Walker's

Marine Mammals of the World

Walker's

Marine Mammals of the World

Ronald M. Nowak

Foreword by John E. Heyning
Introduction by
Randall R. Reeves and
Brent S. Stewart

The Johns Hopkins University Press
Baltimore and London

Printed in the United States of America on acid-free paper
9 8 7 6 5 4 3 2 1

The Johns Hopkins University Press
2715 North Charles Street
Baltimore, Maryland 21218-4363
www.press.jhu.edu

A catalog record for this book is available from the British Library.

Library of Congress Cataloging-in-Publication Data

Nowak, Ronald M.
Walker's marine mammals of the world / Ronald M. Nowak ; foreword by John E. Heyning ; introduction by Randall R. Reeves and Brent S. Stewart.
p. cm.
"Portions of this book have been adapted from Walker's mammals of the world, 6th edition, by Ronald M. Nowak, c1999 . . ."—T.p. verso.
Includes bibliographical references and index. (p.).
ISBN 0-8018-7343-6 (pbk : alk. paper)
1. Marine mammals. 2. Marine mammals—Classification. I. Title: Marine mammals of the world. II. Walker, Ernest P. (Ernest Pillsbury), 1891–1969. Walker's mammals of the world. III. Title.
QL713.2 .N69 2003
599.5—dc21

2002043259

Contents

Foreword

Human interest in marine mammals predates the written record. Stone Age paintings of whales adorn rock walls in the land now called Norway. Over the millennia, innumerable coastal societies have incorporated marine mammals into their cultures and commerce. We see this in the dependence of many on the bounty of meat, fat, furs, and bone from stranded or harvested marine mammals. We see this, too, in enduring ancient iconography. We can look to the heavens and still find constellations named for both dolphins (Delphinus) and whales (Cetus). As we do today, the earliest mariners must have observed, with wonder, bow-riding dolphins.

Our longtime intrigue with marine mammals comes in part from the environment they inhabit. The ocean is an alien world for humans. Although we may walk along its shores, sail on its surface, or a very few of us briefly venture below its waves, we are foreigners in the water. All marine creatures seem mysterious and therefore intriguing to us. We question what they are and wonder how they live. As air-breathing brethren, we are fascinated by these mammals of land-based ancestry that have so successfully adapted to life in the oceans.

We also find ourselves collectively struck by the sheer size of marine mammals—by any and all measures, they are immense. Recent fossil discoveries have unearthed several species of colossal sauropod dinosaurs, some estimated to have reached more than one hundred feet in length. Nonetheless, these long-vanished dinosaurs cannot compete in sheer weight with the living blue whales, the largest of which weighs some 150 tons (136,000 kg)! Large size is not limited to the cetaceans. The largest member of the order Carnivora is the southern elephant seal, the largest mustelid is the sea otter, and the largest bear by average weight is the polar bear. The recently extinct Steller's sea cow, a sirenian, measured up to 27 feet (8.3 meters) in length.

The basis for marine mammals' large size is twofold. First, there is the heat conservation advantage imparted by size—the larger an object is, the less the surface-to-volume ratio. As heat is lost through the animal's surface area and generated by its volume through metabolic activity, a smaller surface-to-volume ratio provides larger animals a better advantage in the cool to frigid oceans and seas. Second, large size can be supported via the water's buoyant properties, thereby allowing animals such as whales to be large without having to develop the architecturally untenable skeletons that would be required to support the same mass on land.

Humans are inherently a wanderlust species. Much of the more recent exploration of the world by ocean-traveling Europeans was undertaken not merely for the joy of exploration, however, but also to seek new marine mammal resources.

Hence, the commercial exploitation of marine mammals has contributed in non-trivial ways to the relatively recent historical phase of world exploration, commerce, and politics. The quest for new fur seal rookeries, in particular, led to much of the earliest marine exploration of the world by both Europeans and Americans. Exemplary of this is the arguably first sighting of the Antarctic continent by the American sealer Nathaniel Palmer. The Russian exploration and settlement of North America from Alaska southward to northern California was primarily in search of sea otters, whose pelts were sold to China. In fact, the severe depletion of sea otter stocks in Alaska was a major contributing factor to Russia's decision to sell this territory to the United States.

Another example of the historical significance of the trade in marine mammal products comes from the U.S. Civil War. The Confederacy attempted to cripple the North's economy by destroying New England's whaling fleet. A trade ship, the *Sea King*, was converted into the warship *Shenandoah*. Captain James I. Waddell was given command of the ship with a mission to "greatly damage and disperse [the Union] fleet, even if [he did] not succeed in utterly destroying it." Without a doubt, the *Shenandoah* succeeded in both dispersing and diminishing the New England whaling fleet. However, the irony of the *Shenandoah* is that the vast majority of the whaling ships she took were captured after General Lee had surrendered in April 1865. While pursuing his enemy, Captain Waddell continued his chase in the distant Bering Sea where, arguably, the last hostile act of the Civil War took place in August 1865.

Origins

The manner in which ancient civilizations explained how marine mammals came to be can seem quite fanciful. According to classic Greek mythology, Dionysus, the god of wine and fertility, took passage on a Tyrrhenian ship. The sailors, however, were pirates and planned to enslave their seemingly powerless passenger. When Dionysus discovered their plot, he avenged himself on the men. As punishment, Dionysus caused the ship's mast to grow as a grapevine, the oars to turn into snakes, and a strange flute to begin playing. These oddities so terrified the pirates that they threw themselves into the sea. They were rescued by the sea god Poseidon, who mercifully changed them into dolphins but demanded their eternal servitude to humankind in return.

The field of evolutionary biology takes a very different approach to understanding the origins and evolution of marine mammals. Today, information is gleaned from the seemingly diverse fields of paleontology, molecular genetics, and morphology. New fossil digs, especially along the Eocene shores of the Tethys Sea (modern-day Egypt, Pakistan, and India), have uncovered the earliest forms of sirenians and cetaceans. Molecular geneticists now have the tools at hand to quickly sequence long sections of DNA from many species at once. Morphologists, molecular geneticists, and paleontologists are now armed with computer programs that can analyze voluminous sets of new data.

Nonetheless, the body shapes of living cetaceans and sirenians differ so profoundly from the typical mammalian *blauplan* that it seems nearly impossible to discern intuitively which, among the other orders of mammals, are their closest kin. Our scientific understanding of cetacean and sirenian origins based on contemporary analyses of fossils, morphology, and molecules may seem as whimsical as that of the ancient Greeks.

Whales and manatees are ungulates, related to cows, horses, and hippos. The extant descendants of archaic ungulates (collectively called condylarths) are the other modern orders of ungulates that radiated in the Paleocene. Traditionally, the fossil evidence suggests that land-dwelling extinct ungulates, called mesonychians, are closely related to the ancestry of whales. Recent insights provided by new fossil finds and DNA sequencing indicate that the Cetacea may have evolved from within the order of even-toed ungulates, the Artiodactyla, or are at least their closest relatives.

The sirenians—manatees, dugongs, and their extinct relatives—are the only fully aquatic herbivorous group of mammals. Sirenians are part of a clade called the Tethytheria. This clade also includes the elephants and their extinct relatives, the extinct marine desmostylians, and, in some classifications, the hyraxes. Some living groups of this clade have a unique form of cheek-tooth replacement in which new teeth originate at the back of the toothrow, slowly move forward, and finally jettison out the worn-down tooth in front.

The oldest and most primitive sirenians are found from the early and middle Eocene of Jamaica and from the middle Eocene of Pakistan, North Africa, and Europe. Ancient sirenians had four limbs and were amphibious, but, as with the cetaceans, modern species have only vestiges of the pelvic girdle tucked within the body contours and are propelled through the water entirely by tail flukes. The fossil record is rich with taxa, but there are only four living species.

The pinnipeds are a group of marine mammals that includes the seals, sea lions, and walruses. Pinnipeds arose from the arctoid (bears, dogs, weasels, etc.) lineage of carnivores. As such, most authorities now include the pinnipeds in the order Carnivora. The terrestrial ancestor of the pinnipeds has been the subject of considerable debate among scientists. There are two alternative hypotheses. One, citing primarily the age and distribution of fossils as evidence, supports a dual origin, attributing the sea lions, fur seals, and walruses to an ursid (bearlike) ancestor evolving in the North Pacific and the true seals to a mustelid (weasels, otters, etc.) ancestor evolving in the North Atlantic, as these are the ocean basins from which the oldest fossils are found. The alternative hypothesis, employing molecular, karyological, and morphological evidence, supports a single origin for all three pinniped families. There is growing agreement among scientists that pinnipeds arose from a single ancestry.

Research: Challenges and Opportunities

Marine mammals, by definition, spend most of their time underwater, far from the watchful eyes of biologists. In addition, they are large and typically fast moving, making below-the-surface viewing difficult in often-turbid water. Further exacerbating the difficulties of investigating marine mammals is our limited ability to study captive animals. With few exceptions, whales are too large to hold in captivity, making most experimentation typically available to elucidate the biology of land mammals unavailable to most marine mammalogists. Thus, such fundamental biological parameters as basic physiology are difficult to study. The study of form and function, morphology, is also difficult with relatively rare animals that are immense. Scholars studying anatomy typically must wait for the random stranding of a dead marine mammal to secure a study specimen. But once that is obtained, logistical challenges still remain. Studying the tongue of a blue whale is comparable, considering the size, to dissecting an entire African elephant but without the aid of skeletal landmarks. Marine mammals are, thus, among the most difficult groups of mammals to study, and our understanding about many aspects of their biology is limited in comparison to that of terrestrial mammals.

The basic biological units of study for whole-organism biologists are species or populations within species. For studies of species diversity and population structure within species, large series of museum specimens are needed to quantify variation—individual, ontogenetic, geographic, and sexual dimorphism. Whereas many museums have excellent holdings of land mammals, especially the species-diverse small mammals, few museums have comparable series of marine mammals for research undertakings. Hence, our understanding of this biological diversity, especially at the population level, has been very limited using classical osteological specimens. New genetic tools are helping in this regard. In the field, biologists can remove tiny plugs of skin from living members of a population in order to increase sample size. Often, populations of marine mammals can be better identified through genetic sampling and analysis, thus providing us with a more complete understanding of within-species population patterning. The same molecular techniques that are so useful for defining species and populations have also successfully been used to identify illegally taken products, such as whale meat from protected species sold in markets as legal catch. The strong conservation biology component of genetic sampling and analysis cannot be overappreciated.

In an ocean world where visibility is limited yet sound travels farther and faster, echolocation—the production of sound and the interpretation of its echoed returns—provides a tremendous advantage in sensing the world. Only the toothed whales, dolphins, and porpoises (suborder Odontoceti) have been experimentally demonstrated to have the ability to echolocate. As we scientists cannot, literally, get into the minds of these animals, we cannot know for sure what details of the environment echolocation provides odontocetes. However, our understanding has been greatly enhanced through integrated research programs involving morphology, acoustics, and field studies. It is known that a blindfolded dolphin can detect a 10-cm sphere at 100 meters. This is nearly the equivalent of our seeing a tennis ball across a football field. Bottlenose dolphins have experimentally demonstrated the ability to distinguish between two objects of the same size but of different densities. This ability may explain how wild dolphins foraging over featureless sandy bottoms can locate the visually cryptic fish buried within the sediments.

Odontocetes make the sounds used in echolocation by moving air in a labyrinth of air sacs off the nasal passages between the blowhole and the skull. This movement of air causes the adjacent soft tissues to vibrate, analogous to what we commonly call a "raspberry." The fatty, lens-shaped melon, or dolphin's "forehead," located in front of the blowhole, focuses the sound. The scientific unraveling of how these sounds are produced took morphologists, physiologists, and acousticians nearly four decades of study.

The science of marine mammalogy has grown exponentially over the past two decades in no small part because of the deployment of new technologies that allow us to answer previously unanswerable questions. Various telemetry devices attached to an animal allow sampling of data unheard of through classic observational methods. Time/depth recorders allow researchers to examine the dive profile of marine mammals and even to sample properties of the water or the animal's blood. Tracking tags, both short-range radio tags and long-range satellite tags, allow scientists to follow the movement patterns of animals for great distances and long periods. Small video cameras can now be attached to marine mammals, thus allowing firsthand viewing of how animals experience their underwater environment. Acoustical recorders can telemeter back to the scientist information as to which animal, within a large social group of dolphins, is making certain sounds and what kind of sounds these are. Such distinctions provide scientists nuanced social context in which to interpret the meaning of these sounds.

However illuminating the attachment of such telemetric devices would seem, the reality has been one of mixed success. For slow-moving animals such as manatees, these devices are incredibly successful because they are relatively easy to attach via tethers around the flukes and divers can even replace the battery packs while the tethers are still attached! Time/depth recorders and satellite tags can also successfully be glued to the pelage of seals and sea lions (thus, e.g., providing records of the amazing diving and migratory patterns of elephant seals). Such tags are naturally shed when the animals next molt their fur. In striking contrast, the large size and smooth skin of cetaceans have made attachment of such devices extremely troublesome. Cetaceans have antifouling reactions to foreign objects attached to their skin, be they barnacles or human-placed tags, so they shed these freeloaders rather quickly.

Social Systems and Behavior

The social systems of marine mammals range nearly the gamut of general mammal systems. Some species are relatively solitary, coming together only for breeding or when aggregating around food resources. Others, such as many species of oceanic dolphins, spend their entire lives within complex, and often large, social groupings.

Polygyny, in which a male may mate with numerous females in a single breeding season, is a predominant mating strategy for mammals. The evolutionarily related phenomenon of pronounced sexual dimorphism (one sex, usually the male, is much larger) is equally widespread. The driving force for this is thought to be the fact that the female mammals, through lactation, provide the sole source of nutrition for the early stages of the offspring's development. In a general evolutionary context, this frees males from the demands of child care, with males increasing their evolutionary fitness by mating with as many females as possible. This in turn leads to increased male-male competition, and such sexual selection drives the evolution of sexual dimorphism. Many species of marine mammals exhibit the greatest sexual dimorphism and degree of polygyny among any mammals. This is especially true for the species of pinnipeds that breed on islands where optimal breeding space is a limiting resource further fueling male-male competition. This is why the male northern fur seal may weigh up to four and a half times as much as the female and why, by controlling the prime pupping sites sought by females, a dominant bull may control breeding access to scores of females. Seals that breed on ice, a more widely spaced resource, tend not to exhibit this extreme degree of sexual dimorphism and polygyny.

Humans and some large dolphins are among the few female mammals known to cease to be reproductive comparably early in life. The hypothesis as to the evolutionary advantage in humans is that these postreproductive females contribute to their own fitness by assisting their daughters with child rearing. This is based on the premise that humans are long lived and that their children require a prolonged development to learn all the skills, including social skills, needed to survive. Interestingly, both pilot whales and killer whales have postreproductive females, complex social structures, and prolonged calf development. The parallels are striking and merit more research.

Sound production is a common form of communication among mammals. It is well known that all marine mammals produce sounds in a wide range of social contexts. Among the varied repertoire of sounds it makes, each bottlenose dolphin also produces its own unique whistle. Biologists believe that these whistles may serve to identify each individual dolphin and thus coined these *signature whistles.* Bottlenose dolphins live in fission-fusion societies, in which small groups may temporarily come together and then split up. However, membership within the smaller groups can be stable for decades. Individuals in the small groups tend to produce signature whistles when they are in distress or out of sight of other members of the subgroup. A newborn bottlenose quickly develops its own whistle, hypothesized as a necessity to becoming a member of the local dolphin society. Male calves may have a signature whistle similar to that of their mother, whereas female calves tend to have more distinct whistles. This is of interest because, on reaching maturity, the males typically disperse from the school. Sperm whales also produce unique individual sounds. Instead of whistles, however, sperm whales produce a unique train of echolocation clicks called *codas.* Biologists know very little about how sperm whales use these codas or if they allow acoustical recognition of individuals, as is hypothesized for the signature whistles of bottlenose dolphins.

In addition to acoustical communication, visual postures are an important form of visual communication among marine mammals. For many species of highly polygynous pinnipeds, a male with its head held high is considered to be threatening toward other males. Field biologists working on the breeding islands of these pinnipeds soon realized that the upright stance of a human closely enough resembles the threat posture of a male pinniped to elicit the unwelcome aggressive response from the dominant bull. These biologists quickly learned that, by crawling though the colony and keeping a low profile, they would be far less likely to incur the wrath of challenged males.

Biologists studying wild spinner dolphins in the tropics have occasionally observed what is thought to be a threat posture as well. During this display, the dolphin strongly arches its back into an S-shaped curve with the flukes up and the flippers held outward and downward. This awkward-looking posture resembles the threat posture of the gray reef shark, which inhabits these same waters. Some biologists believe that the dolphins may be mimicking the threat posture of the shark. It is unclear whether the posture is intended to scare off sharks or is directed toward other dolphins. During this display, the dolphin will also slash its flukes from side to side, the exact opposite of the normal tail movement. Whether this is to mimic the side-to-side movement of the shark's tail or to strike the intended victim with the side of the flukes is also unknown.

It is often mentioned that cetaceans are very intelligent animals. But what do we mean by intelligence, and how do we measure it? One might think, for example, that a way to determine intelligence would be simply to measure the size of the brain. Many of the toothed whales have brains that are larger than that of humans. In fact, the largest brain of any animal is that of the sperm whale, which can weigh up to 21 pounds (9.5 kg). A typical brain weight for a human is about 3 pounds (1.5 kg). However, most dolphins and all whales are much bigger than humans, and brain size tends to be proportional to body size; therefore, they could have larger brains notwithstanding any real measure of intelligence. Another way to measure the brain is as a ratio to body weight. Using this measure, all cetaceans have brains relatively smaller than those of humans but larger than those of most other mammals. It is increasingly evident that no measure of brain size, either absolute or as a ratio of body size, gives any clear measure of intelligence.

There is no universal test to measure intelligence. Even among humans, the conclusions of the tests we have are far from perfect, not to mention controversial. These tests are of limited use for other species that do not read or use their hands as humans do and that have significantly different milieus. What is important for one species to learn may be irrelevant to another species. Based on observations of behavior, we know that many species of dolphins are good at problem solving, a trait we associate with intelligence. Problem solving is usually more developed in carnivorous mammals, including cetaceans, than in herbivorous mammals. Thus, we commonly think of foxes and cats as being "smarter" than cows and sheep. So what about the cetaceans? Most scientists would agree that at least most dolphins and porpoises are good problem solvers and can communicate abstract concepts to one another, both attributes we generally consider to be markers of relatively high intelligence.

Conservation

The rich prizes offered by marine mammals—meat, fat, hides, fur, ivory—made these creatures desirable targets for most coastal communities. Harvest from freshly dead stranded whales and dolphins undoubtedly led to direct hunting of living animals. Such exploitation probably resulted in extinction of at least three species in historic times: the sea mink of the New Brunswick and New England coast, Steller's sea cow of the North Pacific, and the Caribbean monk seal, a species Christopher Columbus observed on his second voyage to North America. Subsistence hunting led to commercial exploitation, usually with little or no management, and eventually to serious depletion of many species, including most fur seals and large whales.

Incidental and indirect destruction of marine mammals often is far more difficult to control than is direct exploitation. For example, the baiji dolphin, restricted to the Chang Jiang (Yangtze) River system of China, teeters at the brink of extinction as its habitat is degraded by the pressures of a burgeoning human population using the river for fishing, travel, and hydroelectric generation. The entanglement and drowning of marine mammals and seabirds in fishing gear, especially gillnets, remains a major worldwide problem. Hundreds of thousands of dolphins have been killed in the eastern tropical Pacific during purse-seining operations for tuna over the past four decades. The highly endangered North Atlantic right whale continues to lose individuals to ship strikes and entanglement in fishing gear. Recent modeling of that population estimates that saving just an additional two reproductive-age female whales a year will result in population recovery, while maintaining the status quo will result in eventual extinction! Pingers, acoustical devices designed to alert marine mammals to the presence of nets, have shown promise in some areas.

Humans also affect marine mammals with pollution, both chemical and acoustical. It has been known for some time that marine mammals in some areas can accumulate very high levels of chemical pollutants such as pesticides and PCBs. These high loads are known to suppress the immune systems of pinnipeds and are suspected of having the same effects on cetaceans. Sounds generated by human activity now permeate the oceans. This acoustical "smog" is created by shipping, boating, oil and mineral exploration, sonar, and warfare. The effects of sound on marine mammals can range from mild disturbance to hearing loss and even death. Recent recognition of these potential problems has led to active research as to how to mitigate them.

The Future

Our understanding of many aspects of the biology of marine mammals is less than that for their terrestrial mammalian kin. However, new technology continues to help us advance our knowledge at a phenomenal clip. Human fascination with marine mammals allows them to serve as environmental ambassadors for marine ecosystems, environments where "out of sight, out of mind" has been the standard for all too long. Informed decision making regarding the wise management of the marine mammals and the ecosystems they inhabit requires an informed public. Thus, the inherent fascination that marine mammals hold for us should compel us to discover more about their intriguing biology. Such knowledge can be gleaned from this most informative book. By providing proper stewardship of our natural heritage, we can ensure that our children's children have the opportunity to catch a glimpse of the largest creature ever to have inhabited the earth.

John E. Heyning
Deputy Director for Research and Collections
Natural History Museum of Los Angeles County
Los Angeles, California

Walker's

Marine Mammals of the World

Marine Mammals of the World: An Introduction

Randall R. Reeves and Brent S. Stewart

The oceans and seas cover three-fourths of the surface of planet Earth. Although life arose in water, the mammalian experiment began on land, perhaps some 200 million years ago. Thus, the existence of mammals fully adapted to life in the sea represents a return to the ancestral home rather than a kind of in situ flowering. This reinvasion of the sea by terrestrial mammals took place along several independent lines of descent. First to enter the sea were the small, primitive, hoofed, probably hippopotamus-like ancestors of the Cetacea (whales, dolphins, and porpoises; Gingerich et al. 2001) and the subungulate ancestors of the Sirenia (manatees, dugongs, and sea cows). Animals recognizable as cetaceans and sirenians were present during the early Eocene epoch more than 50 million years ago (Domning 2002; Fordyce 2002). Much later, possibly in the late Oligocene or early Miocene (15–30 million years ago), the bear- or weasel-like ancestors of the Otariidae (fur seals and sea lions), Odobenidae (walrus), and Phocidae (true seals) began adapting to life in the sea (Berta 2002). Later developments would result in the polar bear, sea otter, and marine otter (Heyning and Lento 2002).

The marine fauna of our present-day world includes an elite array of highly derived mammals. Their remarkable adaptations allow them to exploit the sea's diverse resources—seagrasses and algae (sirenians), zooplankton (baleen whales and some filter-feeding seals), vertebrate and invertebrate nekton (many pinnipeds and toothed cetaceans), benthic invertebrates (walruses and sea otters), and even fellow mammals (killer whales, leopard seals, and polar bears). They are found in waters of abyssal depth as well as in coastal shallows and in a wide range of temperatures and salinities. Several groups—the manatees, river dolphins, and lake seals—have become specially adapted to riverine and lacustrine systems, often far removed from any marine influence.

In fact, deciding which of the mammals are "marine" is more difficult than one might suppose. Some groups are obvious. The orders Cetacea and Sirenia consist of animals with an obligate aquatic existence: they cannot survive out of water for more than a few days. Most of the cetaceans and certain of the sirenians live in saline or brackish waters, yet a few species and populations are full-time inhabitants of freshwater lakes and rivers. Whether they could survive in a marine (saline) environment is uncertain. Seals, sea lions, and the walrus (the pinnipeds) are also counted among the marine mammals without hesitation. Yet pinnipeds are born on land or ice, many are conceived there as well, and they crawl out of the water for periods of their lives to rest and molt. Moreover, as mentioned above, some seal species and populations live an entirely "land-locked" existence in fresh or saline lake systems.

Those are the easiest groups to classify as marine mammals. From this point on, matters get murkier. The U.S. Congress placed several fissiped carnivores, specifically the polar bear (*Ursus maritimus*), sea otter (*Enhydra lutris*), and marine otter (*Lontra felina*), under the authority of the Marine Mammal Protection Act of 1972. That legal decision, based on the fact that these species depend on marine habitat for survival, has generally been accepted and used in other contexts. However, as Rice (1998) points out, the arctic fox (*Vulpes lagopus*) and several local populations of additional otter species (e.g., the Eurasian and American river otters, *Lutra lutra* and *L.* [or *Lontra*] *canadensis,* respectively; the African small-clawed otter, *Aonyx capensis;* see Chanin 1985) forage in marine environments, and a few bats (order Chiroptera) prey on fish and other small aquatic animals. Two bat species (*Myotis vivesi* and *Noctilio leporinus*) regularly fish in marine waters.

While recognizing that any selection of taxa for inclusion is bound to invite controversy at some level, Ronald Nowak and the publishers have chosen what we consider a reasonable representation. The present volume defines *marine mammals* to include the two orders Cetacea and Sirenia, three families of the order Carnivora (Otariidae, Odobenidae, and Phocidae), and three members of the carnivoran families Mustelidae (sea otter and marine otter) and Ursidae (polar bear). In this introduction, most of our attention is given to the cetaceans, pinnipeds, and sirenians. There are some inconsistencies between the names and taxonomy used here and those in the main text that follows mainly because several years have elapsed since the text was written. It must be acknowledged, however, that some of the discrepancies between this introduction and Nowak's text reflect genuine differences of opinion and interpretation. We have tried to make our information current through early 2002.

Taxonomy, Diversity, and Zoogeography

The current standard for marine mammal taxonomy and systematics was outlined by Rice (1998), with a few exceptions based on subsequent analyses (Table 1). Until another comprehensive procedural and substantive evaluation is conducted, we have adopted Rice's scheme with the recent modifications, although we may not fully endorse it (see also Perrin, Würsig, and Thewissen 2002).

CETACEANS

The cetacean skull is *telescoped,* that is, it is elongated, with overlapping bones (Pabst, Rommel, and McLellan 1999). The nares are positioned dorsally, although the blowhole (in odontocetes) or blowholes (in mysticetes) can be anywhere from the front of the rostrum (e.g., in the sperm whale) to the back (e.g., in the mysticetes). In odontocetes, the main bones of the upper jaw have been thrust backward and upward over the eye sockets, and they are usually asymmetrical, skewed to the left. The cranial architecture seems to be geared to accommodate a sophisticated system of transmitting, receiving, and processing sound (Fordyce 1988). A series of sacs (diverticula) in the soft tissue of the nasal passages probably play a role in producing and projecting the diverse high-frequency clicks and other calls of odontocetes, and the thin-walled pan bone (i.e., shaped like a frying pan) in the lower jaw presumably receives sounds from the water and transmits them to the internally placed ear bones via fatty connecting channels (Wartzok and Ketten 1999).

Mainstream cetacean taxonomy has been conservative over the past half-

TABLE 1. TAXONOMY OF MARINE MAMMALS

Order	Suborder	Family	Genus and Species
Carnivora		Ursidae	polar bear, *Ursus maritimus*
		Mustelide	marine otter, *Lutra felina*[1]
			sea otter, *Enhydra lutris*
		Otariidae	northern fur seal, *Callorhinus ursinus*
			Guadalupe fur seal, *Arctocephalus townsendi*
			Juan Fernández fur seal, *Arctocephalus philippii*
			Galápagos fur seal, *Arctocephalus galapagoensis*
			South American fur seal, *Arctocephalus australis*
			subantarctic fur seal, *Arctocephalus tropicalis*
			Antarctic fur seal, *Arctocephalus gazella*
			New Zealand fur seal, *Arctocephalus forsteri*
			Australian/Tasmanian (*A. p. doriferus*) and Cape/South African (*A. p. pusillus*) fur seals, *Arctocephalus pusillus*
			California sea lion, *Zalophus californianus*
			Japanese sea lion, *Zalophus japonicus*
			Galápagos sea lion, *Zalophus wollebaeki*
			New Zealand (Hooker's) sea lion, *Phocarctos hookeri*
			Australian sea lion, *Neophoca cinerea*
			South American sea lion, *Otaria flavescens*
			Steller (northern) sea lion, *Eumetopias jubatus*
		Odobenidae	walrus, *Odobenus rosmarus*
		Phocidae	Hawaiian monk seal, *Monachus schauinslandi*
			Carribean (West Indian) monk seal, *Monachus tropicalis*
			Mediterranean monk seal, *Monachus monachus*
			crabeater seal, *Lobodon carcinophaga*
			leopard seal, *Leptonychotes weddellii*
			Ross seal, *Ommatophoca rossii*
			northern elephant seal, *Mirounga angustirostris*
			southern elephant seal, *Mirounga leonina*
			bearded seal, *Erignathus barbatus*
			hooded seal, *Cystophora cristata*
			gray seal, *Halichoerus grypus*
			harp seal, *Pagophilus groenlandicus*
			ringed seal, *Pusa hispida*
			Baikal seal, *Pusa sibirica*
			Caspian seal, *Pusa caspica*
			spotted seal, *Phoca largha*
			harbor seal, *Phoca vitulina*
			ribbon seal, *Histriophoca fasciata*
Cetacea	Odontoceti	Platanistidae	susu/Ganges river dolphin (*P. g. gangetica*) and bhulan/Indus river dolphin (*P. g. minor*), *Platanista gangetica*
		Lipotidae	baiji (Chinese or Yangtze river dolphin), *Lipotes vexillifer*
		Pontoporiidae	franciscana (La Plata dolphin), *Pontoporia blainvillei*
		Iniidae	boto (Amazon river dolphin), *Inia geoffrensis*
		Monodontidae	beluga, *Delphinapterus leucas*
			narwhal, *Monodon monoceros*
		Phocoenidae	harbor porpoise, *Phocoena phocoena*
			vaquita, *Phocoena sinus*
			Burmeister's porpoise, *Phocoena spinipinnis*
			spectacled porpoise, *Phocoena dioptrica*
			finless porpoise, *Neophocaena phocaenoides*

(*continued*)

TABLE 1. *(Continued)*

Order	Suborder	Family	Genus and Species
			Dall's porpoise, *Phocoenoides dalli*
		Delphinidae	rough-toothed dophin, *Steno bredanensis*
			Indo-Pacific hump-backed dolphin, *Sousa chinensis*[2]
			Atlantic hump-backed dolphin, *Sousa teuszi*
			tucuxi, *Sotalia fluviatilis*
			white-beaked dolphin, *Lagenorhynchus albirostris*
			Atlantic white-sided dolphin, *Lagenorhynchus acutus*
			Pacific white-sided dolphin, *Lagenorhynchus obliquidens*
			dusky dolphin, *Lagenorhynchus obscurus*
			Peale's dolphin, *Lagenorhynchus australis*
			hourglass dolphin, *Lagenorhynchus cruciger*
			Risso's dolphin, *Grampus griseus*
			common bottlenose dolphin, *Tursiops truncatus*
			Indo-Pacific bottlenose dolphin, *Tursiops aduncus*
			spinner dolphin, *Stenella longirostris*
			clymene dolphin, *Stenella clymene*
			striped dolphin, *Stenella coeruleoalba*
			pantropical spotted dolphin, *Stenella attenuata*
			Atlantic spotted dolphin, *Stenella frontalis*
			short-beaked common dolphin, *Delphinus delphis*
			long-beaked common dolphin, *Delphinus capensis*[3]
			Fraser's dolphin, *Lagenodelphis hosei*
			northern right whale dolphin, *Lissodelphis borealis*
			southern right whale dolphin, *Lissodelphis peronii*
			Irrawaddy dolphin, *Orcaella brevirostris*
			Commerson's dolphin, *Cephalorhynchus commersonii*
			Chilean dolphin, *Cephalorhynchus eutropia*
			Heaviside's dolphin, *Cephalorhynchus heavisidii*
			Hector's dolphin, *Cephalorhynchus hectori*
			melon-headed whale, *Peponocephala electra*
			pygmy killer whale, *Feresa attenuata*
			false killer whale, *Pseudorca crassidens*
			killer whale (orca), *Orcinus orca*
			long-finned pilot whale, *Globicephala melas*
			short-finned pilot whale, *Globicephala macrorhynchus*
		Ziphiidae	Arnoux's beaked whale, *Berardius arnuxii*
			Baird's beaked whale, *Berardius bairdii*
			Cuvier's beaked whale, *Ziphius cavirostris*
			Shepherd's beaked whale, *Tasmacetus shepherdi*
			northern bottlenose whale, *Hyperoodon ampullatus*
			southern bottlenose whale, *Hyperoodon planifrons*
			Indo-Pacific (Longman's) beaked whale, *Indopacetus pacificus*
			pygmy beaked whale, *Mesoplodon peruvianus*
			Hector's beaked whale, *Mesoplodon hectori*
			True's beaked whale, *Mesoplodon mirus*
			Gervais's beaked whale, *Mesoplodon europaeus*
			ginkgo-toothed beaked whale, *Mesoplodon ginkgodens*
			Gray's beaked whale, *Mesoplodon grayi*
			Hubbs's beaked whale, *Mesoplodon carlhubbsi*
			Andrews's beaked whale, *Mesoplodon bowdoini*
			Stejneger's beaked whale, *Mesoplodon stejnegeri*

(continued)

TABLE 1. *(Continued)*

Order	Suborder	Family	Genus and Species
			Perrin's beaked whale, *Mesoplodon perrini*[4]
			Sowerby's beaked whale, *Mesoplodon bidens*
			spade-toothed whale, *Mesoplodon traversii*[5]
			Layard's (strap-toothed) beaked whale, *Mesoplodon layardii*
			Blainville's beaked whale, *Mesoplodon densirostris*
		Kogiidae	pygmy sperm whale, *Kogia breviceps*
			dwarf sperm whale, *Kogia sima*
		Physeteridae	sperm whale, *Physeter macrocephalus*
	Mysticeti	Eschrichtiidae	gray whale, *Eschrichtius robustus*
		Neobalaenidae	pygmy right whale, *Caperea marginata*
		Balaenidae[6]	North Atlantic right whale, *Eubalaena glacialis*
			North Pacific right whale, *Eubalaena japonica*
			southern right whale, *Eubalaena australis*
			bowhead (Greenland) whale, *Balaena mysticetus*
		Balaenopteridae	common minke whale, *Balaenoptera acutorostrata*
			Antarctic minke whale, *Balaenoptera bonaerensis*
			Bryde's whales, *Balaenoptera brydei/edeni*[7]
			sei whale, *Balaenoptera borealis*
			fin whale, *Balaenoptera physalus*
			blue whale, *Balaenoptera musculus*
			humpback whale, *Megaptera novaeangliae*
Sirenia		Dugongidae	dugong, *Dugong dugon*
			Steller's sea cow, *Hydrodamalis gigas*
		Trichechidae	Amazonian manatee, *Trichechus inunguis*
			West Indian manatee, *Trichechus manatus*
			West African manatee, *Trichechus senegalensis*

Source: Rice (1998) and Perrin, Würsig, and Thewissen (2002), with a few modifications.
Note: As discussed in the text, marine mammals are classified as such on an informal and somewhat arbitrary basis.
[1]Some mammalogists have assigned the three New World river otters (including the marine otter) to a separate genus (*Lontra*), while reserving *Lutra* for the Old World Eurasian (*Lutra lutra*) and spotted-necked (*Lutra maculicollis*) otters (e.g., Van Zyll de Jong 1987; Koepfli and Wayne 1998).
[2]Although Rice (1998) recognized a third species of *Sousa* as the Indian hump-backed dolphin, *Sousa plumbea*, the IWC Scientific Committee has maintained the conservative position of recognizing only two species, pending further genetic, morphological, and other analyses (IWC, in press). Thus, the putative Indian and Pacific forms are subsumed within *Sousa chinensis*.
[3]Although Rice (1998) recognized a third very long-beaked species, *Delphinus tropicalis*, from the Indian Ocean, Jefferson and Van Waerebeek (2002) concluded after analyzing available specimens that this form is a variant of *D. capensis*.
[4]Described in 2002 by Dalebout et al.
[5]Established by van Helden et al. (2002) as a senior synonym for *Mesoplodon bahamondi*, Bahamonde's beaked whale.
[6]Recent genetic analyses support the concept of three species of right whale: one in the North Atlantic, one in the North Pacific, and one in the Southern Hemisphere (Rosenbaum et al. 2000; IWC 2001b). Also, although Rice (1998) used the genus name *Balaena* for all of the balaenids, the IWC Scientific Committee has decided to retain the genus name *Eubalaena* for the right whales while retaining *Balaena* for the bowhead whale. Rice recognized only one species of right whale, *B. glacialis*, with two subspecies, *B. g. glacialis*, the Northern Hemisphere right whales, and *B. g. australis*, the Southern Hemisphere right whale.
[7]There are at least two morphologically distinct forms of Bryde's whales, very likely different species. The nomenclature of the two known forms is unresolved (see Kato 2002). Perrin, Würsig, and Thewissen (2002) provisionally designated the common Bryde's whale as *Balaenoptera brydei* and the pygmy Bryde's whale as *B. edeni*.

century, dominated by a reluctance to recognize subgeneric distinctions and casting aside almost completely the concept of subspecies. This tendency has begun to give way in recent years, however, as large study samples have become available for some groups, such as the pelagic delphinids, and as molecular genetic analyses have revealed surprising amounts of isolation and divergence, sometimes even on small geographic scales (Dizon, Chivers, and Perrin 1997; Perrin and Brownell 1994). The cetaceans may be unique among mammals in having a global management body, the International Whaling Commission (IWC), with a scientific committee that meets annually, publishes a scholarly journal, and maintains an authoritative taxonomic list of species (which appears as an appendix to the Guide to Authors in the first issue of each volume of the *Journal of Cetacean Research and Management*). At its annual meeting in 2000, the scientific committee agreed to use Rice's (1998) list of marine mammals as a starting point in discussions to revise the IWC's own list of species (International Whaling Commission 2001c).

Mysticeti

There are at least 14 species of mysticetes, or baleen whales, belonging to just four families (Table 1). The Balaenidae, or right whales, consist of three temperate-region species (*Eubalaena* spp.) and an Arctic endemic, the bowhead whale (*Balaena mysticetus*). Molecular genetic analyses have recently clarified specific differences among right whales in the North Atlantic (*Eubalaena glacialis*), North Pacific (*E. japonica*), and Southern Ocean (*E. australis*) (Rosenbaum et al. 2000).

Neobalaenidae and Eschrichtiidae both contain single species. The pygmy right whale (*Caperea marginata*) is confined to the cool temperate Southern Hemisphere. It has rarely been identified at sea, so its distribution is inferred mainly from strandings. The gray whale (*Eschrichtius robustus*) is a Northern Hemisphere species with a primarily coastal distribution. It became extinct in the North Atlantic within the last 300–400 years and survives as two geographically separate populations in the North Pacific. One of these, the eastern or "California" population, migrates annually between the warm lagoon and bay waters of Baja California, Mexico, and the productive shelf waters of the Bering and Chukchi Seas. The other, the western or Asian population, migrates between the South China Sea (winter) and Okhotsk Sea (summer).

Balaenopteridae is the most diverse family of mysticetes, with two well-defined genera and at least eight species. The balaenopterids, or rorquals as they are commonly called (including the humpback whale, *Megaptera novaeangliae*), are distinguished by having a distinct dorsal fin and a series of long, parallel grooves, or pleats, on the throat and chest. Several of the rorquals are cosmopolitan, or at least widely distributed, with populations in all of the major ocean basins: the blue whale (*Balaenoptera musculus*), fin whale (*B. physalus*), sei whale (*B. borealis*), and humpback whale. Two forms of Bryde's whales were long known to exist in tropical and warm temperate regions (Best 1977; Kawamura and Satake 1976), but only in recent years has it become clear that there is more than one species (*Balaenoptera brydei* and *B. edeni;* see Pastene et al. 1997; Rice 1998; Wada and Numachi 1991). Preliminary mitochondrial DNA analyses indicate that the "standard-form" Bryde's whale is more closely related to the sei whale than to the "pygmy" Bryde's whale (Dizon et al. 1997). Among minke whales, a "white-shouldered" form (*B. acutorostrata*) is found throughout the North Atlantic (nominally *B. a. acutorostrata*) and North Pacific (*B. a. scammoni*), and a "dark-shouldered" form (*B. bonaerensis*) is widely distributed in the Southern Hemisphere. Still unresolved is the taxonomic status of a "dwarf" form of minke whale found in portions of the Southern Hemisphere (Arnold, Marsh, and Heinsohn 1987; Best 1985). Since it is genetically and morphologically closer to the Northern Hemisphere form, it is provisionally viewed as an unnamed subspecies of *B. acutorostrata* (Rice 1998).

Although it may seem difficult at first glance to distinguish one blue whale (*Balaenoptera musculus*) from another, the pattern of mottling on the back and sides, together with the appearance of the small dorsal fin, makes it possible for researchers to "photo-identify" these majestic animals individually. Important insights have been gained from such work. Blue whales travel great distances, and their very-low-frequency sounds can be heard hundreds of miles away. Photograph by Thomas A. Jefferson.

Odontoceti

The geographic radiation of toothed whales is perhaps the most extensive among any mammalian group. They have not only adapted to life at most latitudes and managed to penetrate virtually every bay, gulf, and partially enclosed sea, but also colonized major river systems of South America and southern and eastern Asia. Not surprisingly, this suborder is much more diverse than the Mysticeti. It includes 10 families, 34 genera, and more than 70 species (Table 1).

Cosmopolitan Families. Three odontocete families can be thought of as cosmopolitan: Physeteridae, Ziphiidae, and Delphinidae. The sperm whale (*Physeter macrocephalus*), the sole living physeterid, shares with the killer whale (*Orcinus orca;* see later) the distinction of having the most extensive distribution of any marine mammal (Rice 1998). It ranges throughout the world in deep marine waters, although only adult males regularly enter high latitudes.

The deep-diving ziphiids are a fairly diverse group, with 6 genera and at least 20 species. Their zoogeography is complex and poorly known. There are 3 single-species genera. Cuvier's beaked whale (*Ziphius cavirostris*) is the most nearly cosmopolitan species, inhabiting tropical and temperate waters worldwide (Heyning 1989). Shepherd's beaked whale (*Tasmacetus shepherdi*) is known from only a few localities, all in cold temperate waters of the Southern Hemisphere (Mead 2002). The third single-species genus, *Indopacetus,* rested until recently on only two partial crania, one found in Somalia and the other in Queensland, Australia. Numerous sightings of unidentified large ziphiids along the equator and in low latitudes of the North Pacific and northwestern Indian Ocean have been provisionally assigned to this species, and its distribution is assumed to be limited to the tropical Indo-Pacific (Pitman et al. 1999; Pitman 2002a). Two ziphiid genera consist of antitropical species pairs. In each case, one member of the pair is a circumpolar Southern Ocean endemic—Arnoux's beaked whale (*Berardius arnuxii*) and the southern bottlenose whale (*Hyperoodon planifrons*). Baird's beaked whale (*B. bairdii*) is endemic in the cool temperate and subarctic North Pacific, while the northern bottlenose whale (*H. ampullatus*) is confined to the cool temperate and subarctic North Atlantic.

Mesoplodon is the most diverse of all cetacean genera. Mesoplodonts can be "morphologically cryptic"; even though two specimens appear similar, detailed examination, sometimes including gene sequencing and phylogenetic analyses, reveals that they belong to separate species (Dalebout et al. 1998, 2000). They are elusive and difficult to observe and identify at sea; much of what is known about their biology and distribution comes from stranded specimens. Fourteen named species are currently recognized (Mead 1989; Pitman 2002b; Table 1), including two first described in the 1990s—the pygmy beaked whale (*M. peruvianus;* Reyes, Mead, and Van Waerebeek 1991) and the spade-toothed whale (described and given the name Bahamonde's beaked whale, *M. bahamondi,* by Reyes et al. 1996 but recently redescribed by van Helden et al. 2002 as *M. traversii*)—and one in 2002, Perrin's b.w. (*M. perrini;* Dalebout et al. 2002). These three are known only from the Pacific Ocean. Endemism to portions of single ocean basins seems to be a common feature of mesoplodonts: at least one species is restricted to the North Atlantic (*M. bidens*), two to the North Pacific (*M. stejnegeri* and *M. carlhubbsi*), and two to the Indo-Pacific (*M. bowdoini* and *M. ginkgodens*). Two species (*M. layardii* and *M. hectori*) are considered circumpolar in the cold temperate Southern Hemisphere. Gervais's beaked whale (*M. europaeus*) is known only from the tropical and warm temperate Atlantic, including the Gulf of Mexico and Caribbean Sea but apparently not the Mediterranean Sea. True's beaked whale (*M. mirus*) was once thought to be confined to the North Atlantic but has since been found stranded in South Africa and Australia. Conversely, Gray's beaked whale (*M. grayi*) was considered to be limited to the temperate Southern Ocean until a fresh female carcass stranded on the North Sea coast of the Netherlands (Boschma 1950). Only one mesoplodont has an almost cosmopolitan distribution: Blainville's beaked whale (*M. densirostris*) occurs in deep tropical and warm temperate waters worldwide.

The delphinids are the largest and most diverse family of cetaceans, with more than 35 recognized species in 17 families (LeDuc, Perrin, and Dizon 1999; Rice 1998). Only the killer whale is truly cosmopolitan, but its numbers and regularity of occurrence vary geographically. Killer whales can be extremely mobile (transient) or essentially resident, depending on the availability of their preferred prey (Baird 2000; Bigg et al. 1990; Dahlheim and Heyning 1999). Several other genera are almost as widespread. The two pilot whales, long-finned (*Globicephala melas*) and short-finned (*G. macrorhynchus*), together cover much of the globe. Short-finned pilot whales inhabit tropical and warm temperate waters worldwide; long-finned pilot whales occur throughout the Southern Hemisphere, from roughly 20–30°S and southward to the Antarctic, and throughout the temperate and subarctic North Atlantic, including the Mediterranean Sea (Bernard and Reilly 1999; Cañadas and Sagarminaga 2000). Three additional "blackfish," as this blunt-headed, dark-colored group is sometimes called, are pantropical: false killer whale (*Pseudorca crassidens*), pygmy killer whale (*Feresa attenuata*), and melon-headed whale (*Peponocephala electra*). A broadly similar pattern of distribution is shared by several other monotypic genera of small, pelagic odontocetes: rough-toothed dolphin (*Steno bredanensis*), Fraser's dolphin (*Lagenodelphis hosei*), and Risso's dolphin (*Grampus griseus*).

Two delphinid genera, *Stenella* and *Delphinus*, include both widespread species

The clymene dolphin (*Stenella clymene*) is known to occur only in tropical and warm temperate waters of the Atlantic Ocean. It was not well described as a separate species until 1981, and even now its biology and ecology are poorly known in comparison to those of many other pelagic dolphin species. Photograph by Barbara E. Curry.

that inhabit more than one ocean basin and species that are endemic to a single ocean basin. The *Stenella* dolphins are primarily tropical, with three pantropical species: the pantropical spotted dolphin (*S. attenuata*), the spinner dolphin (*S. longirostris*), and the striped dolphin (*S. coeruleoalba*). Substantial variation in form and coloration has led to the establishment of four well-defined spinner dolphin subspecies, including the "dwarf" spinners that are known only from shallow, sheltered waters of southeastern Asia and northern Australia (Perrin, Dolar, and Robineau 1999). The Atlantic spotted dolphin (*S. frontalis*) and the clymene dolphin (*S. clymene*) are both known only from the tropical and warm temperate Atlantic. Dolphins belonging to the genus *Delphinus* occur throughout the world in tropical and warm temperate waters, including the Mediterranean and Black Seas. However, there is considerable uncertainty about how many species should be recognized and which species occurs where. Heyning and Perrin (1994) and Rosel, Dizon, and Heyning (1994) confirmed with morphological and mitochondrial DNA evidence, respectively, that there are at least two valid species. The short-beaked common dolphin (*D. delphis*) is the more abundant and widespread; the long-beaked common dolphin (*D. capensis*) seems to have a more localized, strictly coastal distribution. Although Rice (1998) accepts a third *Delphinus* species, *D. tropicalis*, characterized by an exceptionally long beak, recent evidence supports the position that this is simply a variant of *D. capensis* that occurs in portions of the coastal northern Indian Ocean and the South China Sea (Jefferson and Van Waerebeek 2002).

The delphinid genus *Lagenorhynchus* nominally consists of six antitropical species. The two North Atlantic endemics, the Atlantic white-sided dolphin (*L. acutus*) and the white-beaked dolphin (*L. albirostris*), are similar in external form but have different color patterns. They are partially sympatric in cold temperate to subarctic latitudes, but of the two, white-beaked dolphins occur farther north. There is only one species in the North Pacific, the Pacific white-sided dolphin (*L. obliquidens*). It is abundant across the entire cool temperate zone. The other three "lags"

Dusky dolphins (*Lagenorhynchus obscurus*) are acrobatic at the surface as they socialize and forage over the continental shelf. They occur in three main regions of the Southern Hemisphere—southern South America, southwestern Africa, and New Zealand—with vast expanses of ocean separating the populations. Photograph by Steve Leatherwood.

live in the Southern Hemisphere. The hourglass dolphin (*L. cruciger*) has a circumpolar distribution south of the subtropical convergence, occurring mainly between 45°S and 65°S. Peale's dolphin (*L. australis*) is a nearshore species that occurs along the convoluted coastline of southern South America and around the Falkland (Malvinas) Islands. It is partially sympatric with the much more abundant and widespread dusky dolphin (*L. obscurus*). Dusky dolphins inhabit the South American continental shelf from Peru in the west to northern Argentina in the east, with a low-density zone in the far south (Cape Horn). Disjunct populations are present off southwestern Africa and New Zealand, and dusky dolphins also occur around various oceanic islands, including Gough, Prince Edward, Amsterdam, and St. Paul (Brownell and Cipriano 1999). *Lagenorhynchus*-like dolphins observed in tropical and subtropical waters have given rise to suggestions that there may be an unrecognized species or that Peale's and dusky dolphins sometimes wander far outside their normal ranges (Leatherwood, Grove, and Zuckerman 1991; Van Waerebeek, Goodall, and Best 1997; Van Waerebeek, van Bree, and Best 1995).

The common bottlenose dolphin (*Tursiops truncatus*) is among the most versatile and widespread of the delphinids, with locally resident populations in many coastal and inshore areas along continents and around oceanic islands, as well as populations that migrate over long distances and that range over huge expanses of ocean (Wells and Scott 1999). Its congener, the Indo-Pacific bottlenose dolphin (*T. aduncus*), is a warm-water, primarily coastal species known, thus far, only from the tropical Indian Ocean and western Pacific. The two types of bottlenose dolphin are sympatric in some areas (e.g., around Taiwan; Wang, Chou, and White 1999, 2000).

The other five delphinid genera consist of either tropical or cool-temperate endemics. The tucuxi (*Sotalia fluviatilis*) lives in warm nearshore marine waters from southern Brazil north to Nicaragua (Edwards and Schnell 2001). It is common in bays and estuaries and moves into the lower reaches of many rivers. A small freshwater form lives permanently in the lakes and rivers of the Amazon drainage as far upriver as Ecuador (da Silva and Best 1994). The Irrawaddy dolphin (*Orcaella brevirostris*) of southern Asia has a similarly nearshore and estuarine distribution, from the western Bay of Bengal to the Philippines and northern Australia. Like the tucuxi, it moves far up rivers, including the Irrawaddy, hence its common name (Stacey and Leatherwood 1997). The other tropical genus, *Sousa,* has a patchy, neritic distribution centered in large estuaries and mangrove systems along the coasts of Africa, southern Asia, and Australia (International Whaling Commission, in press; Ross, Heinsohn, and Cockcroft 1994). Although the systematics of this genus are unresolved, the Atlantic form off West Africa is clearly a separate species (*S. teuszi*). The animals occurring from South Africa, along the rim of the Indian Ocean, and into the western Pacific between southern China and eastern Australia are provisionally called Indo-Pacific hump-backed dolphins (*S. chinensis*) (Jefferson and Karczmarski 2001).

The temperate-zone delphinid genus *Cephalorhynchus* consists of four well-defined coastal Southern Hemisphere endemics: Hector's dolphin (*C. hectori*) in New Zealand, the Chilean dolphin (*C. eutropia*) in Chile, Commerson's dolphin (*C. commersonii*) in southern and southeastern South America, and Heaviside's dolphin (*C. heavisidii*) in southwestern Africa. Only one of these, Commerson's dolphin, is known to have populations around oceanic islands, notably the Falklands (Malvinas) in the South Atlantic and the Kerguelen Islands in the southern Indian Ocean (Goodall 1994). The final delphinid genus to consider is *Lissodelphis,* consisting of an antitropical species pair. The northern right whale dolphin (*L. borealis*) is distributed across the cool temperate North Pacific, much like the Pacific white-sided dolphin, and the southern right whale dolphin (*L. peronii*) has a circumpolar distribution in the Southern Hemisphere mainly between 40°S and 55°S.

Porpoises. The terms *porpoise* and *dolphin* make a useful, if somewhat colloquial, distinction. The true porpoises belong to a small family of small animals, the Phocoenidae. They are distinguished by their spade-shaped (spatulate) teeth and lack of a noticeable beak. The Delphinidae may be considered to represent the "true" dolphins, although some of them have little or no beak and theirs is not the only family whose members are called *dolphins* (see below). Of the six modern phocoenid species, four inhabit shallow, nearshore marine waters, while two species are found in deeper offshore waters (Rosel, Haygood, and Perrin 1995). The harbor porpoise (*Phocoena phocoena*) has a northern temperate distribution, with disjunct populations along continental margins of the North Atlantic and North Pacific Oceans and in the Black Sea. The finless porpoise (*Neophocaena phocaenoides*) has a ribbonlike distribution along the tropical and warm temperate margins of the Indo-Pacific. A freshwater population inhabits the Yangtze River system of China. The vaquita (*Phocoena sinus*) has the smallest range of any marine cetacean: it is entirely confined to the upper quarter of the Gulf of California, Mexico. Burmeister's porpoise (*Phocoena spinipinnis*) is a coastal South American endemic that occurs from northern Peru in the west to southern Brazil in the east. Of the two deep-water phocoenids, Dall's porpoise (*Phocoenoides dalli*) is the better known. It is abundant across the cool temperate North Pacific. The spectacled porpoise (*Phocoena dioptrica;* formerly known as *Australophocaena dioptrica*) has rarely been observed alive, but based on strandings and the few at-sea observations, it seems to have a circumpolar distribution in cold temperate waters of the Southern Hemisphere.

Pygmy and Dwarf Sperm Whales. These odd-looking small whales, the Kogiidae, are superficially similar to the much larger sperm whale. They are also reminiscent of the ziphiids in that much of what we think we know about them is inferred from examinations of stranded specimens because they are difficult to detect, identify, and observe at sea. They were not clearly distinguished as separate species—the pygmy *Kogia breviceps* and the dwarf *K. sima*—until the mid-twentieth century (Handley 1966). Both seem to be fairly widely distributed in tropical to temperate latitudes, centered on the continental slopes and offshore.

River Dolphins. Four families, each with only one living genus, comprise the true river dolphins of Asia and South America. China's baiji, or Yangtze dolphin (*Lipotes vexillifer*), was unknown to Western science until the early twentieth century, when a specimen collected in Dongting Lake, a large Yangtze tributary, reached the U.S. National Museum and was described by Miller (1918). In recent times, the baiji has been limited to the middle and lower reaches of the Yangtze River. In the South Asian subcontinent, the blind river dolphins (*Platanista gangetica*) inhabit several large river systems that drain into the Bay of Bengal and one, the Indus, that drains into the Arabian Sea. The species exists as a metapopulation, with no present-day exchange between the Indus and Ganges systems. Since these river dolphins are fairly common in the Sundarbans delta of the Ganges-Brahmaputra-Meghna complex, some may disperse in the Meghna's freshwater plume to the Karnaphuli River of southeastern Bangladesh, where there is a small and at least semi-isolated population (Ahmed 2000; Smith et al. 2001). The boto, or Amazon dolphin (*Inia geoffrensis*), also exists as a metapopulation, with three morphologically distinguishable populations: one in the Orinoco River system of Venezuela and Colombia, another in most of the Amazon River system, and the third in the upper Madeira River drainage of Bolivia. These populations, or subspecies, are isolated by waterfalls or long stretches of rapids (Best and da Silva 1989). Unlike the others in this group of true river dolphins, the franciscana, or La Plata dolphin (*Pontoporia blainvillei*), does not enter fresh water. Rather, it ranges along the Atlantic coast of South America between southeastern Brazil and central Argentina. The franciscana's distribution seems to be influenced by the discharge of large continental rivers, the subtropical Brazil Current, and the availability of large concentrations of juvenile sciaenids in nearshore waters (Crespo, Harris, and González 1998).

Arctic Endemics. One small family of odontocetes, the Monodontidae, is endemic to high latitudes of the Northern Hemisphere. The narwhal (*Monodon monoceros*) has a limited distribution centered in deep, ice-infested basins that branch northward from the western and central North Atlantic (e.g., Davis and Hudson Straits, Baffin Bay, and the Greenland Sea). The white whale, or beluga (*Delphinapterus leucas*), has a more nearly continuous circumpolar distribution, as well as several populations outside the Arctic—for example, in Canada's lower St. Lawrence River (ca. 48°N), in Bristol Bay (Alaska, ca. 57°N), and in Russia's Amur River delta (ca. 53°N) (International Whaling Commission 2000). Both species are ice adapted and at least occasionally follow lanes of open water into the Polar Basin north of 80°N (Belikov, Gorbunov, and Shil'nikov 1984; Dietz et al. 1994; Martin and Smith 1999).

MARINE CARNIVORES

Seals, sea lions, fur seals, the walrus, sea otters and marine otters, and the polar bear are conceptually grouped together as marine carnivores, a nontaxonomic assemblage within the order Carnivora (cf. Reeves et al. 2002; Rice 1998). All

The narwhal (*Monodon monoceros*) is best known for the long, spiraled tusk of the adult male. Females, like the one shown here with a very young calf, not only lack the external tusks of adult males but also have no erupted teeth within the mouth. Native hunters kill hundreds of narwhals each year in northern Canada and Greenland. Females and young narwhals are hunted primarily for their nutritious and tasty skin (muktuk or mattaq), while adult males offer the added incentive of the commercially valuable tusk ivory. Photograph by Randall Reeves.

of these marine carnivores are well adapted for aquatic life and are excellent swimmers and divers, with some of the phocid pinnipeds excelling at long-duration, deep diving (Stewart 2002). All but the polar bear forage exclusively in the water.

Carnivores are known mostly as flesh (or meat) eaters, with *flesh* generally referring to the soft tissues of other mammals rather than to fish and invertebrates such as squid. However, not even all terrestrial carnivores fit that description. Marine carnivores generally have fewer teeth than do terrestrial carnivores, and the carnassial teeth typical of terrestrial carnivores have been substantially modified (King 1983; Reeves et al. 2002). The modifications are, in fact, so extreme that marine carnivores essentially lack carnassial teeth. Rather, their cheek teeth are similar in shape and modified for handling a soft-flesh diet, with little or no crushing surface (cf. Heinrich 2002).

The marine carnivores are often divided into two nontaxonomic categories, known casually as the pinnipeds (seals, sea lions, fur seals, and the walrus) and the marine fissipeds (polar bear, sea otter, and marine otter). For organizational convenience, we treat these two groups under separate headings here.

Pinnipeds

The term *pinniped* is derived from the Latin *pinnipes,* for wing- or fin-footed, and originally from the classical Latin *pinna,* meaning feather or wing, and *pes,* meaning foot (King 1983). These terms refer to the modifications of the appendages into finlike structures. Formerly, the pinnipeds were classified as an order, the Pinnipedia, separate from the order Carnivora (cf. Reeves, Stewart, and Leatherwood 1992; Riedman 1990). The consensus now is that these animals constitute three closely related families of carnivores, all derived from terrestrial carnivore ancestors. As yet, however, there is no consensus on the phylogenetic relationships among these families (cf. Arnason et al. 1995; Cipriano 2002; Fordyce 2002; Lipps and Mitchell 1976; Repenning 1976). Some cladistic taxonomists argue for a monophyletic origin from an ursid arctoid (bearlike) ancestor, whereas others argue for a monophyletic origin from a mustelid arctoid (weasel-like) ancestor (cf. Berta 2002; Heinrich 2002; Wyss 1989). Other analysts, including many numerical taxonomists (i.e., pheneticists), have found support in the fossil record and in the anatomy, physiology, morphology, behavior, distribution, and molecular genetics of extant species pointing toward ursid ancestries for the otariid and obobenid pinnipeds and a mustelid origin for the phocid pinnipeds (Fordyce 2002). The polarized debate seems to derive as much from methodological differences between cladistics and phenetics as from substantive difficulties in collecting unambiguous comparative data on the various important characters. Apart from the many unresolved questions about extinct forms, the taxonomy of living pinnipeds is straightforward and stable compared to that of the cetaceans. Remaining controversy is centered on questions of whether some geographically isolated populations should be regarded as separate species or only subspecies.

The three pinniped families have several features in common: large body size with small surface-to-mass ratio, relatively large eyes, well-innervated facial vibrissae, streamlined body shape (with enclosure of the limb bones in the body), development of insulation (subcutaneous blubber depots or dense fur, or both), modification of the appendicular skeleton resulting in a reduction in the size of the appendages and paddlelike or finlike structures of the hands and feet (e.g., Bryden 1971; Godfrey 1985; Howell 1929, 1930; Hyvårinen 1989; Innes et al. 1990; Jamieson and Fisher 1972; King 1983; Pabst, Rommel, and McLellan 1999; Pardue, Sivak, and Kovacs 1993; Williams and Kooyman 1985), and physiological mechanisms allowing extended apnea while diving and searching for food (e.g., Blix, Grav, and Ronald 1975; Butler and Jones 1997; Kooyman 1973, 1989; Kooyman and Ponganis 1998; Lavigne et al. 1986; Ponganis et al. 1997; Snyder 1983). Pinniped hair is flattened in cross section, compared to round in other carnivores (Yochem and Stewart 2002). This is evidently an adaptation to enhance streamlining and swimming performance. Pinnipeds do not have arrector pili muscles, so the angle of their hair canals cannot be altered and their hairs cannot be made to stand up (as a dog's can, for example). These animals are able to spend substantial amounts of time foraging in water, which has 25 times the heat-conducting capacity and 800 times the density of air. The impressive aquatic adaptations of pinnipeds are conditioned, however, by the animals' need to haul out on land or ice to give birth and molt for several days to weeks at a time.

Pinnipeds have fewer incisors than do terrestrial carnivores. The canines of most pinnipeds are large and conical, often larger and more robust in males than females. Otariid pinnipeds typically have 6 incisors, 2 canines, and 10 to 12 postcanines in the upper jaw and 4 incisors, 2 canines, and 10 postcanines in the mandible. Most true seals (phocid pinnipeds) have 6 incisors, 2 canines, and 10 postcanines in the upper jaw and 4 incisors, 2 canines, and 10 postcanines in the lower jaw. Exceptions are the hooded seal (*Cystophora cristata*), monk seals (*Monachus* spp.), and elephant seals (*Mirounga* spp.), which have only four upper incisors and, in the cases of hooded and elephant seals, only two lower incisors. The cheek teeth of crabeater seals (*Lobodon carcinophaga*) are the most highly modified of mammalian teeth, specialized for straining krill (small, shrimplike crustaceans) from gulps of seawater. Leopard seal (*Hydrurga leptonyx*) teeth are also modified to act as sieves, but they retain some shearing abilities, reflecting

Walruses (*Odobenus rosmarus*) were almost exterminated from the northeastern Atlantic part of their range by commercial hunters. After many decades of protection, their numbers are steadily increasing in Svalbard. Photograph by Ian Gjertz.

this species' more diverse diet. The cheek teeth of other pinniped species, such as elephant seals and the Ross seal (*Ommatophoca rossii*), are greatly reduced in size.

Odobenidae. The walrus (*Odobenus rosmarus*) is the only living representative of what was once a diverse group that included several lineages of non-tusked and sea lion–like walruses (Barnes, Domning, and Ray 1985). Its massive body, sparse hair, heavy and coarse mustache, and robust canine teeth modified into long tusks make the walrus unmistakable. It nevertheless has some characteristics of both otariids and phocids. Like the phocids, the walrus lacks external ear flaps. The foreflippers are like those of the otariids, while the rear flippers are more similar to those of the phocids. Unlike the phocids, though, the walrus can rotate its rear flippers underneath its body and walk on them on ice or land. In the water, walruses use either the front or rear flippers for propulsion. Walruses swim mostly by using the hindflippers to generate propulsive force. The foreflippers are also used like paddles at slower speeds to move forward or backward, to steer, and to provide lift when swimming or gliding fast.

The distribution of the walrus is disjunct circumpolar in the Arctic, with large gaps caused by permanent ice massifs along the central northern coasts of both North America and Eurasia (cf. Fay 1985). Separate subspecies are recognized in the Atlantic (*O. r. rosmarus*) and Pacific (*O. r. divergens*) basins, with a third group centered in the Laptev Sea; this group is less well differentiated but sometimes also accorded subspecific rank as *O. r. laptevi* (Fay 1981; Rice 1998). Eight putative "stocks" of *O. r. rosmarus* are recognized, ranging from Hudson Bay and Foxe Basin in the west to Novaya Zemlya in the east (Born, Gjertz, and Reeves 1995).

Otariidae. There are two main groups of otariids: the sea lions and fur seals (Gentry 2002; Reeves et al. 2002; Rice 1998). Of the 16 otariid species, 9 are fur seals and 7 are sea lions. These "eared" seals have short external flaps of skin as outer-ear appendages. They can walk on both front and rear flippers, with the latter rotated underneath the body. Slow walking involves alternately moving the left and right foreflippers forward. The animals can also gallop or run across the sand and rocks by jumping, with both foreflippers thrust forward simultaneously, while pushing off with the hindfippers (e.g., Beentjes 1990). When in the water, otariids use their long foreflippers for propulsion (pectoral oscillation) and also to steer and provide lift when gliding (Feldkamp 1987; Ponganis et al. 1990). The hindflippers may have some function in steering but are not used for propulsion (English 1976, 1977). The primary morphological adaptations for this type of swimming are shortened proximal limb bones, which are also completely enveloped in the body to reduce drag and increase streamlining, dorsoventral flattening of the radius and ulna, and lengthening of the bones of the manus, giving the appendage a winglike structure (Pabst, Rommel, and McLellan 1999). This design affords a powerful and efficient lever system, with the primary locomotor muscles acting directly on the upper arm. The proximal bones of the rear appendages are also shortened and completely enclosed in the body, serving to reduce drag and to enhance streamlining as the body tapers to a narrow pelvic girdle.

Fur seals have a short, dense layer of hair close to the skin, which traps a layer of air and provides a waterproof insulation (Yochem and Stewart 2002). Longer, less dense hairs form an outer layer. Sea lions have only a sparse layer of shorter hair that is not effective in keeping the skin dry. Both groups have a thin blubber layer. Otariids are found exclusively in the North Pacific and the Southern Hemisphere. They penetrate the South Atlantic only as far north as Angola in the east and southern Brazil in the west.

The sea lions are classified into five genera (*Eumetopias, Zalophus, Otaria, Phocarctos,* and *Neophoca*), all but one of them (*Zalophus*) monotypic. The northern, or Steller, sea lion (*Eumetopias jubatus*) occurs in coastal and shelf waters of the cool temperate North Pacific, northward from Hokkaido in the west and southern California in the east almost to the Bering Strait. The Steller sea lion is the only otariid that regularly hauls out on sea ice (Rice 1998). California sea lions (*Z. californianus*) are familiar to most people because of their role as performers in circuses and amusement parks. They have haul-out sites along the west coast of North America from Vancouver Island in the north to the southern tip of the Baja California peninsula in the south, although only males occur regularly north of California. A genetically isolated population inhabits the Gulf of California (Maldonado et al. 1995). Two even more isolated populations are often regarded as subspecies or species. The Japanese sea lion (*Z. japonicus*) occurred historically in the Sea of Japan and along portions of Japan's Pacific coast. There have been no authenticated observations of Japanese sea lions for several decades, and they may be extinct; however, reports of pinnipeds on and near Takeshima Island as recently as the 1970s suggest that a few may still exist (Rice 1998). The Galápagos sea lion (*Z. wollebaeki*) inhabits the Galápagos archipelago as well as Isla de La Plata, a small island near the coast of Ecuador (Rice 1998). Its residence, and that of the Galápagos fur seal (see later), in equatorial waters is explained by the cool, highly productive Humboldt (Peru) Current, which creates sea conditions more like those typical of temperate latitudes.

The other three sea lions are confined to particular regions in the Southern Hemisphere. The most widespread and abundant is the South American sea lion (*Otaria flavescens* [= *byronia*]). Its rookeries are scattered along the mainland coast from northern Peru in the west, south to Tierra del Fuego and Isla de los Estados, and north to southern Brazil in the east. The New Zealand and Australian

Northern fur seals and California sea lions share beach space during the summer breeding season on San Miguel Island in southern California. The breeding range of the California sea lion extends from islands in the southern Gulf of California northward along the Pacific coast of Baja California to the Farallon Islands, off San Francisco, California. The principal colonies of northern fur seals are in the Bering Sea, but a few pups have been born on the Farallon Islands, and the colony at San Miguel Island has been growing steadily since its establishment in the 1960s. While at sea, these species overlap along the west coast of North America from southern California to British Columbia. However, the fur seals generally occur well offshore, while the sea lions travel and feed within a few miles of the coast. Photograph by Brent S. Stewart.

sea lions (*Phocarctos hookeri* and *Neophoca cinerea,* respectively) are endemic to cool temperate coastal waters of Oceania, including several subantarctic islands. They are much less abundant (probably by at least an order of magnitude) than is the South American sea lion (see Reijnders et al. 1993).

The world's fur seals are classified into one monotypic genus (*Callorhinus*) and one polytypic genus (*Arctocephalus*). The genus *Callorhinus* is restricted to the North Pacific Ocean, where the northern fur seal (*C. ursinus*) occurs from the Sea of Japan northward to the southeastern Sea of Okhotsk, eastward across the southern Bering Sea and northern Gulf of Alaska, and thence southward all the way to northern Baja California. This species supported the large commercial seal hunts at the Commander, Pribilof, and Aleutian Islands in the nineteenth and twentieth centuries (Busch 1985; Scheffer, Fiscus, and Todd 1984). The genus *Arctocephalus* is considerably more diverse and widespread, encompassing all of the southern fur seals (Croxall and Gentry 1987). Only one species, the Guadalupe fur seal (*A. townsendi*), occurs in the Northern Hemisphere. It is endemic to Guadalupe and San Benito Islands off the west coast of Baja California, Mexico, with regular occurrences in recent years on certain of the Southern California Channel Islands (Stewart et al. 1987). Two other species are endemic to single archipelagoes—the Galápagos fur seal (*A. galapagoensis*) and the Juan Fernández fur seal (*A. philippii*). The other five *Arctocephalus* species have considerably wider ranges: the Antarctic, or Kerguelen, fur seal (*A. gazella*) in polar latitudes of the South Atlantic, Indian, and western South Pacific Oceans; the subantarctic, or Amsterdam, fur seal (*A. tropicalis*) in somewhat lower latitudes of the same region; the South American fur seal (*A. australis*) along the coasts of South America and the Falkland Islands; the South Australian and New Zealand fur

Harbor seals occur along the rim of the North Pacific Ocean from Baja California northward, throughout the Aleutian Islands, and to northern Japan. They also range on both sides of the North Atlantic Ocean, and a small population exists in a freshwater lake system in eastern Canada. Harbor seals forage mostly in shallow marine habitats on the bottom, in kelp beds, and in estuaries in some areas. Their body color patterns range from dark backgrounds with lighter rings, predominantly in the southern parts of their range, to light gray or silver backgrounds with scattered spots, in the northern parts. Photograph by Brent S. Stewart.

seals (*A. forsteri*), consisting of two disjunct populations, one along the southern coast of Australia and the other around New Zealand; and the Tasmanian (Australian) and Cape (South African) fur seals (*A. pusillus*), also consisting of two widely disjunct populations, one along the southeastern coast of Australia and Tasmania and the other along the west coast of southern Africa (Warneke and Shaughnessy 1985). As populations of many of the Southern Hemisphere fur seal species have recovered from past exploitation and their ranges have expanded from small, isolated remnant populations, hybridization has occurred at some locations (e.g., Kerley 1983; Shaughnessy, Shaughnessy, and Fletcher 1988).

Phocidae. There are 19 species in 13 genera of these "true" or "earless" seals. Phocids lack external ear flaps and cannot rotate their rear flippers underneath the body. On land, they pull themselves along with inch-worming undulations, often accompanied by rowing movements of the foreflippers against the substrate. The rear flippers trail passively. On ice, some phocids (e.g., crabeater seals, ribbon seals) can move very quickly, making sinusoidal lateral movements to glide, or slither, across the surface. The foreflippers are either held tightly against the sides or used as paddles to gain traction and thus control speed and direction on snow-covered surfaces. Other phocids (e.g., the Ross seal, elephant seals, and the Weddell seal, *Leptonychotes weddellii*) lurch forward slowly, sometimes paddling with their foreflippers. Phocids swim primarily by pelvic oscillation, making propulsive, side-to-side figure-eight movements with their rear flippers. The digits are spread widely, stretching the interdigital webbing to increase the surface area pushing against the water. The foreflippers are used mostly for steering and to provide lift when gliding. They may also be used, together or independently, for backing up or quickly moving forward over short distances as the seal pursues prey on the seafloor or around coral reefs. Some species with especially long foreflippers, like crabeater and leopard seals, may use them for propulsion. As in otariids, the proximal bones of both front and rear flippers are shortened and enveloped within the body; this reduces drag and improves streamlining (Pabst, Rommel, and McLellan 1999).

The tips of the digits of phocid foreflippers are not completely enclosed in skin, but there is some short webbing between them. Phocids have well-developed claws compared with the poorly developed or barely noticeable claws on the foreflippers of otariids. The hair of phocids is short, and they have either a very sparse second layer of hair close to the skin (underfur) or none at all. All phocids have a thick blubber layer that serves to insulate their body core and store energy. They generally fast for longer periods than otariids—up to a month in some species (Boyd 1998).

Four phocid genera live exclusively in the Antarctic, with circumpolar distributions around the Antarctic continent (Boyd 2002a; Reeves et al. 2002). These include the extremely abundant crabeater seal, the widespread but less abundant leopard seal, the fast-ice-inhabiting Weddell seal, and the relatively rare and little known Ross seal. Five more monotypic genera live in polar to cold temperate latitudes of the Northern Hemisphere. Three are confined to the North Atlantic and its appended seas: the coastal gray seal (*Halichoerus grypus*), the much more pelagic hooded seal, and the very abundant, wide-ranging harp seal (*Pagophilus groenlandicus*). The bearded seal (*Erignathus barbatus*) is circumpolar in the pack-ice zone, while the ribbon seal (*Histriophoca fasciata*) has a restricted distribution in the Bering Sea, extending into the northern North Pacific during the ice-free summer period when seals may be pelagic.

Three polytypic genera (*Phoca, Pusa, Monachus*) also live exclusively in the Northern Hemisphere (Reeves et al. 2002; Rice 1998). The harbor seal (*Phoca vitulina*) ranges along the east and west coasts of both the North Pacific and the North Atlantic, while its close relative, the spotted seal (*Phoca largha*), is restricted to the Bering Sea, with occasional movements south into the eastern North Pacific and north into the Beaufort and Chukchi Seas. Some populations of harbor seals inhabit freshwater lake systems in eastern Canada. The ringed seal (*Pusa hispida*) has a circumpolar Arctic and subarctic marine distribution, while its close relatives, the Caspian and Baikal seals (*P. caspica* and *P. sibirica*), are restricted to inland water bodies—the saline Caspian Sea in Asia Minor and the freshwater Lake Baikal in eastern Russia. Small populations of ringed seals inhabit freshwater lake systems in Finland and western Russia (Kunnasranta et al. 2001, 2002; Sipilä and Hyvårinen 1998).

The monk seals are the most tropical pinnipeds. Hawaiian monk seals (*Monachus schauinslandi*) occur primarily around the northwestern Hawaiian islands, although their numbers have been increasing around the main Hawaiian islands in recent years (Ragen and Lavigne 1999). The Mediterranean monk seal (*M. monachus*) survives only in small, disjunct, remnant populations along the west coast of north Africa, at Madeira, and in the northeastern Mediterranean Sea (Forcada, Hammond, and Aguilar 1999). The Caribbean monk seal (*M. tropicalis*) apparently lived throughout much of the Caribbean basin, in portions of the Gulf of Mexico, and eastward to the Bahamas; it has probably been extinct since the late 1950s (Kenyon 1977). Finally, there are two species of elephant seals, one (*Mirounga angustirostris*) in the eastern North Pacific (Stewart and Huber 1993) and the other (*Mirounga leonina*) in the temperate and subpolar Southern Hemisphere, with numerous disjunct breeding populations (Laws 1994).

Marine Fissipeds

The term *fissiped* refers to the fact that these animals' toes are separated from each other, while the digits of pinnipeds are joined together either by skin webbing or cartilage. All members of the order Carnivora, except the pinnipeds, were formerly included in the suborder Fissipedia. Although that taxonomic grouping has been abandoned, it is still often convenient to distinguish between the fissiped marine carnivores and the pinniped marine carnivores.

Mustelidae is the most diverse family of carnivores. The mustelids are long, thin-bodied animals with short legs adapted for pushing through dense underbrush or maneuvering in tunnels on land. The body plan is also well suited to an aquatic lifestyle. Although several otter species forage at least occasionally in marine and estuarine environments, only the sea otter and marine otter are intimately and obligately tied to marine ecosystems.

Sea otters are unique among carnivores in having no sharp cutting edges on their teeth; even the canines are blunt and rounded. Their dentition is adapted for crushing invertebrate prey, many of which have strong exoskeletons, rather than for cutting or shearing flesh. The hind feet are large and flipperlike, and the pelage has extraordinary insulative properties. A layer of air trapped in the fur prevents excessive heat loss across the body surface. A principal function of grooming, which is often accomplished as the otters raft on their backs at the sea surface, is aeration of the fur.

The sea otter lives along the coasts of northern Japan and Russia in the western Pacific, throughout the Aleutian Islands, and along the North American coast from the Gulf of Alaska south to Baja California (Estes and Bodkin 2002). This extensive range, however, is fragmented because of a combination of ecological variability and the effects of past hunting. Deep, wide passes that separate islands from the mainland and each other help prevent dispersal. Three subspecies are distinguishable based on geographic variation in skull characters: *Enhydra lutris lutris* from the Kuril to Commander Islands in the western Pacific, *E. l. kenyoni* in the Aleutian Islands south to Washington, and *E. l. nereis* in California waters (Wilson et al. 1991). Translocation programs have reestablished sea otters in portions of southeastern Alaska, British Columbia, and Washington, where they had been extirpated or reduced to very low numbers. An attempt at reintroduction in Oregon proved a failure.

The marine otter, or chungungo, lives along the west coast of South America from south-central Peru south to the Cape Horn region. It is morphologically similar to the neotropical river otters but distinct both phylogenetically and ecologically.

The polar bear has a circumpolar distribution in the Arctic (Derocher et al. 1998). Polar bears occur along the margins of land masses and roam offshore onto the stable and drifting pack ice in search of their pinniped prey, mainly the ubiquitous ringed seal. Adult females in most areas are tied to coastal habitat for denning, but radio-tracking studies in Alaska have revealed that a large proportion of the maternity dens in that region are in the offshore multiyear pack ice of the Beaufort Sea (Amstrup and DeMaster 1988). Adult males, in particular, cover vast distances by walking, swimming, and riding on the drifting pack ice far from land.

SIRENIANS

The order Sirenia consists of two modern families: the dugongs, Dugongidae, and the manatees, Trichechidae. In comparison to the cetaceans and pinnipeds, the modern sirenians are a small and homogeneous assemblage. This was not always true. Sirenians were much more widespread and diverse in the past, reaching a peak in abundance and diversity during the Miocene (5–25 million years ago) when, not incidentally, the earth reached its own peak of climatic tropicality

Polar bears occur exclusively in northern polar and subpolar latitudes, in close association with sea ice. They subsist mostly on ringed seals, although they also eat other seal species, walruses, belugas, and occasionally narwhals. Recent climatic warming evidently has begun to have significant effects on the reproduction and survival of polar bears (especially cubs), caused by changes in the timing of ice formation and breakup and in the availability of ringed seals. Polar bears near Churchill, Manitoba, on the west coast of Hudson Bay, support a sizable tourist business. On the outskirts of Churchill, people are able to approach and photograph bears in relative safety. Photograph courtesy of U.S. Fish and Wildlife Service.

(Barnes, Domning, and Ray 1985). It has been suggested that, when they were more abundant than they are today, sirenians could have exerted important influences on the evolution of marine plant communities (e.g., Domning 1989; Reynolds and Odell 1991).

The modern sirenians consist of only five species (Reynolds and Odell 1991). The family Dugongidae has two monotypic genera: the dugong (*Dugong dugon*) and the extinct Steller's sea cow (*Hydrodamalis gigas*). The family Trichechidae has a single genus with three species: the West Indian manatee (*Trichechus manatus*), West African manatee (*T. senegalensis*), and Amazonian manatee (*T. inunguis*). The four living sirenian species all fall within the size range of about 2.5–4.5 meters and 250–1,500 kg, whereas the sea cow grew up to 8 meters and could weigh as much as 3,500 kg.

The dugong is an Indo-Pacific marine species whose historical range encompassed virtually all tropical coastal and island waters with sizable seagrass communities from eastern Africa in the west to Melanesia and the Ogesawara (Ryukyu) Islands of southern Japan in the east. Although the dugong's capabilities for dispersal across wide, deep expanses of ocean should not be underestimated, numerous populations seem to have been isolated at least partially by natural factors and only secondarily by the effects of human hunting, fishing, and habitat modification (see Nishiwaki et al. 1979; Nishiwaki and Marsh 1985). For example, there is a sizable dugong population in the Persian Gulf, but the nearest known populations outside the Gulf are in the Red Sea to the west and the Gulf of Kutch to the east.

By the time of its discovery around the Commander Islands of the western Bering Sea in 1741, Steller's sea cow had a very limited range. It was completely extirpated by fur hunters within less than 30 years (Anderson and Domning 2002). This sea cow was the surviving relict of a genus that apparently had occurred along the rim of the North Pacific from Japan to southern California during the Pliocene and possibly into the Pleistocene. Steller's sea cow and its close relatives were the only cold-adapted sirenians, and they flourished by reliance on the luxuriant kelp beds along North Pacific shorelines (Barnes, Domning, and Ray 1985). It has been suggested that, under pristine conditions, the matching ranges of sea otters and sea cows may have reflected an ecological relationship, with the latter benefiting from the former's habit of feeding on algivorous sea urchins (Dayton 1975).

The West Indian manatee has two well-defined subspecies based on cranial morphometrics (Domning and Hayek 1986). The Florida subspecies, *T. m. latirostris,* ranges along the southeastern coast of North America, seasonally reaching Maryland in the north and Louisiana in the west. Its core distribution is the Florida peninsula. The Antillean subspecies, *T. m. manatus,* may have ranged in the past almost continuously from the United States–Mexico border south along the Central and South American coasts to eastern Brazil. It was also once abundant in portions of the Greater Antilles (Cuba, in particular) but is now scarce in the Caribbean except possibly in parts of Puerto Rico, Jamaica, and Cuba (Lefebvre et al. 2001). A recent phylogeographic analysis using mitochondrial DNA indicated three distinct lineages, broadly corresponding to (1) Florida and the West Indies, (2) the Gulf of Mexico to South American rivers flowing into the Caribbean Sea, and (3) the northeastern Atlantic coast of South America (Garcia-Rodriguez et al. 1998). This analysis also suggested limited long-distance gene flow, implying that stretches of open water and unsuitable coastal habitat have served as effective biogeographic barriers to manatee dispersal.

The Amazonian manatee is the smallest of the three manatee species and is distinguished by the lack of nails on the flippers. Its natural distribution seems to be limited to the Amazon drainage, where it occurs from Ecuador and Peru all the way to the Atlantic Ocean. The distributions of the West Indian and Amazonian manatees apparently abut and may even overlap slightly in the Amazon estuaries (Domning 1981; Lefebvre et al. 2001; Rosas 1994). Translocation of Amazonian manatees into West Indian manatee habitat in Guyana and Panama, ostensibly for aquatic weed control, may have led to a certain amount of hybridization (Garcia-Rodriguez et al. 1998).

The West African manatee inhabits rivers, estuaries, and lagoons from Mauritania in the north to the Cuanza River in Angola in the south. Its occurrence in coastal marine waters is little known. However, manatees are widely distributed and relatively common in the Bijagós archipelago off Guinea-Bissau (Silva and Araújo 2001).

Habitat and Ecology

CETACEANS

As intelligent, highly mobile, and adaptive mammals, cetaceans exploit a diversity of habitats. Their ecological roles can vary by species, geographic area, season, age class or stage, and in some cases sex. Some mysticete populations are strongly migratory, with a general pattern of seasonal movement between productive high-latitude feeding grounds and low-latitude breeding grounds (Stevick, McConnell, and Hammond 2002). This phenomenon is manifest most spectacularly by gray and humpback whales, whose annual migrations are the longest of any mammal species. Although less impressive in terms of distance traversed, the seasonal movements of right whales and bowhead whales are often similarly predictable.

A series of hypotheses have been offered to explain mysticete migration (Corkeron and Connor 1999): (1) it optimizes energy budgets by taking the animals away from thermally stressful regions during periods when production there is low; (2) it is vestigial behavior, with no evident selection value; (3) the whales are tracking prey concentrations; or (4) movement into low latitudes is driven by benefits to calf growth and survivorship. The last of these could involve thermoregulatory considerations (e.g., the calf is able to expend more of its energy for growth rather than heat production), the availability of calmer surface waters (reducing energy demands and risks of "drowning"), or avoidance of killer whale predation. None of these hypotheses is entirely satisfactory, but Corkeron and Connor (1999) advocate the hypothesis that, at least for gray and humpback whales, it is the selective advantage conferred on calves that drives the migration. In favor of the predation hypothesis, they suggest that mammal-eating pods of killer whales are linked primarily to pinnipeds, which remain year-round in temperate and higher latitudes (Connor and Corkeron 2001). Thus, young baleen whales would be viewed, from the killer whales' perspective, as highly desirable but only episodically available prey (i.e., primarily during their first poleward migration). Clapham (2001), in contrast, champions the idea that energy conservation is the selective force underlying mysticete migration. Thus, individual adult whales (humpbacks in his example) increase their own reproductive success by migrating to warmer waters, where their offspring can devote most of their energy to growth and development during the first months of life.

In general, odontocetes are less prone than mysticetes to seasonal feasting and fasting strategies (they are "income breeders" rather than "capital breeders"; Bowen, Beck, and Austin 2002; and see below). Many odontocete species, especially those living in tropical or temperate latitudes, do not migrate in the usual sense. River dolphins must attune their movements and reproduction to seasonally variable hydrological conditions. For example, in the Brazilian Amazon, the boto and tucuxi experience annual changes in water level of about 10 meters, and they respond in differing ways. Tucuxis give birth mainly in October and November, when water levels are lowest, while botos give birth between May and July, when water levels are high or beginning to decline. Best and da Silva (1984) propose that the seasonality of calving in both dolphin species is linked to food availability, the idea being that calves should be born at "the most opportune time of year for increased feeding." Tucuxis prey mainly on pelagic schooling fishes that become spatially concentrated and thus vulnerable to predation when water levels are low. Botos, in contrast, prey on solitary benthic organisms (as well as pelagic species) and readily take advantage of opportunities to forage in the inundated grasslands and rain forest. Interestingly, botos living in the Orinoco River system of Venezuela give birth near the end of the low-water season and during the period of rising water, which puts them out of phase with botos in the central and lower Amazon (McGuire and Winemiller 1998). This leads to an alternative hypothesis. If fishes are indeed larger and easier to catch as water levels recede, perhaps it is beneficial for the female and her offspring to experience this bounty four to six months after the calf's birth rather than near the time of birth (ibid.). An intensive, ongoing study of the ecology and population dynamics of botos in Brazil (da Silva and Martin 2000) should soon make it possible to reevaluate these apparently competing hypotheses.

The beaked whales and sperm whales feed in the aphotic zone, largely but not

The boto, also known as the Amazon river dolphin (*Inia geoffrensis*), is endemic to river and lake systems of northern South America. It is the most abundant of the obligate freshwater cetaceans. Local people have traditionally regarded these long-beaked, pink-tinged animals with superstitious dread, and this has meant that they are not hunted in most areas. Photograph by Elizabeth L. Zúñiga.

entirely on cephalopods (e.g., Hooker and Baird 1999; Jaquet, Dawson, and Slooten 2000). All beaked whales, except Shepherd's beaked whale, have reduced dentition, usually consisting of only one or two pairs of erupted mandibular teeth. The sperm whale has teeth only in its narrow lower jaw, with matching sockets in the upper jaw to receive the teeth when the mouth is closed. This trend toward fewer functional teeth is typical of squid specialists (Clarke 1986; Gaskin 1982). Anatomical and other evidence suggests that these whales use suction to capture their prey (Heyning and Mead 1996). As the tongue is retracted through muscular action, the throat becomes distended, creating a sudden drop in pressure inside the mouth. All beaked whales and sperm whales have at least one pair of anteriorly convergent throat grooves, which seem to facilitate gular expansion. Suction feeding would explain why squid are often found intact in the stomachs of ziphiids and sperm whales: the strong force of suction obviates the need for the whale to grasp or pierce the body of its prey before ingestion. An alternative explanation has been suggested by Norris and Møhl (1983), who propose that some odontocetes might use loud, impulsive sounds to stun and debilitate prey, making them easier to hold and swallow. The two hypotheses, suction feeding on one hand and "acoustic stunning" on the other, are not mutually exclusive. While the former is generally accepted, however, the latter remains in the "interesting but untested" category (Tyack and Miller 2002).

Insulated as they are, with a dense hypodermal layer of blubber, cetaceans manage to maintain a muscle core temperature of 36–37°C while living in much colder environments (seawater can be as cold as −1.7°C). This homeothermy is possible because of not only the insulation provided by blubber but also heat-conserving features, such as the low ratios of body surface to body mass (or volume) (this applies particularly to the large whales) and the body streamlining (e.g., loss of hind limbs, internal positioning of genitalia and mammary glands) (Elsner 1999; Fish 1993; Howell 1930; Pabst, Rommel, and McLellan 1999; Rommel et al. 1995; Williams and Worthy 2002). Cetaceans also have efficient countercurrent heat exchange systems that regulate blood flow to the extremities (flippers, flukes, and dorsal fin), conserving or dissipating heat as necessary (Scholander and Schevill 1955). In some species, particularly the migratory baleen whales that fast for months at a time, the blubber content and thickness change dramatically, depending on life stage and season. In a classic study of gray whales, Rice and Wolman (1971) established that southbound whales in the autumn were in much better nutritive condition than northbound whales in the spring. Although considerable amounts of lipid are stored in muscle, visceral fat, bone, and organs as well, the blubber is specially suited to this role. In fin and sei whales, the blubber can constitute as much as a fifth to a quarter of the entire body mass (Lockyer 1987). The blubber layer can be up to 50 cm thick in bowhead whales (Lowry 1993).

The ecological role of cetaceans is much debated (Clapham and Brownell 1996; Plagányi and Butterworth 2002). In the past, culling of some odontocetes (e.g., killer whales, belugas, and Black Sea dolphins) was pursued as a way of reducing perceived competition with fisheries, and in recent years Norwegian and Japanese interests have sought to justify more whaling of mysticetes (minke and Bryde's whales) by citing their consumption of harvestable fishery resources (e.g., International Whaling Commission 2002; Víkingsson and Kapel 2000). The roles of cetaceans in nutrient cycling, maintaining commensal and parasitic faunas, and providing food for scavengers have been noted but little explored (Bowen 1997; Katona and Whitehead 1988; Smith et al. 1989). One of the more interesting examples of physical structuring of habitat involves gray whales. Their disturbance of the benthos while feeding on dense beds of amphipods may help maintain the sand substrate by suspending fine particles, resulting in higher densities of prey over large spatial scales (Oliver and Slattery 1985).

MARINE CARNIVORES

Pinnipeds

The seasonal movements of pinnipeds range from virtually none at all to the extensive migrations of elephant seals (e.g., Bonadonna, Lea, and Guinet 2000; Bowen and Siniff 1999; Campagna, Fedak, and McConnell 1999; Folkow, Martensson, and Blix 1996; Lacoste and Stenson 2000; Merrick and Loughlin 1997; Stewart 1997; Stewart and DeLong 1993, 1995). Most northern phocids that occupy pack-ice habitats engage in seasonal migrations to some extent, moving between breeding sites and productive foraging areas. Weddell, crabeater, and Ross seals move between fast ice and pack ice. Weddell seals evidently migrate seasonally between those fast-ice or interior pack-ice habitats and offshore pelagic areas, and some seals may actually remain offshore for several years (e.g., Stewart et al. 2000, 2001; Testa 1994). Perhaps the most interesting recent data are for the Ross seal, which was previously thought to live year-round in the interior pack ice, where it is almost inaccessible for study. Ross seals evidently leave the pack ice after they molt, forage for at least several months in pelagic habitats of the Southern Ocean, and return to the interior pack ice to breed the following summer (cf. Boveng and Bengtson 2001; Nordoy and Blix 2001).

All pinnipeds shed their hair once a year, usually during a brief period in summer or autumn, although the timing varies among species (Ling 1970, 1974; Scheffer 1964; Yochem and Stewart 2002). Single hairs are shed, except by the elephant seals, monk seals, and perhaps the Ross seal. These species shed their

Ross seals were formerly considered rather rare Antarctic seals, but recent studies suggest that they are fairly common in some areas, such as the Weddell and Ross Seas. They breed in dense, interior pack-ice habitats, which research vessels cannot penetrate in early summer, and then spend most of the rest of the year in open water north of the pack-ice zone. During late summer, when they are molting, their range may overlap those of Adelie and emperor penguins (*Pygoscelis adeliae* and *Aptenodytes forsteri*, respectively), particularly in the eastern Ross Sea. Photograph by Brent S. Stewart.

hair in large sheets, with the upper layer of epidermis still attached. This type of shedding or molt is called "catastrophic." Effective replacement of hair during only several weeks requires the almost continuous transport of nutrients to the skin surface by the blood. Adaptations for heat conservation when in the water and oxygen conservation while diving, however, result in routine restriction of blood flow to the body surface (Elsner 1999). Consequently, most pinnipeds spend substantially more time hauled out when molting. For example, elephant seals spend up to several weeks continuously hauled out on beaches. These interruptions in foraging may require the animals to return over great distances to islands and stable ice habitats, having migrated to distant feeding grounds after the breeding season (e.g., Stewart and DeLong 1995).

With few exceptions (e.g., occasional cannibalism on pups by otariid males, seal eating by walruses, and predation by fur seals on penguins ashore), pinnipeds forage in aquatic habitats on fish and squid, crustaceans, euphausiids, and other invertebrates (e.g., Bowen, Boness, and Iverson 1999; Bowen, Read, and Estes 2002; Bradshaw, Lalas, and McConkey 1998; Dellinger and Trillmich 1999; Holst, Stirling, and Hobson 2001; Lowry and Fay 1984; Lowry, Burns, and Nelson 1987; Lowry, Testa, and Calvert 1996; Nilssen, Haug, and Lindblom 2001). The diets of most species tend to be fairly broad but are usually dominated by five or fewer key species, often varying with season, sex, and age. The leopard seal's diet is exceptionally broad, encompassing fish, squid, and krill but also other Antarctic seals, penguins, seabirds, and virtually anything else that lives in the Southern

Ocean (e.g., Lowry, Testa, and Calvert 1996; Siniff and Stone 1985). Newly weaned pups of several phocid species prey on crustaceans for a few months, until the skills needed to capture other prey are sufficiently developed (e.g., Nilssen, Haug, and Lindblom 2001). When two or more pinniped species live in the same region and the composition of their diet overlaps, they tend to focus their foraging efforts on different species, forage at slightly different depths or in otherwise spatially separate areas, or perhaps hunt at different times of the day.

The preferred types of foraging habitat vary by species, age and sex, and geographic location (cf. Bowen and Siniff 1999; Bowen, Read, and Estes 2002; Sjoberg and Ball 2000). Some phocids (e.g., the bearded seal and monk seals) forage exclusively in shallow benthic and epibenthic habitats, while others (e.g., elephant seals, the hooded seal, and the harp seal) forage primarily in pelagic waters or perhaps in either, depending on season (e.g., the Weddell seal). Adult northern elephant seals segregate geographically by sex when foraging at sea (Stewart and DeLong 1995), and adult males and females of several otariid species migrate differentially. Fur seals generally forage on fish and squid in offshore mesopelagic habitats, often in areas of upwelling. Some species, such as the Antarctic fur seal, forage in shallower pelagic habitats on krill. The picture for sea lions is less clear. Although it has long been claimed that they prey mainly on schooling organisms in the water column in neritic or pelagic habitats, recent studies have demonstrated that New Zealand, Australian, and South American sea lions forage in epibenthic coastal habitats, at least in some seasons. Walruses feed principally in shallow benthic habitats, where they prey on mollusks, and occasionally small fish, that live either in the muddy sediment of the seabed or just above it (Fay 1981).

Although most pinnipeds congregate on land or sea ice during brief periods of the year to breed and molt, they are usually solitary when foraging at sea. Like other marine mammals, they are ultimately tied to the sea surface to breathe, yet natural selection has clearly operated to minimize the time needed there and to maximize the amount of time that can be spent submerged. Foraging bouts may last, with minor interruptions, for several weeks to several months, punctuated for periods of several days to weeks on land or ice when no feeding occurs (e.g., Folkow and Blix 1999; Frost, Simpkins, and Lowry 2001; Gentry and Kooyman 1986; Georges, Tremblay, and Guinet 2000; Gjertz, Lyderson, and Wiig 2001; Kelly and Wartzok 1996; Kraft et al. 2000; Mattlin, Gales, and Costa 1998; Nordoy, Folkow, and Blix 1995; Stewart et al. 1996). Among phocids in particular, little time is spent at the surface breathing between successive dives (e.g., Bengtson and Stewart 1992, 1997; Boyd and Croxall 1996; DeLong and Stewart 1991; Simpkins, Kelly, and Wartzok 2001; Stewart 2002). Some otariids and phocids forage principally in the water column (elephant seals, California sea lions, northern fur seals, Antarctic fur seals) in areas of upwelling, downwelling, or ocean current shears, whereas others seem to spend most of their foraging effort on the seafloor (e.g., harbor seals, New Zealand sea lions) (Stewart 2002).

Tropical otariids live in variable, relatively unproductive environments (e.g., Bowen and Siniff 1999). Their survival is tied to local upwelling systems, which are influenced by small-scale annual fluxes in sea surface temperature, as well as periodic, or cyclic, large-scale climatic oscillations. Thus, survival rates of sympatric Galápagos fur seals and sea lions fluctuate in response to El Niño–Southern Oscillation (ENSO) events. For example, in 1982–83 fur seal numbers declined by 50–70% and sea lion numbers by about 30% (Trillmich and Dellinger 1991). An interesting feature of these species is that, even though their diving capabilities appear similar, their feeding behavior and diet do not overlap (Dellinger and Trillmich 1999). Fur seals in the Galápagos forage at night on squid and vertically migrating shoaling fish (e.g., myctophids and bathylagids), while sea lions forage during daytime on pelagic or demersal shoaling fish (e.g., sardines). This niche separation applies even in areas such as Fernandina Island, where the two species use rookeries in about the same location, forage at about the same distance from shore, and spend approximately the same amount of time at sea per trip.

Marine Fissipeds

Sea otters dive and forage mostly in shallow nearshore waters, taking bottom-dwelling mollusks, crustaceans, echinoderms, and fish (Bowen, Read, and Estes 2002; Estes and Bodkin 2002; Ralls, Hatfield, and Siniff 1995). Dives may occur in bouts lasting several hours during the day and night, interrupted by periods at the surface to groom, process food, or rest. Juvenile males often dive in deeper water, for longer periods, and farther from shore than do juvenile and adult females.

The sea otter's role in coastal kelp ecosystems has been the subject of much study and speculation (e.g., Estes, Duggins, and Rathbun 1989). Recreational and commercial fishermen who exploit clams, abalones, and shellfish have vehemently resisted the reestablishment of sea otters in much of their historic range, viewing them as competitors for these economic resources. It has been shown, however, that sea otter populations in Alaska transformed nearshore reefs from two- to three-trophic-level systems by limiting herbivorous sea urchins (through predation), thereby promoting kelp forest development, with associated increases in abundance of kelp-bed fishes (Estes and Duggins 1995). A further restructuring of the coastal ecosystem from a three- to a four-tiered system may have occurred in the 1990s if, as hypothesized by Estes et al. (1998), predation by killer whales contributed to the reduction of sea otter populations over large areas of the Aleutians. Central to their hypothesis is the idea that a major decline in the regular prey base of some killer whale pods in the Aleutian region—specifically, the formerly large populations of northern sea lions and harbor seals—has forced these predators to devote more of their attention to sea otters as prey. Although the cause or causes of the regional declines of pinniped populations are uncertain, overexploitation of their prey by industrial fishing is probably a contributing factor. The long-term study of sea otter ecology and population dynamics in western Alaska has provided exceptional insights into processes that apparently link oceanic top carnivores with nearshore plant and invertebrate communities via a cascading, top-down series of interactions (Bowen, Read, and Estes 2002).

Marine otters (chungungos) live along exposed, rocky coasts and typically den in the spaces between boulders or in vertical crevices on cliff faces. Their dens often have underwater entrances. Crabs and fish predominate in their diet. Marine otters generally dive for about half a minute, although they may remain submerged for more than a minute (Ostfeld et al. 1989).

Quick, powerful, and clever, polar bears are supreme predators. Although ringed seals are their staple prey, they also hunt other northern phocids. The bears in certain areas, notably Wrangel Island in the western Chukchi Sea and Coats Island in northern Hudson Bay, are adept at capturing young walruses. Belugas and narwhals sometimes become confined to small cracks or pools of open water in heavy ice, and this can be a boon for polar bears (as well as human hunters). The bears manage to kill the whales and drag them onto the ice; as surplus killers, bears sometimes kill many more whales than they can consume (Lowry, Burns, and Nelson 1987). Bowhead whale carcasses that strand on shore or float near ice platforms can attract bears from great distances all around (Larsen 1986). There are intriguing historical descriptions of white bears catching salmon from a river in Labrador, much as brown bears (*Ursus arctos*) are known to do in Alaska (Stirling 1988).

SIRENIANS

The modern sirenians are all tropical animals. The average lower limit of thermal neutrality for West Indian manatees is approximately 24°C (Irvine 1983). Their ability to survive in temperate latitudes along the east coast of North America is dependent on warm-water winter refugia, some of which are natural springs and others of which are artificial, most notably power plant effluents (Hartman 1979; Shane 1983). Dugongs similarly show a strong preference for warm water and are rarely seen in water colder than about 18°C. Both West Indian manatees and dugongs are known to migrate seasonally in response to water temperature changes, although the distances may be relatively short and some individuals are essentially resident (Wells, Boness, and Rathbun 1999).

Dugongs are ecologically tied to shallow marine areas, whether along continental coastlines, over offshore reefs, or in lagoonal areas between islands and barrier reefs. They feed on submerged vegetation, either seagrass beds on the sea bottom to depths of up to 20 meters or in surface canopies. Somewhat like gray whales and walruses, dugongs leave evidence of their foraging on the seafloor. Long, serpentine furrows in seagrass beds indicate where dugongs have uprooted or defoliated the plants (Marsh 2002). Such furrows have been observed at depths of up to 33 meters off northeastern Queensland, Australia (Marsh 2002). Although they can travel to reefs hundreds of kilometers from shore, their preferred habitat is in sheltered waters no deeper than about 6 meters. In Shark Bay, Western Australia, where extensive studies have been carried out, large aggregations are also sometimes found in water 9–15 meters deep, where they forage on the large, fleshy, high-starch-content rhizomes of *Halophila spinulosa* (Anderson 1994; Marsh et al. 1994). Satellite tracking has shown them capable of traveling 200 km in only 2 days (Marsh and Rathbun 1990) and covering distances of up to 600 km in just a few days (Marsh 2002).

Domning (1982) used features of dentition and rostral deflection to infer feeding niches of the three modern manatee species. He concluded that the West African manatee is adapted to feed primarily on plants high in the water column, perhaps reflecting a scarcity of seagrasses and other submerged plants in its habitat, and a consequent reliance on emergent or riverbank plants such as mangroves and *Impatiens*. Amazonian manatees may be specially adapted to exploit the floating meadows of macrophytes that abound in the lake systems of the Amazon basin, although some experimental evidence from captive animals showed a strong preference for plants that could be pulled under the water (Domning 1980). The West Indian manatee is probably the most versatile of the three species, feeding on a wide spectrum of forms that notably includes seagrasses and other benthic plants. In descending order of preference, West Indian manatees feed on submerged, floating, and emergent vegetation (Hartman 1979).

Considering their plant diet and typical occurrence in shallow, nearshore habitats, manatees probably rarely dive deeper than 25–30 meters. Direct observations of free-ranging animals have shown that most dives last less than 5 minutes, although a few have been timed at more than 20 minutes. These longer dives

The state of Florida provides the best opportunities anywhere for observing manatees in natural settings. Here, a diver gets a close look at a Florida manatee (*Trichechus manatus latirostris*) in the clear waters of Homosassa Springs, on Florida's Gulf Coast. Interactions with people are a central feature of life for manatees in Florida, and the state's burgeoning recreational boating industry poses one of many threats to this endangered species. Photograph by Howard Hall.

may involve periods of rest at the bottom rather than feeding or other relatively high-energy activities.

A remarkable feature of the Amazonian manatee's physiology is its ability to fast for long periods. In this respect, it may rival or even exceed the capabilities of the migratory baleen whales. Although Amazonian manatees probably do not undertake long-distance migrations comparable to those of baleen whales, their environment undergoes enormous fluctuations, with water levels in the central Amazon varying by 10–15 meters between the dry and wet seasons. When the level declines, many of the manatees' food sources die or become dormant. In years of prolonged drought, manatees may depend on their low metabolic rate and thick blubber layer to survive (Best 1983).

Social Systems, Behavior, and Life History

CETACEANS

Studying the social systems, behavior, and life history of animals that live in a marine or aquatic environment presents special challenges. Historically, cetacean science was rooted in what could be gleaned from carcasses—animals that either washed ashore in what cetologists call *strandings* or died at the hands of whalers or fishermen. Making inferences from large series of carcasses is what Samuels and Tyack (2000) describe as "making the best of a bad situation." Observations of the living animals in their natural environment were largely anecdotal and were often made under circumstances influenced by chasing and efforts to capture. Some insights began to accumulate as a few dolphin and small whale species were maintained for long periods in captivity, but these were, of course, always suspect because of the artificiality of the context. It was not until the 1970s that systematic studies of cetacean communities in the wild became possible, thanks to new research tools involving individual photographic identification, radio-tracking, and later, genetic analyses. These tools, particularly when employed in long-term studies of populations, have helped transform concepts of these animals at both individual and group levels (see Connor et al. 1998; Mann et al. 2000; Wells, Boness, and Rathbun 1999).

Although odontocetes are generally regarded as the more social of the two major groups of cetaceans (Clapham 2000; Connor et al. 2000; Connor 2002), this impression is at least partly owing to the fact that they occur more regularly in groups, with individuals moving in coordinated fashion and maintaining little interindividual distance. In fact, one of the more difficult aspects of studying the behavior of cetaceans is defining *group*. Closely spaced and apparently close-knit groups of cetaceans are often referred to as *pods*, while large aggregations that appear to coordinate their movements for long periods may legitimately be called *schools* (Norris et al. 1994). Particularly in view of their potential for long-range acoustic communication, mysticetes may be capable of coordinating their activities across spatial scales far beyond what can be observed visually (see Tyack 2000). Thus, the concept of a "group" of baleen whales may require some adjustment in our thinking.

Group composition and size in some species vary, depending on demographic and environmental factors. For example, although sperm whales are typically born into their mother's pod of perhaps 10 related females, young males leave these pods and form "bachelor schools" (Best 1979). As they age, the males become increasingly solitary: schools of small, medium-sized, and large bachelors are progressively smaller. Old bulls live solitary lives, alternately foraging in high latitudes and "roving" between groups of females in low latitudes (Whitehead and Weilgart 2000). Some other odontocetes also seem to segregate according to age or sex. For example, mass strandings of Atlantic white-sided dolphins have involved groups of males and females that were either very young or sexually mature, leading to the inference that maturing juveniles tend to leave or be driven away from their "reproductive herd" (Rogan et al. 1997).

The size of a mysticete group rarely exceeds a few tens of animals, and when large numbers of individuals are in the same area, the aggregation seems to be more adventitious than socially significant. Boisterous courting groups of right whales can grow to include as many as 30 to 35 adult males, jostling one another for access to a focal female (Kraus and Hatch 2001). These bouts of interaction may last for more than an hour before the animals disperse to resume their solitary or small-group lives.

The behavior of cetaceans has been given much attention in popular literature, which is often anthropomorphic and speculative. There are, however, several features in the behavioral repertoire of cetaceans that are truly exceptional. For example, they are among the few nonhuman mammals that exhibit vocal learning. Matrilineal groups of sperm and killer whales use distinctive vocal dialects that are probably culturally acquired (Ford 1991; Weilgart and Whitehead 1997). Sperm whales share many characteristics with elephants, including their large brain size, great longevity, complex social structure, and apparent dependence on old, experienced individuals to maintain and transmit a pool of environmental and other knowledge (Weilgart, Whitehead, and Payne 1996). Similar convergence has been noted in the behavior and social systems of bottlenose dolphins and chimpanzees, both of which live in fission-fusion societies (Connor et al. 2000).

The popular notion that cetaceans are peaceable and nonaggressive is a considerable oversimplification. In many odontocete species, scarring patterns on adult males indicate that males test one another through combat. Most fighting is probably related to the establishment of breeding hierarchies, but the evidence is mostly inferential. For example, the teeth of adult male beaked whales and sperm whales have been shown to match the arrays of long, linear or rakelike scars on their bodies. Heyning (1984) showed that, for Hubbs's beaked whale, in which the two erupted teeth of adult males project from the lower jaw outside the mouth, up along both sides of the rostrum to a height greater than the top of the rostrum, wounds can be inflicted only while the mouth is closed. Apparently, the attacking animal must maneuver to bring the dorsal aspect of its rostrum into contact with the body of its victim to achieve the desired effect. The adult male narwhal has what is perhaps the most striking of all secondary sex characters in the order Cetacea. Its long, straight, spiraled tusk could hardly be used, as some have supposed, to spear prey or root out bottom-dwelling organisms. Rather, it is almost certainly used to intimidate and, when necessary, injure competitors. Gerson and Hickie (1985) used body and tusk dimensions, along with scarring rates on the melons of male narwhals, to infer tusk function. They concluded that, although dominance, fighting success, and reproductive success could not be measured directly for narwhals, "the congruence of narwhal data on physical attributes with those for red deer *[Cervus elaphus]* indicates that the most massive male narwhals with the most robust tusks are likely to be dominant."

Aggressiveness in mysticetes is less obvious. However, male humpback whales can be violently aggressive toward one another when competing for access to adult females on the breeding grounds. At times, bouts of slashing and scraping result in bleeding or abrasion of the dorsal fin and head knobs (Darling, Gibson, and Silber 1983; Tyack and Whitehead 1983). Although apparently less severe, scrape marks on the bodies of right whales also signify aggressive male-male interactions. In fact, it has been suggested that one function of the callosities on the heads of right whales is to inflict painful wounds on conspecifics (Payne and Dorsey 1983).

In addition to predation and breeding, the two contexts in which aggressiveness has a ready and obvious explanation, cetaceans are sometimes aggressive for much more obscure reasons. For example, bottlenose dolphins in Scottish waters are known to attack, and sometimes kill, harbor porpoises. Ross and Wilson (1996) initially suggested that this behavior could be a response to competition for food or a misguided effort to defend young or ill dolphins from a perceived threat. These authors also stated that they could not rule out the possibility that the dolphins were simply playing, practice fighting, or expressing sexual frustration. Subsequently, Patterson et al. (1998) reported finding dead bottlenose dolphin calves on Scottish beaches, with injuries suggesting that they, too, had been killed by adult dolphins. This evidence raises further possibilities. Perhaps infanticide is a factor that shapes the social behavior of some dolphins, as it is for certain other mammals (Wolff 1997). In this light, the attacks on harbor porpoises would appear to be cases of mistaken identity, as their adult body size is roughly equivalent to that of a first-year bottlenose dolphin (Patterson et al. 1998).

Small and medium-sized odontocetes also "mob" large whales for unknown reasons. Pilot whales and bottlenose dolphins have been seen harassing sperm whales and right whales, racing excitedly along and across their heads, sides, and tails, causing the larger animals to take defensive or evasive action (Reeves, personal observations; Weller et al. 1996). Although these interactions are reminiscent of avian mobbing, the selective value of such behavior is less evident. The large whales pose no direct threat to their smaller relatives, nor are they particularly similar to killer whales in either appearance or behavior.

Killer whales (*Orcinus orca*) are apex predators. They attack, kill, and consume a remarkable range of marine organisms, from small schooling fish to large baleen whales. Here, a pod of killer whales is attacking a young blue whale (*Balaenoptera musculus*) off Cabo San Lucas, Baja California, Mexico. Although the observers lost contact with the animals before they could confirm the outcome, the blue whale's wounds were serious and its prospects of escape and survival poor. Photograph by R. French, courtesy of Hubbs-SeaWorld Research Institute and SeaWorld, Inc.

Cetaceans exhibit a fairly impressive diversity of life history strategies (Boness, Clapham, and Mesnick 2002; Whitehead and Mann 2000). At one extreme, harbor porpoises mature rapidly (females can ovulate at 3 years of age) and rarely live beyond their teens (Read 1990). Females in the Gulf of Maine population were found to be simultaneously pregnant and lactating for almost their entire adult lives (Read and Hohn 1995). It is uncertain to what extent this "life in the fast lane" strategy (Read and Hohn 1995) is shared by other phocoenids (e.g., see Hohn et al. 1996) or indeed other families of small odontocetes. Preliminary studies of the franciscana (Brownell 1989; Kasuya and Brownell 1979) suggest that it, like the harbor porpoise, matures rapidly and lives a relatively short but productive life. Among the baleen whales, the humpback whale stands out as particularly productive and capable of rapid increase, although this may have more to do with social and ecological factors than with life history characteristics per se (e.g., see Barlow and Clapham 1997; Clapham 1996; see below). Minke whales, the smallest balaenopterids, are also relatively fast to mature (average of 6–8 years, but variable depending on the population). They may be capable of annual breeding and probably do not live longer than about 30 years (Horwood 1990).

At the other end of the life history continuum are the long-lived social odontocetes and the very long-lived bowhead whale. The former group includes, first and foremost, the sperm whale, which has been characterized as "the epitome of a *K*-selected animal, in the sense of a species that seems to have evolved with populations close to the carrying capacity of the environment" (Whitehead and Weilgart 2000). Sperm whales apparently live for 60–70 years, have low fecundity, and exhibit prolonged parental care. Gestation lasts 14–16 months. Although the calf may begin taking solid food before 1 year of age, it is likely to continue being suckled for at least 5 more years. Females reach sexual maturity at 7–13 years of age, while males do not seem to begin contributing to reproduction (i.e., attain "social maturity") until their late twenties. The maximal calving rate, which would be reached only in a population well below carrying capacity, is about one calf every 5 years (Best, Canham, and Macleod 1984). Fecundity declines with age such that females older than 40 give birth about half as often as 10- to 30-year-olds. At least some populations of short-finned pilot whales (Kasuya and Marsh 1984) and killer whales (Olesiuk, Bigg, and Ellis 1990) exhibit life history parameters similar to those of sperm whales, while long-finned pilot whales appear to be somewhat more productive, with a gestation period of about a year and an average calving interval of about 4 years (Martin and Rothery 1993). The reproductive parameters of other odontocetes generally fall somewhere between the harbor porpoise's and the sperm whale's (Perrin and Reilly 1984).

Focusing principally on lactation, Oftedal (1997) noted interesting parallels between, on one hand, the relatively brief lactation periods of baleen whales and phocid seals (accompanied by fasting) and, on the other hand, the more prolonged lactation of odontocetes and otariids (accompanied by maternal foraging). These opposite strategies have the effect of allowing the different taxa to exploit trophic and other resources in optimal fashion. In general, the baleen whales and many of the phocid seals occupy environments with large seasonal pulses of production, while the toothed cetaceans and otariid pinnipeds tend to live where the plane of nutrition is more even across the seasons (Bowen, Read, and Estes 2002).

MARINE CARNIVORES

Pinnipeds

All pinnipeds must leave the water to give birth and nurse their young for several days to weeks each year (Bartholomew 1970; Bowen, Beck, and Austin 2002; Boyd 1998; Oftedal, Boness, and Tedman 1987). All otariids and some phocids (e.g., elephant seals and the gray seal) also mate on land, although most phocids mate in the water (Bartholomew 1970; Boness 1991; Boness and Bowen 1996; Stirling 1983). The breeding distributions of most species are fairly well known because of the limited availability of suitable terrestrial and ice substrates. Most of the phocids breed in fast-ice or seasonal pack-ice habitats (Boyd 2002a; Reeves et al. 2002). Exceptions are elephant seals, whose breeding colonies are exclusively on land—predominantly offshore islands but also at a few continental mainland sites in California and Argentina (Laws 1994; Stewart and Huber 1993); monk seals, which breed on the beaches of coral atolls (Hawaiian monk seal) or in caves along island or continental coastlines (Mediterranean monk seal; e.g., Ragen and Lavigne 1999); harbor seals, which give birth mainly on land but occasionally in the water near shore and also sometimes on ice floes; and gray seals, which have both ice-breeding and land-breeding colonies in the North Atlantic.

All pinnipeds give birth to a single pup, once a year at most (e.g., Bowen 1991; Bowen, Beck, and Austin 2002; Boyd 2002a). The sex of the offspring is determined by male heterogamy. Twins have sometimes been suspected and have recently been confirmed in the Weddell seal by molecular genetic analyses (Gelatt et al. 2001), but twinning is generally rare. Sex ratios at birth are often slightly male-biased but virtually never statistically different from 50:50.

One likely reason that twinning is rare is the large size of the newborn, which, in phocids, is often a quarter to a third the size of the mother. Phocid mothers make a large investment of energy in their pups (e.g., Arnbom, Fedak, and Boyd

Weddell seals breed primarily in seasonal fast-ice habitats around the Antarctic continent, although there are a few land-breeding colonies in some isolated northern locations (e.g., at South Georgia Island in the South Atlantic Ocean). It has recently been learned that, during the nonbreeding season, many seals leave those colonies to forage in northern pack-ice habitats. Weddell seals dive deeply, down to 800 meters. They feed mostly on bottom-dwelling fish and cephalopods when in fast-ice habitats but perhaps on midwater species when in pack-ice habitats farther offshore. A unique, highly inbred colony of a few dozen Weddell seals lives at White Island in southern McMurdo Sound, where it is trapped and isolated from other colonies by the surrounding Ross Ice Shelf. Photograph by Brent S. Stewart.

1997; Bowen, Beck, and Austin 2002; Boyd 1998, 2002b; Kovacs and Lavigne 1986; Oftedal, Boness, and Tedman 1987; Trillmich 1990, 1996). Their fat-rich milk is transferred during a short lactation period, ranging from 3 to 6 weeks in most species and lasting only 4 days in the hooded seal. Fur seal and sea lion pups are smaller relative to their mothers. However, otariids are less able than phocids to store energy in the form of fat, so mother fur seals and sea lions must go to sea every few days to forage, leaving their pups on shore (e.g., Gentry and Kooyman 1986; Kovacs and Lavigne 1992). Natural selection apparently has favored the strategy of investing energy in a single pup over a relatively long period of time (6 months to 2 years or longer), rather than producing several pups that would require substantially more time to locate and nurse after every foraging trip.

A pinniped female becomes receptive soon after weaning her pup—within a week in some fur seals, several weeks in most phocids and some sea lions (e.g., Boyd 1991; Daniel 1981). In all pinnipeds that have been closely studied, development of the embryo continues for only a short period after the egg has been fertilized. It then remains quiescent before attaching to the uterine wall about 2–3 months later, at which time growth resumes. This phenomenon is called *embryonic diapause* or *delayed implantation* (Boyd, Lockyer, and Marsh 1999). Although the total gestation period of most pinnipeds is about 11 months, fetal growth generally takes place over an 8–9 month period. The Australian sea lion seems to be exceptional among pinnipeds in having a postimplantation period of up to 14 months (Gales et al. 1997).

The testes of phocids and the walrus are inguinal and positioned lateral to the penis; those of otariids are scrotal (Pabst, Rommel and McLellan 1999). All male pinnipeds have a baculum, which is larger relative to overall body size in the wal-

California sea lions have increased steadily in abundance in California waters since the 1940s, when the species numbered only a few thousand after indiscriminate killing for various purposes. Today, the principal colonies at San Nicolas and San Miguel Islands in southern California contain large, robust populations. Females give birth from late May through early July, and most wean their pups within four to six months. Some individuals, however, nurse pups for up to a year. Adult males migrate north after the breeding season ends in August. They have become common sights in marinas and rivers from San Francisco to British Columbia during late autumn and winter, hauled out on artificial jetties, bait barges, boat slips, and private yachts. Females, pups, and juveniles seem to be less migratory. Photograph by Brent S. Stewart.

rus and in phocids than in otariids. The walrus and all otariids have four teats with retractable nipples. Monk seals and the bearded seal also have four teats, whereas all other phocids have only two (King 1983; Reeves et al. 2002).

The breeding season for most pinnipeds is spring or summer (cf. Bowen, Beck, and Austin 2002; Reeves et al. 2002). The four Antarctic phocids give birth between October and December, while most Southern Hemisphere fur seals and sea lions give birth in November and December. The southern elephant seal gives birth in the southern spring (October and November), while the northern elephant seal gives birth in the northern winter (December and January). Several other northern phocids (e.g., harp, hooded, Caspian, and Baikal seals) also give birth in the winter. The breeding season of Hawaiian monk seals is prolonged, and pups can be born any time from March through August and occasionally in September (Ragen and Lavigne 1999). Walruses mate in winter, with pups born in spring after about 15 months of gestation, including a 4- or 5-month embryonic diapause (Fay 1981).

The duration of lactation varies considerably among species but is substantially shorter in phocids, ranging from 4 days (hooded seals) to 6 weeks (Weddell seals), compared with 4 months to 2 years in otariids and 2 years in the walrus (Boyd 1991). Phocids have been referred to as capital breeders. They generally store energy and nutrients in their blubber while foraging intensively for several months, then either fast (elephant seals, monk seals, and the hooded and crabeater seals) or feed very little (the Weddell and harbor seals) while nursing their pups. Otariids, in contrast, have been referred to as "income breeders" (Boyd, Lockyer, and Marsh 1999). Once they have given birth, females may remain ashore for 3–8 days nursing their pups (e.g., Baker and Donohue 1999;

Bowen, Beck, and Austin 2002; Gentry and Kooyman 1986; Horning and Trillmich 1997). They then begin a series of foraging trips lasting one to several days, with brief periods of 1 or 2 days ashore to nurse their pups.

In most pinniped species, females mature fairly rapidly and are often sexually mature when 3 or 4 years old and sometimes as young as 2 years (Boyd 2002a; Laws 1956). They then give birth annually until they die at 20–30 years of age. Males become sexually mature a little later, but in polygynous species, they do not begin breeding for several more years. Thereafter, they are likely to breed successfully for only one or two seasons, and males generally die at much younger ages than do females.

Pinniped mating systems vary from monogamy to extreme polygyny. All otariids are polygynous, with small numbers of males excluding others from mating opportunities with the majority of females. These polygynous systems vary from strong harem systems in which an individual male sequesters and guards a small group of females until they come into estrus and are mated, generally within 10 days after giving birth (e.g., northern fur seal, Antarctic fur seal), to nonharem organization in which dominant males defend areas of the beach or intertidal zone where receptive females depart for or return from foraging trips to sea (e.g., California sea lion). Polygyny in phocids ranges from substantial, as in elephant seals, where dominant males defend opportunities to mate with receptive females by defending general areas rather than individual females (i.e., not strictly harem polygyny), to slight, as in Weddell seals, where males defend underwater territories around breathing holes and may mate with several females during the summer breeding season. Other phocids (e.g., harbor and crabeater seals) are serially monogamous, with males remaining near a postpartum female until she becomes receptive. After mating with one female, the male may search for others that may soon be receptive. All pinnipeds that mate on land are sexually dimorphic in body size, males being several times larger than females, and strongly polygynous. Most phocids that mate in the water or on ice are far less polygynous or are serially monogamous, and males and females are of similar size (harbor, Baikal, Caspian, and ringed seals) or females may be slightly larger than males (e.g., Antarctic phocids; Ralls 1977).

At a few locations over the last several decades, large numbers of pinnipeds have been tagged or branded (e.g., Weddell seals in McMurdo Sound, northern elephant seals in southern California Channel Islands and central California colonies, and southern elephant seals on Macquarie, Heard, and Marion islands). Returns have provided detailed information about the demographic and life history characteristics of these populations and their responses to short- and long-term environmental changes. Similarly, the systematic collection of data on age, sex, size, and reproductive status of hunted seals has made certain species (e.g., harp, hooded, northern fur, and Cape fur seals) among the better-known large wild mammals. Indeed, long-running studies of pinniped populations have become cornerstones of population biology (cf. Bowen and Siniff 1999).

Although pinnipeds are generally regarded as less sophisticated than cetaceans in their acoustic repertoires and behavior (indeed, they are not known to echolocate; Tyack and Miller 2002), many species produce acoustic displays during the breeding season. Some of these are songlike advertisement displays that must have evolved by sexual selection (e.g., Miller 1991; Ray, Watkins, and Burns 1969; Rogers, Cato, and Bryden 1996; Schevill, Watkins, and Ray 1966; Stirling 1973; Stirling, Calvert, and Spencer 1987; Thomas, Zinnel, and Ferm 1983). Seals clearly have the capability of vocal imitation (Ralls, Fiorelli, and Gish 1984). However, the possible role of vocal learning in the development of seal songs has been much less explored than it has for whale songs (especially those of humpback whales).

Marine Fissipeds

Sea otters spend most of their time on the sea surface, often on their backs grooming, resting, or handling and consuming prey (Wells, Boness, and Rathbun 1999). When a prey item such as a clam or abalone must be broken open, the otter uses a stone (occasionally even a bottle or tin can) as either a hammer or an anvil, manipulating the items with its forepaws and using its chest as a pounding platform. A good stone may be stored in one of the animal's axillae while it dives for another prey item.

Unlike other mustelids, sea otters give birth to a single (twins on rare occasion; cf. Williams, Mattison, and Ames 1980), precocial pup, normally in the water. The pup remains dependent on its mother for about 5 months and may remain with her for as long as 8 months. Females generally ovulate once a year, coming into estrus within a few days after weaning or losing a pup. Mating takes place in the water, with the male gripping, and often lacerating, the female's face (especially her nose) with its teeth. Male sea otters exhibit resource defense polygyny; they defend territories and, like some pinnipeds, may sequester and even herd females (Wells, Boness, and Rathbun 1999).

With respect to behavior, life history, and social organization, the marine otter appears to be much more similar to river otters than to the sea otter. Twinning apparently is typical (Sielfeld 1983), and both parents seem to care and provide for the young (Ostfeld et al. 1989).

Polar bears are powerful swimmers. They probably dive while moving among ice floes or along the edges of fast ice, but nothing is known about the depths and durations of such dives. Adult female polar bears enter maternity dens in the autumn (by late November) and give birth to their altricial cubs (average litter size about two) in midwinter (late December or early January). It is important that the den not be disturbed and that the family group be allowed to emerge only after the cubs are large enough to walk and withstand ambient climatic conditions (Amstrup and DeMaster 1988). Even though they are apex predators, the lives of polar bears are full of hazards, not the least of which is cannibalism. Females with cubs generally try to avoid adult males, as a chance encounter can result in the cubs being killed and eaten. Occasionally, the female herself dies while defending the cubs (Stirling 1988). Of course, for thousands of years the polar bear's only serious predators have been humans.

SIRENIANS

Like many cetaceans and some pinnipeds, sirenians are long-lived with low reproductive rates, long generation times, and high investment in each offspring. Female dugongs do not give birth until at least 10 years of age and thereafter produce calves at intervals of 3–7 years (Marsh 2002). They can live for as long as 73 years. Florida manatees seem to be slightly more productive than dugongs, with females giving birth for the first time at a mean age of 5 years and a mean calving interval of about 2.5 years (Reynolds and Powell 2002). They can live for as long as 60 years.

Both dugongs and manatees occur in large, resource-based aggregations but are otherwise mostly asocial. The "mating herds" described for West Indian manatees, involving a focal female in estrus and about 10 (range, 5–22) adult males pushing and maneuvering for access to her (Rathbun et al. 1995), are reminiscent of the "surface active groups" described for right whales (Kraus and Hatch 2001). Wells, Boness, and Rathbun (1999) liken the manatee's mating behavior to what they call "scramble competition polygyny" in otariids. There is no evidence that the focal female manatee vocalizes to advertise her condition, as female right whales apparently do. However, female manatees remain in estrus for as long as 2 or 3 weeks, and they are highly mobile during that time, thus probably widening their choice of mates. So-called cavorting herds of juvenile male manatees engage in intense pushing and shoving as well as sexual interactions, perhaps providing "practice" for later participation in mating herds or alternatively serving to sort out dominance relationships (Wells, Boness, and Rathbun 1999).

A particularly intriguing aspect of aggregative behavior in dugongs is the suggestion that the grazing activity of a large herd modifies the species composition and growth stages of seagrasses, making them more nutritious for the dugongs themselves (Preen 1995). This "cultivation grazing" has been likened to the positive feedback relationship between savanna ungulates and their food plants (Wells, Boness, and Rathbun 1999).

Conservation

Marine mammal conservation has been a high-profile public policy issue in many industrialized countries (see Lavigne, Scheffer, and Kellert 1999). Deliberations and decisions of the International Whaling Commission (IWC) have become front-page news. The Newfoundland and Namibian seal hunts grab headlines and spur controversy. Dolphin and seal die-offs raise awareness about pollution and climate change, creating widespread anxiety about environmental deterioration. Fishermen complain about gear damage and reduced fish catches, which they attribute to recovering marine mammal populations. At the same time, fishermen are vilified by conservationists for causing hundreds of thousands of marine mammals to die each year in fishing gear. Some marine protected areas (parks, reserves, etc.) are premised, at least to some degree, on protecting favored animal species such as whales, dolphins, seals, manatees, and dugongs (Reeves 2000, 2002a). Rights to hunt marine mammals are entrenched in land-claim agreements between aboriginal groups and central governments.

The great public interest in marine mammals and their conservation stems from a combination of factors. Among these are the animals' demonstrated economic and symbolic importance, beliefs about their intelligence and sociability, and a collective sense of guilt about having overexploited their populations and abused countless individuals. In this charged climate, the traditional meaning of conservation has lost force. Advocates for individual whales, dolphins, seals, and other marine mammals insist that conservation means complete protection from "consumptive use" (i.e., killing for products such as meat, skin, and oil). In contrast, promoters of hunting insist that conservation means "sustainable use," which they claim is necessary to ensure an elusive "ecological balance."

We prefer to regard conservation as the effort made by people to maintain the autonomous regeneration of natural populations under natural conditions (see Reeves 2002a; Reeves and Reijnders 2002). The goal of conservation is simply to ensure that the full variety of wild organisms on the planet remains intact indefinitely, influenced only by the effects of natural selection rather than by human-caused mortality and environmental degradation. Obviously, such a concept of conservation is no more than an idealization. In practice, the human

Right whales (*Eubalaena* spp.) were the first whales to be hunted commercially, and their numbers in the North Atlantic had been greatly reduced by the middle of the eighteenth century. Because they remained lucrative targets, however, they were given no respite and continued to be killed at every opportunity until legal protection was conferred in the 1930s. Today, the population of some 300–350 right whales in the western North Atlantic is the focus of intense conservation interest, as its recovery is jeopardized by mortality from entanglement in fishing gear and collisions with ships. Photograph by Robin W. Baird.

presence has had, and will continue to have, a pervasive impact on most marine mammal species.

CETACEANS

Many cetacean populations have been reduced in both abundance and range, and it has usually been clear that the root cause was overkill, that is, more animals were killed by whalers and fishermen than the populations could replace. The often-told history of whaling need not be recounted here. It is sufficient to note that the world's large whales experienced a holocaust of sorts, beginning with the brave exploits of Basque and Norse adventurers late in the first and early in the second millennium and ending with the superefficient factory ships and their fast catcher boats that literally combed the Antarctic and North Pacific during the first three-quarters of the twentieth century (Gambell 1999). What remained of the world's stocks of great whales when large-scale industrial whaling ended in the late 1970s and early 1980s was a mere remnant. At the time of this writing, Norway seemed steadfastly committed to continuing its annual kill of at least several hundred minke whales in the North Atlantic, legitimized by a formal objection to the IWC's global moratorium on commercial whaling that took effect in 1982 in the Antarctic and 1986 in the Northern Hemisphere. Japan was killing about three hundred minke whales in the Antarctic and one hundred more in the North Pacific each year, along with an expanding list of other species, all under the guise of a supposed need for scientific specimens—IWC members are allowed to issue permits for research that supercede protective measures. Finally, local people in some parts of the world continued whaling for food. Aboriginal people or people with lengthy histories of engagement with whaling (e.g., on the West Indies island of Bequia) enjoy an "aboriginal subsistence" exemption under the IWC's regulatory schedule (Reeves 2002b).

Although issues related to commercial whaling continue to resonate in the media and the annual meetings of the IWC still attract global attention and foment controversy, the most immediate threats to cetacean populations are now recognized as being much more diverse than they were a few decades ago. Toxic contaminants are seen by many scientists and conservation activists as a major threat. Noise from shipping and industrial activities and, perhaps most worrisome, military operations has emerged as a serious concern. For at least the last two or three decades, incidental mortality from entanglement in fishing gear has been the primary threat to numerous species, driving some to the very brink of extinction (Perrin, Donovan, and Barlow 1994). Freshwater cetacean populations have been fragmented as their habitat is transformed by hydroelectric, irrigation, and flood-control projects. Not only is much of their habitat now degraded or eliminated, but individuals are also at risk of becoming accidentally trapped in canals and other artificial water bodies with little prospect of survival. Several populations of large whales that were severely depleted by whaling appear to be suffering from the intrinsic effects of low numbers and may be unable to mount any significant recovery. Such populations are now exceedingly vulnerable to low-level mortality from ship strikes (North Atlantic right whales) and "subsistence" whaling (stocks of bowhead whales in the eastern Canadian Arctic). Also, there are concerns about inbreeding, demographic variability, low-density occurrence (difficulty finding mates), and loss of cultural memory in these populations.

There may be reason for cautious optimism about the future of some populations of large whales. Some right whale populations in the Southern Hemisphere, humpback whales in many areas, gray whales in the eastern North Pacific, and blue whales in the eastern North Pacific have shown signs of recovery under protection (Best 1993; International Whaling Commission 2001b). In contrast, the small numbers of right whales in the Northern Hemisphere and parts of the Southern Hemisphere, the continued scarcity of bowhead whales in former strongholds such as the Greenland and Barents Seas, and the lack of evidence for strong recoveries by blue whales and fin whales in the Southern Ocean suggest that the effects of intensive commercial whaling are not entirely reversible (Clapham, Young, and Brownell 1999). In the 1980s and 1990s, direct exploitation was a less immediate threat to most endangered whale populations than was accidental mortality from ship strikes and entanglement in fishing gear (e.g., Knowlton and Kraus 2001). Reduced abundance of prey as a result of overfishing and habitat degradation by humans (Bearzi, Politi, and Notarbartolo di Sciara 1999), climate change (Würsig, Reeves, and Ortega-Ortiz 2001), the direct effects of pollution on health and reproduction (O'Shea, Reeves, and Long 1999; Reijnders, Aguilar, and Donovan 1999), and the disturbance caused by noise from ship traffic and industrial activity (Gordon and Moscrop 1996; Würsig and Richardson 2002) have become additional major concerns in recent decades.

The baiji is regarded as the world's most critically endangered species of cetacean. Although it was probably still common and widely distributed when China's Great Leap Forward began in the autumn of 1958, it soon became the target of intensive hunting for meat, oil, and leather (Zhòu and Zhang 1991). It has been legally protected in China since the early 1980s and accorded the status of a "national treasure" in the same way as the giant panda. However, the baiji has continued to be killed accidentally in fishing operations (by both entanglement and electrical shock), by collisions with powered vessels, and by exposure to underwater blasting during harbor construction. This mortality, combined with the effects of overfishing of prey resources, pollution, industrial and vessel noise, and the damming of Yangtze tributaries, has made Chinese and foreign scientists pessimistic about the baiji's survival (Zhou et al. 1998). At a recent meeting of the IWC Scientific Committee's Standing Subcommittee on Small Cetaceans (International Whaling Commission 2001a), participants were unable to agree on how to proceed. Some argued that efforts to live-capture and relocate dolphins from the Yangtze into a "seminatural reserve," as recommended by an international panel of experts in 1986 (Perrin and Brownell 1989), should be abandoned and that all available resources should be devoted to reducing threats in the river. Other participants expressed the view that the baiji's only hope for survival lies in continuation of the ex situ approach, providing, of course, that the "seminatural reserve" is made to function as envisioned in 1986.

Other species of small cetaceans that are in serious trouble include the vaquita, threatened primarily and most immediately by incidental mortality in gillnets (D'Agrosa, Lennert, and Vidal 2000; Rojas-Bracho and Jaramillo-Legorreta 2002); the Ganges and Indus dolphins, threatened not only by entanglement in fishing gear but also by major development projects that have essentially carved up their habitat into isolated fragments between irrigation dams (Reeves and Chaudhry 1998; Smith et al. 2000); the franciscana, of which at least 1,500 are killed annually in gillnets (Crespo 2002); and Hector's dolphin, also threatened primarily by mortality in fishing gear (Martien et al. 1999). Several geographic populations are critically endangered, including the North Island (New Zealand) population of Hector's dolphin (Dawson et al. 2001; Pichler and Baker 2000) and the Mahakam River (Indonesia) population of the Irrawaddy dolphin (Kreb 1999). There is no doubt about the fact that documentation of "endangerment" lags well behind the ongoing threats, especially to freshwater, inshore, and coastal marine populations of small cetaceans. Some populations of Irrawaddy dolphins, finless porpoises, hump-backed dolphins, and even bottlenose dolphins and harbor porpoises have probably been extirpated or are on the verge of disappearing, but clear-cut scientific evidence to document their status is not yet available.

MARINE CARNIVORES

Pinnipeds

Indigenous people on all continents except Antarctica have hunted pinnipeds for thousands of years (Bonner 1982). The meat has been used for food, the oil from their fat for fuel, their furs and skins for clothing and shelter, and their teeth and bones for making tools and weapons. Some local populations were probably extirpated by human hunters in prehistoric or early historic times. Antarctic seal populations, however, were truly pristine until the late 1700s or early 1800s, when commercial whalers and explorers began to penetrate high southern latitudes. According to one source, more than 1.2 million fur seals had been taken from South Georgia by 1822, rendering the population there virtually extinct (McCann and Doidge 1987). Throughout the nineteenth and early twentieth centuries, markets for oil, hides, and ivory (in the case of the walrus) fueled commercial hunting on a massive scale (Busch 1985). The introduction of firearms meant that large numbers of animals were killed or seriously injured without being secured by the hunters (e.g., Fay et al. 1994; Reeves, Wenzel, and Kingsley 1998). All species of fur seals were brought to near extinction within decades after their discovery (Croxall and Gentry 1987). Elephant seals were relentlessly pursued for their oil; "elephanting" was an adjunct to whaling (Fuller 1980), and both species were brought close to extinction. In contrast, harp and hooded seals

The Guadalupe fur seal is the only member of the genus *Arctocephalus* that occurs in the Northern Hemisphere. It was hunted extensively for its fur in the late 1700s and early 1800s and was thought to be extinct until 1949, when a lone male was seen ashore at San Nicolas Island, off southern California. A small colony of a few more than a dozen seals was subsequently discovered farther south in Mexican waters at Guadalupe Island. The primary breeding colony is still at Guadalupe Island, and it has grown steadily to about ten thousand seals. A small colony recently became established at nearby San Benito Island, and one pup was born in the late 1990s at San Miguel Island, southern California. Also, a few adult males have maintained breeding territories among California sea lions at the Southern California Channel Islands, and they have occasionally come ashore as far north as San Francisco, usually during warm-water, El Niño years. Genetic studies have revealed that a substantial amount of genetic variation was lost during the population bottleneck of the eighteenth and nineteenth centuries. Photograph by Brent S. Stewart.

were subjected to intensive, and often controversial, commercial sealing (Lavigne and Kovacs 1988), yet their numbers remained large enough to support rapid recovery once the scale of killing declined in the 1980s.

Today, Asian markets for pinniped penises and bacula, along with governmental support for culling operations to reduce seal predation on fish, help to sustain the sealing industry (e.g., Bräutigam and Thomsen 1993; Lavigne, Scheffer, and Kellert 1999; Malik et al. 1997). The large "subsistence" takes of walruses in Alaska, Canada, Greenland, and Russia continue to supply raw ivory and carvings to an international market (Bräutigam and Thomsen 1993). These ongoing exploitation regimes may be sustainable in some sense (e.g., see Prescott-Allen and Prescott-Allen 1996) but require constant monitoring and scientific vigilance.

Some otariid species have recovered substantially from past overexploitation, and their populations are now relatively robust. A few number only several thousands to tens of thousands but appear to be stable or increasing (e.g., the Guadalupe and Juan Fernández fur seals). The three Southern Hemisphere sea lion species (Australian, South American, and New Zealand) have probably always been less abundant than fur seals, and their recoveries have been less pronounced and sustained (Childerhouse and Gales 1998; Gales, Shaughnessy, and Dennis 1994; Reijnders et al. 1993). Populations of Steller sea lions throughout the Aleutian Islands and western Gulf of Alaska have crashed during the past three decades and now number only about 10–20% of their peak abundance in the 1950s (Loughlin 2002). Likewise, populations of southern elephant seals at most breeding sites in the Southern Ocean have declined substantially since the 1960s. Exceptions are at Macquarie Island and South Georgia, where the populations may be stable, and on the mainland in southern Argentina, where numbers are still increasing. The causes of these mysterious declines have yet to be identified. Populations of seals have also declined in the Baltic Sea, evidently at least in part because of the effects of marine pollution on reproductive performance.

The two surviving monk seal species are in serious trouble, with estimated total populations of about 1,400 (Hawaiian; Ragen and Lavigne 1999) and a few hundred (Mediterranean; Forcada, Hammond, and Aguilar 1999). They are among the most urgent conservation priorities for pinnipeds because of their demographic and ecological vulnerability (e.g., Brasseur, Reijnders, and Verriopoulos 1997; Lavigne 1999; Ragen and Lavigne 1999). Both species populations are fragmented so that genetic exchange among subpopulations is limited. Interactions with fisheries are ongoing sources of conflict, with fishermen resenting the seals as perceived competitors and pests and with seals frequently dying from entanglement in fishing gear. The small numbers of seals and their relegation to remote and perhaps marginal habitat mean that they have little resilience to withstand the effects of environmental change (e.g., sea level rise) and catastrophic events (e.g., toxic algal blooms and epizootics).

Marine Fissipeds

The sea otter's curse was its luxuriant pelt, which was one of the most desirable of all mammalian furs. Sea otters were hunted remorselessly to supply the Oriental market from the 1780s onward—until very few were left and protection came in 1911 with signing of the Treaty for the Preservation and Protection of Fur Seals. Although anchored nets were sometimes used to catch sea otters, most of the hunting was conducted by men in boats, using lances initially and rifles later on. In California, sea otters were sometimes shot by men standing on shore, and in Washington, shooting towers were erected at the surf line and Indians were employed to swim out and retrieve the carcasses. Aboriginal people in Alaska are still allowed to hunt sea otters as long as the furs are used locally to make clothing or authentic handicraft items. The reported annual kill during the mid- to late 1990s was in the range of 600 to 1,200. There is undoubtedly a very large incipient demand for sea otter furs, as evidenced by poaching in Russia's Kamchatka Peninsula and Kuril Islands and the growing interest in hunting on the part of Alaskan natives (Estes 1994).

Sea otter populations in prime habitat are able to increase at annual rates as high as 17–20% (Estes 1990). Most populations appeared to be recovering well in the mid-1990s, with about 100,000 in Alaska, 2,300 in California, and 300 in Washington in 1995 (Marine Mammal Commission 2001), plus an additional 3,500 in the Commander Islands and an expanding population in British Columbia (Estes and Bodkin 2002). However, sea otters in the west-central Aleutians evidently began declining in the 1970s and declined rapidly in the 1990s. By 2000, there were an estimated 6,000 sea otters left throughout the Aleutian islands, down from about 50,000 to 100,000 otters when the population was at a peak in the 1980s (cf. Estes and Bodkin 2002, Estes et al. 1998). Whether predation by killer whales (perhaps associated with depletion of normal whale prey by human fisheries or other environmental disruption) was the sole or primary cause remained uncertain. Abundance was also declining during the late 1990s in California, where the annual rate of increase had not exceeded about 5% since

This sea otter (*Enhydra lutris*) is rafting on its back at the water's surface and manipulating a sea urchin before ingestion. Sea otters have recovered strongly across much of their historical range. Recent population crashes in portions of the Aleutian Islands, however, as well as a leveling off and possible downturn in the trajectory of the California population in recent years, are cause for continuing concern. Photograph courtesy of U.S. Fish and Wildlife Service.

the late 1930s, when the California subspecies there was "rediscovered." Again, the cause or causes of the recent decline of this population are uncertain. A controversial attempt was made during the 1980s to establish a new population of sea otters in the California Channel Islands to reduce the risk that an oil spill would destroy the mainland population. More than 135 otters were captured and translocated to San Nicolas Island, but by the late 1990s, only about 15 remained. In very recent years, the colony has grown slowly; about 30 sea otters were living there in 2002, and a few pups were being born each year.

Little is known about the abundance or conservation status of the marine otter, although it has long been viewed as in danger of extinction. It has been hunted for its pelt and also because fishermen view it as a pest or competitor.

Although commercial and large-scale sport hunting of polar bears has now ceased, at least several hundred are still killed each year, most of them by Inuit for meat and hides (mainly to be sold for cash), some of them in "self-defense" to protect human life or property (Derocher et al. 1998; IUCN/SSC Polar Bear Specialist Group 1999). Hunt management varies both between and within the range states. While removals are limited by a quota system in much of Canada, there is no limit on how many polar bears can be killed by native hunters in Alaska and Greenland. Illegal hunting and live-captures are known to occur in Russia, but the scale is unknown. The international trade in polar bear gallbladders provides a commercial incentive to the "subsistence" hunt. Also, the existence of a legal supply of bear gallbladders adds to the difficulty of enforcing bans on the trade in these organs from other endangered ursids.

The status of polar bears is monitored under the International Agreement on the Conservation of Polar Bears and Their Habitats, which took effect in 1976 (Prestrud and Stirling 1994). Representatives of the circumpolar signatories—Canada, the United States, Denmark (on behalf of Greenland), Norway, and the Soviet Union (now the Russian Federation)—meet regularly to share information, discuss research needs, and assess polar bear stocks. The current total population is thought to be within the range of 22,000 to 27,000 (Derocher et al. 1998; IUCN/SSC Polar Bear Specialist Group 1999). In recent years, concern has centered on the need for better protection of females and their denning habitat, the particularly high levels of PCBs found in bears in the Atlantic sector of the Arctic, and the effects of global warming on polar bears and their prey (see Stirling, Nunn, and Iacozza 1999).

SIRENIANS

As mentioned earlier, Steller's sea cow was hunted to extinction in the eighteenth century (Anderson and Domning 2002). The extant tropical sirenians have proven more resilient, but only because of their more extensive ranges and, no doubt, greater initial abundance before encountering human hunters. Another factor in their survival may have been their ability to adapt to hunting pressure by selection for reclusive, cryptic behavior (Lefebvre et al. 2001). Nearly all sirenian populations have nevertheless been decimated, or in many cases extirpated, by a combination of hunting, incidental killing in fishing gear, mortality from vessel strikes, and entrapment or entrainment in water regulation devices (Marsh and Lefebvre 1994; Marsh et al. 2002; O'Shea, Lefebvre, and Beck 2001; Reynolds 1999). Most countries with sirenian populations confer legal protection on them, but few apply rigorous enforcement (e.g., Marsh et al. 1995; Reeves et al. 1996; Rosas 1994; Roth and Waitkuwait 1986). Remarkably, even though no group of large mammals could be more placid and inoffensive, manatees have come into serious conflict with humans in parts of West Africa. In Sierra Leone and Guinea-Bissau, for example, they incur the hostility of farmers and fishermen, the former by moving into rice fields during the rainy season to "raid" crops and the latter by damaging nets while removing caught fish (see Powell 1978) or becoming entangled (Reeves, Tuboku-Metzger, and Kapindi 1988; Silva and Araújo 2001).

There has been little active management of human activities to benefit sirenian conservation, except in Australia and the United States. In Australia, the main challenge has been to manage exploitation of dugongs by aboriginal hunters (Marsh, Harris, and Lawler 1997; Smith and Marsh 1990) and to reduce mortality in fishing gear and antishark nets (Marsh 1988; Marsh et al. 1999). In view of the beneficial effects of "cultivation grazing" by dugongs (see above), their extirpation from an area may result in deterioration of the seagrass meadows as dugong habitat (Marsh 2002). In the United States, Florida manatees experience high mortality and morbidity from collisions and entanglement, and their habitat is under constant pressure from coastal development (Reynolds 1999). Modeling has shown, however, that measures to reduce boat activity and boat speeds in 13 key areas of the state would go a long way toward ensuring the species' survival indefinitely (Marmontel, Humphrey, and O'Shea 1997).

Literature Cited

Ahmed, B. 2000. Water development and the status of the shushuk (*Platanista gangetica*) in southeast Bangladesh. *In* R. R. Reeves, B. D. Smith, and T. Kasuya (eds.), Biology and Conservation of Freshwater Cetaceans in Asia, 62–66. Occasional Papers of the IUCN Species Survival Commission No. 23. IUCN, Gland, Switzerland.

Amstrup, S. C., and D. P. DeMaster. 1988. Polar bear *Ursus martitimus*. *In* J. W. Lentfer (ed.), Selected Marine Mammals of Alaska: Species Accounts with Research and Management Recommendations, 39–56. Marine Mammal Commission, Washington, D.C.

Anderson, P. K. 1994. Dugong distribution, the seagrass *Halophila spinulosa,* and thermal environment in winter in deeper waters of eastern Shark Bay, Western Australia. Australian Wildlife Research 21:381–88.

Anderson, P. K., and D. P. Domning. 2002. Steller's sea cow *Hydrodamalis gigas. In* W. F. Perrin, B. Würsig, and J. G. M. Thewissen (eds.), Encyclopedia of Marine Mammals, 1178–81. Academic Press, San Diego.

Arnason, U., K. Bodin, A. Gullberg, C. Ledje, and S. Mouchaty. 1995. A molecular view of pinniped relationships with particular emphasis on the true seals. Journal of Molecular Evolution 40:78–85.

Arnbom, T. R., M. A. Fedak, and I. L. Boyd. 1997. Factors affecting maternal expenditure in southern elephant seals during lactation. Ecology 78:471–83.

Arnold, P., H. Marsh, and G. Heinsohn. 1987. The occurrence of two forms of minke whales in east Australian waters with a description of external characters and skeleton of the diminutive or dwarf form. Scientific Reports of the Whales Research Institute (Tokyo) 38:1–46.

Baird, R. W. 2000. The killer whale: foraging specializations and group hunting. *In* J. Mann, R. C. Connor, P. L. Tyack, and H. Whitehead (eds.), Cetacean Societies: Field Studies of Dolphins and Whales, 127–53. University of Chicago Press, Chicago.

Baker, J. D., and M. J. Donohue. 1999. Ontogeny of swimming and diving in northern fur seal (*Callorhinus ursinus*) pups. Canadian Journal of Zoology 78:100–109.

Barlow, J., and P. J. Clapham. 1997. A new birth-interval approach to estimating demographic parameters of humpback whales. Ecology 78:535–46.

Barnes, L. G., D. P. Domning, and C. E. Ray. 1985. Status of studies on fossil marine mammals. Marine Mammal Science 1:15–53.

Bartholomew, G. A. 1970. A model for the evolution of pinniped polygyny. Evolution 24:546–99.

Bearzi, G., E. Politi, and G. Notarbartolo di Sciara. 1999. Diurnal behavior of free-ranging bottlenose dolphins in the Kvarnerić (northern Adriatic Sea). Marine Mammal Science 15:1065–97.

Beentjes, M. P. 1990. Comparative terrestrial locomotion of the Hooker's sea lion (*Phocarctos hookeri*) and the New Zealand fur seal (*Arctocephalus forsteri*): evolutionary and ecological implications. Zoological Journal of the Linnaean Society 98:307–25.

Belikov, S. E., Y. A. Gorbunov, and V. I. Shil'nikov. 1984. Observations of cetaceans in the seas of the Soviet Arctic. Report of the International Whaling Commission 34:629–32.

Bengtson, J. L., and B. S. Stewart. 1992. Diving and haulout behavior of crabeater seals in the Weddell Sea, Antarctica, during March 1986. Polar Biology 12:635–44.

———. 1997. Diving patterns of a Ross seal (*Ommatophoca rossii*) near the eastern coast of the Antarctic Peninsula. Polar Biology 18:214–18.

Bernard, H. J., and S. B. Reilly. 1999. Pilot whales *Globicephala* Lesson, 1828. *In* S. H. Ridgway and R. Harrison (eds.), Handbook of Marine Mammals. Volume 6: The Second Book of Dolphins and Porpoises, 245–79. Academic Press, San Diego.

Berta, A. 2002. Pinniped evolution. *In* W. F. Perrin, B. Würsig, and J. G. M. Thewissen (eds.), Encyclopedia of Marine Mammals, 921–29. Academic Press, San Diego.

Best, P. B. 1977. Two allopatric forms of Bryde's whale off South Africa. Report of the International Whaling Commission (Special Issue) 1:10–38.

———. 1979. Social organization in sperm whales, *Physeter macrocephalus. In* H. E. Winn and B. L. Olla (eds.), Behavior of Marine Animals: Current Perspectives in Research. Volume 3: Cetaceans, 227–89. Plenum Press, New York.

———. 1985. External characters of southern minke whales and the existence of a diminutive form. Scientific Reports of the Whales Research Institute (Tokyo) 36:1–33.

———. 1993. Increase rates in severely depleted stocks of baleen whales. ICES Journal of Marine Science 50:169–86.

Best, P. B., P. A. S. Canham, and N. Macleod. 1984. Patterns of reproduction in sperm whales, *Physeter macrocephalus.* Report of the International Whaling Commission (Special Issue) 6:51–79.

Best, R. C. 1983. Apparent dry-season fasting in Amazonian manatees (Mammalia: Sirenia). Biotropica 15:61–64.

Best, R. C., and V. M. F. da Silva. 1984. Preliminary analysis of reproductive parameters of the boutu, *Inia geoffrensis*, and the tucuxi, *Sotalia fluviatilis*, in the Amazon River system. Report of the International Whaling Commission (Special Issue) 6:361–69.

———. 1989. Biology, status, and conservation of *Inia geoffrensis* in the Amazon and Orinoco River basins. *In* W. F. Perrin, R. L Brownell Jr., K. Zhou, and J. Liu (eds.), Biology and Conservation of the River Dolphins, 23–34. Occasional Papers of the IUCN Species Survival Commission No. 3, IUCN, Gland, Switzerland.

Bigg, M. A., P. F. Olseiuk, G. M. Ellis, J. K. B. Ford, and K. C. Balcomb. 1990. Social organization and genealogy of resident killer whales (*Orcinus orca*) in the coastal waters of British Columbia and Washington State. Report of the International Whaling Commission (Special Issue) 12:383–405.

Blix, A. S., H. J. Grav, and K. Ronald. 1975. Brown adipose tissue and the significance of the venous plexes in pinnipeds. Acta Physiologica Scandinavica 94:133–35.

Bonadonna, F., M.-A. Lea, and C. Guinet. 2000. Foraging routes of Antarctic fur seals (*Arctocephalus gazella*) investigated by the concurrent use of satellite tracking and time-depth recorders. Polar Biology 23:149–59.

Boness, D. J. 1991. Determinants of mating systems in the Otariidae (Pinnipedia). *In* D. Renouf (ed.), The Behaviour of Pinnipeds, 1–44. Chapman and Hall, London.

Boness, D. J., and W. D. Bowen. 1996. The evolution of maternal care in pinnipeds. Bioscience 46:645–54.

Boness, D. J., P. J. Clapham, and S. L. Mesnick. 2002. Life history and reproductive strategies. *In* A. R. Hoelzel (ed.), Marine Mammal Biology: An Evolutionary Approach, 278–324. Blackwell Science, Oxford.

Bonner, W. N. 1982. Seals and Man: A Study of Interactions. Washington Sea Grant, University of Washington Press, Seattle.

Born, E. W., I. Gjertz, and R. R. Reeves. 1995. Population Assessment of Atlantic Walrus. Meddelelser No. 8. Norsk-Polarinstitutt, Oslo, Norway.

Boschma, H. 1950. Maxillary teeth in specimens of *Hyperoodon rostratus* (Müller) and *Mesoplodon grayi* von Haast stranded on the Dutch coasts. Koninklijke Nederlandse Akademie van Wetenschappen Proceedings 53(6):3–14.

Boveng, P. L., and J. L. Bengtson. 2001. Antarctic pack ice seals: movements reveal diversity of foraging behavior in relation to sea ice and bathymetry. S3O03 in Antarctic Biology in a Global Context. Proceedings of the VIII SCAR International Biology Symposium. Amsterdam, The Netherlands.

Bowen, W. D. 1991. Behavioural ecology of pinniped neonates. *In* D. Renouf (ed.), The Behaviour of Pinnipeds, 66–127. Chapman and Hall, London.

———. 1997. Role of marine mammals in aquatic ecosystems. Marine Ecology Progress Series 158:267–74.

Bowen, W. D., and D. B. Siniff. 1999. Distribution, population biology, and feeding ecology of marine mammals. *In* J. E. Reynolds III and S. A. Rommel (eds.), Biology of Marine Mammals, 423–84. Smithsonian Institution Press, Washington, D.C.

Bowen, W. D., C. A. Beck, and D. A. Austin. 2002. Pinniped ecology. *In* W. F. Perrin, B. Würsig, and J. G. M. Thewissen (eds.), Encyclopedia of Marine Mammals, 911–21. Academic Press, San Diego.

Bowen, W. D., D. J. Boness, and S. J. Iverson. 1999. Diving behaviour of lactating harbour seals and their pups during maternal foraging trips. Canadian Journal of Zoology 77:978–88.

Bowen, W. D., A. J. Read, and J. A. Estes. 2002. Feeding ecology. *In* A. R. Hoelzel (ed.), Marine Mammal Biology: An Evolutionary Approach, 217–46. Blackwell Science, Oxford.

Boyd, I. L. 1991. Environmental and physiological factors controlling the reproductive cycles of pinnipeds. Canadian Journal of Zoology 69:1135–48.

———. 1998. Time and energy constraints in pinniped lactation. American Naturalist 152:717–28.

———. 2002a. Pinniped life history. *In* W. F. Perrin, B. Würsig, and J. G. M. Thewissen (eds.), Encyclopedia of Marine Mammals, 929–34. Academic Press, San Diego.
———. 2002b. Energetics: consequences for fitness. *In* A. R. Hoelzel (ed.), Marine Mammal Biology: An Evolutionary Approach, 247–77. Blackwell Science, Oxford.
Boyd, I. L., and J. P. Croxall. 1996. Dive durations in pinnipeds and seabirds. Canadian Journal of Zoology 74:1696–1705.
Boyd, I. L., C. Lockyer, and H. D. Marsh. 1999. Reproduction in marine mammals. *In* J. E. Reynolds III and S. A. Rommel (eds.), Biology of Marine Mammals, 218–86. Smithsonian Institution Press, Washington, D.C.
Bradshaw, C. J. A., C. Lalas, and S. McConkey. 1998. New Zealand sea lion predation on New Zealand fur seals. New Zealand Journal of Marine and Freshwater Research 32:101–4.
Brasseur, S. M. J. M., P. J. H. Reijnders, and G. Verriopoulos. 1997. Mediterranean monk seal *Monachus monachus*. *In* P. J. H. Reijnders, G. Verriopoulos, and S. M. J. M. Brasseur (eds.), Status of Pinnipeds Relevant to the European Union, 12–26. IBN Scientific Contributions 8, DLO Institute for Forestry and Nature Research, Wageningen, The Netherlands.
Bräutigam, A., and J. Thomsen. 1993. Appendix 2: harvest and international trade in seals and seal products. *In* P. Reijnders, S. Brasseur, J. van der Toorn, et al. (eds.), Seals, Fur Seals, Sea Lions, and Walrus: Status Survey and Conservation Action Plan, 84–87. IUCN, Gland, Switzerland.
Brownell, R. L., Jr. 1989. Franciscana *Pontoporia blainvillei* (Gervais and d'Orbigny, 1844). *In* S. H. Ridgway and R. Harrison (eds.), Handbook of Marine Mammals. Volume 4: River Dolphins and the Larger Toothed Whales, 45–67. Academic Press, London.
Brownell, R. L., Jr., and F. Cipriano. 1999. Dusky dolphin *Lagenorhynchus obscurus* (Gray, 1828). *In* S. H. Ridgway and R. Harrison (eds.), Handbook of Marine Mammals. Volume 6: The Second Book of Dolphins and Porpoises, 85–104. Academic Press, San Diego.
Bryden, M. M. 1971. Anatomical and allometric adaptations in elephant seals, *Mirounga leonina* (L.). Journal of Anatomy 108:208.
Busch, B. C. 1985. The War against the Seals: A History of the North American Seal Fishery. McGill-Queen's University Press, Kingston, Ontario.
Butler, P. J., and D. R. Jones. 1997. Physiology of diving of birds and mammals. Physiological Reviews 77:837–99.
Campagna, C., M. A. Fedak, and B. J. McConnell. 1999. Post-breeding distribution and diving behavior of adult male southern elephant seals from Patagonia. Journal of Mammalogy 80:1341–52.
Cañadas, A., and R. Sagarminaga. 2000. The northeastern Alboran Sea, an important breeding and feeding ground for the long-finned pilot whale (*Globicephala melas*) in the Mediterranean Sea. Marine Mammal Science 16:513–29.
Chanin, P. 1985. The Natural History of Otters. Facts on File, New York.
Childerhouse, S., and N. Gales. 1998. Historical and modern distribution and abundance of the New Zealand sea lion *Phocarctos hookeri*. New Zealand Journal of Zoology 25:1–16.
Cipriano, F. 2002. Evolutionary biology, overview. *In* W. F. Perrin, B. Würsig, and J. G. M. Thewissen (eds.), Encyclopedia of Marine Mammals, 404–7. Academic Press, San Diego.
Clapham, P. J. 1996. The social and reproductive biology of humpback whales: an ecological perspective. Mammal Review 26:27–49.
———. 2000. The humpback whale: seasonal feeding and breeding in a baleen whale. *In* J. Mann, R. C. Connor, P. L. Tyack, and H. Whitehead (eds.), Cetacean Societies: Field Studies of Dolphins and Whales, 173–96. University of Chicago Press, Chicago.
———. 2001. Why do baleen whales migrate? A response to Corkeron and Connor. Marine Mammal Science 17:432–36.
Clapham, P. J., and R. L. Brownell Jr. 1996. The potential for interspecific competition in baleen whales. Report of the International Whaling Commission 46:361–67.
Clapham, P. J., S. B. Young, and R. L. Brownell Jr. 1999. Baleen whales: conservation issues and the status of the most endangered populations. Mammal Review 29:35–60.

Clarke, M. R. 1986. Cephalopods in the diet of odontocetes. *In* M. M. Bryden and R. Harrison (eds.), Research on Dolphins, 281–321. Clarendon Press, Oxford.

Connor, R. C. 2002. Ecology of group living and social behaviour. *In* A. R. Hoelzel (ed.), Marine Mammal Biology: An Evolutionary Approach, 353–87. Blackwell Science, Oxford.

Connor, R. C., and P. J. Corkeron. 2001. Predation past and present: killer whales and baleen whale migration. Marine Mammal Science 17:436–39.

Connor, R. C., J. Mann, P. L. Tyack, and H. Whitehead. 1998. Social evolution in toothed whales. Trends in Ecology and Evolution 13:228–32.

Connor, R. C., R. S. Wells, J. Mann, and A. J. Read. 2000. The bottlenose dolphin: social relationships in a fission-fusion society. *In* J. Mann, R. C. Connor, P. L. Tyack, and H. Whitehead (eds.), Cetacean Societies: Field Studies of Dolphins and Whales, 91–126. University of Chicago Press, Chicago.

Corkeron, P. J., and R. C. Connor. 1999. Why do baleen whales migrate? Marine Mammal Science 15:1228–45.

Crespo, E. A. 2002. Franciscana *Pontoporia blainvillei*. *In* W. F. Perrin, B. Würsig, and J. G. M. Thewissen (eds.), Encyclopedia of Marine Mammals, 482–85. Academic Press, San Diego.

Crespo, E. A., G. Harris, and R. González. 1998. Group size and distributional range of the franciscana, *Pontoporia blainvillei*. Marine Mammal Science 14:845–49.

Croxall, J. P., and R. L. Gentry (eds.). 1987. Status, Biology, and Ecology of Fur Seals: Proceedings of an International Symposium and Workshop, Cambridge, England, 23–27 April 1984. NOAA Technical Report NMFS 51:1–212.

D'Agrosa, C., C. E. Lennert, and O. Vidal. 2000. Preventing the extinction of a small population: vaquita (*Phocoena sinus*) fishery mortality and mitigation strategies. Conservation Biology 14:1110–19.

Dahlheim, M. E., and J. E. Heyning. 1999. Killer whale *Orcinus orca* (Linnaeus, 1758). *In* S. H. Ridgway and R. Harrison (eds.), Handbook of Marine Mammals. Volume 6: The Second Book of Dolphins and Porpoises, 281–322. Academic Press, San Diego.

Dalebout, M. L., A. Van Helden, K. Van Waerebeek, and C. S. Baker. 1998. Molecular genetic identification of southern hemisphere beaked whales (Cetacea: Ziphiidae). Molecular Ecology 7:687–94.

Dalebout, M. L., J. G. Mead, K. Van Waerebeek, J. C. Reyes, and C. S. Baker. 2000. Molecular genetic discovery of a new beaked whale (Ziphiidae). Document No. SC/52/SM 11. International Whaling Commission, Cambridge, U.K.

Dalebout, M. L., J. G. Mead, C. S. Baker, A. N. Baker, and A. L. van Helden. 2002. A new species of beaked whale *Mesoplodon perrini* sp. n. (Cetacea: Ziphiidae) discovered through phylogenetic analyses of mitochondrial DNA sequences. Marine Mammal Science 18:577–608.

Daniel, J. C. 1981. Delayed implantation in the northern fur seal (*Callorhinus ursinus*) and other pinnipeds. Journal of Reproduction and Fertility (Suppl.) 29:35–50.

Darling, J. D., K. M. Gibson, and G. K. Silber. 1983. Observations on the abundance and behavior of humpback whales (*Megaptera novaeangliae*) off west Maui, Hawaii, 1977–79. *In* R. Payne (ed.), Communication and Behavior of Whales, 201–22. Westview Press, Boulder, Colo.

da Silva, V. M. F., and R. C. Best. 1994. Tucuxi *Sotalia fluviatilis* (Gervais, 1853). *In* S. H. Ridgway and R. Harrison (eds.), Handbook of Marine Mammals. Volume 5: The First Book of Dolphins, 43–69. Academic Press, London.

da Silva, V. M. F., and A. R. Martin. 2000. A study of the boto, or Amazon River dolphin (*Inia geoffrensis*), in the Mamirauá Reserve, Brazil: operation and techniques. *In* R. R. Reeves, B. D. Smith, and T. Kasuya (eds.), Biology and Conservation of Freshwater Cetaceans in Asia, 121–31. Occasional Papers of the IUCN Species Survival Commission No. 23. IUCN, Gland, Switzerland.

Dawson, S., F. Pichler, E. Slooten, K. Russell, and C. S. Baker. 2001. The North Island Hector's dolphin is vulnerable to extinction. Marine Mammal Science 17:366–71.

Dayton, P. K. 1975. Experimental studies of algal canopy interactions in a sea otter–dominated kelp community at Amchitka Island, Alaska. Fishery Bulletin 73:230–37.

Dellinger, T., and F. Trillmich. 1999. Fish prey of the sympatric Galápagos fur seals and sea lions: seasonal variation and niche separation. Canadian Journal of Zoology 77:1204–16.

DeLong, R. L., and B. S. Stewart. 1991. Diving patterns of northern elephant seal bulls. Marine Mammal Science 7:369–84.

Derocher, A. E., G. W. Garner, N. J. Lunn, and Ø. Wiig (eds.). 1998. Polar Bears: Proceedings of the Twelfth Working Meeting of the IUCN/SSC Polar Bear Specialist Group, 3–7 February 1997, Oslo, Norway. Occasional Papers of the IUCN Species Survival Commission No. 19. IUCN, Gland, Switzerland.

Dietz, R., M. P. Heide-Jørgensen, E. W. Born, and C. M. Glahder. 1994. Occurrence of narwhals (*Monodon monoceros*) and white whales (*Delphinapterus leucas*) in East Greenland. Meddelelser om Grønland, Bioscience 39:69–86.

Dizon, A. E., S. J. Chivers, and W. F. Perrin (eds.). 1997. Molecular Genetics of Marine Mammals. Special Publication No. 3. Society for Marine Mammalogy, Lawrence, Kans.

Dizon, A. E., C. A. Lux, R. G. LeDuc, J. Urbán-R., M. Henshaw, C. S. Baker, F. Cipriano, and R. L. Brownell. 1997. Molecular phylogeny of the Bryde's/sei whale complex: separate species status for the pygmy Bryde's form? Report of the International Whaling Commission 47:398.

Domning, D. P. 1980. Feeding position preference in manatees (*Trichechus*). Journal of Mammalogy 61:544–47.

———. 1981. Distribution and status of manatees *Trichechus* spp. near the mouth of the Amazon River, Brazil. Biological Conservation 19:85–97.

———. 1982. Evolution of manatees. Journal of Paleontology 56:599–619.

———. 1989. Kelp evolution: a comment. Paleobiology 15:53–56.

———. 2002. Sirenian evolution. *In* W. F. Perrin, B. Würsig, and J. G. M. Thewissen (eds.), Encyclopedia of Marine Mammals, 1083–86. Academic Press, San Diego.

Domning, D. P., and L. C. Hayek. 1986. Interspecific and intraspecific morphological variation in manatees (Sirenia: *Trichechus*). Marine Mammal Science 2:87–144.

Edwards, H. H., and G. D. Schnell. 2001. Status and ecology of *Sotalia fluviatilis* in the Cayos Miskito Reserve, Nicaragua. Marine Mammal Science 17:445–72.

Elsner, R. 1999. Living in water: solutions to physiological problems. *In* J. E. Reynolds III and S. A. Rommel (eds.), Biology of Marine Mammals, 73–116. Smithsonian Institution Press, Washington, D.C.

English, A. W. 1976. Functional anatomy of the hands of fur seals and sea lions. American Journal of Anatomy 147:1–18.

———. 1977. Structural correlates of forelimb function in fur seals and sea lions. Journal of Morphology 151:325–52.

Estes, J. A. 1990. Growth and equilibrium in sea otter populations. Journal of Animal Ecology 59:385–401.

———. 1994. Conservation of marine otters. Aquatic Mammals 20:125–28.

Estes, J. A., and J. L. Bodkin. 2002. Otters. *In* W. F. Perrin, B. Würsig, and J. G. M. Thewissen (eds.), Encyclopedia of Marine Mammals, 842–58. Academic Press, San Diego.

Estes, J. A., and D. O. Duggins. 1995. Sea otters and kelp forests in Alaska: generality and variation in a community ecological paradigm. Ecological Monographs 65:75–100.

Estes, J. A., D. O. Duggins, and G. B. Rathbun. 1989. The ecology of extinctions in kelp forest communities. Conservation Biology 3:252–64.

Estes, J. A., M. T. Tinker, T. M. Williams, and D. F. Doak. 1998. Killer whale predation on sea otters linking oceanic and nearshore ecosystems. Science 282:473–76.

Fay, F. H. 1981. Walrus *Odobenus rosmarus* (Linnaeus, 1758). *In* S. H. Ridgway and R. J. Harrison (eds.), Handbook of Marine Mammals. Volume 1: The Walrus, Sea Lions, Fur Seals, and Sea Otter, 1–23. Academic Press, London.

———. 1985. *Odobenus rosmarus.* Mammalian Species 238:1–7.

Fay, F. H., J. J. Burns, S. W. Stoker, and J. S. Grundy. 1994. The struck-and-lost factor in Alaskan walrus harvests, 1952–1972. Arctic 47:368–73.

Feldkamp, S. D. 1987. Forelimb propulsion in the California sea lion *Zalophus californianus.* Journal of Zoology (London) 212:4333–57.

Fish, F. E. 1993. Influence of hydrodynamic design and propulsive mode on mammalian swimming. American Zoologist 36:628–41.
Folkow, L. P., and A. S. Blix. 1999. Diving behavior of hooded seals (*Cystophora cristata*) in the Greenland and Norwegian Seas. Polar Biology 22:61–74.
Folkow, L. P., P.-E. Martensson, and A. S. Blix. 1996. Annual distribution of hooded seals (*Cystophora cristata*) in the Greenland and Norwegian Seas. Polar Biology 16:179–89.
Forcada, J., P. S. Hammond, and A. Aguilar. 1999. Status of the Mediterranean monk seal *Monachus monachus* in the western Sahara and the implications of a mass mortality event. Marine Ecology Progress Series 188:249–61.
Ford, J. K. B. 1991. Vocal traditions among resident killer whales (*Orcinus orca*) in coastal waters of British Columbia. Canadian Journal of Zoology 69:1454–83.
Fordyce, E. 1988. Anatomy. *In* R. Harrison and M. M. Bryden (eds.), Whales, Dolphins, and Porpoises, 102–9. Facts on File, New York.
Fordyce, R. E. 2002. Cetacean evolution. *In* W. F. Perrin, B. Würsig, and J. G. M. Thewissen (eds.), Encyclopedia of Marine Mammals, 214–20. Academic Press, New York.
Frost, K. J., M. A. Simpkins, and L. F. Lowry. 2001. Diving behavior of subadult and adult harbor seals in Prince William Sound, Alaska. Marine Mammal Science 17:813–34.
Fuller, J. J. 1980. Master of Desolation: The Reminiscences of Capt. Joseph J. Fuller, edited by B. C. Busch. Mystic Seaport Museum, Mystic, Conn.
Gales, N. J., P. D. Shaughnessy, and T. E. Dennis. 1994. Distribution, abundance and breeding cycle of the Australian sea lion *Neophoca cinerea*. Journal of Zoology (London) 234:353–70.
Gales, N. J., P. Williamson, L. V. Higgins, M. A. Blackberry, and I. James. 1997. Evidence for a prolonged postimplantation period in the Australian sea lion (*Neophoca cinerea*). Journal of Reproduction and Fertility 111:159–63.
Gambell, R. 1999. The International Whaling Commission and the contemporary whaling debate. *In* J. R. Twiss Jr. and R. R. Reeves (eds.), Conservation and Management of Marine Mammals, 179–98. Smithsonian Institution Press, Washington, D.C.
Garcia-Rodriguez, A. I., B. W. Bowen, D. Domning, A. A. Mignucci-Giannoni, M. Marmontel, R. A. Montoya-Ospina, B. Morales-Vela, M. Rudin, R. K. Bonde, and P. M. McGuire. 1998. Phylogeography of the West Indian manatee (*Trichechus manatus*): How many populations and how many taxa? Molecular Ecology 7:1137–49.
Gaskin, D. E. 1982. The Ecology of Whales and Dolphins. Heineman Education Books, London.
Gelatt, T. S., C. S. Davis, D. B. Siniff, and C. Strobeck. 2001. Molecular evidence for twinning in Weddell seals (*Leptonychotes weddellii*). Journal of Mammalogy 82:491–99.
Gentry, R. L. 2002. Eared seals, Otariidae. *In* W. F. Perrin, B. Würsig, and J. G. M. Thewissen (eds.), Encyclopedia of Marine Mammals, 348–51. Academic Press, San Diego.
Gentry, R. L., and G. L. Kooyman (eds.). 1986. Fur Seals: Maternal Strategies on Land and at Sea. Princeton University Press, Princeton.
Georges, J.-Y., Y. Tremblay, and C. Guinet. 2000. Seasonal diving behaviour in lactating subantarctic fur seals on Amsterdam Island. Polar Biology 23:59–69.
Gerson, H. B., and J. P. Hickie. 1985. Head scarring on male narwhals (*Monodon monoceros*): evidence for aggressive tusk use. Canadian Journal of Zoology 63:2083–87.
Gingerich, P. D., M. ul Haq, I. S. Zalmout, I. H. Khan, and M. S. Malkani. 2001. Origin of whales from early artiodactyls: hands and feet of Eocene Protocetidae from Pakistan. Science 283:2239–42.
Gjertz, I., C. Lydersen, and O. Wiig. 2001. Distribution and diving of harbour seals (*Phoca vitulina*) in Svalbard. Polar Biology 24:209–14.
Godfrey, S. J. 1985. Additional observations of subaqueous locomotion in the California sea lion (*Zalophus californianus*). Aquatic Mammals 11:53–57.
Goodall, R. N. P. 1994. Commerson's dolphin *Cephalorhynchus commersonii* (Lacépède 1804). *In* S. H. Ridgway and R. Harrison (eds.), Handbook of Marine Mammals. Volume 5: The First Book of Dolphins, 241–67. Academic Press, London.
Gordon, J., and A. Moscrop. 1996. Underwater noise pollution and its significance for whales and dolphins. *In* M. P. Simmonds and J. D. Hutchinson (eds.), The Conservation

of Whales and Dolphins: Science and Practice, 281–319. John Wiley & Sons, Chichester, U.K.

Handley, C. O., Jr. 1966. A synopsis of the genus *Kogia* (pygmy sperm whales). *In* K. S. Norris (ed.), Whales, Dolphins, and Porpoises, 62–69. University of California Press, Berkeley and Los Angeles.

Hartman, D. S. 1979. Ecology and behavior of the manatee (*Trichechus manatus*) in Florida. American Society of Mammalogists, Special Publication 5.

Heinrich, R. E. 2002. Carnivora. *In* W. F. Perrin, B. Würsig, and J. G. M. Thewissen (eds.), Encyclopedia of Marine Mammals, 197–200. Academic Press, San Diego.

Heyning, J. E. 1984. Functional morphology involved in intraspecific fighting of the beaked whale, *Mesoplodon carlhubbsi*. Canadian Journal of Zoology 62:1645–54.

———. 1989. Cuvier's beaked whale *Ziphius cavirostris* G. Cuvier, 1823. *In* S. H. Ridgway and R. Harrison (eds.), Handbook of Marine Mammals. Volume 4: River Dolphins and the Larger Toothed Whales, 289–308. Academic Press, London.

Heyning, J. E., and G. M. Lento. 2002. The evolution of marine mammals. *In* A. R. Hoelzel (ed.), Marine Mammal Biology: An Evolutionary Approach, 38–72. Blackwell Science, Oxford.

Heyning, J. E., and J. G. Mead. 1996. Suction feeding in beaked whales: morphological and observational evidence. Contributions in Science, Natural History Museum of Los Angeles County 464:1–12.

Heyning, J. E., and W. F. Perrin. 1994. Evidence for two species of common dolphins (genus *Delphinus*) from the eastern North Pacific. Contributions in Science, Natural History Museum of Los Angeles County 442:1–35.

Hohn, A. A., A. J. Read, S. Fernández, O. Vidal, and L. T. Findley. 1996. Life history of the vaquita, *Phocoena sinus* (Phocoenidae, Cetacea). Journal of Zoology (London) 239: 235–51.

Holst, M., I. Stirling, and K. A. Hobson. 2001. Diet of ringed seals (*Phoca hispida*) on the east and west sides of the North Water polynya, northern Baffin Bay. Marine Mammal Science 17:888–908.

Hooker, S. K., and R. W. Baird. 1999. Deep-diving behaviour of the northern bottlenose whale, *Hyperoodon ampullatus* (Cetacea: Ziphiidae). Proceedings of the Royal Society of London B, Biological Sciences 266:671–76.

Horning, M., and F. Trillmich. 1997. Ontogeny of diving behaviour in the Galapagos fur seal. Behaviour 134:1211–57.

Horwood, J. W. 1990. Biology and Exploitation of the Minke Whale. CRC Press, Boca Raton.

Howell, A. B. 1929. Contributions to the comparative anatomy of the eared and earless seals (genera *Zalophus* and *Phoca*). Proceedings of the U.S. National Museum 73:1–142.

———. 1930. Aquatic Mammals: Their Adaptations to Life in the Water. Charles C Thomas, Springfield, Ill.

Hyvärinen, H. 1989. Living in darkness: whiskers as sense organs of the ringed seal (*Phoca hispida saimensis*). Journal of Zoology (London) 218:663–78.

Innes, S. G., G. A. J. Worthy, D. M. Lavigne, and K. Ronald. 1990. Surface area of phocid seals. Canadian Journal of Zoology 68:2531–38.

International Whaling Commission. 2000. Report of the Sub-committee on Small Cetaceans. Journal of Cetacean Research and Management 2 (Suppl.):235–63.

———. 2001a. Report of the Standing Sub-committee on Small Cetaceans. Journal of Cetacean Research and Management 3 (Suppl.):263–91.

———. 2001b. Report of the Workshop on the Comprehensive Assessment of Right Whales: A Worldwide Comparison. Journal of Cetacean Research and Management (Special Issue) 2:1–60.

———. 2001c. Annex U. Report of the Working Group on Nomenclature. Journal of Cetacean Research and Management 3 (Suppl.):363–67.

———. 2002. Statements concerning scientific permit catches. Journal of Cetacean Research and Management 4 (Suppl.):395–403.

———. In press. Report of the Standing Sub-committee on Small Cetaceans. Journal of Cetacean Research and Management 5 (Suppl.).

Irvine, A. B. 1983. Manatee metabolism and its influence on distribution in Florida. Biological Conservation 25:315–34.

IUCN/SSC Polar Bear Specialist Group. 1999. Global status and management of the polar bear. *In* C. Servheen, S. Herrero, and B. Peyton (eds.), Bears: Status Survey and Conservation Action Plan, 255–70. IUCN, Gland, Switzerland.

Jamieson, G. S., and H. D. Fisher. 1972. The pinniped eye: a review. *In* R. J. Harrison (ed.), Functional Anatomy of Marine Mammals, 1:245–61. Academic Press, London.

Jaquet, N., S. Dawson, and E. Slooten. 2000. Seasonal distribution and diving behaviour of male sperm whales off Kaikoura: foraging implications. Canadian Journal of Zoology 78:407–19.

Jefferson, T. A., and L. Karczmarski. 2001. *Sousa chinensis.* Mammalian Species 655:1–9.

Jefferson, T. A., and K. Van Waerebeek. 2002. The taxonomic status of the nominal dolphin species *Delphinus tropicalis* van Bree, 1971. Marine Mammal Science 18(4):787–818.

Kasuya, T., and R. L. Brownell Jr. 1979. Age determination, reproduction, and growth of the franciscana dolphin, *Pontoporia blainvillei.* Scientific Reports of the Whales Research Institute (Tokyo) 31:45–67.

Kasuya, T., and H. Marsh. 1984. Life history and reproductive biology of the short-finned pilot whale, *Globicephala macrorhynchus,* off the Pacific coast of Japan. Report of the International Whaling Commission (Special Issue) 6:259–310.

Kato, H. 2002. Bryde's whales *Balaenoptera edeni* and *B. brydei. In* W. F. Perrin, B. Würsig, and J. G. M. Thewissen (eds.), Encyclopedia of Marine Mammals, 171–77. Academic Press, San Diego.

Katona, S., and H. Whitehead. 1988. Are Cetacea ecologically important? Oceanography and Marine Biology Annual Review 26:553–68.

Kawamura, A., and Y. Satake. 1976. Preliminary report on the geographical distribution of the Bryde's whale in the North Pacific with special reference to the structure of the filtering apparatus. Scientific Reports of the Whales Research Institute (Tokyo) 28:1–35.

Kelly, B. P., and D. Wartzok. 1996. Ringed seal diving behavior in the breeding season. Canadian Journal of Zoology 74:1547–55.

Kenyon, K. W. 1977. Caribbean monk seal extinct. Journal of Mammalogy 58:97–98.

Kerley, G. I. H. 1983. Relative population sizes and trends, and hybridization of fur seals *Arctocephalus tropicalis* and *A. gazella* at the Prince Edward Islands, Southern Ocean. South African Journal of Zoology 18:388–92.

King, J. E. 1983. Seals of the World, 2d edition. British Museum (Natural History), London.

Knowlton, A. R., and S. D. Kraus. 2001. Mortality and serious injury of northern right whales (*Eubalaena glacialis*) in the western North Atlantic. Journal of Cetacean Research and Management (Special Issue) 2:193–208.

Koepfli, K.-P., and R. K. Wayne. 1998. Phylogenetic relationships of otters (Carnivora: Mustelidae) based on mitochondrial cytochrome *b* sequences. Journal of Zoology (London) 246:401–16.

Kooyman, G. L. 1973. Respiratory adaptations in marine mammals. American Zoologist 13:457–68.

———. 1989. Diverse Divers: Physiology and Behavior. Springer-Verlag, Berlin.

Kooyman, G. L., and P. J. Ponganis. 1998. The physiological basis of diving to depth: birds and mammals. Annual Review of Physiology 60:19–32.

Kovacs, K. M., and D. M. Lavigne. 1986. Maternal investment and neonatal growth in phocid seals. Journal of Animal Ecology 55:1035–51.

———. 1992. Maternal investment in otariid seals and walruses. Canadian Journal of Zoology 70:1953–64.

Kraft, B. A., C. Lydersen, K. M. Kovacs, I. Gjertz, and T. Haug. 2000. Diving behaviour of lactating bearded seals (*Erignathus barbatus*) in the Svalbard Area. Canadian Journal of Zoology 78:1408–18.

Kraus, S. D., and J. J. Hatch. 2001. Mating strategies in the North Atlantic right whale (*Eubalaena glacialis*). Journal of Cetacean Research and Management (Special Issue) 2:237–44.

Kreb, D. 1999. Observations on the occurrence of Irrawaddy dolphins, *Orcaella brevirostris*, in the Mahakam River, East Kalimantan, Indonesia. Zeitschrift für Säugetierkunde 64:54–58.

Kunnasranta, M., H. Hyvärinen, T. Sipilä, and M. Nikolai. 2001. Breeding habitat and lair structure of the ringed seal (*Phoca hispida ladogensis*) in northern Lake Ladoga in Russia. Polar Biology 24:171–74.

Kunnasranta, M., H. Hyvärinen, J. Hakkinen, and J. T. Koskela. 2002. Dive types and circadian behaviour patterns of Saimaa ringed seals, *Phoca hispida saimensis*, during open-water season. Acta Theriologica 47:63–72.

Lacoste, K., and G. B. Stenson. 2000. Winter distribution of harp seals (*Phoca groenlandica*) off eastern Newfoundland and southern Labrador. Polar Biology 23:805–11.

Larsen, T. 1986. Population biology of the polar bear (*Ursus maritimus*) in the Svalbard area. Norsk Polarinstitutt, Oslo, Skrifter 184:1–55.

Lavigne, D. M. 1999. The Hawaiian monk seal: management of an endangered species. *In* J. R. Twiss Jr. and R. R. Reeves (eds.), Conservation and Management of Marine Mammals, 246–66. Smithsonian Institution Press, Washington, D.C.

Lavigne, D. M., and K. M. Kovacs. 1988. Harps and Hoods: Ice-breeding Seals of the Northwest Atlantic. University of Waterloo Press, Waterloo, Ontario.

Lavigne, D. M., V. B. Scheffer, and S. R. Kellert. 1999. The evolution of North American attitudes toward marine mammals. *In* J. R. Twiss Jr. and R. R. Reeves (eds.), Conservation and Management of Marine Mammals, 10–47. Smithsonian Institution Press, Washington, D.C.

Lavigne, D. M., S. Innes, G. A. J. Worthy, K. M. Kovacs, O. J. Schmitz, and J. P. Hickie. 1986. Metabolic rates of seals and whales. Canadian Journal of Zoology 64:279–84.

Laws, R. M. 1956. Growth and sexual maturity in aquatic mammals. Nature 178:193–94.

———. 1994. History and present status of southern elephant seal populations. *In* B. J. Le Boeuf and R. M. Laws (eds.), Elephant Seals: Population Ecology, Behavior, and Physiology, 49–65. University of California Press, Berkeley and Los Angeles.

Leatherwood, S., J. S. Grove, and A. E. Zuckerman. 1991. Dolphins of the genus *Lagenorhynchus* in the tropical South Pacific. Marine Mammal Science 7:194–97.

LeDuc, R. G., W. F. Perrin, and A. E. Dizon. 1999. Phylogenetic relationships among the delphinid cetaceans based on full cytochrome *b* sequences. Marine Mammal Science 15:619–48.

Lefebvre, L. W., M. Marmontel, J. P. Reid, G. B. Rathbun, and D. P. Domning. 2001. Status and biogeography of the West Indian manatee. *In* C. A. Woods and F. E. Sergile (eds.), Biogeography of the West Indies: Patterns and Perspectives, 425–74. CRC Press, Boca Raton.

Ling, J. K. 1970. Pelage and molting in wild mammals with special reference to aquatic forms. Quarterly Review of Biology 45:16–54.

———. 1974. The integument of marine mammals. *In* R. J. Harrison (ed.), Functional Anatomy of Marine Mammals, 2:1–44. Academic Press, London.

Lipps, J. H., and E. Mitchell. 1976. Trophic model for the adaptive radiations and extinctions of pelagic marine mammals. Paleobiology 2:147–55.

Lockyer, C. 1987. Cetacean bioenergetics. *In* A. C. Huntley, D. P. Costa, G. A. J. Worthy, and M. A. Castellini (eds.), Approaches to Marine Mammal Energetics, 183–203. Special Publication No. 1. Society for Marine Mammalogy, Lawrence, Kans.

Loughlin, T. R. 2002. Steller's sea lion, *Eumetopias jubatus*. *In* W. F. Perrin, B. Würsig, and J. G. M. Thewissen (eds.), Encyclopedia of Marine Mammals, 1181–85. Academic Press, San Diego.

Lowry, L. F. 1993. Foods and feeding ecology. *In* J. J. Burns, J. J. Montague, and C. J. Cowles (eds.), The Bowhead Whale, 201–38. Special Publication No. 2. Society for Marine Mammalogy, Lawrence, Kans.

Lowry, L. F., and F. H. Fay. 1984. Seal eating by walruses in the Bering and Chukchi Seas. Polar Biology 3:11–18.

Lowry, L. F., J. J. Burns, and R. R. Nelson. 1987. Polar bear, *Ursus maritimus,* predation on belugas, *Delphinapterus leucas,* in the Bering and Chukchi seas. Canadian Field-Naturalist 101:141–46.

Lowry, L. F., J. W. Testa, and W. Calvert. 1996. Notes on winter feeding of crabeater and leopard seals near the Antarctic Peninsula. Polar Biology 8:475–78.

Maldonado, J. E., F. O. Davila, B. S. Stewart, E. Geffen, and R. K. Wayne. 1995. Intraspecific genetic differentiation in California sea lions (*Zalophus californianus*) from southern California and the Gulf of California. Marine Mammal Science 11:46–58.

Malik, S., P. J. Wilson, R. J. Smith, D. M. Lavigne, and B. N. White. 1997. Pinniped penises in trade: a molecular-genetic investigation. Conservation Biology 11:1365–74.

Mann, J., R. C. Connor, P. L. Tyack, and H. Whitehead (eds.). 2000. Cetacean Societies: Field Studies of Dolphins and Whales. University of Chicago Press, Chicago.

Marine Mammal Commission. 2001. Annual Report to Congress 2000. Marine Mammal Commission, Bethesda, Md.

Marmontel, M., S. R. Humphrey, and T. J. O'Shea. 1997. Population viability analysis of the Florida manatee (*Trichechus manatus latirostris*), 1976–1991. Conservation Biology 11:467–81.

Marsh, H. 1988. An ecological basis for dugong conservation in Australia. *In* M. L. Augee (ed.), Marine Mammals of Australasia: Field Biology and Captive Management, 9–21. Royal Zoological Society, Mosman, New South Wales, Australia.

———. 2002. Dugong *Dugong dugon. In* W. F. Perrin, B. Würsig, and J. G. M. Thewissen (eds.), Encyclopedia of Marine Mammals, 344–47. Academic Press, San Diego.

Marsh, H., and L. W. Lefebvre. 1994. Sirenian status and conservation. Aquatic Mammals 20:155–70.

Marsh, H., and G. B. Rathbun. 1990. Development and application of conventional and satellite radio-tracking techniques for studying dugong movements and habitat usage. Australian Wildlife Research 17:83–100.

Marsh, H., A. N. M. Harris, and I. R. Lawler. 1997. The sustainability of the indigenous dugong fishery in Torres Strait, Australia/Papua New Guinea. Conservation Biology 11:1375–86.

Marsh, H., R. I. T. Prince, W. K. Saalfeld, and R. Shepherd. 1994. The distribution and abundance of the dugong in Shark Bay, Western Australia. Australian Wildlife Research 21:149–61.

Marsh, H., G. B. Rathbun, T. J. O'Shea, and A. R. Preen. 1995. Can dugongs survive in Palau? Biological Conservation 72:85–89.

Marsh, H., C. Eros, P. Corkeron, and B. Breen. 1999. A conservation strategy for dugongs: implications of Australian research. Marine and Freshwater Research 50:979–90.

Marsh, H., H. Penrose, C. Eros, and J. Hugues. 2002. Dugong status report and action plans for countries and territories. United Nations Environment Programme, Nairobi, Early Warning and Assessment Report Series 1, UNEP/DEWA/RS.02–1.162 pp.

Martien, K. K., B. L. Taylor, E. Slooten, and S. M Dawson. 1999. A sensitivity analysis to guide research and management for Hector's dolphin. Biological Conservation 90:183–91.

Martin, A. R., and P. Rothery. 1993. Reproductive parameters of female long-finned pilot whales (*Globicephala melas*) around the Faroe Islands. Report of the International Whaling Commission (Special Issue) 14:263–304.

Martin, A. R., and T. G. Smith. 1999. Strategy and capability of wild belugas, *Delphinapterus leucas,* during deep, benthic diving. Canadian Journal of Zoology 77:1783–93.

Mattlin, R. H., N. J. Gales, and D. P. Costa. 1998. Seasonal dive behaviour of lactating New Zealand fur seals (*Arctocephalus forsteri*). Canadian Journal of Zoology 76:50–60.

McCann, T. S., and D. W. Doidge. 1987. Antarctic fur seal, *Arctocephalus gazella. In* J. P. Croxall and R. L. Gentry (eds.), Status, Biology, and Ecology of Fur Seals: Proceedings of an International Symposium and Workshop, Cambridge, England, 23–27 April 1984, pp. 5–8. NOAA Technical Report NMFS 51.

McGuire, T. L., and K. O. Winemiller. 1998. Occurrence patterns, habitat associations, and potential prey of the river dolphin, *Inia geoffrensis,* in the Cinaruco River, Venezuela. Biotropica 30:625–38.

Mead, J. G. 1989. Beaked whales of the genus *Mesoplodon. In* S. H. Ridgway and R. Harrison (eds.), Handbook of Marine Mammals. Volume 4: River Dolphins and the Larger Toothed Whales, 349–430. Academic Press, London.

———. 2002. Shepherd's beaked whale *Tasmacetus shepherdi. In* W. F. Perrin, B. Würsig, and J. G. M. Thewissen (eds.), Encyclopedia of Marine Mammals, 1078–81. Academic Press, San Diego.

Merrick, R. L., and T. R. Loughlin. 1997. Foraging behavior of adult female and young-of-the-year Steller sea lions in Alaskan waters. Canadian Journal of Zoology 75:776–86.

Miller, E. H. 1991. Communication in pinnipeds, with special reference to non-acoustic signalling. *In* D. Renouf (ed.), The Behaviour of Pinnipeds, 128–235. Chapman and Hall, London.

Miller, G. S., Jr. 1918. A new river-dolphin from China. Smithsonian Miscellaneous Collections 68(2486):1–12.

Nilssen, K. T., T. Haug, and C. Lindblom. 2001. Diet of weaned pups and seasonal variations in body condition of juvenile Barents Sea harp seals *Phoca groenlandica.* Marine Mammal Science 17:926–36.

Nishiwaki, M., and H. Marsh. 1985. Dugong *Dugong dugon* (Müller, 1776). *In* S. H. Ridgway and R. Harrison (eds.), Handbook of Marine Mammals. Volume 3: The Sirenians and Baleen Whales, 1–31. Academic Press, London.

Nishiwaki, M., T. Kasuya, N. Miyazaki, T. Tobayama, and T. Kataoka. 1979. Present distribution of the dugong in the world. Scientific Reports of the Whales Research Institute (Tokyo) 31:133–41.

Nordoy, E. S., and A. S. Blix. 2001. The previously pagophilic Ross seal is now rather pelagic. *In* Antarctic Biology in a Global Context: Proceedings of the VIII SCAR International Biology Symposium, S5O14. Amsterdam, The Netherlands.

Nordoy, E. S., L. Folkow, and A. S. Blix. 1995. Distribution and diving behaviour of crabeater seals (*Lobodon carcinophagus*) off Queen Maud Land. Polar Biology 15:261–68.

Norris, K. S., and B. Møhl. 1983. Can odontocetes debilitate prey with sound? American Naturalist 122:85–104.

Norris, K. S., B. Würsig, R. S. Wells, and M. Würsig. 1994. The Hawaiian Spinner Dolphin. University of California Press, Berkeley and Los Angeles.

Oftedal, O. T. 1997. Lactation in whales and dolphins: evidence of divergence between baleen and toothed species. Journal of Mammary Gland Biology and Neoplasia 2:205–30.

Oftedal, O. T., D. J. Boness, and R. A. Tedman. 1987. The behavior, physiology, and anatomy of lactation in the Pinnipedia. Current Mammalogy 1:175–245.

Olesiuk, P. F., M. A. Bigg, and G. M. Ellis. 1990. Life history and population dynamics of resident killer whales (*Orcinus orca*) in the coastal waters of British Columbia and Washington State. Report of the International Whaling Commission (Special Issue) 12:209–43.

Oliver, J. S., and P. N. Slattery. 1985. Destruction and opportunity on the sea floor: effects of gray whale feeding. Ecology 66:1965–75.

O'Shea, T. J., L. W. Lefebvre, and C. A. Beck. 2001. Florida manatees: perspectives on populations, pain, and protection. *In* L. A. Dierauf and F. M. D. Gulland (eds.), CRC Handbook of Marine Mammal Medicine, 2d ed, 31–43. CRC Press, Boca Raton.

O'Shea, T. J., R. R. Reeves, and A. K. Long (eds.). 1999. Marine Mammals and Persistent Ocean Contaminants: Proceedings of the Marine Mammal Commission Workshop, Keystone, Colorado, 12–15 October 1998. Marine Mammal Commission, Bethesda, Md.

Ostfeld, R. S., L. Ebensperger, L. L. Klosterman, and J. C. Castilla. 1989. Foraging, activity budget, and social behavior of the South American marine otter *Lutra felina* (Molina 1782). National Geographic Research 5:422–38.

Pabst, D. A., S. A. Rommel, and W. A. McLellan. 1999. The functional morphology of marine mammals. *In* J. E. Reynolds III and S. A. Rommel (eds.), Biology of Marine Mammals, 15–72. Smithsonian Institution Press, Washington, D.C.

Pardue, M. T., J. G. Sivak, and K. M. Kovacs. 1993. Corneal anatomy of marine mammals. Canadian Journal of Zoology 71:2282–90.

Pastene, L. A., M. Goto, S. Itoh, S. Wada, and H. Kato. 1997. Intra- and inter-oceanic patterns of mitochondrial DNA variation in the Bryde's whale, *Balaenoptera edeni*. Report of the International Whaling Commission 47:569–74.

Patterson, I. A. P., R. J. Reid, B. Wilson, K. Grellier, H. M. Ross, and P. M. Thompson. 1998. Evidence for infanticide in bottlenose dolphins: an explanation for violent interactions with harbour porpoises? Proceedings of the Royal Society of London B 265:1167–70.

Payne, R., and E. M. Dorsey. 1983. Sexual dimorphism and aggressive use of callosities in right whales (*Eubalaena australis*). *In* R. Payne (ed.), Communication and Behavior of Whales, 295–329. Westview Press, Boulder, Colo.

Perrin, W. F., and R. L. Brownell Jr. (eds.). 1989. Report of the workshop. *In* W. F. Perrin, R. L. Brownell Jr., K. Zhou, and J. Liu (eds.), Biology and Conservation of the River Dolphins, 1–22. Occasional Papers of the IUCN Species Survival Commission No. 3. IUCN, Gland, Switzerland.

Perrin, W. F., and R. L. Brownell Jr. 1994. A brief review of stock identity in small marine cetaceans in relation to assessment of driftnet mortality in the North Pacific. Report of the International Whaling Commission (Special Issue) 15:393–401.

Perrin, W. F., and S. B. Reilly. 1984. Reproductive parameters of dolphins and small whales of the family Delphinidae. Report of the International Whaling Commission (Special Issue) 6:97–125.

Perrin, W. F., M. L. L. Dolar, and D. Robineau. 1999. Spinner dolphins (*Stenella longirostris*) of the western Pacific and Southeast Asia: pelagic and shallow-water forms. Marine Mammal Science 15:1029–53.

Perrin, W. F., G. P. Donovan, and J. Barlow (eds.). 1994. Gillnets and cetaceans. Report of the International Whaling Commission (Special Issue) 15:1–629.

Perrin, W. F., B. Würsig, and J. G. M. Thewissen (eds.). 2002. Marine mammal species. *In* W. F. Perrin, B. Würsig, and J. G. M. Thewissen (eds.). Encyclopedia of Marine Mammals, 1335–37. Academic Press, San Diego.

Pichler, F. B., and C. S. Baker. 2000. Loss of genetic diversity in the endemic Hector's dolphin due to fisheries-related mortality. Proceedings of the Royal Society of London B 267:97–102.

Pitman, R. L. 2002a. Indo-Pacific beaked whale *Indopacetus pacificus*. *In* W. F. Perrin, B. Würsig, and J. G. M. Thewissen (eds.), Encyclopedia of Marine Mammals, 615–17. Academic Press, San Diego.

———. 2002b. Mesoplodont whales *Mesoplodon* spp. *In* W. F. Perrin, B. Würsig, and J. G. M. Thewissen (eds.), Encyclopedia of Marine Mammals, 738–42. Academic Press, San Diego.

Pitman, R. L., D. M. Palacios, P. L. R. Brennan, B. J. Brennan, K. C. Balcomb III, and T. Miyashita. 1999. Sightings and possible identity of a bottlenose whale in the tropical Indo-Pacific: *Indopacetus pacificus?* Marine Mammal Science 15:531–49.

Plagányi, É. E., and D. S. Butterworth. 2002. Competition with fisheries. *In* W. F. Perrin, B. Würsig, and J. G. M. Thewissen (eds.), Encyclopedia of Marine Mammals, 268–73. Academic Press, San Diego.

Ponganis, P. J., E. P. Ponganis, K. V. Ponganis, G. L. Kooyman, R. L. Gentry, and F. Trillmich. 1990. Swimming velocities in otariids. Canadian Journal of Zoology 68:2105–15.

Ponganis, P. J., G. L. Kooyman, E. A. Baranov, P. H. Thorson, and B. S. Stewart. 1997. The aerobic submersion limit of Baikal seals, *Phoca sibirica*. Canadian Journal of Zoology 75:1323–27.

Powell, J. A. 1978. Evidence of carnivory in manatees (*Trichechus manatus*). Journal of Mammalogy 59:442.

Preen, A. 1995. Impacts of dugong foraging on seagrass habitats: observational and experimental evidence for cultivation grazing. Marine Ecology Progress Series 124:201–13.

Prescott-Allen, R., and C. Prescott-Allen (eds.). 1996. Assessing the Sustainability of Uses of Wild Species: Case Studies and Initial Assessment Procedure. Occasional Papers of the IUCN Species Survival Commission No. 12. IUCN, Gland, Switzerland.

Prestrud, P., and I. Stirling. 1994. The International Polar Bear Agreement and the current status of polar bear conservation. Aquatic Mammals 20:113–24.

Ragen, T. J., and D. M. Lavigne. 1999. The Hawaiian monk seal: biology of an endangered species. *In* J. R. Twiss Jr. and R. R. Reeves (eds.), Conservation and Management of Marine Mammals, 224–45. Smithsonian Institution Press, Washington, D.C.

Ralls, K. 1977. Sexual dimorphism in mammals: avian models and unanswered questions. American Naturalist 111:917–38.

Ralls, K., P. Fiorelli, and S. Gish. 1984. Vocalizations and vocal mimicry in captive harbor seals, *Phoca vitulina*. Canadian Journal of Zoology 63:1050–56.

Ralls, K., B. B. Hatfield, and D. B. Siniff. 1995. Foraging patterns of California sea otters as indicated by telemetry. Canadian Journal of Zoology 73:523–31.

Rathbun, G. B., J. P. Reid, R. K. Bonde, and J. A. Powell. 1995. Reproduction in free-ranging Florida manatees. *In* T. J. O'Shea, B. B. Ackerman, and H. F. Percival (eds.), Population Biology of the Florida Manatee, 135–56. U.S. Department of the Interior, National Biological Service, Information and Technology Report 1.

Ray, C., W. A. Watkins, and J. J. Burns. 1969. The underwater song of *Erignathus* (bearded seal). Zoologica 54:79–83.

Read, A. J. 1990. Age at sexual maturity and pregnancy rates of harbour porpoises *Phocoena phocoena* from the Bay of Fundy. Canadian Journal of Fisheries and Aquatic Sciences 47:561–65.

Read, A. J., and A. A. Hohn. 1995. Life in the fast lane: the life history of harbor porpoises from the Gulf of Maine. Marine Mammal Science 11:423–40.

Reeves, R. R. 2000. The Value of Sanctuaries, Parks, and Reserves (Protected Areas) as Tools for Conserving Marine Mammals. Contract Report to Marine Mammal Commission, Bethesda, Md.

———. 2002a. Conservation efforts. *In* W. F. Perrin, B. Würsig, and J. G. M. Thewissen (eds.). Encyclopedia of Marine Mammals, 276–97. Academic Press, San Diego.

———. 2002b. The origins and character of "aboriginal subsistence" whaling: a global review. Mammal Review 32:71–106.

Reeves, R. R., and A. A. Chaudhry. 1998. Status of the Indus River dolphin *Platanista minor*. Oryx 32:35–44.

Reeves, R. R., and P. J. H. Reijnders. 2002. Conservation and management. *In* A. R. Hoelzel (ed.), Marine Mammal Biology: An Evolutionary Approach, 388–415. Blackwell Science, Oxford.

Reeves, R. R., B. S. Stewart, and S. Leatherwood. 1992. The Sierra Club Handbook of Seals and Sirenians. Sierra Club Books, San Francisco.

Reeves, R. R., D. Tuboku-Metzger, and R. A. Kapindi. 1988. Distribution and exploitation of manatees in Sierra Leone. Oryx 22:75–84.

Reeves, R. R., G. W. Wenzel, and M. C. S. Kingsley. 1998. Catch history of ringed seals (*Phoca hispida*) in Canada. *In* M. P. Heide-Jørgensen and C. Lydersen (eds.), Ringed Seals in the North Atlantic, 100–129. North Atlantic Marine Mammal Commission Scientific Publications 1, Tromsø, Norway.

Reeves, R. R., S. Leatherwood, T. A. Jefferson, B. E. Curry, and T. Henningsen. 1996. Amazonian manatees, *Trichechus inunguis*, in Peru: distribution, exploitation, and conservation status. Interciencia 21:246–54.

Reeves, R. R., B. S. Stewart, P. J. Clapham, and J. A. Powell. 2002. National Audubon Society Guide to Marine Mammals of the World. A. A. Knopf, New York.

Reijnders, P. J. H., A. Aguilar, and G. P. Donovan (eds.). 1999. Chemical pollutants and cetaceans. Journal of Cetacean Research and Management (Special Issue) 1:1–273.

Reijnders, P., S. Brasseur, J. van der Toorn, P. van der Wolf, I. Boyd, J. Harwood, D. Lavigne, and L. Lowry. 1993. Seals, Fur Seals, Sea Lions, and Walrus: Status Survey and Conservation Action Plan. IUCN, Gland, Switzerland.

Repenning, C. A. 1976. Adaptive evolution of sea lions and walruses. Systematic Zoology 25:375–99.

Reyes, J. C., J. G. Mead, and K. Van Waerebeek. 1991. A new species of beaked whale *Mesoplodon peruvianus* sp. n. (Cetacea: Ziphiidae) from Peru. Marine Mammal Science 7: 1–24.

Reyes, J. C., K. Van Waerebeek, J. C. Cárdenas, and J. L. Yáñez. 1996. *Mesoplodon bahamondi* sp. n. (Cetacea: Ziphiidae), a new living beaked whale from the Juan Fernández Archipelago, Chile. Boletín del Museo Nacional de Historia Natural, Chile 45: 31–44.

Reynolds, J. E., III. 1999. Efforts to conserve the manatees. *In* J. R. Twiss Jr. and R. R. Reeves (eds.), Conservation and Management of Marine Mammals, 267–95. Smithsonian Institution Press, Washington, D.C.

Reynolds, J. E., III, and D. K. Odell. 1991. Manatees and Dugongs. Facts on File, New York.

Reynolds, J. E., III, and J. A. Powell. 2002. Manatees *Trichechus manatus, T. senegalensis,* and *T. inunguis. In* W. F. Perrin, B. Würsig, and J. G. M. Thewissen (eds.), Encyclopedia of Marine Mammals, 709–20. Academic Press, San Diego.

Rice, D. W. 1998. Marine Mammals of the World: Systematics and Distribution. Special Publication Number 4. Society for Marine Mammalogy, Lawrence, Kans.

Rice, D. W., and A. A. Wolman. 1971. The Life History and Ecology of the Gray Whale (*Eschrichtius robustus*). American Society of Mammalogists, Special Publication 3.

Riedman, M. 1990. The Pinnipeds: Seals, Sea Lions, and Walruses. University of California Press, Berkeley and Los Angeles.

Rogan, E., J. R. Baker, P. D. Jepson, S. Berrow, and O. Kiely. 1997. A mass stranding of white-sided dolphins (*Lagenorhynchus acutus*) in Ireland: biological and pathological studies. Journal of Zoology (London) 242:217–27.

Rogers, T., D. Cato, and M. Bryden. 1996. Behavioral significance of underwater vocalizations of captive leopard seals, *Hydrurga leptonyx*. Marine Mammal Science 12:414–27.

Rojas-Bracho, L., and A. Jaramillo-Legorreta. 2002. Vaquita *Phocoena sinus. In* W. F. Perrin, B. Würsig, and J. G. M. Thewissen (eds.), Encyclopedia of Marine Mammals, 1277–80. Academic Press, San Diego.

Rommel, S. A, G. A. Early, K. A. Matassa, D. A. Pabst, and W. A. McLellan. 1995. Venous structures associated with thermoregulation of phocid seal reproductive organs. Anatomical Record 243:390–402.

Rosas, F. C. W. 1994. Biology, conservation and status of the Amazonian manatee *Trichechus inunguis.* Mammal Review 24:49–59.

Rosel, P. E., A. E. Dizon, and J. E. Heyning. 1994. Genetic analysis of sympatric morphotypes of common dolphins (genus *Delphinus*). Marine Biology 119:159–67.

Rosel, P. E., M. G. Haygood, and W. F. Perrin. 1995. Phylogenetic relationships among the true porpoises (Cetacea: Phocoenidae). Molecular Phylogenetics and Evolution 4:463–74.

Rosenbaum, H. C., R. L. Brownell Jr., M. W. Brown, C. Schaeff, V. Portway, B. N. White, S. Malik, L. A. Pastene, N. J. Patenaude, C. S. Baker, M. Goto, P. B. Best, P. J. Clapham, P. Hamilton, M. Moore, R. Payne, V. Rowntree, C. T. Tynan, and R. DeSalle. 2000. Worldwide genetic differentiation of *Eubalaena:* questioning the number of right whale species. Molecular Ecology 9:1793–1802.

Ross, G. J. B., G. E. Heinsohn, and V. G. Cockcroft. 1994. Humpback dolphins *Sousa chinensis* (Osbeck, 1765), *Sousa plumbea* (G. Cuvier, 1829) and *Sousa teuszii* (Kukenthal, 1892). *In* S. H. Ridgway and R. Harrison (eds.), Handbook of Marine Mammals. Volume 5: The First Book of Dolphins, 23–42. Academic Press, London.

Ross, H. M., and B. Wilson. 1996. Violent interactions between bottlenose dolphins and harbour porpoises. Proceedings of the Royal Society of London B 263:283–86.

Roth, H. H., and E. Waitkuwait. 1986. Répartition et status des granes espèces de mammifères en Côte-d'Ivoire III. Lamantins. Mammalia 50:227–42.

Samuels, A., and P. Tyack. 2000. Flukeprints: a history of studying cetacean societies. *In* J. Mann, R. C. Connor, P. L. Tyack, and H. Whitehead (eds.), Cetacean Societies: Field Studies of Dolphins and Whales, 9–44. University of Chicago Press, Chicago.

Scheffer, V. 1964. Hair patterns in seals (Pinnipedia). Journal of Morphology 115:291–304.

Scheffer, V. B., C. H. Fiscus, and E. I. Todd. 1984. History of scientific study and management of the Alaskan fur seal, *Callorhinus ursinus*, 1786–1964. NOAA Technical Report NMFS SSRF-780:1–70.

Schevill, W. E., W. A. Watkins, and G. C. Ray. 1966. Analysis of underwater *Odobenus* calls with remarks on the development and function of the pharyngeal pouches. Zoologica 51:103–6.

Scholander, P. F., and W. E. Schevill. 1955. Counter-current vascular heat exchange in the fins of whales. Journal of Applied Physiology 8:279–82.

Shane, S. H. 1983. Abundance, distribution, and movements of manatees (*Trichechus manatus*) in Brevard County, Florida. Bulletin of Marine Science 33:1–9.

Shaughnessy, P. D., G. L. Shaughnessy, and L. Fletcher. 1988. Recovery of the fur seal population at Macquarie Island. Papers and Proceedings of the Royal Society of Tasmania 122:177–87.

Sielfeld, W. 1983. Mamíferos Marinos de Chile. Ediciiones de la Universidad de Chile, Santiago.

Silva, M. A., and A. Araújo. 2001. Distribution and current status of the West African manatee (*Trichechus senegalensis*) in Guinea-Bissau. Marine Mammal Science 17:418–24.

Simpkins, M. A., B. P. Kelly, and D. Wartzok. 2001. Three-dimensional diving behaviors of ringed seals (*Phoca hispida*). Marine Mammal Science 17:909–25.

Siniff, D. B., and S. Stone. 1985. The role of the leopard seal in the tropho-dynamics of the Antarctic marine ecosystem. *In* W. R. Siegfried, P. R. Condy, and R. M. Laws (eds.), Antarctic Nutrient Cycles and Food Webs, 555–59. Springer-Verlag, Berlin.

Sipilä, T., and H. Hyvärinen. 1998. Status and biology of Saimaa (*Phoca hispida saimensis*) and Ladoga (*Phoca hispida ladogensis*) ringed seals. *In* M. P. Heide-Jørgensen and C. Lydersen (eds.), Ringed Seals in the North Atlantic, 83–99. NAMMCO Scientific Publications 1. North Atlantic Marine Mammal Commission, Tromsø, Norway.

Sjoberg, M., and J. P. Ball. 2000. Grey seal, *Halichoerus grypus*, habitat selection and haulout sites in the Baltic Sea: bathymetry or central-place foraging. Canadian Journal of Zoology 78:1661–67.

Smith, A., and H. Marsh. 1990. Management of traditional hunting of dugongs [*Dugong dugon* (Müller, 1776)] in the northern Great Barrier Reef, Australia. Environmental Management 14:47–55.

Smith, B. D., R. K. Sinha, K. Zhou, A. A. Chaudhry, R. Liu, D. Wang, B. Ahmed, A. K. M. Aminul Haque, R. S. L. Mohan, and K. Sapkota. 2000. Register of water development projects affecting river cetaceans in Asia. *In* R. R. Reeves, B. D. Smith, and T. Kasuya (eds.), Biology and Conservation of Freshwater Cetaceans in Asia, 22–39. Occasional Papers of the IUCN Species Survival Commission No. 23. IUCN, Gland, Switzerland.

Smith, B. D., B. Ahmed, M. Edrise, and G. Braulik. 2001. Status of the Ganges river dolphin or shushuk *Platanista gangetica* in Kaptai Lake and the southern rivers of Bangladesh. Oryx 35:61–72.

Smith, C. R., H. Kukert, R. A. Wheatcroft, P. A. Jumars, and J. W. Deming. 1989. Vent fauna on whale remains. Nature 341:27–28.

Snyder, G. K. 1983. Respiratory adaptations in diving mammals. Respiration Physiology 54:269–94.

Stacey, P. J., and S. Leatherwood. 1997. The Irrawaddy dolphin, *Orcaella brevirostris:* a summary of current knowledge and recommendations for conservation action. Asian Marine Biology 14:195–214.

Stevick, P. T., B. J. McConnell, and P. S. Hammond. 2002. Patterns of movement. *In* A. R. Hoelzel (ed.), Marine Mammal Biology: An Evolutionary Approach, 185–216. Blackwell Science, Oxford.

Stewart, B. S. 1997. Ontogeny of differential migration and sexual segregation in northern elephant seals. Journal of Mammalogy 78:1101–16.

———. 2002. Diving behavior. *In* W. F. Perrin, B. Würsig, and J. G. M. Thewissen (eds.), Encyclopedia of Marine Mammals, 333–39. Academic Press, San Diego.

Stewart, B. S., and R. L. DeLong. 1993. Seasonal dispersion and habitat use of foraging northern elephant seals. Symposium, Zoological Society of London 66:179–94.

———. 1995. Double migrations of the northern elephant seal, *Mirounga angustirostris*. Journal of Mammalogy 76:196–205.

Stewart, B. S., and H. R. Huber. 1993. *Mirounga angustirostris*. Mammalian Species 449:1–10.

Stewart, B. S., P. K. Yochem, R. L. DeLong, and G. A. Antonelis. 1987. Interactions between Guadalupe fur seals and California sea lions at San Nicolas and San Miguel islands, California. NOAA Technical Report NMFS 51:103–6.

Stewart, B. S., E. A. Petrov, E. A. Baranov, A. Timonin, and M. Ivanov. 1996. Seasonal movements and dive patterns of juvenile Baikal seals, *Phoca sibirica*. Marine Mammal Science 12:528–42.

Stewart, B. S., P. K. Yochem, T. S. Gelatt, and D. B. Siniff. 2000. First-year movements of Weddell seal pups in the western Ross Sea, Antarctica. *In* W. Davison, C. H. Williams, and P. Broady (eds.), Antarctic Ecosystems: Models for Wider Ecological Understanding, 71–76. New Zealand Natural Sciences, Canterbury University.

Stewart, B. S., P. K. Yochem, T. S. Gelatt, and D. B. Siniff. 2001. Ecological roles of foraging Weddell seals in autumn and winter pack-ice and polynya ecosystems of the western Ross Sea. *In* Antarctic Biology in a Global Context: Proceedings of the VIII SCAR International Biology Symposium, S5P52. Amsterdam, The Netherlands.

Stirling, I. 1973. Vocalization in the ringed seal (*Phoca hispida*). Journal of the Fisheries Research Board of Canada 30:1592–94.

———. 1983. The evolution of mating systems in pinnipeds. *In* J. F. Eisenberg and D. G. Kleinman (eds.), Advances in the Study of Mammalian Behavior, 489–527. American Society of Mammalogists, Allen Press, Lawrence, Kans.

———. 1988. Polar Bears. University of Michigan Press, Ann Arbor.

Stirling, I., W. Calvert, and C. Spencer. 1987. Evidence of stereotyped underwater vocalizations of male Atlantic walruses (*Odobenus rosmarus rosmarus*). Canadian Journal of Zoology 65:2311–21.

Stirling, I., N. J. Nunn, and J. Iacozza. 1999. Long-term trends in the population ecology of polar bears in western Hudson Bay in relation to climatic change. Arctic 52:294–306.

Testa, J. W. 1994. Over-winter movements and diving behavior of female Weddell seals (*Leptonychotes weddellii*) in the southwestern Ross Sea, Antarctica. Canadian Journal of Zoology 72:1700–1710.

Thomas, J. A., K. C. Zinnel, and L. M. Ferm. 1983. Analysis of Weddell seal (*Leptonychotes weddelli*) vocalizations using underwater playbacks. Canadian Journal of Zoology 61:1448–56.

Trillmich, F. 1990. The behavioral ecology of maternal effort in fur seals and sea lions. Behaviour 114:3–20.

———. 1996. Parental investment in pinnipeds. Advances in the Study of Behavior 25:533–77.

Trillmich, F., and T. Dellinger. 1991. The effects of El Niño on Galápagos pinnipeds. *In* F. Trillmich and K. A. Ono (eds.), The Ecological Effects of El Niño on Otariids and Phocids: Responses of Marine Mammals to Environmental Stress, 66–74. Springer-Verlag, Berlin.

Tyack, P. L. 2000. Functional aspects of cetacean communication. *In* J. Mann, R. C. Connor, P. L. Tyack, and H. Whitehead (eds.), Cetacean Societies: Field Studies of Dolphins and Whales, 270–307. University of Chicago Press, Chicago.

Tyack, P. L., and E. H. Miller. 2002. Vocal anatomy, acoustic communication and echolocation. *In* A. R. Hoelzel (ed.), Marine Mammal Biology: An Evolutionary Approach, 142–84. Blackwell Science, Oxford.

Tyack, P. L., and H. Whitehead. 1983. Male competition in large groups of wintering humpback whales. Behaviour 83:131–54.

van Helden, A. L., A. N. Baker, M. L. Dalebout, J. C. Reyes, K. Van Waerebeek, and C. S. Baker. 2002. Resurrection of *Mesoplondon traversii* (Gray, 1874), senior synonym of

M. bahamondi Reyes, Van Waerebeek, Cárdenas, and Yáñez, 1995 (Cetacea: Ziphiidae). Marine Mammal Science 18:609–21.

Van Waerebeek, K., R. N. P. Goodall, and P. B. Best. 1997. A note on evidence for pelagic warm-water dolphins resembling *Lagenorhynchus.* Report of the International Whaling Commission 47:1015–17.

Van Waerebeek, K., P. J. H. van Bree, and P. B. Best. 1995. On the identity of *Prodelphinus petersii* (Lütken, 1889) and records of dusky dolphin, *Lagenorhynchus obscurus* (Gray, 1828), from the southern Indian and Atlantic Oceans. South African Journal of Marine Science 16:25–35.

Van Zyll de Jong, C. G. 1987. A phylogenetic study of the Lutrinae (Carnivora; Mustelidae) using morphological data. Canadian Journal of Zoology 65:2536–44.

Víkingsson, G. A., and F. O. Kapel (eds.). 2000. Minke Whales, Harp and Hooded Seals: Major Predators in the North Atlantic Ecosystem. North Atlantic Marine Mammal Commission, Tromsø, Norway. NAMMCO Scientific Publications 2. 132 pp.

Wada, S., and K. Numachi. 1991. Allozyme analyses of genetic differentiation among the populations and species of *Balaenoptera.* Report of the International Whaling Commission (Special Issue) 13:125–54.

Wang, J. Y., L.-S. Chou, and B. N. White. 1999. Mitochondrial DNA analysis of sympatric morphotypes of bottlenose dolphins (genus *Tursiops*) in Chinese waters. Molecular Ecology 8:1603–12.

———. 2000. Osteological differences between two sympatric forms of bottlenose dolphins (genus *Tursiops*) in Chinese waters. Journal of Zoology (London) 252:147–62.

Warneke, R. M., and P. D. Shaughnessy. 1985. *Arctocephalus pusillus,* the South African and Australian fur seal: taxonomy, evolution, biogeography, and life history. *In* J. K. Ling and M. M. Bryden (eds.), Studies of Sea Mammals in South Latitudes, 53–77. South Australian Museum, Adelaide.

Wartzok, D., and D. R. Ketten. 1999. Marine mammal sensory systems. *In* J. E. Reynolds III and S. A. Rommel (eds.), Biology of Marine Mammals, 117–75. Smithsonian Institution Press, Washington, D.C.

Weilgart, L., and H. Whitehead. 1997. Group-specific dialects and geographical variation in coda repertoire in South Pacific sperm whales. Behavioral Ecology and Sociobiology 40:277–85.

Weilgart, L., H. Whitehead, and K. Payne. 1996. A colossal convergence. American Scientist 84:278–87.

Weller, D. W., B. Würsig, H. Whitehead, J. C. Norris, S. K. Lynn, R. W. Davis, N. Clauss, and P. Brown. 1996. Observations of an interaction between sperm whales and short-finned pilot whales in the Gulf of Mexico. Marine Mammal Science 12:588–94.

Wells, R. S., and M. D. Scott. 1999. Bottlenose dolphin *Tursiops truncatus* (Montagu, 1821). *In* S. H. Ridgway and R. Harrison (eds.), Handbook of Marine Mammals. Volume 6: The Second Book of Dolphins and the Porpoises, 137–82. Academic Press, San Diego.

Wells, R. S., D. J. Boness, and G. B. Rathbun. 1999. Behavior. *In* J. E. Reynolds III and S. A. Rommel (eds.), Biology of Marine Mammals, 324–422. Smithsonian Institution Press, Washington, D.C.

Whitehead, H., and J. Mann. 2000. Female reproductive strategies of cetaceans. *In* J. Mann, R. C. Connor, P. L. Tyack, and H. Whitehead (eds.), Cetacean Societies: Field Studies of Dolphins and Whales, 219–46. University of Chicago Press, Chicago.

Whitehead, H., and L. Weilgart. 2000. The sperm whale: social females and roving males. In J. Mann, R. C. Connor, P. L. Tyack, and H. Whitehead (eds.), Cetacean Societies: Field Studies of Dolphins and Whales, 154–72. University of Chicago Press, Chicago.

Williams, T. D., J. A. Mattison, and J. A. Ames. 1980. Twinning in a California sea otter. Journal of Mammalogy 61:575–76.

Williams, T. M., and G. L. Kooyman. 1985. Swimming performance and hydrodynamic characteristics of harbor seals *Phoca vitulina.* Physiological Zoology 58:158–78.

Williams, T. M., and G. A. J. Worthy. 2002. Anatomy and physiology: the challenge of aquatic living. *In* A. R. Hoelzel (ed.), Marine Mammal Biology: An Evolutionary Approach, 73–97. Blackwell Science, Oxford.

Wilson, D. E., M. A. Bogan, R. L. Brownell Jr., A. M. Burdin, and M. K. Maminov. 1991. Geographic variation in sea otters, *Enhydra lutris.* Journal of Mammalogy 72:22–36.

Wolff, J. O. 1997. Population regulation in mammals: an evolutionary perspective. Journal of Animal Ecology 66:1–13.

Würsig, B., and W. J. Richardson. 2002. Effects of noise. *In* W. F. Perrin, B. Würsig, and J. G. M. Thewissen (eds.), Encyclopedia of Marine Mammals, 794–802. Academic Press, San Diego.

Würsig, B., R. R. Reeves, and J. G. Ortega-Ortiz. 2001. Global climate change and marine mammals. *In* P. G. H. Evans and J. A. Raga (eds.), Marine Mammals: Biology and Conservation, 589–608. Kluwer Academic/Plenum Press, New York.

Wyss, A. R. 1989. Flippers and pinniped phylogeny: has the problem of convergence been overrated? Marine Mammal Science 5:343–60.

Yochem, P. K., and B. S. Stewart. 2002. Hair and fur. *In* W. F. Perrin, B. Würsig, and J. G. M. Thewissen (eds.), Encyclopedia of Marine Mammals, 548–49. Academic Press, San Diego.

Zhou, K., and X. Zhang. 1991. Baiji: The Yangtze River Dolphin and Other Endangered Animals of China. Stone Wall Press, Washington, D.C.

Zhou, K., J. Sun, A. Gao, and B. Würsig. 1998. Baiji (*Lipotes vexillifer*) in the lower Yangtze River: movements, numbers, threats and conservation needs. Aquatic Mammals 24:123–32.

Pinnipeds

Seals, Sea Lions, and Walrus: Pinnipedia

This order of aquatic mammals occurs along ice fronts and coastlines, mainly in polar and temperate parts of the oceans and adjoining seas of the world but also in some tropical areas and in certain inland bodies of water. The Pinnipedia traditionally are regarded as a full order, and some authorities, including Corbet (1978), Corbet and Hill (1991), and E. R. Hall (1981), continue to treat them as such. Other authorities, such as Schliemann (*in* Grzimek 1990) and Simpson (1945), have considered the Pinnipedia to be only a suborder of the order Carnivora. Recently there was a growing consensus, based mainly on morphological evidence, that the pinnipeds belong within the arctoid division of the Carnivora and are biphyletic in origin, the families Otariidae (eared seals, sea lions) and Odobenidae (walrus) having arisen from bearlike ancestors and the family Phocidae (earless seals) being an early offshoot of the line leading to the otters (McLaren 1960; Rice 1977; Stains 1984; Tedford 1976; Wozencraft 1989). At the same time, however, there were immunological and chromosomal data that supported a monophyletic origin for the Pinnipedia (Arnason 1974; Sarich 1969).

Studies of cranial and postcranial skeletal material, both fossil and Recent, now indicate that the pinnipeds are indeed monophyletic (Berta, Ray, and Wyss 1989; Berta and Wyss 1990, 1994; Wiig 1983; Wyss 1987, 1988*a*). There is general agreement that the group is an evolutionary offshoot of the Carnivora. However, although morphological evidence supports an origin from the same ancestral line that gave rise to the Ursidae (Wyss and Flynn 1993), some DNA analyses suggest that the closest living relatives of the pinnipeds are the Mustelidae (Arnason and Ledje 1993). There also remains a question, based partly on whether cladistic principles are followed (see account of family Hominidae), whether the pinnipeds warrant ordinal rank or would best be retained within the Carnivora. The former course has been followed here, with recognition of 3 Recent pinniped families (in the sequence suggested by Wyss 1987), 18 genera, and 34 species. However, many authorities now consider the pinniped families to be components of the carnivore suborder Caniformia and/or the superfamily Arctoidea (Barnes 1989; Berta and Wyss 1994; Jones et al. 1992; Wozencraft *in* Wilson and Reeder 1993; Wyss and Flynn 1993).

Pinnipeds are measured in a straight line from the tip of the nose to the tip of the tail. Total length varies from 120 cm to 600 cm; a short or vestigial tail between the hind limbs grows very little after birth. Adults weigh from about 35 kg to 3,700 kg, with *Phoca* containing some of the smallest species and *Mirounga* the largest. Pinnipeds have a streamlined, torpedo-shaped body, with all four limbs modified into flippers. The arm and leg bones are similar to those in the Carnivora, but the bases of the limbs, to or beyond the elbows and the knees, are deeply enclosed within the body. The hands and feet are long and flattened; hence the name Pinnipedia, which means "feather-footed." Each limb has five broadly webbed, oarlike digits, which form the flipper. In most species the head is flattened, and the face shortened, to aid in rapid propulsion through the water. External ears are small or entirely lacking, and the nostrils are slitlike; the ears and the nose can be tightly closed when the animal is underwater. The eyes, which are set in deep protective cushions of fat, are also adapted for underwater use. The cornea is flattened, and the pupil is capable of great enlargement to enable better sight in dark water. The neck in pinnipeds is generally thick and muscular yet quite flexible. A reduction in the interlocking processes of the vertebrae enables these animals to bend backward to a greater degree than most other mammals. The overall design of the body is fluid, with great power and grace evident in the movements. This allows the animal to absorb the shock of the impact of ocean waves, to haul out on ice or rocky coasts, or to execute agile maneuvers in order to capture prey at sea.

Pinnipeds are less modified for aquatic life than the wholly aquatic cetaceans. Like the cetaceans, pinnipeds have a thick coat of subcutaneous blubber to provide energy, buoyancy, and insulation, but most also have a hairy coat to protect them from sand and rocks when ashore. All pinnipeds have whiskers and a hairy covering (though this is almost lacking in *Odobenus*), which is kept lubricated by secretions from sebaceous glands. Although all have a coarse coat of guard hairs, the fur seals *(Callorhinus* and *Arctocephalus)* also possess a dense layer of underfur. This underfur traps small bubbles and keeps the skin dry, while the stiffer guard hairs protect the body from abrasion. Most pinnipeds are born with a woolly coat called "lanugo," which is white in some species and jet black in others. Molting in pinnipeds usually occurs after the breeding season and is most spectacular in the elephant seals and monk seals, whose outer layer of skin is shed in patches along with the fur.

Pinnipeds are clumsy on land, but in the water they are skillful divers and swimmers. They swim by means of the flippers and by sinuous movements of the trunk. In the Otariidae and the Odobenidae locomotion is accomplished mainly by use of the forelimbs; in the Phocidae the hind limbs provide most of the thrust.

Expert diving by pinnipeds is dependent upon the ani-

A. Guadalupe fur seal *(Arctocephalus philippii),* photo by Warren J. Houck. B. Leopard seal *(Hydrurga leptonyx),* photo by Michael C. T. Smith. C. A young northern elephant seal *(Mirounga angustirostris)* on a sandy beach, the remarkable modification of the hind limbs and body, which adapts it for an aquatic existence, being well illustrated; the front flippers are partially hidden by the loose sand; photo by Julio Berdegué.

mals' efficient use of oxygen, by means of which they remain submerged longer than terrestrial mammals without sustaining brain damage or the "bends." According to J. E. King (1983), just before diving the seal exhales; when breathing stops, the heartbeat slows, thus conserving oxygen. Adult seals can slow their heart rate from a normal speed of 55–120 beats per minute to 4–15 beats per minute. This phenomenon, known as "bradycardia," develops more rapidly and lasts longer as the seal grows older. During a dive the seal's peripheral blood vessels are constricted and circulation is reserved for the heart and brain, thus reducing oxygen consumption by one-third. In addition, pinnipeds have a high tolerance for carbon dioxide and lactic acid buildup in the blood. Within 5–10 minutes of surfacing the seal regains its normal heartbeat, and the blood is reoxygenated by the increased beat and a few deep breaths of air. This efficient use of oxygen enables pinnipeds to make long, deep dives. Maximum depth is known to be nearly 900 meters in some species, and submergence time may reach 73 minutes.

Pinnipeds, unlike cetaceans, must keep some sort of link to the land since they can mate and give birth to the young only on shore or on ice. The habitats in which they gather for mating and pupping vary from floe ice in the Arctic and Antarctic to ragged cliffs, sandy beaches, and lava caves elsewhere. The major necessity appears to be isolation from humans and other predators (Haley 1978).

Pinnipeds are carnivorous and consume a wide variety of animal matter, ranging from krill and other crustaceans to mollusks and fish. They usually eat the common seafood of an area, swallowing moderate-sized species whole and headfirst. Larger catches are shaken into bite-sized pieces (Haley 1978). Some pinnipeds seek food at night, and certain polar seals feed in total darkness for four months of the year. A 100-kg seal eats approximately 5–7 kg of food daily when not fasting.

Some species, such as the Ross seal *(Ommatophoca)*, live alone during the winter, but most pinnipeds are much more gregarious than land carnivores. A breeding colony of pinnipeds ranges from a few individuals to more than 1 million animals within a radius of 50 km. These mammals tend to frequent small, isolated breeding grounds and are polygamous (Otariidae and Odobenidae) or mostly monogamous (Phocidae). All give birth ashore, on land or ice, and mate once a year. The period of pregnancy is 8–15 months, with delayed implantation occurring in many species. Delayed implantation may represent an adaptation that allows the births to take place at approximately the same time of the year, an important feature for colonial and, in some cases, migratory species. Single births are the rule; twins are the exception. Newborn pinnipeds can swim, but the pups of some species do not have enough blubber to provide buoyancy and insulation until they are several weeks old. Growth during the nursing period is rapid, for the mother's milk is particularly rich, about 50 percent fat. The adult pelage is usually acquired near the end of the first summer. Pinnipeds are sexually mature at 2–5 years and may live to 40 years in the wild. Predators include large sharks, killer whales *(Orcinus)*, leopard seals *(Hydrurga)*, and polar bears *(Ursus)*.

Pinnipeds long have been greatly valued by humans because of their fur, oil, and ivory and for use as food or fertilizer. They generally have been easy targets because of their tendency to congregate in large numbers in localized areas for breeding (Haley 1978). Pinnipeds have been hunted commercially for hundreds of years, and sealing expeditions have been responsible for the slaughter of millions of animals. With the decline of seal populations and the worldwide rise of conservation and humanitarian movements, commercial exploitation has been brought under control or stopped altogether. One of the last major organized hunts was the annual take of the valuable soft pelts of young harp seals ("white coats") and hooded seals ("blue coats") in the North Atlantic. Barzdo (1980) reported that 208,759 seals of both species were killed during the 1977 season. Public protests and official import bans subsequently led to a sharp curtailment of the harvest. The United States has prohibited the take of pinnipeds in its waters (except for certain specified purposes) since the passage of the Marine Mammal Protection Act of 1972.

For purposes of fossil history, Berta and Wyss (1994) restricted the term Pinnipedia to the most recent common ancestor of the group comprising the Otariidae, the Odobenidae, the Phocidae, and all of the other descendants of that ancestor. The term Pinnipedimorpha was used to designate the group comprising the Pinnipedia plus more primitive relatives, the oldest known of which is the fossil genus *Enaliarctos*. Other authorities, however, have argued that this genus is ancestral only to the Otariidae (Barnes 1989, 1992; Repenning 1990). A nearly complete skeleton of *Enaliarctos*, showing many primitive characters that would be expected in a common ancestor of all pinnipeds, recently was discovered in late Oligocene or early Miocene deposits of California (Berta, Ray, and Wyss 1989). *Enaliarctos* also has been found in the late Oligocene of Oregon (Berta 1991). Otherwise the known geological range of the Pinnipedia is early Miocene to Recent in North America, Pliocene to Recent in South America and Europe, late Pliocene in Egypt, and Pleistocene to Recent in New Zealand, Australia, and Japan.

PINNIPEDIA; **Family OTARIIDAE**

Eared Seals, Fur Seals, and Sea Lions

This family of 7 Recent genera and 14 species occurs along the coasts of northeastern Asia, western North America, South America, southern Africa, southern Australia, New Zealand, and many, predominantly southern, oceanic islands. The fur seals are in the genera *Callorhinus* and *Arctocephalus*, and the sea lions are in *Zalophus, Phocarctos, Neophoca, Otaria*, and *Eumetopias*. According to Repenning and Tedford (1977), the lineage of *Callorhinus* evidently separated from that of *Arctocephalus* in the late Miocene, whereas the sea lions did not diverge from *Arctocephalus* until the late Pliocene or early Pleistocene. Warneke and Shaughnessy (1985) suggested that the species *Arctocephalus pusillus* is intermediate to the other fur seals and the sea lions. And Bonner (1984*b*) considered *Arctocephalus* to be more closely related to the sea lions than to *Callorhinus* despite its resemblance to the latter in certain characters. These three viewpoints, along with Barnes's (1989) arrangement of the sea lions, form the basis for the sequence of genera presented herein. Berta and Deméré (1986) supported the affinity of *Callorhinus* and *Arctocephalus* and placed both in the subfamily Arctocephalinae, while putting the sea lions in the subfamily Otariinae; however, they indicated that the latter is phylogenetically nearer to the sea lions than is the former. Subsequently, Berta and Wyss (1994) suggested that *Arctocephalus* actually is nearer to the Otariinae than it is to *Callorhinus*. Barnes (1989) included both *Arctocephalus* and *Callorhinus* in the subfamily Otariinae and regarded the Odobenidae (walruses) as only a subfamily of the Otariidae. E. R. Hall (1981) used the name Rosmarinae for the latter subfamily.

New Zealand sea lion *(Phocarctos hookeri)*, photo by D. J. Griffiths.

Total length in otariids is 120–350 cm and weight is about 27–1,100 kg, males always being much larger than females. The body form is slender and elongated, and the tail is small but always distinct. The external ears are small and entirely cartilaginous. The long, oarlike flippers bear rudimentary nails. The flippers, which are thick and cartilaginous, are thickest at the leading edge and have a smooth, leathery surface. In both the Otariidae and the Odobenidae the hind flippers can be turned forward to help support the body, so that all four limbs can be used for traveling on land. Members of these two families walk or run on land in a somewhat doglike fashion. In the Phocidae the hind flippers cannot be moved ahead, and the animals must wiggle and hunch to travel on land. The swimming mechanism of the Otariidae is centered near the forepart of the body, and locomotion in water is accomplished mainly by use of the forelimbs. Phocids swim primarily by strokes of the hind flippers.

Sea lions have a blunt snout and a coat of short, coarse guard hairs covering only a small amount of underfur. Fur seals have a more pointed snout and very thick underfur, which may be of considerable commercial value. The pelage of newborn otariids is silky, never woolly. Adult coloration varies from yellowish or red-brown to black; there generally are no stripes or sharp markings. Females usually have two pairs of mammae. In males the testes are scrotal and the baculum is well developed.

The otariid skull is somewhat elongate and rounded, though rather bearlike in overall appearance. J. E. King (1983) pointed out a number of distinguishing characters. For example, the otariid skull has supraorbital processes and only slightly inflated tympanic bullae, the Odobenidae lack supraorbital processes and have moderately inflated bullae, and the Phocidae lack supraorbital processes and have well-inflated bullae. The normal dental formula of the Otariidae is: (i 3/2, c 1/1, pm 4/4, m 1–2/1) × 2 = 34–36. The first and second upper incisors are small and divided by a deep groove into two cusps; the third (outer) upper incisor is caninelike, especially in the sea lions; the canine teeth are large, conical, pointed, and recurved; and the premolars and molars are similar, with one main cusp. The number of upper molars varies within and among genera.

Eared seals inhabit arctic, temperate, and subtropical waters. Their breeding habitat is exclusively marine, never freshwater. They shelter along seacoasts, in quiet bays, and on rocky, isolated islands. They may be active by both day and night. J. E. King (1983) explained that like all pinnipeds, the Otariidae have acute vision and good hearing underwater and apparently depend on olfaction to distinguish individuals. These seals protect themselves by tearing an adversary with their canine teeth, by hurling their weight against the adversary, or by diving and swimming away. They feed mainly on fish but also eat cephalopods and crustaceans. Dominant bulls generally fast during the breeding season.

Otariids are highly gregarious, especially during the reproductive season. The males arrive first on the breeding grounds, where dominant individuals establish territories. The females come later. The males are polygamous and may associate with a group of more than 50 females. Mating occurs on land, soon after the females give birth to the young conceived during the previous season. Thus, the total time of pregnancy is nearly a year, but in some species it is known to include a period of delayed implantation. There normally is a single pup, which is cared for only by the mother. The young usually do not swim for at least two weeks; weaning occurs after 3–36 months.

The known geological range of this family is Miocene to Recent in Pacific North America, Europe, and Asia; Pliocene to Recent in South America; Pleistocene to Recent in Africa, Australia, New Zealand, and Japan; and Recent in other parts of the current range (Stains 1984). If the Enaliarctinae are treated as a basal subfamily of the Otariidae, as was done by Barnes (1989, 1992) and Repenning (1990), the history of the family could be extended back into the late Oligocene. However, Berta (1991) and Berta and Wyss (1994) did not consider the Enaliarctinae to be a natural group and indicated that the earliest members (including *Enaliarctos*) diverged from the Pinnipedia long before the development of the Otariidae.

PINNIPEDIA; OTARIIDAE; **Genus CALLORHINUS**
Gray, 1859

Northern Fur Seal

The single species, *C. ursinus*, occurs in a great arc across the North Pacific, from the Sea of Japan, through the Sea of Okhotsk and the Bering Sea, to the Channel Islands off southern California (Rice 1977; U.S. National Marine Fisheries Service 1981). The southerly limits of the winter–spring migration are at about 35° N on the Japanese side, and at San Diego, California, 33°10′ N, on the American side. On very rare occasions young individuals have been recorded from the arctic coast, as far east as Letty Harbor, Northwest Territories (J. E. King 1983). One specimen was reported from the coast of northeastern China (Zhou 1986).

There is striking sexual dimorphism in size. Fully mature males have a length of about 213 cm and a weight of 181–272 kg; mature females measure 142 cm and weigh 43–50 kg (Baker, Wilke, and Baltzo 1970). Adult males have dark gray to brown upper parts, usually grayish shoulders and foreneck, a short mane, and reddish brown

A northern fur seal *(Callorhinus ursinus)*, member of the family Otariidae, which, like the Odobenidae, is able to move its foreflippers with great freedom and to turn its hind limbs forward and thus is able to assume an erect posture. Photo by Victor B. Scheffer through U.S. Fish and Wildlife Service.

underparts and flippers. Adult females and immature males have grayish brown upper parts, reddish brown underparts, and a pale area on the chest.

Like *Arctocephalus, Callorhinus* is characterized by a sharper snout than that found in sea lions as well as abundant underfur. Unlike that of *Arctocephalus,* the fur of the foreflipper of *Callorhinus* extends only to the wrist, where it terminates in a sharp, straight line. The facial angle of the skull is always less than 125° in *Callorhinus* but more than 125° in *Arctocephalus.* As a result, the rostrum of *Callorhinus* is shorter and down-curved. *Callorhinus* also is distinguished from other otariids by longer ear pinnae, longer hind flippers, and certain characters of the premaxillary bones, baculum, and pelage (Gentry 1981; J. E. King 1983; Repenning and Tedford 1977).

Except as noted, the remainder of this account is based largely on Gentry (1981) and J. E. King (1983). *Callorhinus* is more pelagic than other otariids of the Northern Hemisphere, with only neonates spending more than 60–70 days per year on land. There appear to be no regular landing sites other than the breeding islands. Individuals concentrate in areas of upwelling over seamounts and along continental slopes and thus are rarely found close to shore. They are most common 48–100 km offshore in waters of 6°–11° C, probably because of the availability there of preferred prey. Dierauf (1984) reported the unusual occurrence of a northern fur seal approximately 144 km upstream from the Pacific in the Sacramento River.

The vast oceanic range of the northern fur seal contrasts with the relatively few and tiny islands where the genus concentrates for reproduction. Although such sites probably once extended along much of the Pacific coasts of North America and Asia, the only major breeding colonies today are on St. George and St. Paul islands of the Pribilofs in the eastern Bering Sea, Copper and Bering islands of the Commanders in the western Bering Sea, Robben Island in the Sea of Okhotsk, the central Kuril Islands, and San Miguel Island off southern California. Colonies at the last two sites have become reestablished only recently. Lloyd, McRoy, and Day (1981) reported the discovery of another small breeding group on Bogoslof Island in the eastern Bering Sea near the central Aleutians. Stein, Herder, and Miller (1986) reported the birth of one pup on the mainland coast of northern California in 1983 and cited a record of another birth on the Washington coast in 1959.

Although the entire population of San Miguel Island probably remains in California waters all year, some *Callorhinus* make an annual round trip of more than 10,000 km, the most extensive migration of any pinniped's. Seasonal movements vary according to the sex, age, and breeding site of the individuals. Adult males leave the breeding grounds in early August and go to sea; most of those from the Pribilof Islands apparently winter south of the Aleutian Islands and eastward into the Gulf of Alaska. Adult females and juveniles of both sexes begin to leave the Pribilofs in October and are followed by pups of the year. By

Northern fur seals *(Callorhinus ursinus):* A. Harem bull with females and young; B. Courtship between fur seals. Photos by Victor B. Scheffer.

December most of the seals have left the islands, headed southeast through the Aleutian passes, and spread as far south as San Francisco. From January to April they may be found anywhere along the migration route from southeastern Alaska to the California-Mexico border. Generally, adult females deploy farther to the south than do the young seals. Northward migration begins in April, and by May large numbers of the seals are back in the Gulf of Alaska. From June to October most are back in the vicinity of the Pribilofs, though younger animals, not reproductively active, are still widely scattered.

In the western Pacific most of the fur seals from Robben Island winter in the Sea of Japan, and most of those from the Commander Islands migrate toward Japan. The animals are first seen off Hokkaido in October, their numbers increasing there until December, and they then move south to Honshu. Northward migration begins later than on the American side, with the last seals still in Hokkaido waters in July. Tag-recapture studies cited by Lander and Kajimura (1982) indicate that thousands of seals born on the Pribilof Islands migrate into Japanese waters and intermingle with Asian seals and that a significant number return north to the Asian rookeries. Movement from the western to the eastern Pacific also occurs but is not as common.

While the seals are at sea most activity occurs during the evening, night, and early morning. The animals commonly sleep in the middle of the day, floating on one side. On the island rookeries activity continues unabated day and night. *Callorhinus* travels well on land but must rest frequently. At sea it is a skillful swimmer and expert diver. For short distances it can keep ahead of a ship moving at about 16–24 km/hr (Baker, Wilke, and Baltzo 1970).

Gentry, Kooyman, and Goebel (1986) carried out studies utilizing an instrument called the "time-depth recorder," which they attached to fur seal mothers captured in the Pribilofs. These instrumented and released females were found to go to sea for 5.5–10.0 days at a time and to make 122–362 dives per trip. About 69 percent of dives occurred at night, and 80 percent of resting took place during daylight. Dives averaged 2.6 minutes, with some lasting 5–7 minutes. Mean depth was 68 meters, and maximum depth was 207 meters. Some individuals made only shallow dives, others made only deep dives, and still others made both. The most frequent depths attained were 50–60 meters and 175 meters. Female *Callorhinus* made relatively long trips and large numbers of dives compared with females of tropical species. Gentry et al. (1986) suggested that this contrast reflects the need of subpolar female seals to obtain a large amount of energy (through feeding) on each trip so that their young can be quickly brought to weaning weight.

Observations of noninstrumented females indicate that the first feeding trip is made about 7 days after giving birth and that the trips usually last 4–9 days, alternating with 2-day periods of suckling the young. Nonsuckling females make longer visits to shore but spend less total time on land and move to sea and back in an unpredictable pattern (Gentry and Holt 1986). Reproductively active males may remain on land for the entire breeding season, not feeding for up to two months and surviving on stored fat reserves. When *Callorhinus* does eat, it takes a wide variety of food, especially small, schooling fish. The diet consists mainly of juvenile pollock in continental shelf areas of the Bering Sea and of squid in oceanic areas. Other important prey species, among the 75 known, are anchovies, capelin, herring, and rockfish. *Callorhinus* sometimes is accused of being detrimental to commercial fisheries, but salmon occurred in only 239 of 9,580 seal stomachs containing food, collected from 1958 to 1966 in the northeastern Pacific (Baker, Wilke, and Baltzo 1970). Even off Japan, where the overlap between human and fur seal diet seems greatest, 70–90 percent of the volume of take by *Callorhinus* is lantern fish, which are not harvested by people.

Overall population density in favorable wintering waters can range up to 26/sq km. A concentration of food often will attract a group of 6–20 seals, and a loose assembly of about 100 has been observed (Baker, Wilke, and Baltzo 1970). Otherwise *Callorhinus* usually is seen in smaller groups or alone while at sea. Despite the vast aggregations that form at rookeries, it is not a particularly sociable animal. There apparently are no lasting bonds other than those between mothers and suckling young.

The breeding season begins when adult males arrive at the rookeries, which is early June in the Pribilof Islands. The males may return year after year to the site of their own birth to breed. They establish small territories on the beaches, with the animals often no more than a few meters apart. These areas are constantly defended through fights, threats, and roars. Immature and senescent males are forced away from the breeding grounds, often farther inland, where they form groups and maintain a size-related dominance hierarchy with respect to resting sites.

The adult females begin to arrive on the Pribilofs in mid-June. They are attracted to one another and form dense groups in favorable locations, often near the water. These groups comprise from 1 to more than 100 females, the average being about 40 (Baker, Wilke, and Baltzo 1970). Although the females come together within the territory of a bull, the presence of the latter is incidental and there is no actual harem formation. The males do not control the females and cannot prevent movement to the sea or across territorial boundaries. Adult females are aggressive toward one another as well as toward the bulls, immature animals of both sexes, and neonates other than their own. Toward the end of the breeding season the younger females are able to come ashore, and at least some of them mate with the bachelor males.

Adult females give birth about 2 days after they come ashore. They then mate again about 6 days later during an estrus of less than 48 hours. After fertilization the zygote develops to the blastocyst stage and enters the uterus. Implantation then is delayed for 3.5–4.0 months. The total gestation period is approximately 11.75 months.

Most pups are born from 20 June to 20 July. There normally is a single young. Twins are rare but have been recorded more often in *Callorhinus* than in any other otariid (Spotte 1982). At birth, males average 66 cm in length and 5.4 kg in weight, and females average 63 cm and 4.5 kg. The newborn has coarse black hair, but after about 8 weeks this coat is shed for one that is steel gray dorsally and creamy white underneath. After giving birth a female remains with her pup and is constantly attentive for about a week. She then begins extensive feeding trips at sea, returning to nurse the pup for only about 2 days a week. At other times the pups form peer groups of their own, wander about the rookery, and spend much time sleeping and playing. Although they are able to swim at birth, they usually do not enter the water during the first month. They are not taught to swim by their mother, and they generally are ignored by their father. Females evidently use scent to distinguish their pups from among a large group. Weaning occurs after 3–4 months and is initiated by the young. Afterwards there evidently is no contact between mother and pup (Gentry and Holt 1986).

Females reach sexual maturity at 3–7 years and may then give birth once a year until age 23; the highest pregnancy rates occur in females 8–16 years old. Males become sexually fertile at 5–6 years but are unable to maintain a breeding territory until 10–12 years. Bulls breed until 15–20 years. Age determinations of up to 26 years have been made, and maximum life span is thought to be about 30 years (Baker, Wilke, and Baltzo 1970).

The thick underfur of *Callorhinus* makes its pelt the most valuable of any pinniped's. There are about 57,000 hairs per sq cm, which is half the density found in the sea otter. The search for the latter species brought Russian explorers to the waters of the North Pacific in the early eighteenth century and led to the discovery of the fur seal rookeries on the Commander Islands. As the sea otter became scarce, there was an increasing take of the fur seal. The Pribilof Islands herd was found in 1786, and the Russians brought Aleut Indians there to hunt the seals (Bonner 1982*a;* Busch 1985). It is estimated that there originally were 1.5–2.0 million fur seals in the Commander Island rookeries and 2.0–2.5 million on the Pribilofs (Lander and Kajimura 1982). Many more occurred at the Kuril and Robben island rookeries and probably at numerous other sites along the coast of western North America. Some 130,000 skins taken by sealers in 1808 and 1809 at the Farallon Islands, off central California, probably were from *Callorhinus* (J. E. King 1983).

By the early nineteenth century a drastic depletion of the Pribilof herd was evident to the Russians. Controls were implemented, including a restriction of the harvest to immature males from 1835 to 1867. By the latter year, when the United States purchased Alaska, including the Pribilofs, the herd there seemed to have recovered to near its original size. An estimated 4 million northern fur seals had been taken through 1867, with the annual kill on the Pribilofs in the 1860s numbering 30,000–40,000. When the United States took over the Pribilofs, there was a failure to immediately reestablish hunting restrictions, and the kill there during 1868 and 1869 alone was about 329,000. Controls were then restored, but they were far more liberal than under the Russians, and the kill of males on land averaged more than 100,000 annually through 1889 (Bonner 1982*a;* Busch 1985; Lander and Kajimura 1982).

A new commercial lease and more stringent quotas reduced the average kill in the Pribilof rookeries to only about 17,000 annually over the next 20 years. By then, however, a severe new threat had developed in the form of pelagic sealing. Unlike the harvest on land, in which nonbreeding males could be separated easily from other animals, killing at sea involved shooting or harpooning any seals that were encountered. The individuals most affected by this new procedure were lactating females on their lengthy foraging trips. The death of one of these females also meant the loss of the unborn she carried and of her suckling young at the rookery. More than 1 million animals, 60–80 percent of them females, were taken by pelagic sealers from 1868 to 1911, and at least as many were killed and not recovered. The resulting decline in the herds led to a series of diplomatic crises, generally pitting the major rookery owners, the United States and Russia, against the two main pelagic sealing nations, Canada and Japan. These conflicts never were fully resolved, but they eventually moderated simply because there were so few seals left. By the early twentieth century the Kuril rookeries had been completely destroyed and those on the Commander and Robben islands nearly eliminated. Estimates of the number of seals then remaining in the Pribilof herd vary from about 130,000 to 300,000 (Baker, Wilke, and Baltzo 1970; Bonner 1982*a;* Busch 1985; Lander and Kajimura 1982).

In 1911 the United States, Russia, Japan, and Great Britain (representing Canada), through the North Pacific Fur Seal Convention, agreed to prohibit pelagic sealing except by aboriginal peoples using primitive equipment. Under the agreement, Japan and Canada would each receive 15 percent of the seals taken on the Pribilofs and 15 percent of those taken on the Commander Islands. Canada, Russia, and the United States would each receive 10 percent of the skins taken on Robben Island (then Japanese-owned). From 1912 to 1917 all killing of the Pribilof seals, which had come under direct control of the U.S. government, was prohibited, except for local subsistence purposes. Subsequently the authorized harvest there, consisting almost entirely of young males, climbed along with a dramatic increase of the seal herd. The annual kill averaged about 40,000 from 1918 to 1940. By the latter year the Pribilof herd was thought to be near its original size of 2.0–2.5 million animals (Bonner 1982*a;* Busch 1985; Lander and Kajimura 1982).

In October 1941 Japan terminated the 1911 agreement and resumed pelagic sealing on the grounds that the seals were damaging fisheries. There evidently was no substantial loss to the herds, however, and in 1957 a new convention was established by the United States, Canada, Japan, and Russia (now in control of Robben Island and the Kurils). Pelagic sealing again was prohibited, an international commission was established to coordinate management and research, and Canada and Japan were each provided 15 percent of the commercial harvest by the United States and Russia. There seemed to be every reason for optimism, and the story of the northern fur seal was held up as a model of conservation. The annual commercial harvest on the Pribilofs, still composed almost entirely of young males, averaged about 66,000 from 1941 to 1955, but the herds appeared to be thriving. The seals on the Russian rookeries also were increasing under strict protection. By about 1970 there were an estimated 265,000 on the Commanders, 165,000 on Robben Island, and 33,000 on the Kurils, this last population having become reestablished only in the 1950s. A breeding group returned to San Miguel Island, off southern California, in 1968 (Baker, Wilke, and Baltzo 1970; Bonner 1982*a;* Busch 1985).

There were some problems, however. By the mid-1950s mortality on the Pribilofs was considered to be too great. It appeared that high population densities were facilitating the spread of disease. A decision was made to reduce the size of the herd through a major commercial harvest of females. From 1956 to 1968 approximately 321,000 females were killed, along with an annual average of 52,000 males (Bonner 1982*a;* Lander and Kajimura 1982).

The expected improvement did not materialize, and a general deterioration of status continued on both sides of the North Pacific (Bonner 1982*a;* Busch 1985; Fowler 1985, 1987; Lander and Kajimura 1982; Marine Mammal Commission 1994; Trites 1992; Trites and Larkin 1989; U.S. National Marine Fisheries Service 1984, 1985, 1986, 1987, 1989, 1994). The situation has been blamed partly on what now generally is considered to have been an excessive kill of females. Another critical factor seems to have been development of the pollock fishery in the North Pacific, which led to reduced prey availability for the seals, particularly for the young during their autumn migration southward through the Aleutians. As many as 50,000 fur seals annually also have perished as a result of entanglement and drowning in discarded fish nets. The yearly kill on the Pribilofs, again restricted largely to males, averaged 34,000 from 1969 to 1975 and about 25,000 from 1976 to 1983. There has been no commercial harvest since 1973 on St. George and none since 1983 on St. Paul. The estimated number of pups born on St. Paul, where most of the rookeries are located, dropped from 450,000 per year in the mid-1950s to 253,000 in 1992. The total number of seals present on the Pribilofs was estimated at 815,000 in 1987, less than half as many as in the 1940s and early 1950s. The count was 982,000 in 1992, but the increase was due primarily to the presence of more males, as a result of the cessation of the commercial harvest, rather than to any reversal of the general downward population trend. There also were about 4,000 seals on San Miguel and 360,000 in the Russian rookeries. Based on the sharp recent decline, the IUCN now classifies *C. ursinus* as vulnerable. The U.S. National Marine Fisheries Service has designated it as a depleted species pursuant to the Marine Mammal Protection Act of 1972 but has rejected a petition to add it to the U.S. List of Endangered and Threatened Wildlife. Whatever classifications are applied, it is evident that *Callorhinus* is undergoing its most serious crisis since the era of pelagic sealing a century ago.

PINNIPEDIA; OTARIIDAE; **Genus ARCTOCEPHALUS**
E. Geoffroy St.-Hilaire and F. Cuvier, 1826

Southern Fur Seals

There are eight species (Bonner 1981*b*, 1984*a;* Boyd 1993; Carr, Carr, and David 1985; David, Mercer, and Hunter

Southern fur seal *(Arctocephalus forsteri)*, photo by John Warham.

1993; Gales, Coughran, and Queale 1992; J. E. King 1983; Reijnders et al. 1993; Shaughnessy and Fletcher 1987; Smithers 1983):

A. townsendi, Guadalupe Island and occasionally other islands off southern California and Baja California;
A. philippii, Juan Fernandez Islands off central Chile, San Felix and San Ambrosio islands off northern Chile;
A. galapagoensis, Galapagos Islands;
A. australis, coasts of South America from central Peru and southern Brazil to Tierra del Fuego, Falkland Islands;
A. tropicalis, primarily to the north of the Antarctic Convergence on Tristan, Inaccessible, Nightingale, Gough, Marion, Prince Edward, Crozet, Amsterdam, and St. Paul islands but wandering individuals also recorded from southern Brazil, South Georgia, Angola, South Africa, the Comoro Islands, Heard Island, much of the southern coast of Australia, Tasmania, New Zealand, Macquarie Island (where breeding now is occurring), and the Juan Fernandez Islands;
A. gazella, primarily to the south of the Antarctic Convergence on South Shetland, South Orkney, South Sandwich, South Georgia, Bouvet, Kerguelen, Heard, and McDonald islands but also breeds on Marion, Crozet, and Macquarie islands, and wandering individuals found on Prince Edward Island and Tierra del Fuego;
A. forsteri, coastal waters of Western Australia, South Australia, Tasmania, and New Zealand, and many subantarctic islands east and south of New Zealand;
A. pusillus, coastal waters of Angola, Namibia, South Africa, southeastern Australia, and Tasmania, and a wandering individual recorded from Marion Island.

The species vary considerably in size, ranging from the comparatively small *A. galapagoensis* to the very large *A. pusillus.* Otherwise all southern fur seals are quite similar in overall appearance and generally have grizzled dark gray-brown upper parts and slightly paler underparts. Like *Callorhinus,* but unlike the sea lions, *Arctocephalus* is characterized by dense underfur. Unlike that of *Callorhinus,* the fur of the foreflipper of *Arctocephalus* extends distally past the wrist and descends to a sinuous line over the metacarpals. The facial angle of the skull of *Arctocephalus* always is greater than 125°. The shape of the snout varies from very short in *A. galapagoensis* to very long in *A. philippii,* and the rhinarium may be smooth and inconspicuous, as in *A. gazella* and *A. galapagoensis,* or inflated and bulbous, as in *A. philippii* and *A. forsteri.* Based on the simple crowns of their cheek teeth, the island species appear to be the most primitive of the southern fur seals (Bonner 1981*b;* J. E. King 1983; Repenning and Tedford 1977). Additional information is provided separately for each species.

Arctocephalus townsendi

Adult males are estimated to be 200 cm long and to weigh 140 kg; females are about 135 cm long and weigh 50 kg (U.S. National Marine Fisheries Service 1984). In color the males are dusky black, the head and shoulders appearing grayish because of the lighter tips of the guard hairs and the animals appearing a grizzled gray when dry. Both *A. townsendi* and *A. philippii* are distinguished from other species of *Arctocephalus* by their extremely long, pointed nose (J. E. King 1983).

Unlike many seals, *A. townsendi* seldom, if ever, lands on open sandy beaches but generally occurs on shores characterized by solid rock and large lava blocks, usually at the base of towering cliffs. Perhaps the most unusual feature of its behavior, however, is the tendency to frequent caves and rocky recesses while on land. It sometimes is found at least 25 meters back from the cave entrances. Based on these observations, Peterson et al. (1968) suggested that during the era of commercial sealing in the nineteenth century seals occurring in open rookeries were quickly slaughtered but that those with secretive behavior survived and passed on their traits to the current population.

Smoothly polished rock extending up to 30 meters above sea level indicates the former occurrence of great numbers of fur seals on much of Guadalupe Island, off Baja California, Mexico. *A. townsendi* also evidently once bred on the Channel Islands off southern California. Even today, individuals range widely at sea, sometimes appearing on San Miguel and San Nicolas in the Channel Islands and at Cedros Island, 300 km southeast of Guadalupe. Two specimens recently appeared on the coast of central California (Webber and Roletto 1987). In contrast to the seasonal landings of other pinnipeds, however, *A. townsendi* can be found on shore year-round and shows strong site tenacity. Males are territorial, defending a cave or recess and barking and puffing at other bulls. Associated with each territorial male is a group of females, usually only 2 or 3 but occasionally as many as 10. Mating and the birth of young conceived the previous year take place from May to July (Peterson et al. 1968; Thornback and Jenkins 1982).

Estimates of the original population on Guadalupe alone range from 20,000 to 200,000 individuals, and there would have been many more on the Channel Islands and at other sites. Most of the seals were killed for their valuable skins around 1800–1820, and the last commercial catches were made from 1876 to 1894. *A. townsendi* was considered to be extinct from 1895 to 1926, but in 1928 two males were received by the San Diego Zoo after capture by fishermen who were aware of a small population on Guadalupe. Subsequently it was thought that the fishermen had killed off the population and that the species really was extinct. However, in 1949 a lone bull was seen on San Nicolas Island, and in 1954 a small breeding colony again was found on Guadalupe. Under complete protection by the laws of Mexico and the United States, the population appears to have increased substantially and currently contains about 6,000 seals (Reijnders et al. 1993). *A. townsendi* is classified as vulnerable by the IUCN and as threatened by the USDI and is on appendix 1 of the CITES.

Arctocephalus philippii

Adult males are estimated to be 150–200 cm long and to weigh about 140 kg; females are about 140 cm long and weigh 50 kg (Bonner 1981*b*). Males have a slim, pointed snout, a heavy mane of silver-tipped guard hairs, and shiny blackish brown fur on the posterior body parts and belly (Hubbs and Norris 1971).

This species, like *A. townsendi,* comes ashore mainly on solid lava rock at the base of cliffs, on ledges, and in caves and recesses. When it enters the sea, it tends to stay rather close to the rocks. In a characteristic behavior, also shared with *A. townsendi,* this seal often inverts itself, with the head straight down in the water, exposing and gently waving the spread rear flippers (Hubbs and Norris 1971). *A. philippii* may go to sea in the cold waters of the Humboldt Current in winter and spring. Breeding occurs in late spring and summer (November–January). The diet includes fish and cephalopods (Thornback and Jenkins 1982).

It is likely that there were more than 4 million *A. philippii* in the late seventeenth century, prior to the start of commercial exploitation (Torres N. 1987). There was an estimate of 2–3 million fur seals on Isla Alejandro Selkirk in the Juan Fernandez Archipelago in 1797. By that year, however, intensive slaughter by American and British sealers was under way, primarily to obtain skins for trade in China. An estimate made in 1798 put the number of surviving seals on the same island at 500,000–700,000. The take on this island from 1793 to 1807 may have exceeded 3.5 million skins. At times crews from as many as 15 vessels were simultaneously engaged in killing the seals. Great numbers were taken in the same period on other of the Juan Fernandez Islands and farther north on San Felix and San Ambrosio. The seals seem to have been largely extirpated by 1824, though a few were killed in the late nineteenth century. Subsequently *A. philippii* generally was considered to be extinct (Hubbs and Norris 1971).

In 1965 the species was discovered to still occur in small numbers on Isla Alejandro Selkirk, in 1968 it also was found on nearby Isla Robinson Crusoe, and in 1970 two individuals were recorded on San Ambrosio. A census in 1983–84 counted 6,300 animals throughout the Juan Fernandez Archipelago (Torres N. 1987). More recent surveys estimated the total population size for the 1990–91 breeding season at 12,000 (Reijnders et al. 1993). The species is completely protected by Chilean law, but there is some poaching and harassment by fishermen (Thornback and Jenkins 1982). *A. philippii* is classified as vulnerable by the IUCN and is on appendix 2 of the CITES.

Arctocephalus galapagoensis

This is the smallest and least sexually dimorphic of the Otariidae. Length is about 154 cm in males and 120 cm in females, and weight is about 64 kg in males and 27 kg in females (Bonner 1984*b*). The upper parts are grizzled gray-brown and the underparts, muzzle, and ears are light tan (J. E. King 1983). The skull is very small and lightly constructed, and the snout is very short (Bonner 1981*b*).

The Galapagos fur seal is found in an area unusually warm for an otariid. When ashore, it selects rugged terrain with caves or overhanging lava ledges where it can shelter from the sun. In contrast, *Zalophus californianus,* another otariid in the Galapagos Islands, prefers gently sloping, open beaches. All fur seal colonies have access to deep water (Bonner 1984*a*). There is no migration, but the animals move into the sea to take advantage of the cold and productive waters of the Humboldt Current. The diet consists of squid and fish (Trillmich 1987*b*).

In a study of lactating females instrumented with a time-depth recorder, Kooyman and Trillmich (1986) found trips to sea to average 16.4 hours. There was an average of 79 dives per trip, 95 percent of them made at night. Most dives measured less than 30 meters, but maximum depth was 115 meters and maximum duration was 7.7 minutes.

On shore, 6–10 females may occupy an area of about 100 sq meters, a relatively low density compared with that of most other fur seals (Trillmich 1987*b*). Breeding males in turn establish comparatively large territories, of up to 200 sq meters, encompassing a number of females. Because of this size, as well as the broken terrain, a bull has difficulty defending his entire territory, and a rival male may sometimes invade and mate with a female. All territories have access to the sea, so that the bulls, overheated from threatening and fighting, can cool off at midday. However, during their entire territorial tenure, up to 51 days, they do not actually feed (Bonner 1984*a*).

Some unusual reproductive adaptations reflect the equatorial habitat of the Galapagos fur seal (Bonner 1984*a*; Trillmich 1986, 1987*b*). Breeding occurs during the cool season, August–November, when there is likely to be less heat stress and greater availability of prey. The peak of births comes in the first week of October, and there is a single pup per female. The female enters estrus and mates about 8 days after giving birth. Because of delayed implantation, the total period of pregnancy lasts nearly a year. The mother initially remains with the newborn for 5–10 days and then establishes a routine of 1–3 days feeding at sea alternating with 1–2 days ashore with the pup. Foraging trips have been found to last 50–70 hours at the time of the new

Southern fur seal *(Arctocephalus galapagoensis)*, photo by H. Hoeck.

moon but only 10–20 hours at the time of the full moon. Lactation may last 2–3 years or even longer, a remarkable period for the Otariidae. Although all females mate during the postpartum estrus, only 15 percent give birth the following year if they are still feeding a pup. Of females without dependent young, about 70 percent give birth. If a pup is born to a female that is feeding a year-old individual, the newborn usually starves or, occasionally, is killed by the yearling. If the older sibling is already 2 years old, the newborn has a 50 percent chance of survival. The extended lactation evidently is associated with diminishing food availability to the nonmigratory seals when the warm season begins in December. At that time the females must forage for 4–6 days and spend only 1 day ashore. The new pups, now weighing about 7 kg, do not have the strength to find their own food and are able to grow very little during the warm season. A year later they do some hunting on their own but still are not self-sufficient. A female has her first young at about 5 years and probably can raise only 5 offspring if she lives to age 15. Clark (1985) estimated the age of one female specimen to be 22 years.

A. galapagoensis was taken regularly by commercial sealers during the nineteenth century and was thought to be extinct by the early twentieth century. A small colony was rediscovered in 1932–33, and numbers have increased substantially since then. Censuses in 1977–78 and 1988–89 yielded population estimates of about 30,000–40,000 (Reijnders et al. 1993; Trillmich 1987*a*). The species occurs on at least 15 islands of the Galapagos and is fully protected by Ecuadorean law, though some colonies may be jeopardized by feral dogs (Thornback and Jenkins 1982). *A. galapagoensis* is classified as vulnerable by the IUCN and is on appendix 2 of the CITES.

Arctocephalus australis

Adult males are about 189 cm long, weigh about 150–200 kg, and are blackish gray in color with longer hairs on the neck and shoulders. Adult females are about 143 cm in length, weigh 30–60 kg, and are usually grayish black above and lighter below. There is some evidence that animals on the mainland of South America are larger than those on the Falkland Islands (Bonner 1984*b;* Gentry and Kooyman 1986; J. E. King 1983).

In a study on the coast of Peru, Trillmich and Majluf (1981) found *A. australis* to inhabit rocky, vertically structured slopes that provided some areas of shade for most of the day. Some seals lived in a huge cave where they had to climb about 15 meters up and down. There was much local movement because of changing temperature. During the hot part of the day the animals concentrated near the sea or in tide pools. At sea *A. australis* may range widely but is not migratory (J. E. King 1983). Most hunting is done at night, and dives have been found to average 29 meters in depth and to reach a maximum of 170 meters (Trillmich et al. 1986). The diet consists mostly of fish, cephalopods, and crustaceans (Bonner 1984*b*).

Trillmich and Majluf (1981) found an overall population density of 0.5–1.5/sq meter. Bulls established breeding territories averaging about 50 sq meters, but males whose territories lacked access to water had to abandon these areas during midday, and some low-lying territories were lost each day when the tide came in. Nonterritorial males gathered on beaches where no females were present. Bulls drove other males away by threatening displays and a call that began with a growl and rose to a higher frequency. Males attempted to herd females, but the latter moved about freely in accordance with their own needs. The breeding season lasted from October to January, with the great majority of births in November–December. According to J. E. King (1983), pups weigh 3–5 kg and have soft black fur. Mating occurs 6–8 days after parturition, but implantation is delayed about 4 months, so that total gestation lasts about 11.75 months. Mothers and young locate one another through vocalizations and scent. During a period of poor feeding conditions on the coast of Peru lactating females were found on the average to alternate 4.7 days of foraging with 1.3 days of suckling the young. As in *A. galapagoensis,* lactation is unusually long. Weaning occurs after 1–2 years, and a female may simultaneously suckle young of different ages (Trillmich and Majluf 1981; Trillmich et al. 1986). Sexual maturity is attained at 7 years by males and at 3 years by females (Bonner 1981*b*). Females evidently attain full size at around 10 years of age and may live for nearly 30 years (Lima and Páez 1995).

A. australis was used as a source of food and hides by prehistoric Indians. Commercial exploitation began on the coast of Uruguay shortly after 1515. For many years the kill was controlled, but by the 1940s a decline was evident. Subsequent management, restricting the take to young males, has allowed an increase. The annual authorized kill in Uruguay had been around 12,000 seals, but fell to about 5,000 in 1987–91, from a total population of about 252,000. The Falkland Islands population of *A. australis,* which was greatly reduced by American and British sealers in the late eighteenth century, seems never to have fully recovered and now contains around 15,000 seals. There also are about 3,000 in Argentina, 40,000 in Chile, and 20,000 in Peru (Bonner 1981*b,* 1982*a;* Gentry and Kooyman 1986; J. E. King 1983; Majluf 1987; Reijnders et al. 1993). *A. australis* is on appendix 2 of the CITES.

Arctocephalus tropicalis

Adult males are 150–80 cm long and weigh about 100–150 kg. Adult females are 120–45 cm long and weigh about 50 kg. This is the only fur seal with a clear color pattern. The upper parts are dark grayish brown, the belly is more ginger in color, and the chest, nose, and face are white to orange. There is a conspicuous crest on the top of the head of adult males, formed from longer guard hairs. The foreflippers are proportionately shorter in *A. tropicalis* than in *A. gazella* and *A. pusillus* (Bonner 1984*b;* J. E. King 1983; Smithers 1983).

This fur seal lives mainly on isolated islands to the north of the oceanic convergence where the cold currents of the Antarctic region sink beneath warmer waters. The convergence is not a total barrier, however, as wandering individuals occasionally cross to the south. Some evidently moved from Amsterdam Island across the Antarctic Sea to New Zealand, a distance of about 5,000 km. Other wanderers have reached the coasts of Africa and South America. Aside from females that are feeding young, most individuals spend the winter and early spring (June–September) at sea (J. E. King 1983). This period corresponds with when most wandering seals have appeared on the coast of South Africa (Shaughnessy and Ross 1980).

There may be two main periods of activity on land, during summer for breeding and during autumn for molting. On Marion and Gough islands these numerical peaks ashore were found to occur, respectively, in December and March–April (Bester 1981; Kerley 1983*b*). On Amsterdam Island, however, there is a marked increase only of adult males ashore during the autumn peak (Roux and Hes 1984). *A. tropicalis* prefers to haul out in rugged, uneven terrain on the windward side of islands. The cooling effects of wind and spray are important in reducing heat stress, especially during the peak of breeding activity. Females may give birth in caves and rocky recesses (Smithers 1983). In a study on Gough Island, Bester (1982) found all habitat on the westward, windward side to be occupied by breeding colonies, provided that there was easy access and protection from high seas. Nonbreeding colonies occupied most parts of the leeward, eastern coast. The diet of *A. tropicalis* has

Southern fur seals *(Arctocephalus tropicalis),* photo by John Visser.

been estimated to comprise 50 percent squid, 45 percent fish, and 5 percent krill but also has been reported to include penguins (Smithers 1983).

J. E. King (1983) wrote that the general breeding biology of *A. tropicalis* seems to be much the same throughout its range. The first adult males start to arrive on the breeding grounds in September, and the rest of the seals arrive in October and November. The pups are born from the end of November to February. Smithers (1983) provided additional details, especially with regard to Gough Island. Colonies at established breeding sites consist almost exclusively of territorial males, adult females, and newborn. Other colonies contain immature animals and adult males that have been unable to establish territories. The latter have been found to spend 94 percent of their time in total inactivity.

The number of breeding males on Gough Island increases rapidly during November. These animals engage in vicious fighting that may leave the contestants exhausted and badly injured. Territories are eventually established and are maintained by rushing to the boundaries with open-mouthed, guttural threats and slashing briefly at the opponent's face and chest. Territories may vary from 5 to 63 sq meters in size and tend to be delineated by natural topographical features. Most of the adult females arrive about a week after the territorial males. They tend to space themselves out but can be highly aggressive toward one another and toward pups other than their own. They enter estrus soon after giving birth to the young conceived the previous season and are then actively herded by the territorial males. On Amsterdam Island the average number of females associated with a territorial bull has been reported to be 6–8, with a maximum of 14 (Bonner 1968).

According to J. E. King (1983), the females give birth 5–6 days after coming ashore. Mating occurs 8–12 days after parturition. There is a period of delayed implantation, so that total gestation is approximately 11.75 months. The pup is about 60 cm long, weighs 4.5 kg, and has a black and chestnut coat, which is shed 8–12 weeks later for pelage more like that of the adult.

The peak pupping season on Gough Island is about mid-December. The female remains with her young for about a week and then begins to make trips into the water. Upon returning to land she gives a call that may attract a number of young; she then distinguishes her own by scent. Pups begin to associate with one another at about 2 weeks; most congregate at the back of the beach by the time they are 4–5 weeks. Bester (1981) reported that the young first enter the surf zone and rock pools at Gough Island at about 6 weeks. Weaning occurs at 8–11 months, and the young may not leave the breeding beaches until October. Normally a single young is produced by a female each year, but Bester and Kerley (1983) reported that one mother on Marion Island apparently raised twins to weaning age.

All of the islands inhabited by *A. tropicalis* were regularly raided by commercial sealers during the late eighteenth and early nineteenth centuries. By the 1830s the breeding colonies had been greatly reduced, but a partial recovery led to renewed hunting in the late nineteenth and early twentieth centuries. Subsequent lessening of demand for seal skins and oil, plus legal restrictions, allowed increases in some areas. The most remarkable comeback has been that of the Gough Island population, which grew from around 300 individuals in 1892 to 13,000 in the 1950s and to 200,000 in 1978. The increase subsequently continued, albeit at a slower rate, on Gough and various other islands (Bester 1980, 1987, 1990; Bonner 1981*b*; Kerley 1987; J. E. King 1983; Roux 1987; Wilkinson and Bester 1990). The total world population of *A. tropicalis* was recently estimated to be 322,000 (U.S. National Marine Fisheries Service 1994). The species is on appendix 2 of the CITES.

Arctocephalus gazella

Except as noted, the information for the account of this species was taken from Bonner (1968, 1982*b*) and J. E. King (1983). Adult males are 172–97 cm long and weigh 126–60 kg. Adult females are 113–39 cm long and weigh 30–51 kg. The back and sides are gray to slightly brownish, and the neck and underparts are creamy. The male has a well-developed mane with many white hairs, which give it a grizzled appearance. Compared with *A. tropicalis,* with which it overlaps in range, *A. gazella* lacks the conspicuous yellow chest and has an apparently longer, less bulky body, a slenderer neck, relatively longer foreflippers, and smaller eyes.

A. gazella is found primarily on islands to the south of the Antarctic Convergence but also occurs to the north, especially on Marion and Prince Edward islands. These islands are shared by larger populations of *A. tropicalis,* and limited hybridization has been reported between the two species there. Southernmost limits are not well known; individuals have been recorded in winter on the pack ice southwest of Bouvet Island (Smithers 1983), and *A. gazella* may be capable of traveling long distances over ice, but the species does not seem to be well adapted to such habitat. There is a general movement away from the breeding islands and out to sea during winter (May–November), but where the animals go, and whether they have a directional migration or simply a dispersal, is not known. Some adult males and juveniles are found ashore or in the vicinity of the breeding islands throughout the year (Doidge, McCann, and Croxall 1986). In a survey of *A. gazella* on Marion Island, Kerley (1983*a*) found peak numbers during summer breeding in December and autumn molting in March. These two peaks were, respectively, earlier and later for *A. gazella* than for *A. tropicalis* on the same island.

Breeding colonies of *A. gazella* prefer rocky stretches of beach with some protection from the sea and with access to the interior. On South Georgia, where the largest colonies occur, there is a lush growth of tussock grass inland from the beaches. Numerous seals move into this area and frequently lie on top of the tussocks. The seals have a surprising agility in land travel, sometimes progressing over slippery rocks or through dense tussock considerably faster than a human. On a smooth surface they can gallop at around 20 km/hr. They probably can exceed this speed when swimming.

Kooyman, Davis, and Croxall (1986) captured 20 lactating female *A. gazella* at South Georgia Island, instrumented them with time-depth recorders and radio transmitters, and released them. The seals were found to forage at sea for 1–13 days at a time. Average trip length was 5.3 days for these individuals, compared with 4.3 days for noninstrumented seals. The first dive was made after an average journey of 57 km and 8 hours. Maximum range was about 150 km over 4–5 days. The mean number of dives per trip was 414, with 81 percent being made at night, when krill rose in the water. Dives averaged 1.9 minutes in length and 30 meters in depth, with respective maximums of 4.9 minutes and 101 meters. In another study in the same area but using more advanced electronic equipment, Boyd and Croxall (1992) obtained data on 11 lactating females. Results basically confirmed the earlier investigation, but dives to depths as great as 181 meters were recorded, and it was found that some individuals dived much deeper and more frequently than others. Krill *(Euphausia superba),* comprising small shrimplike crustaceans, is the staple diet of *A. gazella.* Fish and squid are taken occasionally. Males do not feed while occupying a breeding territory ashore.

The adult males come ashore on South Georgia from late October to early December. They establish and maintain territories through fierce ritualized displays, high-pitched whimpering vocalizations, and fierce fighting. Opponents face each other and make ponderous slashes with open mouths. Most of the blows are received on the chest and sides of the neck, which are protected by a heavy mane, and thus cause no serious injury. Eventually a dominance hierarchy is established, with the most successful fighters winning territories on the beach, near the water's edge but well above the high-water mark. Next in status are males occupying areas higher up on the beach. Lower-ranking males are forced onto territories that are well inland or partly submerged, and some bulls become fully aquatic. Younger males move onto beaches not used for breeding or into areas of tussock vegetation behind the breeding beaches. Unlike the juveniles of some other otariids, they do not form groups but tend to avoid one another. Some young females also occupy these areas.

In a study at South Georgia, McCann (1980) found mean territory size to decline from about 60 sq meters in mid-November to 22 sq meters in December, as males continued to arrive and fight for breeding space. Territorial tenure lasted from less than 1 day to more than 53 days, averaging 34 days.

Adult females arrive on South Georgia mainly in late November and early December. They form groups in favorable areas that correspond to bull territories. The average number of females associated with each territorial male is about 15; McCann (1980) reported a range of 1–27. Bulls generally can prevent a single female from leaving a territory but have little control over a group. Studies by Doidge, McCann, and Croxall (1986) show that within an average of 1.8 days of arrival females give birth to the young conceived the previous year. They enter estrus after another 6 days, mate, and then return to the sea about 6.9 days after parturition. The mean length of foraging trips was found to be 4.3 days, and the mean length of visits ashore to suckle the pup was 2.1 days.

Total gestation is about 11.75 months and probably includes a period of delayed implantation. Most pups are born in early December. They are about 65 cm long, weigh about 6 kg, and have a dark brown or black coat, which is shed for a more silvery yearling pelage in January or February. After the initial departure of the female the pups begin to roam about and associate with one another. When the mother returns, she attracts the young by calling and then confirms identification by scent. As the season progresses, most mothers and young move inland, and suckling often occurs while the female lies on top of a tussock. By early January some pups already are venturing into the sea, but they do not swim competently until early March. Weaning, initiated by the pup, occurs at about 117 days (Doidge, McCann, and Croxall 1986). There normally is a single young, but Doidge (1987) reported that two sets of twins were raised to weaning on South Georgia. Sexual maturity is attained at 3–4 years, but males are unable to hold a territory until they are at least 8 years old.

The skin of *A. gazella,* with a density of about 40,000 hairs per sq cm (Bonner 1985), has a value comparable to that of *Callorhinus.* The colonies on South Georgia were discovered by Captain Cook in 1775, and commercial hunting for skins began there in the 1790s. At the peak of exploitation on South Georgia, in the season of 1800–1801, 17 American and British vessels took 112,000 skins. In 1819 the last great fur seal colonies were located on the South Shetland Islands, off the coast of the Antarctic Peninsula, and 250,000 skins were taken there during the 1820–21 season. The smaller colonies at the South Orkney, South Sandwich, and Bouvet islands were subsequently devastated. By 1830 *A. gazella* was commercially extinct, but enough individuals survived to allow a partial recovery and a resumption of limited commercial sealing from the 1870s to the 1920s.

In the 1930s *A. gazella* again was at a low ebb, there being only about 100 individuals left in the vicinity of South Georgia. A remarkable recovery then began in that area, possibly in association with human extermination of baleen whales in the Southern Hemisphere during the twentieth century. These whales, like *A. gazella,* feed primarily on krill. The reduction of the whales to only about 16 percent of their original biomass may have resulted in a far greater availability of krill for the seals. Under legal protection the seals began to increase at an annual rate of around 16.8 percent. By 1957 there were an estimated 15,000 at South Georgia, and in 1976 there were about 369,000. McCann and Doidge (1987) listed estimates of 1.2 million for South Georgia and about 15,000 in the remainder of the range. A recent comprehensive census at South Georgia resulted in an estimate of 1.55 million (Boyd 1993). Wanderers from there may have reestablished colonies on some of the other islands formerly inhabited by *A. gazella,* but the current population on Bouvet (9,000) apparently represents the original stock. Following the spread of *A. gazella* to Marion Island, cases of hybridization between that species and *A. tropicalis,* which also occurs there, were reported (Kerley and Robinson 1987; Wilkinson and Bester 1990).

Given current conditions, and considering old records of abundance, the South Georgia population can be expected to reach a peak of 2 million seals; large colonies may also grow on the South Shetlands and Kerguelen, and smaller groups (numbering in the tens of thousands) will develop on some of the other islands of the range. There already are about 15,000 on Heard Island (Shaughnessy and Goldsworthy 1990) and 16,000 on South Shetland Island (Boyd 1993). *A. gazella* is on appendix 2 of the CITES and is protected by the laws of all the nations in which it occurs and also, south of 60° S, by the Antarctic Treaty and the Convention for the Conservation of Antarctic Seals. There is concern, however, that krill-harvesting technology, now being developed by several countries, may result in competition between human interests and the fur seals. Another problem is that the increasing seal population at South Georgia is leading to destruction of tussock vegetation, erosion, and jeopardy to certain bird species that depend on this habitat (Bonner 1985). Severe environmental damage also has been reported on Signy Island in the South Orkneys (Smith 1988).

Arctocephalus forsteri

Except as noted, the information for this account was taken from Crawley (*in* Strahan 1983) and Crawley and Wilson (1976). Adult males are 150–250 cm long, weigh 120–85 kg, and have a massive neck and thick mane. Adult females are 130–50 cm long and weigh 40–70 kg. The upper parts are dark gray-brown and the underparts are paler.

This fur seal is found along rocky coasts and on islands. It generally tends to stay close to land and does not make long-distance migrations, but there is a seasonal population shift in New Zealand waters. Adult males and some subadults begin to leave the rookeries in January, move northward for the austral winter, and start back south in August. Nonbreeding colonies are established ashore as far north as Three Kings Island, off the northern tip of New Zealand. Three young individuals were found on New Caledonia in 1972–73 (J. E. King 1983). During the nonbreeding season, March–September, the rookery sites may be occupied by

Southern fur seals *(Arctocephalus forsteri)*, photo by Graham J. Wilson.

pups, yearlings, and small subadults of both sexes. Adult females divide their time during this season between the rookeries and the sea. Rookeries usually are on the exposed western coast of islands. They are characterized by rocky, irregular topography; reefs or outlying boulders that offer protection from the open sea; and areas above the splash zone where mothers and young pups can take shelter. *A. forsteri* feeds mainly on squid and octopus and also occasionally takes fish and penguins. Adult males may fast for up to 10 weeks while defending breeding territories on land.

Adult males establish a size-based dominance hierarchy while ashore in nonbreeding colonies but are strictly territorial in the rookeries. They begin to arrive on the breeding grounds in October, their numbers increasing steadily until mid-November and then more slowly until late December. Agonistic behavior is ritualized and involves guttural and barking vocalizations and threatening displays that emphasize the size of the neck. Only about 30 percent of encounters need to be resolved by physical confrontations. When fighting, opponents push with their chests, wave their necks from side to side, and attempt to inflict bites on the face, neck, or shoulder. Despite regular challenges and intrusions, the bulls spend about 75 percent of their time lying down. Territory size declines as the breeding season progresses and more bulls arrive; at the peak of the season few territories exceed 100 sq meters. Young males and other nonbreeding animals congregate on beaches near the breeding colonies and may move into the rookeries in January, when the bulls begin to leave.

Adult females arrive at the rookeries in large numbers in late November and December. They come together in groups of about six in favorable locations near the water. They are highly agonistic and constantly threaten one another regarding the slightest disturbance. The nearest territorial bull attempts to control a group of females but has little success, and the latter move freely across territorial boundaries. Miller (1975*a*) found that after an average of 2.1 days of first coming ashore, females give birth to the young conceived the previous breeding season. They enter estrus after another 7.9 days and are receptive for up to 14 hours. Gestation lasts about 11.75 months (J. E. King 1983) and includes a 4-month period of delayed implantation.

Pups are born from late November to mid-January, mostly around mid-December. At birth they weigh 3.5 kg, are 55 cm long, and have black fur. The mother remains with her single pup for about 10 days, then goes to sea to forage for 3–5 days, and then returns to suckle the young for 2–4 days. As the pup grows, the mother's foraging trips become longer. At a few weeks of age the pups form small groups, or pods, of 4–5 individuals. Pups continue to suckle until August or September, when they and the females leave the rookeries. Females attain sexual maturity at 4–6 years of age, but males are not able to win breeding territories until they are 10–12 years old.

According to Warneke (1982), the pelt of *A. forsteri* was considered more valuable than that of the partly sympatric *A. pusillus.* Exploitation began about 1798 in the Bass Strait, between Australia and Tasmania, where *A. forsteri* formerly occurred. Sealers soon shifted east to New Zealand, and by 1825 all significant and accessible colonies had been destroyed or greatly reduced. Macquarie Island was discovered in 1810, 57,000 seal skins were taken in the 1810–11 season, and the seal population was thought to have been exterminated by 1815. Limited exploitation continued until the late nineteenth century, when both Australia and New Zealand began to apply regulations on sealing. There subsequently has been a modest general recovery, but there has been no return of *A. forsteri* to the Bass Strait. The populations in New Zealand and nearby islands are estimated to number 50,000 individuals and are thought to be increasing (Mattlin 1987). A small population has returned to Macquarie Island, at the southern edge of the range, probably through immigration from other islands (J. E. King 1983). Richards (1994) suggested that Macquarie and the Antipodes Islands were the primary hauling grounds for juveniles in the New Zealand region and that the devastation of this age class by sealers in the nineteenth century has resulted in a very slow recovery of the overall population of that region, which originally may have numbered 1.5–2.0 million. For Australia, Ling (1987) reported that there were an estimated 5,000 or fewer *A. forsteri* and that population trends were not clear. However, a survey in southwestern Australia in 1989–90 led to an estimate of 27,000 and indicated that numbers were increasing (Reijnders et al. 1993). *A. forsteri* is on appendix 2 of the CITES.

Arctocephalus pusillus

Except as noted, the information for this account was taken from J. E. King (1983) and Warneke and Shaughnessy (1985). Adult males are 184–234 cm long and weigh 134–363 kg; adult females are 136–76 cm long and weigh 36–122 kg. In the African subspecies, *A. p. pusillus,* adult males have dark blackish gray upper parts and are lighter below, and females are brownish gray above and lighter brown ventrally. In the Australian subspecies, *A. p. doriferus,* adult males are dark gray-brown all over, except for the paler mane, and females have silvery gray upper parts, a creamy yellow throat and chest, and a chocolate brown abdomen. Despite the great geographical difference between the two subspecies, they are almost identical in cranial characters.

Both subspecies are confined largely to waters of the continental shelf and their immediate vicinity. Maximum known range offshore is 220 km in Africa and 112 km in Australia; however, a wandering individual appeared on subantarctic Marion Island in 1982 (Kerley 1983*c*). *A. pusillus* generally remains near the breeding grounds throughout the year and does not make regular migrations. Some individuals following cold water currents along the West African coast have been recorded as far north as 11°19′ S. Preferred breeding beaches are characterized by bare rock, boulders, or ledges, but in Africa *A. pusillus* sometimes breeds on sandy beaches. The diet consists mostly of fish and also includes cephalopods and crustaceans. In a study off South Africa, Kooyman and Gentry (1986) found lactating females to make foraging trips of 5–6 days. Their dives averaged 2.1 minutes in length and 45 meters in depth, with respective maximums of 7.5 minutes and 204 meters.

Southern fur seal *(Arctocephalus pusillus)*, photo by John Visser.

Adult males begin to concentrate on shore and compete for territories in October. Territory size is 10–20 sq meters in Africa and 20–140 (averaging 62) sq meters in Australia. The larger size in the latter region may reflect an unnaturally low population density. *A. pusillus* differs from all other species of *Arctocephalus* in that the bulls tolerate the presence of some juveniles over a year old. Most young males and older bulls that cannot hold a territory congregate on beaches near the breeding sites.

Adult females form groups within the male territories. These groups contain averages of about 7 individuals in Africa (Reijnders et al. 1993) and 9 in Australia, though they may number as many as 66. In both Africa and Australia females give birth from late October to late December, mostly around 1 December. They mate 5–7 days after parturition. Total gestation lasts 51 weeks and includes a period of delayed implantation of about 4 months. According to David and Rand (1986), females make a brief trip to sea between parturition and mating. The mean duration of foraging trips subsequently increases to about 4 days by the end of the third month after parturition.

Pups weigh 4.5–7.0 kg at birth and are 60–70 cm long. Lactation generally lasts 11–12 months but sometimes continues into the second or third year of life. By June or July the young are usually foraging effectively on their own. Sexual maturity is attained at about 4–5 years, but males usually are not capable of winning territories until they are about 11 years old. They seldom can hold territories for more than another 2 years. Maximum longevity is at least 18 years.

Commercial exploitation began in South Africa in 1610 and has continued to the present. Fur seal populations were reduced to very low levels by the end of the nineteenth century, when regulations were implemented. Numbers subsequently increased slowly until the 1930s and then began to grow more rapidly. Gentry and Kooyman (1986) reported that the African population of *A. pusillus* numbered 1.1 million individuals, was growing at an annual rate of 3.9 percent, and sustained a commercial harvest of about 75,000 per year. Reijnders et al. (1993) cited a population estimate of as many as 2 million and an annual growth rate of 3 percent but noted that the average yearly kill had been reduced to about 34,000.

Sealing began in the Bass Strait region of Australia in 1798. By 1825 about 200,000 *A. pusillus* had been killed and populations were at a minimum. Less intensive exploitation continued until the late nineteenth century, when protective regulations were established. The populations slowly recovered to about 20,000–25,000 individuals in the 1940s. Subsequent growth has been small, perhaps because many seals are deliberately shot by fishermen or accidentally drowned in nets. In 1991 the total population was tentatively estimated at 30,000–50,000 (Reijnders et al. 1993). The species is on appendix 2 of the CITES.

PINNIPEDIA; OTARIIDAE; **Genus ZALOPHUS**
Gill, 1866

California Sea Lion

The single species, *Z. californianus,* occurs as three isolated subspecies—*Z. c. japonicus,* on the coasts of Japan and Korea; *Z. c. californianus,* on the Pacific coast of North

America from Vancouver Island to Nayarit (western Mexico); and *Z. c. wollebaeki*, in the Galapagos Islands. Except as noted, the information for this account was taken from Odell (1981) and Peterson and Bartholomew (1967).

Adult males are 200–250 cm long, weigh 200 to about 400 kg, and are generally brown in color; adult females are 150–200 cm long, weigh 50–110 kg, and are tan. J. E. King (1983) pointed out that adult males have a very noticeably raised forehead because of the extremely high sagittal crest on the skull and that the hair over the top of the crest may be lighter in coloration.

Zalophus is essentially a coastal animal, frequently hauling out on shore throughout the year. J. E. King (1983) noted that it is rarely found more than 16 km out to sea. Breeding in the subspecies *californianus* occurs from Mazatlan and the Tres Marías Islands, at the southern edge of the range, north through the Gulf of California and along the coast of Baja California, to the Channel Islands off southern California. A few pups are born on smaller islands farther north, including the Farallons, near San Francisco. When the breeding season ends, many of the adult and subadult males migrate north along the coast as far as British Columbia. One individual has been reported in the Gulf of Alaska. Another was seen recently in Acapulco Bay, far south of the normal breeding range (Gallo-R. and Ortega-O. 1986). Movements of the females and young are not fully understood, though at least some remain in the vicinity of the rookeries all year, and some may migrate southward.

On the Channel Islands off California *Zalophus* breeds mainly on flat, open, sandy beaches but sometimes in rocky areas. Sites of rookeries are not permanently fixed, and the sea lions may shift location if disturbed. Eibl-Eibesfeldt (1984) reported that in the Galapagos *Zalophus* may use sandy beaches for resting but seldom for breeding. Trillmich (1986), however, stated that the Galapagos sea lion prefers flat beaches, either sandy or rocky, where there is easy access to relatively calm waters and where it can spend the hot hours around tide pools or in the shade of vegetation.

On land *Zalophus* may walk slowly, move at a rapid gallop, or stride over smooth surfaces using only the front flippers. When swimming it frequently leaps from the water in a shallow arc and reenters headfirst (porpoising). Groups of 5–20 young sometimes swim in file, porpoising one after another. In California, activity on land and sea has been observed regularly both by day and by night. In the Galapagos *Zalophus* is basically diurnal, with peaks of activity in the morning and late afternoon (Eibl-Eibesfeldt 1984). An investigation in the Galapagos utilizing the time-depth recorder found 75 percent of dives by lactating females to be made at night. Foraging trips averaged 15.7 hours, there were 85–198 dives per trip, average depth was 37 meters, and maximum was 186 meters (Kooyman and Trillmich 1986). A study of the same kind in California waters found trips to average 52.8 hours and 199 km, with about 500 dives per trip. Diving took place at all times, there being an average of 3.5 bouts of diving per day with a mean duration of 3.3 hours each. Most dives were less than 3 minutes in duration and 80 meters in depth, but the longest was 9.9 minutes and the deepest was an estimated 274 meters (Feldkamp, DeLong, and Antonelis 1989). Captive individuals have been trained to retrieve objects from depths as great as 250 meters. Several captives flown to islands off California and released were able to find their way back to their pen in San Diego, 115–270 km away, in 2–7 days (Ridgway and Robinson 1985). The diet of *Zalophus* consists mainly of cephalopods and small fish.

During the nonbreeding season there is a conspicuous but not complete sexual segregation. When ashore at this time, individuals form temporary dominance hierarchies based on size, with smaller animals, for example, being forced from favorable resting sites. Nonetheless, *Zalophus* is highly gregarious; the animals pack themselves tightly together even when empty space is available. According to J. E. King (1983), about 13,000 *Zalophus* congregate during autumn and winter on Año Nuevo Island, an area of 6.5 ha.; on some of the more crowded beaches only about 0.6 sq meter is available for each animal.

On the California Channel Islands territorial behavior by adult males occurs from May to August and is especially intense in late June and early July. In contrast to some other otariids, male *Zalophus* do not establish territories until females, and some pups, are already present on the breeding beaches. Fighting for space occurs both on land and in water and includes chest-to-chest pushing and quick slashing bites. Once territories are established, physical combat is reduced and boundaries are maintained by ritualized displays such as oblique stares, head shaking, and lunging without making contact. Bulls patrol their areas on land and swim back and forth along the water side, barking incessantly. Some territories are partly or mostly aquatic. In the Channel Islands they generally are arranged one deep along the beach, have access to the water for cooling, are not separated by topographical features, and are about 10–15 meters wide. In a rocky area territories were determined to average 130 sq meters in size. In the Galapagos, Eibl-Eibesfeldt (1984) found territorial activity to occur throughout the year and diameter to be 40–100 meters. Male *Zalophus* hold their territories an average of only 27 days. Some territories are occupied by a succession of different males in the course of a breeding season, though sometimes the original owner may return to reclaim his space after an interval of feeding at sea. Males that are too young or too weak to win a territory form groups outside of the breeding zone, though occasionally one of these individuals attempts to slip into a territory and remain undetected.

Adult females average 16 per territorial male on the breeding beaches. Although occasionally blocked by the males, the females pay little attention to boundaries, and their movements may cause the males to follow suit. They are attracted to one another but defend their individual spaces with high-pitched barks and threatening gestures. Within a few days of coming ashore for the breeding season they give birth to the young conceived during the previous season. They enter estrus about 3 weeks later (Odell 1984; Trillmich 1986) and actively solicit the male. Total gestation thus would be just over 11 months and probably includes about a 3-month period of delayed implantation.

In California births occur from mid-May to late June. Eibl-Eibesfeldt (1984) stated that in the Galapagos births take place in every month except April and May and that there is commonly a peak in August–October. Trillmich (1986) noted that the Galapagos breeding season varies in onset and duration from year to year but usually lasts 16–40 weeks between June and December. Adult females there were found to spend an average of 6.8 days ashore after giving birth and then to return to sea before mating again. Subsequent visits to suckle the pup averaged 0.6 days, and foraging trips averaged 0.5 days. Females returned to their pups almost every night.

Each female usually produces a single pup per year, though twins have been reported in captivity (Spotte 1982). The pup is about 75 cm long, weighs about 6 kg, and is chestnut brown in color. The mother is very attentive for several days and then makes increasingly long trips to forage at sea. Adult males seem to take a greater interest in the young than do the males of other otariids, and they have been seen apparently joining in an effort to block the ap-

California sea lion *(Zalophus californianus)*, photo by Victor B. Scheffer. Inset: photo by Daniel K. Odell.

proach of sharks toward groups of swimming pups (Eibl-Eibesfeldt 1984). The young are capable of swimming awkwardly at birth and walk with coordination within 30 minutes. At 2–3 weeks they form pods of 5–200 individuals, which move and play together. When a female returns to land, she vocalizes, inspects any pup that gives a responding bleat, and confirms that it is her own through visual and olfactory inspection prior to allowing it to nurse. As time progresses, the mother and pup spend increasing periods of time together at sea. The young probably obtain some food for themselves before they are 5 months old, but complete weaning is a slow process (Trillmich 1986). Usually lactation terminates before 11–12 months, but occasionally a female suckles both a newborn pup and a yearling. Sexual maturity is reached at about 9 years by males and at 6–8 years by females.

Zalophus is the "seal" commonly seen in circuses and animal acts. According to J. E. King (1983), circus training is accomplished by continually rewarding the animal with fish. Ball balancing may be taught by constantly throwing a ball at the sea lion until it is accidentally balanced or by holding a ball on the animal's nose until the objective is understood. It may take a year of training before a trick is ready to be shown to the public, but *Zalophus* has a good memory and will be able to demonstrate it perfectly later, even after a complete rest of three months. The performing life of a sea lion may last 8–12 years. According to Marvin L. Jones (Zoological Society of San Diego, pers. comm., 1995), captive individuals have been reported to live as long as 34 years.

In addition to being captured for the circus trade, *Zalophus* once was exploited for its skin and oil. It still was common in the late nineteenth century, but subsequently it was intensively hunted for the so-called trimmings trade (Busch 1985). Certain of its internal parts were valuable in Oriental medicine, and its whiskers were used as pipe cleaners. This trade led to the extermination of entire herds. The population on the Channel Islands was so reduced that only a single animal could be found there in 1908. A survey in 1938 indicated the presence of 2,020 individuals along the whole California coast. Protection then allowed partial recovery of the subspecies *Z. c. californianus:* by 1967 there were estimated to be 40,000 in California, primarily on the Channel Islands, and another 40,000 in Mexico; the respective figures for 1987 were 74,000 and 83,000, and for 1990, about 110,000 and 92,000. There are problems, however. As the breeding population increases to the south, so also does the autumn and winter migration of males into the coastal waters off Washington, Oregon, and British Columbia. In this region *Zalophus* is reported to be damaging the commercial salmon fishery both by eating the fish and by tearing nets (U.S. National Marine Fisheries Service 1978, 1984, 1987, 1994). Despite some retaliatory killing, together with accidental mortality in gill nets and entanglement in debris, numbers continue to increase rapidly (Reijnders et al. 1993). Poaching for meat and oil is occurring in the Gulf of California, though the population there has increased to about 20,000, 35 percent higher than in 1966 (Le Boeuf et al. 1983).

The subspecies in the Sea of Japan, *Z. c. japonicus*, may already be extinct because of persecution by fishermen. It evidently once occurred along the coasts of Honshu, Kyushu, Shikoku, and Korea, as well as on several small nearby islands. The last possible survivors may be on the South Korean island of Dokto, though there have been no documented reports since the late 1950s (Reijnders et al. 1993). The subspecies is classified as extinct by the IUCN and as endangered by the USDI.

The Galapagos subspecies, *Z. c. wollebaeki*, recovered from sealing activity at the turn of the century and now occurs throughout the archipelago. The population has been estimated to be as high as 50,000. However, numbers declined drastically after an epidemic in the 1970s and possibly again after adverse weather phenomena (El Niño) reduced food supplies in 1982–83 (Eibl-Eibesfeldt 1984; Gentry and Kooyman 1986). A recent estimate put numbers at only 20,000 (U.S. National Marine Fisheries Service 1994). The subspecies now is classified as vulnerable by the IUCN.

PINNIPEDIA; OTARIIDAE; **Genus PHOCARCTOS**
Peters, 1866

Hooker's, New Zealand, or Auckland Sea Lion

The single species, *P. hookeri*, now occurs along the coast of South Island of New Zealand and on Stewart, Snares, Auckland, Campbell, and Macquarie islands to the south. In historical time it also occurred and probably bred on North Island of New Zealand. This species formerly sometimes was placed in the genus *Neophoca* and was even considered to be conspecific with *N. cinerea*, but based on morphological and behavioral distinctions most authorities now treat *Phocarctos* as a full genus. Except as noted, the information for this account was taken from J. E. King (1983) and Walker and Ling (1981*b*).

Adult males are 200–250 cm long, are black or blackish brown in color, and have a dark mane of long, coarse hair. Adult females are 160–200 cm long and are silvery gray above and creamy below. Weight has not been reliably recorded but apparently is less than in *Neophoca*. The facial profile of *Phocarctos* is blunter and more rounded than that of *Neophoca*, and the muzzle appears to be shorter. The skull of *Phocarctos* can be distinguished by the posterior prolongation of the tympanic bullae and the deeply concave palate. The postorbital process of the zygomatic arch is distinct in *Phocarctos* but almost entirely absent in *Neophoca*.

Phocarctos does not make lengthy migrations but may move far out to sea or well up the coast of New Zealand during the nonbreeding season. The current breeding range is restricted to the Auckland, Snares, and Campbell islands. Sandy beaches are preferred for breeding, but individuals also have been observed to move up to 2 km inland to rest in the forest or in grass on top of high cliffs. Fossil bones about 10,000 years old found in caves at the northern tip of South Island indicate that some animals traveled up to 5 km inland (Worthy 1992). *Phocarctos*, like *Neophoca*, moves about well on land and may porpoise while swimming. In an analysis of video film, Beentjes (1990) found the terrestrial locomotion of *Phocarctos* to be similar to that of land vertebrates, in which the limbs on opposite sides of the body are moved in sequence, alternately and independently. In contrast, *Arctocephalus forsteri* was observed to move its hind limbs in unison, an adaptation less suitable for sandy and other surfaces without rocks. The diet of *Phocarctos* consists of fish, cephalopods, crustaceans, and penguins.

Breeding has been reported to begin as early as October, but on Enderby Island in the Aucklands, where most scientific studies have been carried out, the adults of both sexes arrive to breed at the beginning of December. At the height of the season there are about 1,000 individuals present—200 males, 400 females, and 400 pups. Only 19 percent of these males, however, actually hold a territory with females. The defended area is no more than a 2-meter circle, from which the bull makes aggressive charges and which shifts location in relation to the movement of the females. Unlike those of *Neophoca*, male *Phocarctos* defend a territory even if no females are present. They hold these positions throughout the breeding season and do not leave to forage. Defense is largely ritualized, and serious fights are few. The males do not herd the females, which form compact groups on their own, and are much more gentle with them than are male *Neophoca*.

A few days after arrival the females give birth to the young conceived in the previous breeding season. Estrus and mating occur 6–7 days after parturition. Thus total gestation lasts about 11.75 months and probably includes a period of delayed implantation. Most young are born on Enderby Island in December and early January. The single pup is about 75–80 cm long, about 7 kg in weight, and chocolate brown in color. The young remain quietly with their mothers for 2–3 days but then become much more active than the pups of *Neophoca*. At about 1 month the young are brought to the vegetated area behind the beach, where they remain while the mothers forage at sea. Upon returning to land, mother-pup contact is made by mutual vocalization. When the females are away, the young form large pods and may enter streams and rock pools. At 2 months they can swim strongly and are able to visit neighboring islands with their mother. Lactation probably continues for up to a year, but the young can obtain some of their own food before being weaned. Males reach sexual maturity at 6 years (Hayssen, Van Tienhoven, and Van Tienhoven 1993). Maximum longevity is at least 23 years (Reijnders et al. 1993).

Phocarctos, like *Neophoca*, is relatively uncommon. Little is known of its original status, though scattered remains and historical records suggest that it formerly bred throughout New Zealand and that it was greatly reduced in numbers by human exploitation for its skin, meat, and oil. The Auckland Islands were discovered in 1806, and most of the sea lions there had been eliminated by 1830. Some exploitation continued until New Zealand established legal protection in 1881. Partial recovery subsequently occurred, and the Auckland Islands, now uninhabited by people, have been made a reserve. Only about 10,000–15,000 individuals are estimated to exist, and they are regularly subject to accidental capture by commercial squid fisheries near major rookeries (Reijnders et al. 1993). *Phocarctos* is classified as vulnerable by the IUCN.

New Zealand sea lions *(Phocarctos hookeri):* Top, adult male; Bottom, adult female. Photos by Basil Marlow.

PINNIPEDIA; OTARIIDAE; **Genus NEOPHOCA**
Gray, 1866

Australian Sea Lion

The single species, *N. cinerea,* now occurs from Shark Bay, on the west-central coast of Western Australia, south and east to Bridgeport, at the southeastern tip of South Australia. Until about 1800 the species also was present farther east, in the area of Bass Strait, between Victoria and Tasmania. Fulton (1990) reported two recent observations near Sydney on the coast of New South Wales. Except as noted, the information for this account was taken from J. E. King (1983) and Walker and Ling (1981*a*).

Adult males are 200–250 cm long, weigh up to 300 kg, are rich chocolate brown in general color, and have a white to yellowish mane of long, coarse hair. Adult females are 132–81 cm long, weigh 61–104 kg, and are silvery gray to fawn above and creamy below.

Neophoca does not usually make lengthy migrations and may spend its entire life near the beach where it was born. However, some bachelor males on the coast of Western Australia move up to 280 km to the north during the breeding season (Gales et al. 1992). Tagging studies of breeding colonies on Kangaroo Island, South Australia, also have shown that some females and their 4-month-old pups move to other sites 20–40 km distant. The longest movement recorded was by a 24-month-old juvenile found at least 300 km away from the site of tagging. *Neophoca* evidently does not spend long periods at sea, and many individuals can be found ashore at any time of year. Resting sites include flat, sandy beaches, but females sometimes give birth in rocky areas. Most breeding colonies are found on small offshore islands. The current breeding range extends from Houtman Abrolhos, rocky islets off west-central Western Australia, to the Pages, a group of rocks just east of Kangaroo Island in South Australia. Breeding now occurs on at least 50 islands in this region (Gales, Shaughnessy, and Dennis 1994). *Neophoca* is adept on land, can move at a surprisingly fast gallop, has a remarkable ability to climb steep cliffs, and has been found up to 9.7 km inland. It is a strong swimmer, makes series of leaps clear of the water (porpoising), and has been reported to dive to depths of about 40 meters. The diet probably consists mostly of fish and cephalopods but also has been reported to include crayfish, lobsters, and penguins.

Nonbreeding individuals form groups of as many as 15, which may include all age and sex classes but seldom more than a single adult male. Bachelor males sometimes congregate in colonies during the nonbreeding period (Gales et al. 1992). There is a dominance hierarchy, especially in adult males, which is manifested by older animals forcing younger ones from favorable rest sites. Breeding colonies characteristically include fewer than 100 individuals, the largest having 500–600. Male intolerance increases with age and the approach of the breeding season. The bulls establish and maintain territories through ritualized postures and violent combat. Agonistic encounters are accompanied by guttural threats, growls, and high-pitched barking. Unlike other male otariids, male *Neophoca* do not remain on their territories for long periods without eating. A bull occasionally stays for more than 2 weeks but usually leaves more frequently to forage at sea for 2–3 hours. Dominant individuals generally are able to regain their space upon returning to land. They are, however, harassed by groups of subadult males that move about the colony and attempt to mate with the females. The existence of a territory is based on the presence of females, and location may shift as the females move in response to environmental conditions.

Females are gregarious to some extent, but they also defend a small space with hisses, roars, and fighting. As many as 8 females are associated with each territorial male. They are actively herded, and if in or near estrus they sometimes can be prevented from leaving a territory. The males are extremely rough in such activity and may even enter another territory to retrieve a female. In such behavior *Neophoca* shows close resemblance to *Otaria* (Campagna and Le Boeuf 1988). Females have been reported to be hostile to pups other than their own, but human investigators have seen some exhibit "nanny" behavior. A single cow appeared to be in charge of a group of pups, including her own, and would defend the entire group against the humans. Later, she would be relieved by the mother of another pup in the group.

Recent field studies in both South Australia and Western Australia (Gales et al. 1992; Gales, Shaughnessy, and Dennis 1994; Higgins 1993) have confirmed earlier observations indicating that a period of 17–18 months elapses between the start of consecutive pupping seasons and that the breeding cycle of *Neophoca* thus is substantially longer than the usual annual pinniped cycle. The exact length of the cycle averages about 17.5 months, so the timing of births and other reproductive events steadily shifts with respect to the annual calendar and seems independent of environmental factors. Numerous individually marked females have had interbirth intervals of 512–76 days, though some females do not produce young in each consecutive season. Births occur over a period of about 4–6 months in any given colony, and this period varies from colony to colony. Such flexibility in breeding may be associated with the mild climate that prevails throughout the range of *Neophoca* and may serve to mitigate competition between lactating females in what are some of the most biologically depauperate waters in the world.

Direct observations also have shown that females enter estrus 6–7 days after giving birth. Therefore, either there is subsequently a greatly extended period of delayed implantation prior to the normal pinniped active gestation of 8–9 months or there is a normal delay of about 3 months followed by a remarkably long active gestation. Observations made in 1979 suggest the validity of the former alternative—a delay in implantation of the blastocyst for up to 10–11 months. More recently, however, Reijnders et al. (1993) cited evidence that the delay is only about 5–6 months and the placentation phase is about 12 months.

Pregnant females come ashore about 2 days before giving birth. Gales et al. (1992) reported most young to be born in small clearings under shrubbery on the edge of the beach or further inland. Higgins and Gass (1993) observed mothers moving their young away from the natal site after about 30 days. The single pup is 62–68 cm long, weighs 6–8 kg, and has a chocolate brown coat that is replaced after 2 months by pelage resembling that of the adult female. The mother remains with her young for about 10 days and then enters the sea to forage. She returns at intervals of about 2 days to nurse the pup, the average stay on land being 33 hours (Higgins and Gass 1993). Upon first coming ashore, she emits a "moo" that elicits a high-pitched call from the pup, and identification is confirmed by olfactory inspection. As the young grow older they form small groups and begin swimming in rock pools. Mothers lead their pups into the sea, and by the time they are 3 months old they accompany their mothers continuously, both on land and in the water. Weaning normally occurs prior to the birth of the next pup, but females have been seen simultaneously suckling both a yearling and a newborn. In South Australia, Higgins and Gass (1993) found females to nor-

Australian sea lions *(Neophoca cinerea):* A. Photo by Vincent Serventy; B. Photo by Eric Lindgren.

mally suckle their pups for 15–18 months, with weaning occurring an average of 26 days before the next birth. Females first mate at 3 years, males at over 6 years (Hayssen, Van Tienhoven, and Van Tienhoven 1993).

Neophoca is a relatively rare pinniped, there probably being no more than 10,000–12,000 individuals in existence (Reijnders et al. 1993). It may never have been very abundant, but limited evidence indicates that it was more numerous prior to European colonization of Australia and that early sealing activity eliminated the genus in the Bass Strait area. It now is completely protected, and populations seem stable.

PINNIPEDIA; OTARIIDAE; **Genus OTARIA**
Peron, 1816

Southern Sea Lion, or South American Sea Lion

The single species, *O. flavescens,* occurs in the coastal waters of South America from northern Peru and southeastern Brazil south to Tierra del Fuego and the Falkland Islands (Vaz-Ferreira 1981). Oliva (1988) presented a detailed case supporting use of the name *O. byronia* for this species. She has been widely followed in that regard (Corbet and Hill 1991; Reijnders et al. 1993; Wozencraft *in* Wilson and Reeder 1993), but Rodriguez and Bastida (1993) recently argued convincingly in favor of *O. flavescens.* In 1973 an individual was found dead in the Galapagos Islands (Wellington and De Vries 1976), and in 1977 one was killed on the Pacific coast of Panama (Méndez and Rodriguez 1984). Except as noted, the information for the remainder of this account was taken from J. E. King (1983) and Vaz-Ferreira (1981).

Adult males are 216–56 cm long and weigh 200–350 kg. They generally have dark brown upper parts, a slightly paler mane, and dark yellow underparts but are sometimes orange or pale gold in color. Adult females are about 180–200 cm long, weigh about 140 kg, are brownish orange to yellow in color, and often have paler markings on the face, neck, or back of the head. *Otaria* can be distinguished from other sea lions by its blunt, short, broad, and high, slightly upturned, erectile muzzle and high, wide lower jaw. The widened neck and heavy mane reinforces the massive appearance of the head and foreparts of the male.

Otaria is not migratory, but males may go to sea for extensive periods after the breeding seasons. Some individuals move as far north as Rio de Janeiro on the Brazilian coast, about 1,600 km from the nearest breeding colonies, in Uruguay. Some also may enter freshwater rivers. Breeding sites are usually on open sand or pebble beaches or on flat zones of rock, but occasionally they are on steep beaches or in places with large boulders. The diet consists mainly of fish, squid, and crustaceans and also includes fur seals. When seeking schooling prey, *Otaria* sometimes feeds in groups. During the nonbreeding period individuals of any age and sex may be found in the same area.

Studies on the Valdez Peninsula of Argentina (Campagna 1985; Campagna and Le Boeuf 1988; Campagna, Le Boeuf, and Cappozzo 1988) found that adults of both sexes began to assemble on the breeding beaches in the second week of December. Males reached a numerical peak during 15–21 January, and by the first week of February 90 percent of them had departed. Female numbers peaked at the end of January. Males compete for and maintain territories through vocalizations, ritual displays, and physical combat.

South American sea lions *(Otaria flavescens)*, male and female, photo by Lothar Schlawe.

As the season progresses, territorial emphasis shifts from defense of a specific area to defense of a group of females. Fighting, which is most common at the beginning of the breeding season, involves the two opponents' meeting face to face, moving their heads sideways, and attempting to bite and tear. The vocal challenge of the adult male consists of a roar followed by a series of grunts. Successful contenders may spend up to 2 months defending a territory, during which time they do not eat and get little sleep. Territories are about 3–10 meters in diameter and seem to be based on space and the presence of females rather than on topographical features. Defeated individuals and young males gather at the fringes of the breeding beaches.

There are as many as 18 females in each male territory, though the average is only 2.8, considerably fewer than in most other otariids (Campagna and Le Boeuf 1988). Post-estrous individuals are free to move across boundaries or into the sea, but pre-estrous and estrous females frequently are blocked and even carried and tossed about by the males. This sequestering of females is much more strenuous than in other otariids. Also unlike the situation in other genera, the nonterritorial males of *Otaria* do not always remain outside of the breeding zones or limit themselves to individual, and largely ineffective, challenges. New studies (Campagna and Le Boeuf 1988; Campagna, Le Boeuf, and Cappozzo 1988; Conway 1986) indicate that the territorial bulls are constantly hard-pressed by the others and may not be able to hold more than 2 or 3 females at a time. The bachelor males, including subadults, form temporary groups with an average of 10 and a maximum of about 40 animals. They then regularly invade the territories, attempting to split off the females or seizing them in their jaws and carrying them off. They even kidnap pups in an apparent effort to lure the mothers away. Most raids are eventually repulsed by the territorial bulls, but some adult bachelor males are able to retain females in areas on the periphery of the main breeding area, to mate with them, and to successfully defend them from other males. Bulls that lose all their females sometimes leave the breeding area. As a result, a relatively high proportion of male *Otaria* are able to participate in the mating process.

Within a few days of arriving on the breeding grounds adult females give birth to the young conceived the previous year. Births occur from mid-December to early February, mostly in January. There is evidence that the period of pupping on the Pacific coast is earlier and longer than on the Atlantic side. Total gestation is 11.75 months and probably includes a period of delayed implantation. There normally is a single pup, which is 85 cm long and weighs 10–15 kg. The natal coat is black and fades to a dark chocolate color after a month. Following birth, the mothers remain with their young for a few days and then mate and go to sea to forage. They return at intervals to suckle the young, initially calling to the latter, receiving an answering bleat, and then confirming identification by olfaction. When left alone the young form groups or pods of their own. Campagna (1985) reported that pups enter water for the first time at 3–4 weeks, in groups of 20–50 individuals. This activity occurs mainly during middle and low tides, when pools and channels are exposed. Lactation often continues until the mother gives birth the following year and sometimes lasts even longer. Sexual maturity probably is attained by males at 6 years and by females at 4 years. Rosas, Haimovici, and Pinedo (1993) found both sexes to reach full size by 8 years and found that the maximum age in the wild was 16 years. According to Jones (1982), a captive was estimated to have lived to an age of 24 years and 10 months.

Otaria was exploited heavily for its skin, meat, and oil until well into the twentieth century. About 250,000 individuals were killed in Argentina from 1917 to 1953, with a peak harvest of 15,000 in 1946 (Conway 1986). There still was a population of 400,000 animals in the Falkland Islands in 1937, but by 1966 only 30,000 survived and in the late 1980s only 15,000 were counted (Reijnders et al. 1993). The reasons for this decline are not known; however, the reduction of a population on Isle Verde, Uruguay, from 2,400 in 1953 to only 28 in 1977 was the result of deliberate shooting by fishermen, who blamed the sea lions for damaging nets (Pilleri and Gihr 1977*c*). The U.S. National Marine Fisheries Service (1987) estimated that there were 228,000 *Otaria* on the Pacific side of South America and 45,000 on the Atlantic side.

PINNIPEDIA; OTARIIDAE; **Genus EUMETOPIAS**
Gill, 1866

Northern Sea Lion, or Steller Sea Lion

The single species, *E. jubatus*, occurs across the North Pacific from Hokkaido, through the Sea of Okhotsk and the Bering Sea, and down the west coast of North America to the Channel Islands off southern California. Zhou (1986) reported a specimen from the coast of Jiangsu Province in eastern China. Except as noted, the information for this account was taken from J. E. King (1983), Loughlin, Perez, and Merrick (1987), and Schusterman (1981).

Eumetopias is the largest otariid and exhibits pronounced sexual dimorphism. Adult males average about 300 cm in length and 1,000 kg in weight, with some reaching 325 cm and 1,120 kg. Adult females average about 240 cm in length and 270 kg in weight, with some attaining 350 kg. Both sexes are slightly variable in color, ranging from light buff to reddish brown, the chest and abdomen being a little darker. Adult males develop a massive neck with a heavy mane of long, coarse hairs. Females have 2–6, usually 4, retractable mammae. *Eumetopias* differs from *Zalophus* in being much larger in size and having a conspicuous diastema between the upper fourth and fifth postcanine teeth. The head of *Eumetopias* is bearlike, and its nose is short and straight, not upturned as in *Otaria*.

The Steller sea lion is not known to migrate but does disperse widely during the nonbreeding period. Loughlin, Rugh, and Fiscus (1984) listed 51 active breeding colonies. The largest are in the Aleutians and on the Gulf of Alaska. Others occur off Sakhalin, on the Kurils and islands in the Sea of Okhotsk, along the east coast of Kamchatka, on the Pribilof Islands, in southeastern Alaska, on the coast of British Columbia and Oregon, on Año Nuevo and the southeastern Farallon Islands near San Francisco, and on San Miguel Island off southern California. The breeding season ends in August throughout the range of the genus, and Alaskan males may then move both southward and as far north as St. Lawrence Island. California males tend to move northward along the coast.

Rookeries, as well as nonbreeding haul-out sites, are located mainly on remote and rocky coasts and islands. They characteristically have access to the open sea and to abundant food resources. *Eumetopias* commonly is seen in waters about 200 meters deep and is known to dive to depths of 183 meters. It often feeds at night but may hunt schooling prey by day. The diet consists primarily of a wide variety of fish, mostly with no commercial value, and also includes substantial amounts of octopus and squid, bivalve mollusks, and crustaceans. Young fur seals, ringed seals, and

Steller sea lions *(Eumetopias jubatus):* A. Rookery, photo by Victor B. Scheffer; B & C. Male, female and pup, photos by Karl W. Kenyon.

sea otters sometimes are taken. When searching for schooling prey *Eumetopias* may hunt and dive in large groups. Aggregations of several hundred to several thousand sea lions sometimes depart from land and disperse into groups of fewer than 50 at 8–24 km offshore. Group feeding apparently helps to control the movement of schooling fish and squid.

Eumetopias is gregarious and polygynous. Adult males begin to return to the rookeries in large numbers in early May, reaching a maximum by early July. They are highly aggressive toward one another and compete for territories, sometimes inflicting severe wounds. Disputes involve roaring, hissing, and chest-to-chest confrontations with mouths open. Once territories have been established, boundaries can be maintained through ritualized displays with relatively little fighting. Boundaries follow topographical features, such as cracks and ridges in rocks. The most favored areas are semiaquatic, sloping down into the sea. Territories average about 225 sq meters in size and are held an average of 43 days, during which time the bull remains ashore and does not feed. Males have been known to return to the same territory for 7 consecutive breeding seasons. Males that are too young or too weak to hold a territory may congregate on beaches near the rookeries, along with other nonbreeding animals.

Adult females assemble on the breeding grounds about 3 days before they are ready to give birth to the young conceived the previous year. They compete for the most favorable birth sites, areas of sloping rock just above high tide. They are highly aggressive and readily bite, push, and threaten one another and pups other than their own. Groups of 10–30 females form within the territory of each successful male (U.S. National Marine Fisheries Service 1978). Estrus occurs 10–14 days after parturition, and females actively solicit the male through vocalization and displays. The blastocyst implants in September or October after a delay of 3–4 months, and the total gestation period is about 11.5 months.

The young, normally one per female, are born from mid-May to mid-July, mostly about 5–16 June, in both Alaska and California. The pups weigh 18–22 kg and are dark brown to blackish in color. Their weight may double after 7 weeks. The mother remains with the pup for 5–13 days and then spends 9–40 hours foraging at sea. She continues to alternate feeding trips with visits to suckle the young. Pups are capable of swimming at birth. For several weeks the mother encourages the pup and stays with it in shallow water while it develops swimming skills. At 10–14 days the pups form groups of their own and play and sleep together while the mothers forage. Upon a female's return, contact is made with the young through mutual vocalizations and olfaction. Lactation continues at least until the female again gives birth, though the pups also do some foraging on their own. Occasionally a female suckles a young individual into its second or even third year. Sexual maturity is attained at between 3 and 8 years, but males generally cannot win breeding territories until they are 10 years old. Females may breed until they are in their early twenties and may live for 30 years.

Because of its massive size and relatively aggressive nature, *Eumetopias* rarely is seen in captivity and seldom is trained to perform tricks. Its skin formerly was used by the Aleutian natives for boat coverings, harness, clothing, and boats. It now is less in demand, though the very thick hide may be used for leather and the meat may be used as food for ranched mink and fox. There were commercial harvests in the Aleutians as recently as 1972, and there still is a limited take by native peoples. Bigg (1988) stated that *Eumetopias* is well known throughout its range because of the damage it allegedly causes to commercial fish and fishing gear and that as a result there have been organized control programs and bounties.

The Pribilof Islands population, which originally numbered well over 15,000 individuals, was reduced to a few hundred animals by uncontrolled exploitation after the U.S. purchase of Alaska in 1867. Skins were sold and traded for sea otter pelts to Aleutian natives. Protection was established in 1914, and by 1960 the population was near 10,000 (Kenyon 1962).

Whether other populations underwent a comparative cycle of decline and recovery is not known, but the U.S. National Marine Fisheries Service (1978) indicated that *Eumetopias* had increased considerably throughout Alaska since the early twentieth century and was at or near the carrying capacity of the ecosystem. The total world estimate in the late 1950s and early 1960s was 240,000–300,000 individuals, of which 200,000 were in Alaska. Loughlin, Rugh, and Fiscus (1984) obtained about the same estimate based on surveys from 1975 to 1980 but noted that there had been increases in some areas and serious declines in others. The Pribilof population, for example, had fallen again to about 2,000 animals. In British Columbia numbers had dropped from 12,000 to 5,000, partly through commercial harvests and a deliberate control program. The California population, which once numbered nearly 7,000 individuals, had declined to 3,000, and there were only 20 breeding animals left on San Miguel Island. It was suggested that numerical changes in some areas might have been caused by a shift of populations. Braham, Everitt, and Rugh (1980) reported a decline of more than 50 percent in the population of the eastern Aleutians.

Recent reports (Alverson 1992; Bigg 1988; Loughlin, Perlov, and Vladimirov 1992; Marine Mammal Commission 1994; Merrick, Loughlin, and Calkins 1987; Pascual and Adkison 1994; Reijnders et al. 1993; U.S. National Marine Fisheries Service 1986, 1987, 1994) indicate that a general decline is continuing in the eastern Aleutians and spreading across most of the Alaskan region and into Canadian and Asian waters as well. New surveys show that there has been no population shift and that numbers are dropping sharply everywhere. Overall numbers in the Aleutians and southwestern Alaska fell from 140,000 in the late 1950s to about 68,000 in 1985 and fewer than 30,000 in 1989. The actual cause still is unknown, but factors under investigation include disease, loss of prey to greatly expanding commercial fisheries, entanglement and drowning of the sea lions in commercial fishing nets, and deliberate killing by fishermen. In 1989 the total world estimate was approximately 116,000 *Eumetopias*, of which about 82,000 were in Alaska and 17,000 were in the rest of North America. The Russian population comprised only about 17,000 individuals, compared with around 60,000 prior to the recent decline. Counts in the early 1990s indicate a continued decline at an annual rate of nearly 10 percent. On 10 April 1990 the USDI issued an emergency regulation classifying the Steller sea lion as a threatened species. The U.S. National Marine Fisheries Service followed with various regulations and projects aimed at conservation of the species and its habitat, but the effective implementation thereof was questioned by the Marine Mammal Commission. The IUCN now has classified the species as endangered.

PINNIPEDIA; **Family ODOBENIDAE; Genus ODOBENUS**
Brisson, 1762

Walrus

The single living genus and species, *Odobenus rosmarus*, occurs primarily in coastal regions of the Arctic Ocean and adjoining seas. Within historical time the species has been present regularly as far south as Kamchatka, the Aleutians, southern Hudson Bay, Nova Scotia, southern Greenland, northern Norway, and the White Sea. There also are historical records of wandering individuals from Honshu (Japan), the Sea of Okhotsk, Yakutat Bay (southeastern Alaska), coastal Massachusetts, Iceland, Ireland, southern England, northern Spain, France, Belgium, the Netherlands, and coastal Germany (Allen 1942; Duguy 1986; Fay 1981, 1985; J. E. King 1983; Nores and Pérez 1988; Reeves 1978; Reijnders 1982). Barnes (1989) and E. R. Hall (1981) considered the Odobenidae to be only a subfamily of the Otariidae. Hall used the name Rosmarinae for this subfamily and used the name *Rosmarus* Brünnich, 1772, in place of *Odobenus*. In contrast, Wyss (1987) pointed out that many characters indicate a close relationship between the Odobenidae and the Phocidae. Berta and Wyss (1994) included both the Odobenidae and the Phocidae, but not the

Walruses *(Odobenus rosmarus):* A. A herd of walruses, photo from U.S. Navy; B. Walrus bulls, photo by Leonard Lee Rue III; C. Young walrus, photo from Zoologisk Have, Copenhagen.

Otariidae, in a pinniped subgroup, the Phocomorpha. The remainder of this account draws heavily from the extensive studies of Fay (1981, 1982, 1985).

Adult males are 270–356 cm long and up to 150 cm in height and weigh 800–1,700 kg. The smaller females are 225–312 cm long and weigh 400–1,250 kg. There is no external tail. The body is covered with short, coarse hair, most dense on young animals and least dense in old males. During their annual molt in June and July the males shed most of their hair and may appear to be naked. The older males often have numerous raised nodules on the neck and shoulders, which are almost entirely hairless. The overall color is cinnamon-brown, but there is variation. Old males have pale skin and hair and may appear almost white when immersed in cold water, or pinkish when lying in warm sun. All individuals have a conspicuous mustache consisting of

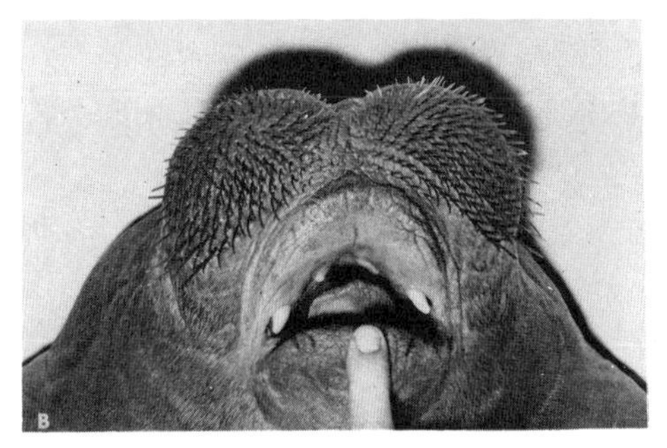

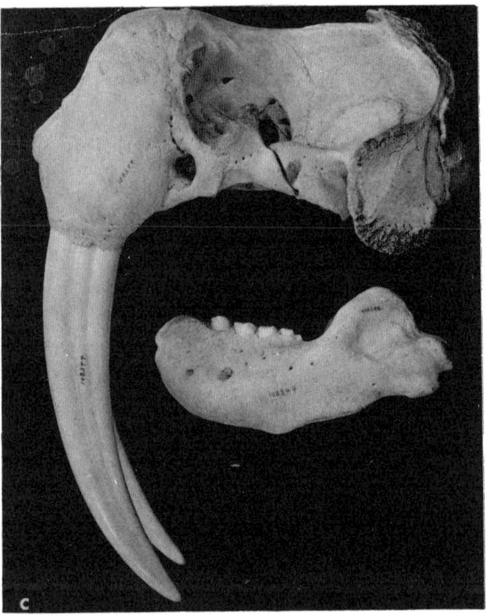

A. Mounted head of an adult male walrus from eastern Siberia *(Odobenus rosmarus divergens)*, photo by Grancel Fitz through New York Zoological Society. B. Yearling male walrus *(O. rosmarus rosmarus)*, photo by Victor B. Scheffer. C. Skull of Pacific walrus *(O. rosmarus divergens)*, photo from U.S. National Museum of Natural History. D. Walrus bulls *(O. rosmarus)*, photo by Leonard Lee Rue III.

about 450 thick bristles, the roots of which are richly supplied with blood vessels and nerves. The tough, wrinkled skin of the walrus is about 2–4 cm thick over most of the body, being thickest on the neck and shoulders of adult males. An underlying layer of blubber is up to about 10 cm thick on the chest.

The walrus has a swollen body form, a rounded head and muzzle, and a short neck that is especially thick in old males. The eyes are small and piglike, and the external ears are represented only by a low wrinkle of skin. The foreflippers are nearly as wide as they are long, while the hind flippers are more triangular. All the flippers are thick and cartilaginous, being thickest on the forward or leading edge, and all have five digits. The palms and soles are bare, rough, and warty for traction on ice. As in the Otariidae, the hind limbs can be turned forward and used together with the forelimbs for maneuvering on land. The baculum of the male walrus, up to 63 cm in length, is the largest of any mammal's in both absolute and relative size. The male also has a pair of pouches that arise from the pharynx and pass backward along the insides of the neck. These sacs can be greatly expanded with air from the lungs and are thought to function principally as resonance chambers for production of a bell-like sound, usually made while underwater; they also are used as a flotation device. Females usually have four mammae, sometimes more.

The massive skull is approximately rectangular, and the halves of the lower jaw are solidly fused. The maximum number of teeth is 38, but more than half are rudimentary and occur with less than 50 percent frequency. The most common permanent dental formula is: (i 1/0, c 1/1, pm 3/3, m 0/0) $\times$ 2 = 18. The upper canines are an outstanding feature of the adult walrus. These teeth grow throughout life in both sexes, developing into tusks that may attain a length of more than 100 cm in males and about 80 cm in females. About 80 percent of the tusk is exposed above the gumline. The tusk is peculiar in that it bears enamel only at the tip for a short time after eruption; the entire crown in the adult consists of dentin (ivory). A single tusk in an old male may weigh 5.35 kg. The tusks of females are more recurved in the middle and more slender than those of males. Skulls with three tusks are occasionally found. Tusks function mainly in intraspecific agonistic interaction and also are used for interspecific defense, cutting through ice, hooking over ice for stability while sleeping in water, and helping to pull the body out of the water. There is no evidence that the tusks are used for digging food from the ocean floor.

The walrus is principally an inhabitant of moving pack ice over shallow waters of the continental shelf. It depends for subsistence on benthic organisms that dwell at depths of not more than 80–100 meters. Most populations appear to be migratory, moving south with the arctic ice in the winter and then north as the ice recedes in the spring. There are currently four main populations. One, the Pacific population, is found mainly in the Bering Sea in winter (December–January), migrates north through Bering Strait in spring (April–June), spends the summer (July–September) along the edge of the permanent polar ice cap in the Chukchi Sea, and then returns to the Bering Sea again in autumn. Some individuals travel more than 3,000 km in the course of the year. Several thousand males of the Pacific population do not participate in the northward migration but remain in the Bering Sea during the summer. The other three populations are found from Hudson Bay to western Greenland, from eastern Greenland to Svalbard and the Kara Sea, and in the Laptev Sea off northern Siberia. Based on limited radio-tracking studies, Born and Knutsen (1992) suggested that the eastern Greenland group may be separate from that in Svalbard and eastward, but Born and Gjertz (1993) subsequently reported some movement of individuals between the two regions.

During winter individuals tend to concentrate where the ice is relatively thin and dispersed and to avoid extensive areas of thick ice. In summer the males may utilize isolated coastal beaches and rocky islets. Females and young usually remain in the pack ice, especially on floes with a surface area of 100–200 sq meters. Ice seems to be preferred for resting, molting, and bearing young, but land is used if no ice is available. An individual can use its head to break through ice up to 20 cm thick and then maintain and enlarge the hole by abrading the perimeter with its tusks. Walruses are active both by day and by night. Terrestrial locomotion is quadrupedal and resembles that of the Otariidae but is accomplished with much less facility and speed. In the water the walrus propels itself by alternating strokes of the hind flippers, in a manner similar to that of the Phocidae. The forelimbs may be used as paddles at low speed but otherwise are held against the body or used only for steering. The normal swimming speed is about 7 km/hr, and the maximum is at least 35 km/hr.

Foraging dives usually last 2–10 minutes. Most feeding is done at depths of 10–50 meters; the deepest known dive was about 80 meters. An animal locates food by swimming headfirst along the ocean bottom, the sensitive vibrissae on the snout constantly in motion and in contact with the floor. Organisms that reside deep in the sediments probably are unearthed by piglike rooting. Captives were observed to first root into the substrate with the upper edge of their snout and then stir up the sand by jetting water out of their mouth (Kastelein and Mosterd 1989). Prey is examined and manipulated with the vibrissae, and soft-bodied organisms are swallowed whole. Organisms with shells are held in the lips, the fleshy parts removed by powerful suction and ingested, and the shells rejected. This remarkable process may account for thousands of small clams during a single meal. The diet consists mainly of bivalve mollusks (clams and mussels) but also includes a wide variety of other benthic invertebrates. Fish and seals are captured occasionally. Adult males evidently eat little during winter, when they accompany the females and young.

The walrus is gregarious, tending to haul out on land or ice in herds of up to several thousand individuals lying in close physical contact. In such groups, which are largely sexually segregated during the nonbreeding season, there are dominance hierarchies based on body and tusk size. In a study of an aggregation of males, Miller (1975*b*) found most social interactions to be agonistic and concluded that "walruses are bullies." When, for example, a large individual comes out of the water to seek a resting spot, it might throw back its head and point its tusks at a smaller animal. If this threatening display is not sufficient, and it often is not, there is striking with the tusks and frequently bloodshed, but serious injury is rare.

During the breeding season the animals congregate in traditional favored areas of up to several hundred square kilometers. The females and young form groups of 20–50 individuals and are followed by a number of males. When the groups stop to rest on ice, the males compete for nearby locations in the water, about 7–10 meters apart. About 10 percent of the males are able to associate with the females, the younger and weaker males remaining nearby. The successful males display by making a variety of clicking and bell-like sounds underwater and then raising their head above the water and emitting a series of sharp clucks and whistles. When a female is attracted, she joins the male and mating takes place underwater. When nearly ready to give birth, a female often separates from her herd, but afterwards she joins a group of other mothers and young. A young female tends to remain in her mother's group. Young males drift away at about 2–3 years to join a male herd.

Mating occurs mainly during winter, especially in January and February. Implantation of the blastocyst is delayed for 4–5 months, usually until June or July, and birth occurs 10–11 months thereafter, mainly from mid-April to mid-June; thus, the total gestation period is 15–16 months. At birth the single calf is about 113 cm long, about 63 kg in weight, grayish in color, and evidently capable of swimming. There is no external evidence of teeth, but the permanent tusks have formed internally and will erupt about a year later. The social bond between mother and calf is apparently stronger than in any other pinniped. The females are extremely solicitous and protective, and there are no lengthy periods of separation as in the Otariidae. There also is evidence of community care and even adoption of orphaned calves. Lactation usually continues for at least 2 years, though the young are capable of finding some of their own food long before final weaning. Females are sexually mature at 6–7 years and full grown at 10–12 years. Males are fertile at 8–10 years but cannot successfully compete for mating privileges until they are at least 15 and full grown. Fecundity is greatest in females at 9–11 years, when they can produce a calf every other year. The interval is longer in older animals. Maximum known longevity in the wild is just over 40 years.

For thousands of years the native peoples of the North regarded the walrus as having supernatural powers and human attributes, but at the same time they depended on the species as a major source of food, fuel, and materials for making tools, shelters, boats, sleds, and clothing. Such indigenous hunting probably had little effect on overall populations; however, serious declines began with European commercial exploitation for ivory, oil, and hides. Such activity began in the Middle Ages and became especially intensive on both sides of the North Atlantic in the sixteenth and seventeenth centuries. The original size of the populations in this region is not known, but given the magnitude of the kill, there must have been large numbers. Thousands of individuals, for example, were killed annually in the Gulf of St. Lawrence and off Nova Scotia during this period (Reeves 1978). Even in the late nineteenth century more than 1,000 were being killed each year at Svalbard, and in 1887 that same number was taken off Novaya Zemlya. Shortly thereafter the catch dropped to an insignificant level in the eastern North Atlantic, reflecting the near disappearance of the species. In the West, the walrus had been exterminated south of Labrador by the mid–nineteenth century and substantially depleted farther north. Ronald, Selley, and Healey (1982) indicated that the last great populations in the Canadian Arctic, those around Baffin Island, were devastated from 1925 to 1931, when 175,000 individuals were killed. However, Richard and Campbell (1988) showed this figure to be grossly exaggerated, though they agreed that the former large populations in the region had been depleted.

Protection and declining demand eventually prevented the total disappearance of the Atlantic subspecies *(O. r. rosmarus)*, but there has been no substantial recovery. The continuing indigenous harvest is about equal to the annual recruitment rate of 6–10 percent. There are now about 25,000 individuals in the population that occurs from Hudson Bay to western Greenland and only 1,000–2,000 from eastern Greenland to the Kara Sea. Numbers seem to have increased recently in Svalbard and the Barents Sea, with a consequent rise in observations along the coast of Norway (Born and Gjertz 1993; Gjertz et al. 1993; Gjertz and Wiig 1994). The population in the Laptev Sea, sometimes considered a separate subspecies *(O. r. laptevi)*, remained largely undisturbed by modern exploitation until the late nineteenth century. It too subsequently declined in the face of excessive hunting and now numbers about 4,000–5,000 individuals. Russia classifies *O. r. rosmarus* as vulnerable and *O. r. laptevi* as rare.

The North Pacific population *(O. r. divergens)* had come under commercial pressure by the eighteenth century. More than 10,000 individuals were taken in some years through the late nineteenth century. In a two-year period about 16,000 kg of ivory was obtained on the Pribilof Islands alone. In 1909, when the population was at a low ebb, commercial harvesting was prohibited in Alaska, and a recovery began. From 1931 to 1957, however, Russia carried out intensive pelagic commercial hunting, and the population declined to only about 40,000–50,000 individuals. Subsequent Russian and U.S. regulations ended commercial use and allowed another recovery. By the early 1980s the Pacific population was estimated to contain nearly 250,000 walruses and may have been close to its original size. However, Fay, Kelly, and Sease (1989) warned that the population again is being mismanaged through excessive subsistence harvests in the United States and Russia and by lack of coordination between the two countries. The total kill of 10,000–15,000 individuals annually, plus natural mortality, was thought to be leading to a decline that could halve the population during the 1990s. A survey in 1990 produced a rangewide estimate of 201,039 Pacific walruses, but there was no clear indication of population trends (Marine Mammal Commission 1994). Recently poaching has intensified as the demand for walrus ivory accelerated following the widespread destruction of Africa's elephant herds (see account of *Loxodonta*). It is feared that excessive numbers are being killed in Alaska under the guise of the subsistence take, the carcasses discarded, and the tusks sold in the Far East to replace elephant ivory (*Oryx* 24 [1990]: 195). There also is concern that expansion of the clam-dredging industry into the North Pacific, as has been proposed, would seriously reduce the walrus's food supply (Ronald, Selley, and Healey 1982).

The walrus has a long geological history, and its family once was far more varied than it is today (Berta and Wyss 1994; Deméré 1994; Repenning and Tedford 1977). The Odobenidae evidently had evolved in the North Pacific by at least the middle Miocene, from the same pinniped group that gave rise to the Phocidae. In the late Miocene and Pliocene there were at least seven genera of odobenids in Pacific North America. The family then disappeared in the Pacific, but a branch evidently had entered the Atlantic in the late Miocene, and it is represented in the Pliocene, Pleistocene, and Recent of both Europe and Atlantic North America. The Odobenidae reentered the North Pacific in the late Pleistocene. There are late Pleistocene records as far south as San Francisco, Michigan, and North Carolina.

PINNIPEDIA; **Family PHOCIDAE**

True, Earless, or Hair Seals

This family of 10 Recent genera and 19 species occurs along ice fronts and coastlines mainly in polar and temperate parts of the oceans and adjoining seas of the world but also in some tropical areas and in certain inland bodies of water. The sequence of genera presented here is based on a review of Burns and Fay (1970), J. E. King (1983), and McLaren

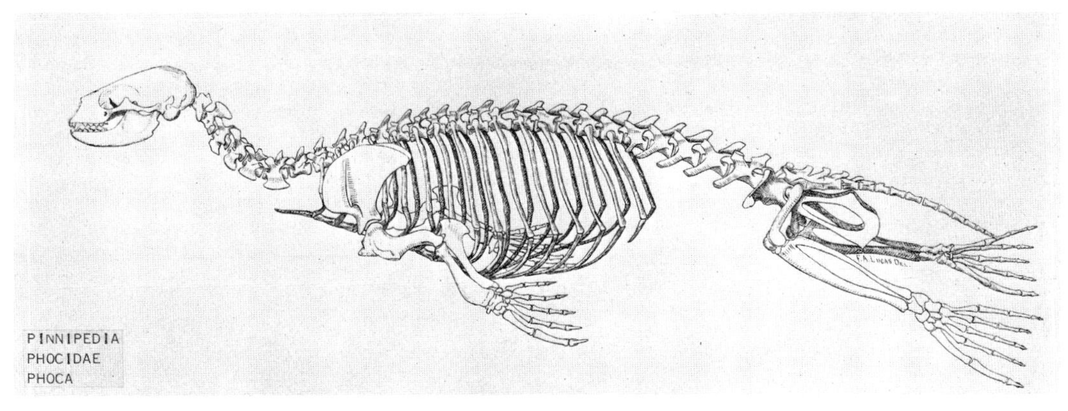

Skeleton of harbor seal *(Phoca vitulina)*, photo from American Museum of Natural History.

(1984), which indicates recognition of two subfamilies: the Monachinae, the southern phocids, with the genera *Monachus, Lobodon, Hydrurga, Leptonychotes, Ommatophoca,* and *Mirounga;* and the Phocinae, the northern phocids, with *Erignathus, Cystophora, Halichoerus,* and *Phoca. Monachus* is considered to be the most primitive of all living phocids, and *Erignathus,* the most primitive member of the Phocinae. Based on a detailed morphological analysis, Wyss (1988*b*) doubted that the Monachinae comprise a monophyletic group. Berta and Wyss (1994) accepted the Phocinae and also indicated that the genera of the Monachinae form a basically valid group but suggested that the species *Monachus schauinslandi* does not belong in that group. Some authorities, including E. R. Hall (1981) and Stains (1984), place *Cystophora* in a third subfamily, the Cystophorinae. Still other authorities, such as Corbet (1978), Rice (1977), and Wozencraft (*in* Wilson and Reeder 1993), do not use subfamilial designations.

Total length in phocids is 120–600 cm and weight is 65–3,700 kg. In most genera males are about the same size as females, but in *Mirounga* and *Cystophora* the males are much larger. The body is streamlined and the tail is small but obvious. External ears are represented only by a faint wrinkling of skin, and as in the Odobenidae, there is no supporting cartilage. The foreflippers, which are placed far forward, are smaller than the hind flippers and make up much less than one-fourth of the body length. The flexible flippers, of nearly uniform thickness, have five digits. In the subfamily Phocinae the metacarpals of the first and second digits are approximately the same size, and there are large claws on the fore- and hind flippers; in the Monachinae the first metacarpal is noticeably larger than the others, and the claws on the hind flippers are reduced (J. E. King 1983). The hind limbs cannot be turned forward, so phocids wriggle and hunch in order to travel on land. Such locomotion is rather laborious, so whenever possible the animals roll or slide; they can move fast when necessary. In water phocids propel themselves by moving the hind flippers in a vertical plane. The swimming mechanism is centered near the hind part of the body, not close to the forebody as in the Otariidae. Phocids frequently swim on their backs, at least in captivity, and regularly stand upright in the water, maintaining their position by "treading" with their foreflippers.

The adult pelage is stiff and lacks appreciable amounts of underfur. Eyebrow vibrissae are well developed, and mustache vibrissae are often beaded. A number of species have spotted color patterns, and some of the species of *Phoca* are the only pinnipeds with a banded pattern and with sexual differences in this pattern. Some phocids have three distinct coats—newborn, subadult, and adult. The newborn are covered with a dense, soft, woolly, often white coat. There is an extensive amount of blubber. Females generally have four mammae. In males the testes are internal and the baculum is well developed.

According to J. E. King (1983), the phocid skull differs from that of the Otariidae in having well-inflated tympanic bullae, no supraorbital processes, and nasal bones that extend well posteriorly between the frontal bones. In the subfamily Phocinae the lateral swelling of the mastoid process forms an oblique ridge at an angle of about 60° from the long axis of the mastoid bone as a whole; in the Monachinae this ridge is absent. The dental formula is: (i 2–3/1–2, c 1/1, pm 4/4, m 0–2/0–2) × 2 = 26–36. The upper incisors have a simple pointed crown, the canines are elongate, and the postcanines usually have three or more distinct cusps.

Phocids shelter on rocky or sandy coasts, islands, and ice floes. Some stay on land or ice much of the time. Some migrate or disperse over a large area, and others make local movements corresponding to fluctuations of the ice. Activity is both diurnal and nocturnal. J. E. King (1983) indicated that phocids see and hear well underwater and that some phocids may have the most efficient hearing of pinnipeds in the air. Olfaction may be important in communicating intraspecific information. Phocids defend themselves by opening their mouths, uttering menacing cries, and advancing against the enemy, or they may flee to the water and dive. Most species eat fish, shellfish, and cephalopods. *Hydrurga* is the only pinniped that preys regularly on penguins and other seals. Some species can fast for long periods.

Phocids do not usually congregate in large rookeries as do otariids. Males may establish territories on land during the breeding season. Most species live in pairs during this period, but the males of some species, such as *Halichoerus* and *Mirounga,* may mate with a number of females. The total time of pregnancy is 270–350 days and usually includes a period of delayed implantation. At birth the single pup weighs 5–40 kg. In *Phoca* newborn can swim immediately, but in some other genera they often shun the water for several weeks. Lactation, lasting about 10–80 days, is generally much shorter than in the Otariidae.

The known geological range of the Phocidae is middle Miocene to Recent in North America and Europe, early Pliocene to Recent in South America, late Miocene or early Pliocene to Recent in Africa, and Recent in nearly all seas and oceans (Berta and Wyss 1994; J. E. King 1983).

A. Southern elephant seals *(Mirounga leonina),* a nonbreeding group assembled on Macquarie Island. In such places they are helpless against human predators, and this group is composed of the descendants of a few that remained after the species was almost exterminated in the Antarctic up to 1918. Photo from Australian News and Information Bureau. B & C. Weddell seals *(Leptonychotes weddelli).* About 14 days after birth of the pup, the mother tries to get it in the water by going in herself and calling it. While the pup gains courage, she saws a ramp in the ice with her teeth so that the pup can slide into the water. Photos from U.S. Navy.

PINNIPEDIA; PHOCIDAE; **Genus MONACHUS**
Fleming, 1822

Monk Seals

There are three species (E. R. Hall 1981; J. E. King 1983; Kinzelbach and Boessneck 1992; Rice 1977):

M. schauinslandi (Hawaiian monk seal), Hawaiian Islands;
M. tropicalis (Caribbean monk seal), originally found throughout the West Indies and along the coasts of Florida, Yucatan, and eastern Central America;
M. monachus (Mediterranean monk seal), originally found throughout the Mediterranean and Black seas, on the Atlantic coast of North Africa from Morocco to Mauritania, and in the Madeira, Canary, and Cape Verde islands.

Wyss (1988*b*) stated that *M. monachus* and *M. tropicalis* seem to be more closely related to other phocids than to *M. schauinslandi* and that generic distinction for the latter seems justified. The more traditional view was set forth by J. E. King (1983), who observed that monachines were the dominant seals of the North Atlantic in the late Miocene and Pliocene and that certain populations presumably gave rise to *Monachus*, which then diverged into various species. However, it is agreed that *M. schauinslandi* has several skeletal features that are more primitive than those of the oldest known fossil monachine, the latter having lived about 14.5 million years ago. Therefore, *M. schauinslandi* apparently separated from the North Atlantic lineage of *Monachus* more than 15 million years ago, when an ancient seaway between North and South America closed. The cranial characters of *M. schauinslandi* are more primitive than those of the other monk seals but are closer to those of *M. tropicalis* than to those of *M. monachus*. In general, *Monachus* is distinguished from other living monachines by having the nasal processes of the premaxillary broadly in contact with the nasals, rather than barely or not touching; having wide, heavy, crushing-type postcanines; having four, rather than two, mammae in females; having smooth, rather than faintly beaded, vibrissae; and having jet black, rather than white or grayish, neonatal pelage. Unlike in some other monachines, the adult pelage of *Monachus* is never spotted; the uniform brown or grayish dorsal color, supposedly resembling that of a monk's robe, gives the genus its common name. Additional information is provided separately for each species.

Monachus schauinslandi (Hawaiian monk seal)

Except as noted, the information for the account of this species was taken from Kenyon (1981) and J. E. King (1983). Females average larger than males and become much heavier when pregnant. One adult male was 214 cm long and weighed 173 kg, and a probably pregnant adult female was 234 cm long and weighed 272 kg. Coloration is slate gray dorsally and silvery gray ventrally. *M. schauinslandi* resembles *M. tropicalis* but has a more primitive bone structure of the inner ear. Another primitive and unique feature is that in *M. schauinslandi* the fibula is not fused to the proximal end of the tibia.

The Hawaiian monk seal is now largely restricted to the small, rocky Leeward Islands. Breeding now occurs regularly on five of these islets and atolls, and individuals tend to remain in the vicinity of their natal beaches. There is, however, much movement between islands and out to sea, and occasional wanderers reach the main Hawaiian Islands, to the southeast. Marked individuals have been found to move up to 1,165 km from the nearest breeding site. To give birth, females venture well up on sandy beaches, near shrubs and away from the tide line. Although *M. schauinslandi* is the only completely tropical phocid, it has no obvious physiological adaptations for warm habitat, and its blubber content is about the same as that of polar seals. It evidently avoids heat stress by remaining inactive during the day, lying in the shade or wet sand. It feeds at night, carrying out lengthy series of dives in shallow lagoons. During a period of 4.4 hours one individual made 25 dives lasting 5–14 minutes each. Another individual was observed to remain underwater for 20 minutes. The diet consists primarily of reef- or floor-dwelling species of eels, other fish, cephalopods, and crustaceans.

Small, loose groups may assemble in environmentally favorable locations, but *M. schauinslandi* is basically solitary. During the spring and summer adult males cruise constantly along favorite basking beaches in search of receptive females. Since the males outnumber the females by about three to one, the latter frequently are disturbed. Squabbles between rival males or between males and upset females may be accompanied by bubbling and bellowing vocalizations, open-mouthed threats, and sometimes fighting. According to the Marine Mammal Commission (1994), mobbing by groups of aroused males frequently results in death or injury of females and pups. In an effort to facilitate conservation, some males are being removed or treated with a testosterone-suppressing drug to reduce their libido.

Mating takes place in the water. The young are born from late December to mid-August, mostly from mid-March to late May. The single pup weighs 16–18 kg, is about 100 cm long, and is covered with soft black hair. Lactation lasts about 35–40 days, during which time the pup reaches a weight of about 64 kg. The mother remains with the pup, fasting during this whole period, and loses about 90 kg of her own weight. The pup can swim weakly at birth and is taken into the water each day to practice under close maternal supervision. After weaning, the female returns to the sea, and the young must then fend entirely for itself. Its weight may fall to 45 kg by the time it is a year old. Weight then begins to increase again, but full size is not attained until at least 4 years (U.S. National Marine Fisheries Service 1978). Females can first bear young at about 5 years (McLaren 1984). Most give birth every other year, though some do so annually. Maximum known longevity in the wild is 30 years.

The Hawaiian monk seal evolved in an environment totally free of people and other sources of terrestrial predation and harassment. It thus unfortunately became both easily approachable and easily disturbed. It may once have bred throughout the Hawaiian Islands, but it disappeared from most beaches long ago. It probably survived in the Leewards because the Polynesians established no lasting settlements there. Remaining populations were largely eliminated by commercial sealing operations in the early nineteenth century, and the species was considered to be extinct by 1824. Some individuals survived, however, and in 1859 a single vessel took 1,500 skins. The species then remained generally undisturbed, though in very low numbers, until World War II. Subsequent establishment of military facilities and increased human presence led to new declines. The breeding colony on Midway had disappeared by 1968. The main problems are that when pregnant females are disturbed by people or dogs, they may move to unfavorable areas to give birth, areas where the young cannot be properly sheltered, and that when lactating females are disturbed they cannot adequately feed their young. There also is concern that increasing fishing activity in the Leeward Islands may lead to competition between the seals and human interests and to entanglement and drowning of the seals in fishing gear. The U.S. National Marine Fisheries Service (1987) estimated the total number of *M. schauinslandi* at 500–1,500 individuals but subsequently (1994) reported that its status was rapidly deteriorating. The species is classified as endangered by the IUCN and the USDI and is on appendix 1 of the CITES.

Monachus tropicalis (Caribbean monk seal)

According to J. E. King (1983), adults are about 200–240 cm long and grayish brown on the back, shading to yellowish white ventrally. Kenyon (1981) noted that very little is known about the biology of this species but that in many respects it appears to have been similar to the other species of *Monachus*. An exception is that the peak of the pupping season was probably December.

European exploitation began during the voyages of Columbus and continued relentlessly as the seals were killed for their skins and oil. The species already was scarce in the mid–nineteenth century. More recently *M. tropicalis* has been persecuted extensively by fishermen, who view the animal as a competitor. An aerial survey in 1973 found fishing activity throughout the former habitat of *M. tropicalis* and led Kenyon (1981) to conclude that the species was extinct. This position was upheld by Le Boeuf, Kenyon, and Villa-Ramirez (1986) following a cruise through the Gulf of Mexico and around the Yucatan Peninsula. The last reliable records are of a small colony at Seranilla Bank, about midway between Jamaica and Honduras, in 1952. There have been a few more recent reports, however, and some authorities hope that *M. tropicalis* may yet survive (Mignucci Giannoni 1986). The species is classified as endangered by the USDI and is on appendix 1 of the CITES but is regarded as extinct by the IUCN.

Monachus monachus (Mediterranean monk seal)

Except as noted, the information for this account was taken from Kenyon (1981). Length was 238 cm and 262 cm in two adult males and 235 cm and 278 cm in two adult females. Usual adult weight is probably about 250–300 kg. Reported maximum weight is 400 kg. J. E. King (1983) wrote that coloration is generally dark brown to black above and a little lighter below and that a whitish gray patch occurs on the belly of some individuals.

Early reports indicate that *M. monachus* originally occurred on open sandy beaches (Bareham and Furreddu 1975). However, the species may always have depended mainly on beaches backed by extensive deserts or cliffs and thus inaccessible to terrestrial predators. Increasing human presence over the last few thousand years may have allowed this seal to survive primarily on small, rocky, waterless islands. Most births now occur in caves and grottoes, some of them with underwater entrances, but the preferred breeding habitat may be beaches under cliffs. Foraging apparently is done mostly in waters less than 30 meters deep, though one individual was taken on fishing tackle set at 75 meters. The diet consists of fish and octopus.

Even though *M. monachus* has been in close contact with Western civilization far longer than any other pinniped, surprisingly little is known of its ecology and behavior. A polygynous social structure is suggested by one observation of a large male visiting a group of five other seals and finding one receptive female. This mating oc-

Hawaiian monk seals *(Monachus schauinslandi)*: A. Photo from San Diego Zoological Garden; B. Photo by Karl W. Kenyon.

curred underwater off Deserta Grande Island, in the Madeiras, on 1 August 1976. According to Marchessaux and Pergent-Martini (1991), the birth season extends from April to December, with a peak in September, and is asynchronous with respect to different colonies. Parturition takes place in caves after a gestation period of 9–10 months. The single pup weighs about 15–20 kg at birth, is about 94 cm long, and has a black woolly coat. Weaning occurs by about 6 weeks, and at least some females evidently reach sexual maturity at 4 years. Jones (1982) reported that a captive lived 23 years and 8 months.

According to old reports and place names, *M. monachus* once was much more common than it is today. Hunting and disturbance by people resulted in major declines, though the species survived over much of its range into modern times. The rocky coasts of the Mediterranean, with numerous small islets and sea caves, offered an abundance of secluded breeding sites. However, recent increases in human presence, fishing activity, and motorized vessels have brought the seal under severe pressure. Although it is legally protected throughout its range, it is killed regularly by fishermen, who consider it to be a competitor and dislike it for damaging nets. The seals also may become entangled in nets and drown, and pregnant females disturbed by people may abort their young.

Since the beginning of the twentieth century *M. monachus* has disappeared from the mainland coasts of Spain, southern France, the Crimea, Palestine, and Egypt. It also has been wiped out in the Canary Islands, but a group of about 10 individuals survives at the Desertas Islands in the Madeiras (Panou, Jacobs, and Panos 1993), and the remains of two animals were found on Sal in the Cape Verde Islands in 1990 (Kinzelbach and Boessneck 1992). Reijnders et al. (1993) reported that the total remaining number is around 500 and is continuing to decline. There are still breeding groups along the Atlantic coast of Africa, particularly at Cap Blanc, where Western Sahara borders Mauritania; along the North African coast from Morocco to Tunisia; and in the eastern Mediterranean. There also may still be a few on the Black Sea coasts of Turkey and Bulgaria (Panou, Jacobs, and Panos 1993). A recent report indicated that the Cap Blanc colony, previously thought to contain only about 100 seals, may actually have 350 (*Oryx* 27 [1993]: 135). Otherwise, the largest concentration, perhaps 300 animals, is found among the Greek islands of the Aegean Sea and along the adjacent Turkish coast (Marchessaux 1989). However, Goedicke (1981) predicted that considering current trends, the species would disappear from Greek territory by the year 2000. *M. monachus* is classified as critically endangered by the IUCN and as endangered by the USDI and is on appendix 1 of the CITES.

PINNIPEDIA; PHOCIDAE; **Genus LOBODON**
Gray, 1844

Crabeater Seal

The single species, *L. carcinophagus,* is found primarily along the coasts and pack ice of Antarctica, but wandering individuals occasionally reach South Africa, Heard Island, southern Australia, Tasmania, New Zealand, and the Atlantic coast of South America as far north as Rio de Janeiro. Except as noted, the information for this account was taken from J. E. King (1983), Kooyman (1981*a*), and Laws (1984).

Adult males, which on average are slightly smaller than females, are usually 203–41 cm long, with a reported maximum of 257 cm. Adult females are usually 216–41 cm long, with a maximum of 262 cm. Few weights have been recorded, but the probable adult range is 200–300 kg. Following a summer molt in January–February, coloration is mainly dark brown dorsally, grading to blonde ventrally. On the back and sides are large chocolate brown markings interrupted by lighter brown. The flippers are the darkest parts of all. The coat fades throughout the year, reaching an almost uniform blonde or creamy white color by summer. The body is relatively slim and the snout is long. The cheek teeth are perhaps more complex than those of any other pinniped or carnivore. Spaces between the prominent tubercles intrude deeply into each tooth. The main cusps of the upper and lower teeth fit between each other, so that when the mouth is closed the only gaps are those between the tubercles.

The crabeater seal occurs largely in the vicinity of the Antarctic pack ice. During the maximum extent of the ice in the winter the seal's range extends over about 22 million sq km. In summer it moves south and is confined to six regions of residual pack ice totaling 4 million sq km. Although concentrated along the edge of the drifting ice, the species sometimes moves onto the shore of Antarctica. One individual was found 113 km from open water, and another was found on a glacier 1,100 meters above sea level, perhaps an elevational record for pinnipeds.

Lobodon has remarkable agility on snow or ice and may be the fastest of pinnipeds there. Speeds of up to 25 km/hr have been reported. When it sprints the forelimbs are thrust alternately against the snow, the pelvis is moved from side to side while also thrusting against the snow, and the hind flippers are held together and off the snow. Stirling and Kooyman (1971) reported that when *Lobodon* is caught, it rolls over many times. Such may be an instinctive reaction developed to escape the killer whale *(Orcinus)* or leopard seal *(Hydrurga)*. Many adult *Lobodon* are scarred from attacks by these other two genera. Despite its name, the crabeater seal feeds almost exclusively on krill, small shrimplike animals of the family Euphausiidae. It catches these animals by swimming into a school of them with its mouth open, sucking them in, and then sieving the krill by forcing the water through the spaces between the tubercles of the postcanine teeth. Most feeding probably takes place at night. Other invertebrates and small fish are also eaten.

Population densities of around 1–7/sq km have been reported. The animals usually are seen alone or in small groups. Siniff et al. (1979) reported that concentrations of 50–1,000, composed primarily of immature individuals, sometimes form on favorable areas of fast ice. Otherwise, *Lobodon* is found in small social units that apparently originate in the spring breeding season, when pregnant females move onto suitable ice floes to give birth. Just before and after parturition each female is joined by a male. The male protects the female and pup from other males as well as from potential predators. An area of up to 50 meters from the female is so defended. The female, however, savagely keeps the male at bay until her pup is weaned and she enters estrus, about 4 weeks after giving birth. The male then becomes aggressive and forces the mother and pup apart. Mating takes place from October to December and occurs on the ice, rather than in the water as in most polar phocids. The male and female remain in close contact for a few days, then break up. As a result of fights with both males and females, the old males are heavily scarred about the head and neck. Agonistic interaction is accompanied by hissing and blowing sounds. *Lobodon* also emits a deep groaning sound underwater (Stirling and Siniff 1979).

Births occur from September to November, with a syn-

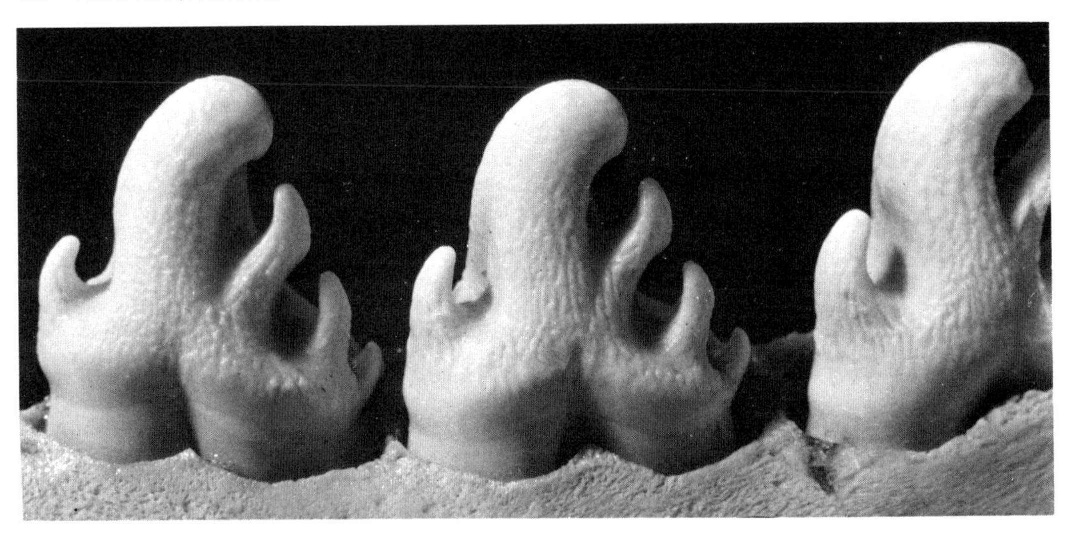

Teeth of crabeater seal, postcanines 3, 4, and 5, right mandible, lingual aspect, photo by Victor B. Scheffer.

chronized peak in early to mid-October. The total gestation time is about 11 months and probably includes a period of delayed implantation. The single pup is about 150 cm long and has a soft, woolly, grayish brown coat. In the 4-week period of lactation it grows from 25 to 120 kg. The young are nearly full grown after 2 years, but some growth continues for years afterwards. Sexual maturity is attained at about 3–6 years (Hayssen, Van Tienhoven, and Van Tienhoven 1993). More than 80 percent of females give birth each year. The life span may be up to 39 years.

Lobodon is the most numerous pinniped and may be the most abundant large mammal on earth. Population estimates range up to 40 million individuals. Numbers apparently increased in response to the greater availability of krill following the commercial extermination of the Antarctic baleen whales. It also has been reported that the age

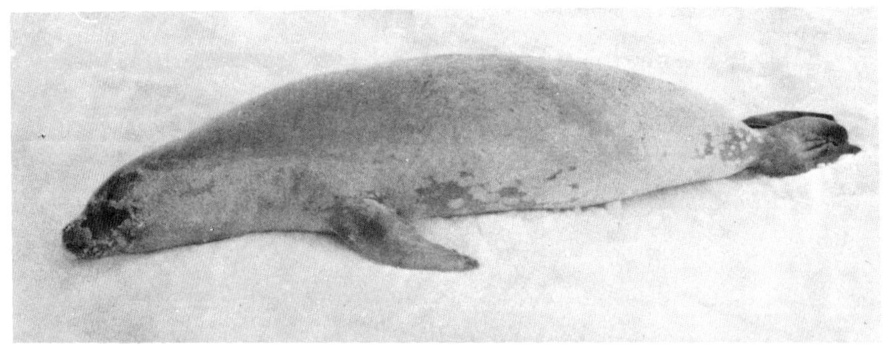

Crabeater seal *(Lobodon carcinophagus)*, photo by University of Minnesota Antarctic Seal Research Program.

of sexual maturity declined from 4 years in 1950 to only 2.5 years by about 1980. Bengtson and Siniff (1981), however, found that this trend may be changing in response to adjustments in Antarctic ecosystems and reported an average age at sexual maturity of 3.8 years in a sample of 94 females. They also cautioned that development of a major commercial krill fishery could further alter the ecosystems. Siniff (1991) indicated that *Lobodon* may be particularly vulnerable to loss of krill both to human harvest and recovering whale populations and cited evidence of a recent decline in seal numbers. Reijnders et al. (1993) considered 11–12 million to be a realistic estimate of total numbers.

Leopard seals *(Hydrurga leptonyx)*: A. Photo by U.S. Office of Territories through the National Archives; B. Photo by John Warham.

PINNIPEDIA; PHOCIDAE; **Genus HYDRURGA**
Gistel, 1848

Leopard Seal

The single species, *H. leptonyx*, occurs along the coasts and pack ice of Antarctica and also is regularly present on most subantarctic islands. Wandering individuals have reached South Africa, southern Australia, Tasmania, New Zealand, Lord Howe Island, the Cook Islands, Tierra del Fuego, and the Atlantic coast of South America nearly to Buenos Aires. Except as noted, the information for this account was taken from J. E. King (1983), Kooyman (1981*b*), and Siniff and Stone (1985).

Adult males, on average smaller than females, are up to 250–320 cm long and weigh 200–455 kg; adult females are 241–338 cm long and weigh 225–591 kg (Reijnders et al. 1993). Coloration is dark gray to almost black dorsally, paler on the sides, and silvery below. There are gray spots on the throat, shoulders, and sides. The body is streamlined but massive, the head seems disproportionately large and reptilian, and the foreflippers are comparatively long. The skull and canine teeth are unusually long, and the postcanine teeth, having three prominent tubercles with narrow clefts between them, are second in complexity to those of *Lobodon*.

Mature animals normally occur in the outer fringe of the pack ice. Young animals move throughout the subantarctic islands in the winter and early spring (June–October). Animals about 3–9 years old may reach the Antarctic coast in summer. There is some question whether these movements are regular migrations or occasional dispersals, but some islands are visited annually. There is a year-round presence on South Georgia and Heard Island. Juveniles occur seasonally on Kerguelen Island, with a peak in September and October, perhaps in association with the presence of breeding colonies of penguins (Borsa 1990).

The leopard seal has a variable diet and is the only pinniped that regularly preys on warm-blooded animals. The biomass taken has been estimated at 45 percent krill, 35 percent seals, 10 percent penguins, and 10 percent fish and cephalopods. Krill is consumed in much the same manner as described for *Lobodon*, with the organisms being filtered from the water by the complex cheek teeth. *Lobodon* itself is a major prey item, especially during the summer, when pups are available. Most crabeater seals bear scars from attacks by *Hydrurga*. Adult penguins are captured mostly in the water, but chicks are seized on the ice. Probably only a few individual leopard seals regularly take penguins.

A population density of about 0.12/sq km of pack ice has been reported. *Hydrurga* seems to be solitary, and unlike in *Lobodon*, males have not been observed together with females and young. Bester and Roux (1986), however, reported groups of 2–5 adults, including both sexes, hauled out in the vicinity of abundant prey at Kerguelen Island. A long, deep droning sound is produced underwater.

Mating occurs from November to February, apparently in the water. Gestation lasts about 11 months, and implantation evidently is delayed for about 2 months. Births take place from September to January, mostly in late October and November. The single pup is about 160 cm long and weighs about 30 kg. The natal coat is soft and thick, the dorsal color is dark gray with a darker central stripe, and the sides and ventral surface are almost white, irregularly spotted with black. Lactation may last about 4 weeks. Males apparently reach sexual maturity in their fourth year, and females, in their third year. Maximum known longevity in the wild is 26 years.

Population estimates for *Hydrurga* vary but seem to center around 400,000 individuals. The genus is highly adaptable and is not exploited, but it may face problems in the future. Young individuals evidently depend heavily on krill but are less efficient in foraging than are krill-feeding specialists such as *Lobodon*. Therefore, *Hydrurga* could be adversely affected by development of a commercial krill fishery.

PINNIPEDIA; PHOCIDAE; **Genus LEPTONYCHOTES**
Gill, 1872

Weddell Seal

The single species, *L. weddelli*, is found in areas of fast ice around Antarctica and some subantarctic islands, including the South Shetlands, the South Orkneys, and South Georgia. Wandering individuals have reached Heard, Kerguelen, Macquarie, and the Auckland, Juan Fernandez, and Falkland islands, as well as southeastern Australia, New Zealand, Patagonia, and Uruguay. Except as noted, the information for this account was taken from J. E. King (1983), Kooyman (1981*c*, 1981*d*), and Stirling (1971).

Adult males, which average slightly smaller than females, are about 250 cm long, with a reported maximum of 297 cm; adult females are about 260 cm long, with a reported maximum of 329 cm. In the spring both sexes commonly weigh around 400–450 kg. Coloration varies, but typically the upper parts are bluish black with white streaks and splashes that increase toward the rear and bottom of the animal. The lower parts are gray with white streaks. During the summer the coat fades to rusty grayish brown, but it never bleaches to the whiteness often found in *Lobodon*. The head of *Leptonychotes* is relatively small, the muzzle is very short, and the eyes and canine teeth are relatively large. The upper incisor teeth are markedly unequal, the outer being about four times longer than the inner. The cheek teeth are strong and fully functional but lack the complexity of those of *Hydrurga* and *Lobodon*.

The Weddell seal is perhaps the most southerly of mammals. It normally is found on fast ice in coastal areas within sight of land, and not on moving pack ice or isolated floes. There apparently are no major migrations, but there is a general northward movement before the onset of winter as the ice expands and a summer concentration to the south where fast ice remains. Some individuals remain in the most southerly parts of the range throughout the year, though they are less obvious in winter, when they presumably stay in the water, where the temperature is constant. *Leptonychotes* depends heavily on holes through the ice for access to the water and for breathing. Holes are made initially by cutting with the incisor teeth and then are en-

Weddell seals *(Leptonychotes weddelli)*, photo from U.S. Navy.

larged and maintained by abrading the ice with the canines. Cutting does not begin in areas where the ice is more than 10 cm thick, but once a hole is made, it may be maintained even if the ice increases to a thickness of 100–200 cm. Holes generally are made near natural cracks in the ice, and *Leptonychotes* evidently uses the radiating pattern of cracks as a guide to find its way to the holes when underwater. Its underwater vision is excellent and is thought to be the primary means of navigation. Dives are deeper and more frequent in daylight. During the long period of winter darkness dives tend to be shallow and probably are made in areas previously familiar to the animals.

The underwater activity of the Weddell seal has been investigated by capturing some individuals, attaching a time-depth recorder to them, and then releasing them. Animals tend to make an initial series of shallow dives to familiarize themselves with an area and then begin to go farther and deeper. Horizontal swimming speed is around 8–12 km/hr, and average rate of descent is 35 meters/minute. Foraging usually involves a series of dives lasting 8–15 minutes each and alternating with periods of 2–4 minutes at the surface hole. The entire series may go on for 8–9 hours without stopping. It is no wonder that when *Leptonychotes* finally emerges to rest or sleep on the ice, it may give the impression of being lethargic and not easily aroused. It is, however, one of the supreme divers among pinnipeds and in this regard even outclasses some cetaceans. Its dives usually reach depths of 200–400 meters, but some have been as deep as 600 meters and lasted up to 73 minutes. Its foraging trips from an access hole may cover a round-trip distance of 10–12 km. The diet consists mostly of fish, both sluggish bottom dwellers and active midwater species. The Antarctic cod, which may weigh more than 30 kg, frequently is taken. Cephalopods and crustaceans also are eaten.

Densities of 15–35/sq km have been recorded in favorable areas of the Weddell Sea. *Leptonychotes* is not gregarious, however, and adults keep apart when hauled out on the ice. Holes through the ice must be shared by a number of seals and sometimes are the scene of much aggressive activity and serious fighting. Younger animals may be forced to disperse into areas where the water is not completely covered by ice, and they sometimes form large aggregations. Dominant adult males establish territories below the ice that vary in size from about 15 by 50 meters to 50 by 400 meters and spend almost all their time from October to December defending these areas. During the same period the females haul out on the ice above to give birth to the young conceived during the previous breeding season. The females form loose groups at suitable pupping sites but keep well apart and may snap at one another. A sex ratio of 10 females to 1 male has been observed at breeding sites. After the season, subadults are admitted into these areas and the distance between individuals decreases. *Leptonychotes* is among the noisiest of seals and produces a continuous and varied stream of underwater sounds, including whistles, buzzes, tweets, and chirps. Some of these sounds probably are for territorial defense, and others are used to attract mates. Thomas and Stirling (1983) reported considerable variation in the vocalizations of different seal populations.

Mating takes place in December, evidently in the water. Implantation of the blastocyst in the uterus is delayed until the period from mid-January to mid-February. Births occur mostly in October in Antarctica but may be as early as late August at South Georgia. The single newborn is about 150 cm long, averages 29 kg in weight, and has a grayish coat that is replaced after 4 weeks by adultlike pelage. The mother remains with the pup almost constantly for about 2 weeks, then begins to spend about a third of her time foraging. Upon her return to the ice she establishes vocal contact with her pup and confirms identification by sniffing. She may bring the pup into the water to practice swimming when it is only 8–10 days old. Lactation lasts 6–7 weeks, after which the mother mates again and departs. By then the young averages 113 kg in weight and is able to dive to a depth of 92 meters. Females may reach sexual maturity as early as 2 years, but Croxall and Hiby (1983) reported that females in the South Orkney Islands first gave birth at an average age of 4–5 years. Up to 80 percent of adult females become pregnant each year. At McMurdo Sound, Antarctica, both sexes are recruited into the adult population at around 5 years, and annual reproductive rates are 46–79 percent (Testa 1987; Testa and Siniff 1987). Maximum known longevity is 25 years (McLaren 1984).

Unlike *Lobodon* and *Hydrurga,* the Weddell seal is easily approached and captured by people. Excessive hunting to supply Antarctic bases with meat for people and sled dogs has eliminated or reduced some breeding colonies. There is fear that if intensive sealing begins in Antarctica, *Leptonychotes* will be especially vulnerable because of its tendency to congregate on stable ice during the summer. Until now, however, there have been no major declines, and current numbers are estimated at 500,000 to 1 million.

PINNIPEDIA; PHOCIDAE; **Genus OMMATOPHOCA**
Gray, 1844

Ross Seal

The single species, *O. rossi,* is circumpolar in pack ice of the Antarctic. Wandering individuals have reached South Australia and Heard Island. Except as noted, the information for this account was taken from J. E. King (1983) and Ray (1981).

Adult males, on average smaller than females, are 168–208 cm long and weigh 129–216 kg; adult females are 190–250 cm long and weigh 159–204 kg. After the summer molt, in January, coloration is dark gray above and silvery white below, with streaks of pale gray on the sides of the head, shoulders, throat, and flanks. In the course of the year the coat fades to tan or brownish. The body hair and vibrissae are the shortest of any phocid's. The head is large and the neck is thick, but the body is slenderer than that of *Leptonychotes.* The snout is very short, and the mouth appears to be relatively small. The eyes are proportionately larger than those of any other seal, though the slits are not unusually large. The front limbs, which are more specialized and flipperlike in shape than those of most phocids, have reduced claws and greatly elongated terminal phalanges. The hind flippers are proportionately longer than those of any other phocid. The skull is distinguished by the short rostrum and the enormous size of the eye sockets. The incisors and canine teeth, which are small compared with those of *Leptonychotes,* are sharp and recurved. The cheek teeth are relatively small and weak.

Ommatophoca is among the least known of pinnipeds. It generally is considered to prefer the heavy consolidated pack ice, but some observations indicate that it also is found on smaller floes. It is slow and clumsy on the ice but is thought to be a swift swimmer and a very capable maneuverer and to pursue its prey under the ice, where its large eyes can perceive movements in the dimly lit waters. The diet consists mostly of cephalopods, including some large squid. Other invertebrates and fish also are taken.

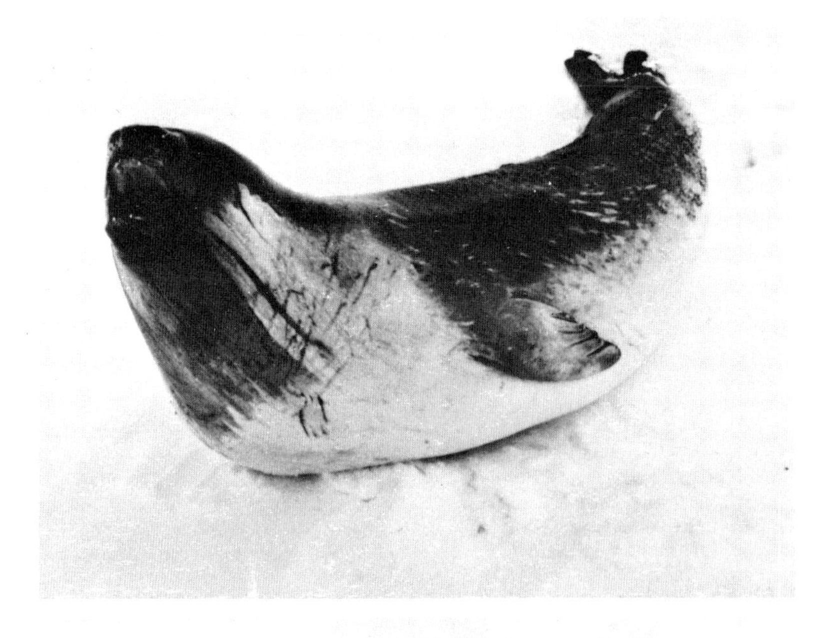

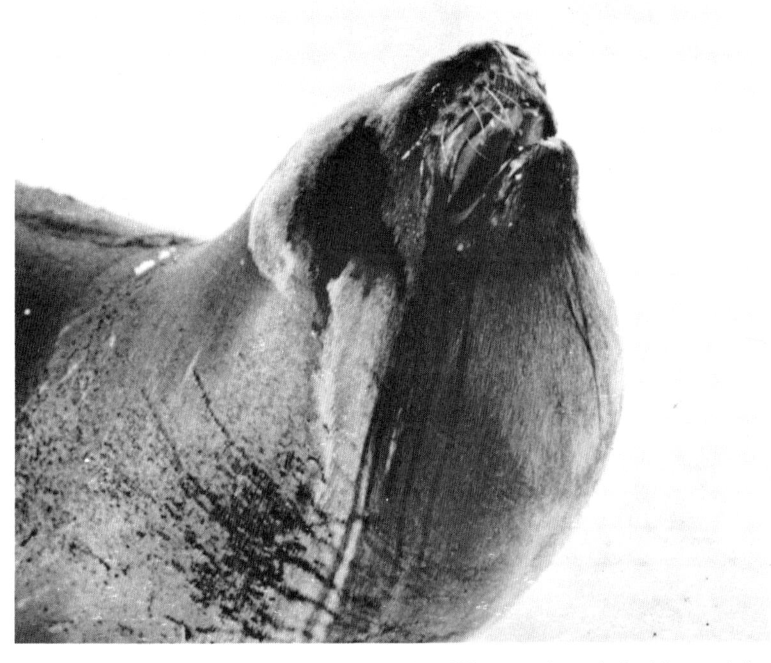

Ross seal *(Ommatophoca rossi)*, photos by University of Minnesota Antarctic Seal Research Program.

Although the mouth is small externally, it can be opened remarkably wide and then closed forcefully. Swallowing large prey is facilitated by an expanded trachea and powerfully developed muscles of the tongue and pharynx. The hard palate is short, but the soft palate is very long, extending far posteriorly to the level of the occipital condyles. The soft palate can be inflated with air, and both it and the expanded trachea seem to act as resonating chambers for the production of sound. When approached by people *Ommatophoca* may lift its front end, open its mouth widely, inflate its soft palate to meet the back of the raised tongue, and emit a series of trills and thumping noises. It shows so little fear that it can be closely photographed and even touched during this display. A variety of birdlike chirps and other sounds are produced both in the air and underwater. The process occurs internally in the throat, and no air is expelled from the body until the end of the display.

The vocal capabilities of *Ommatophoca* suggest that it may establish underwater territories like those of *Leptonychotes* and maintain contacts over considerable distances. However, available evidence indicates that it is generally solitary and that individuals only occasionally come together in small groups. The highest recorded population density is 2.9/sq km. Some individuals are scarred on the neck and shoulders, but it is not known whether these marks result from intraspecific fighting or from attacks by *Orcinus* and *Hydrurga*. Females haul out on the ice in November to give birth to the young conceived during the previous breeding season. Mating apparently takes place in late December, after the young are weaned. Implantation of the blastocyst is delayed for 2–3 months, and the total gestation period is about 11 months. There seems to be a birth peak from 3 to 18 November. The single newborn is about 97 cm long, weighs 17 kg, has a coat similar to that of an adult, and is capable of swimming almost immediately. Weights of up to 75 kg are attained after 15 days, and lactation ceases after approximately 4 weeks. Most females become pregnant by the time they are 3 years old, and males may reach sexual maturity at 3–4 years. Maximum known longevity is 21 years.

Ommatophoca was not discovered until 1840, the latest of any pinniped, and fewer than 50 records accumulated over the next 100 years. Subsequent investigators have suggested that the genus is not rare but have provided an incomplete basis for assessing its status. Overall population estimates of 100,000–650,000 have been made but are not considered reliable. Reijnders et al. (1993) cited a minimal estimate of 130,000 and reported no known threats to the species.

PINNIPEDIA; PHOCIDAE; **Genus MIROUNGA**
Gray, 1827

Elephant Seals

There are two species (Johnson 1990; J. E. King 1983; Ling and Bryden 1981; Rice 1977; Stewart and Huber 1993):

M. angustirostris (northern elephant seal), coastal areas from southeastern Alaska to Baja California, much of the northeastern Pacific Ocean, isolated records from Midway Island and Japan;
M. leonina (southern elephant seal), most subantarctic islands between 40° S and 62° S, coast of southern Argentina, waters south to edge of antarctic ice at 78° S, occasionally on coasts of southern Africa and Australia and as far north as St. Helena (16° S), Mauritius (20° S), and Lima (14° S), and one recent record from Oman (18° N).

According to J. E. King (1983), the ancestor of *Mirounga* probably descended from the same stock as that leading to *Monachus* and also passed through an ancient seaway from the Caribbean to the Pacific. Many characters of the skull, dentition, and growth suggest that the northern elephant seal *(M. angustirostris)* is more primitive than *M. leonina.* Both species are characterized by large overall size and the trunklike, inflatable proboscis of adult males. The proboscis reaches full size in animals about eight years old. Its tip then overhangs the mouth in front, so that the nostrils open downwards. An enlargement of the nasal cavity, internally it is divided into two parts by the nasal septum. Externally the proboscis is flattened and less obvious in the nonbreeding season. When breeding, *Mirounga* can erect this organ, partly by blood pressure, assisted by inflation, to form a high, bolster-shaped cushion on top of the snout. It may act as a resonating chamber to amplify the roar of the bull. The narial basin of the huge skull is increased in length and there is a high frontonasal area to provide space for attachment of the muscles that move the proboscis. The canine teeth of the adult male are relatively large and continue to grow for at least 12 years. The cheek teeth are much reduced. In color the males are dark gray, being a little lighter ventrally just after the molt, but through the year the coat fades to a rusty grayish brown. Fighting between bulls leads to intensive scarring of the neck region, and the scar tissue makes the skin of the chest extremely thick, tough, and cracked. The females are browner and generally darker than the males and have a light-colored yoke around the neck caused by the many small scars resulting from bites during mating. The hair is short, stiff, and harsh, and there is no underfur. Additional information is provided separately for each species.

Mirounga angustirostris (northern elephant seal)

Except as noted, the information for the account of this species was taken from J. E. King (1983) and McGinnis and Schusterman (1981). Adult males are 400–500 cm long and weigh 2,000–2,700 kg; the much smaller adult females are 200–300 cm long and weigh about 600–900 kg. From *M. leonina, M. angustirostris* differs in having a smaller and less flexible body, less sexual dimorphism, a narrower skull with a longer rostrum and a greater degree of suture closure with age, and a larger proboscis in the adult male. This proboscis is very long, hanging down over the mouth about 30 cm when relaxed and showing a very deep transverse groove that almost divides it in two when inflated. Inflation causes it to curve down between the jaws and directs the nostrils toward the pharynx; thus snorts are directed down into the open mouth and pharynx, which act as a resonating chamber.

The northern elephant seal currently breeds on numerous islands, from Cedros, off the west coast of central Baja California, north to the Farallons, near San Francisco. It also uses a few mainland areas (see below). Preferred breeding sites are good sand or gravel beaches remote from human activity. According to Condit and Le Boeuf (1984), there are certain established rookeries where the seals breed and molt. When not at these sites, the seals are feeding at sea, mainly in waters from northern Baja California to northern Vancouver Island. Juveniles move northward from their rookeries during the summer by an average of 900–1,000 km and return to haul out in the autumn. During the winter, when adults are breeding, juveniles again go to sea. Adult males migrate northward during the spring, haul out in July and August to molt, move north again in the autumn, and return to the rookeries to breed during the winter. Adult females are at sea for just over 10 weeks during the spring and again for about 7 or 8 months during the summer and autumn. Individual seals have been found as far away as Midway Island and the central Gulf of California. DeLong, Stewart, and Hill (1992) recorded migrations from the San Miguel Island rookery, off southern California, to the eastern Aleutian Islands and Gulf of Alaska, and back.

Using radio tracking via an earth satellite, as well as a time-depth recorder, Stewart and DeLong (1995) determined that adult seals from San Miguel make a full double migration each year, the first ever documented for any animal. In February or early March, after the breeding season, there is a round trip averaging about 73 days and 6,300 km for females, and 124 days and 12,000 km for males. After returning to San Miguel in May (females) and July (males), and remaining ashore for 3–4 weeks to fast and molt, the seals make a second circuit, the females covering about 12,200 km in 234 days and the males covering about 9,600 km in 126 days. Both outward movements take the animals to the same general foraging sites, which are in the

Northern elephant seals *(Mirounga angustirostris):* A. Male, photo from Allen Hancock Foundation; B. Father and baby, photo by Julio Berdegué; C. Seal shedding its skin, photo from Colorado Museum of Natural History through Alfred M. Bailey and R. J. Niedrach; D. Adult male with proboscis turned downward into mouth, photo by Julio Berdegué; E. Group of females, animal on left shedding, photo by Rupert R. Bonner; F. South Atlantic elephant seal *(M. leonina),* photo from Zoological Garden Berlin-West.

northeastern Pacific, as described below for the study at Año Nuevo. As in that study, the seals from San Miguel are sexually segregated and dive continuously during their migration. Their individual total annual movements of 18,000–21,000 km are the greatest yet reported for mammals.

Based on studies of seals from the Año Nuevo rookery in central California, Le Boeuf (1994) reported that adult males migrate to specific foraging areas along the continental margin off southern Alaska and the Aleutians. Females generally do not move as far to the north and west, but they disperse more widely in the open ocean and forage en route. Seals of both sexes and all ages dive deep, long, and continuously while they are at sea. Using a time-depth recorder, Le Boeuf et al. (1988, 1989) monitored movements of females during their entire period of foraging after lactation, averaging 74.2 days. They were found to dive almost continuously during this period, at a rate of 2.5–3.3 dives per hour. Modal dive duration per female was 17.1–22.5 minutes. The longest dive was 62 minutes. Modal dive depth per female was 500–700 meters. The maximum was estimated at 1,250 meters. Le Boeuf et al. (1992) determined that a female had a mean speed of about 1.5 meters per second while descending and 1.2 meters per second at the beginning of her ascent. DeLong and Stewart (1991) recorded data on adult males that were instrumented on San Miguel Island and then went to sea for 107–45 days during spring and early summer. They were submerged about 86 percent of the time, rarely spending more than 5 minutes between dives at the surface. Dives averaged 22.6 minutes, and most reached depths of 350–450 meters. The longest lasted 77 minutes and the deepest was 1,529 meters. DeLong, Stewart, and Hill (1992) subsequently reported a dive of 1,581 meters for a male elephant seal foraging off Alaska, a record for all air-breathing vertebrates (including whales and sea turtles).

The diet consists mostly of deepwater and bottom-dwelling organisms, particularly squid and elasmobranch and teleost fish. The studies cited above indicate that dives tend to be shallower at night. They suggest also that *M. angustirostris* feeds mostly in the deep scattering layer of the mesopelagic zone and preys on organisms that migrate vertically upward at night. Most fish with such behavior are bioluminescent, as are several species of squid taken by the elephant seal. This factor may help the seal to find its prey at great depths.

Adult males arrive on the breeding beaches in late November and early December and probably do not feed again until they depart in March. Adult females arrive in late December or January and also do not feed until their departure, about 34 days later. The females come together in compact groups, with individuals usually not more than 1–2 meters apart. In newly established colonies with relatively few animals a dominant male may be able to defend a discrete group of females, but this pattern breaks down as more crowded conditions develop. Generally, a single male cannot control more than about 40–50 females. If a colony has a larger number of females, there will be several bulls present. The latter compete for access to the females and eventually form a dominance hierarchy. There is then no defense of a territorial space or a particular harem of females; rather, the highest-ranking males are able to mate the most, and subordinate males are forced to the periphery of the group. According to Reiter, Panken, and Le Boeuf (1981), the largest female group on Año Nuevo Island, southwest of San Francisco, contains 1,100 individuals and is associated with 10–25 males. A successful male in this colony mates with about 100 females during the season.

Males establish dominance through stares, gestures, vocalizations, and physical combat. A threat display may involve elevating the anterior part of the body to about 90°, inflating the proboscis, slamming the body to the ground, and emitting a series of snorts and low-pitched guttural sounds. If a fight develops, the opponents confront one another chest to chest and rock back and forth seeking an opportunity to deliver a downward blow with the canine teeth. Serious wounds occur frequently on the chest and proboscis but rarely lead to death. The roaring vocalization is rhythmic, resonant, and metallic and carries for nearly 1 km. Distinctions have been detected between the sounds produced by the males of different islands. Shipley, Hines, and Buchwald (1986) found that the threat calls of individual males could be readily identified.

Females have a loud and prolonged threat vocalization. After giving birth, they become aggressive toward one another and usually to pups other than their own, but some adoptions have been noted. The young conceived during the previous breeding season is usually born 6–7 days after the female's arrival. Toward the end of the lactation period females enter estrus for 3–5 days. Most mating occurs in the third week of February. Gestation lasts for about 11 months but includes a delayed implantation period of 2–3 months (Stewart and Huber 1993). The peak of births occurs from 20 January to 1 February. The newborn pup is about 127 cm long, weighs 30–45 kg, and has grayish black pelage. At the time of weaning, 27 days later, it weighs about 131 kg (Le Boeuf and Laws 1994), and the coat is silvery. Subsequently the young gather in small pods and practice swimming and diving but do not feed. When they leave the natal beaches in April or May, they are able to remain underwater for 15 minutes. Females are sexually mature in their third year, though most do not give birth until their fourth or fifth year. Nearly all adult females become pregnant each year and have the potential of producing 10–12 young during their lifetime, but individuals 6 or more years in age are much more likely to raise their pups to weaning (Reiter, Panken, and Le Boeuf 1981). Males are fertile at 5 years but usually are not able to win mating privileges until they are 8–9 years old. Longevity in the wild is relatively short, being 14 years for males and 18 years for females (Reijnders et al. 1993).

A 4-meter-long elephant seal can yield about 325 liters of fine oil. Commercial exploitation of this resource began on the West Coast around 1818 and intensified during the California gold rush. By the 1860s, after perhaps 250,000 individuals had been killed (Busch 1985), there were almost too few individuals left to make hunting worthwhile. Stewart et al. (1994) noted that there were three separate periods when the species was generally considered completely extinct, but small groups were rediscovered each time. By 1892 *M. angustirostris,* which once had bred from Point Reyes, north of San Francisco, to Magdalena Bay, in southern Baja California, had been reduced to a single herd of about 100 animals on Guadalupe Island, off northern Baja.

A subsequent recovery resulted from an end to commercial interest and the establishment of complete legal protection by Mexico in 1922. From Guadalupe, which still has one of the largest populations, the species spread both south and north. Breeding colonies were reestablished on the California Channel Islands in the 1930s, in the Farallon Islands in 1972, and at Año Nuevo Point, on the California mainland, in 1975. Since 1981 a small breeding group also has occurred seasonally on the Point Reyes Peninsula, north of San Francisco (Allen, Peaslee, and Huber 1989). The seals now breed on 15 islands and 3 mainland beaches, with more than half of the births occurring on San Miguel Island and about a fifth on Guadalupe (Stewart et al. 1994). Most of the original range now has been reoccupied. Cooper and Stewart (1983) reported that numbers were increasing at a rate of 14 percent annually. The U.S. National Marine Fisheries Service (1989, 1994) estimated overall numbers at 100,000, the same figure that Busch (1985) had calculated for the period prior to exploitation. However, Stewart et al. (1994) reported that total numbers in 1991 were about 127,000, including 28,000 pups of the year, and that there was considerable room for further expansion; numbers were continuing to increase by more than 6 percent annually. Notwithstanding these encouraging trends, current elephant seal populations show a remarkable lack of genetic diversity because of the great numerical reduction in the late nineteenth century and may thus be vulnerable to changing conditions.

Mirounga leonina (southern elephant seal)

Except as noted, the information for the account of this species, the largest of pinnipeds, was taken from J. E. King (1983), Ling and Bryden (1981), and Smithers (1983). Adult males are usually 450–600 cm long and weigh up to 3,700 kg. Old records suggest that some males attained a length of 900 cm and a weight of 5,000 kg. The much smaller females are 200–300 cm long. Adult female weight seems to be the subject of some confusion, having been reported as 359 kg (Ling and Bryden 1981), 400 kg (McCann 1985), 680 kg (McLaren 1984), and 900 kg (Bonner 1982*a;* J. E. King 1983). However, Le Boeuf and Laws (1994) explained that females range widely in mass, with most of them weighing around 400–600 kg shortly after giving birth. *M. leonina* has a more flexible body than does *M. angustirostris* and can bend backwards and touch its tail with its head. The proboscis of the adult male is not so pendulous as in *M. angustirostris,* overhanging the mouth only about 10 cm. When it is erected its transverse grooves are much less deep. It hangs down, so that the nostrils open in front of the mouth, but it does not actually go into the mouth. It functions as a resonator for sounds produced in the mouth.

The southern elephant seal breeds on most subantarctic islands and on the Valdez Peninsula, on the coast of southern Argentina. The three major populations are centered at South Georgia in the southwestern Atlantic, Kerguelen and Heard islands in the South Indian Ocean, and Macquarie Island south of New Zealand. Nonbreeding individuals regularly land at many points, including coastal Antarctica. Gales and Burton (1989) reported that groups haul out to molt at the Vestfold Hills and Windmill Islands along the coast of Greater Antarctica. M. D. Murray (1981) reported establishment of a small breeding colony at Peterson Island, just off the Indian Ocean side of Antarctica. Heimark and Heimark (1986) reported the same for Palmer Station on the Antarctic Peninsula. The annual life cycle involves lengthy stays at sea alternating with periods ashore for breeding, molting, and resting. The pelagic phases are not well known; the longest movement of a tagged animal was from South Georgia to South Africa, a distance of 4,800 km. Immature individuals, along with females that have not recently given birth, come ashore to molt from November to January. Cows that gave birth during the previous breeding season molt in January and February, and adult males do so from February to May. Molting takes 3–40 days, and the animals do not feed during this period. They then go to sea again, returning to the breeding grounds in August or September. Preferred beaches are flat and smooth with a substrate of sand, rounded stones, or pebbles. Some animals move into areas of tussock grass, and molting females often seek muddy wallows. Like *M. angustirostris, M. leonina* has flexible flippers and sometimes scoops sand over itself to lower its body temperature. At the southern edge of its range *M. leonina* hauls out on snow or ice. It can make access holes by butting though the ice with its head, and it may travel 20 km under the ice, utilizing air pockets along the way. Maximum swimming speed is about 25 km/hr. The diet consists almost entirely of cephalopods and fish.

Like the northern elephant seal, *M. leonina* is capable of very long, deep dives and has been studied in this regard by use of time-depth recorders (Hindell, Slip, and Burton 1991; Hindell et al. 1992; Slip, Hindell, and Burton 1994). The instruments were attached to both males and females on Macquarie Island just before they departed to forage and then recovered when they returned. The animals were at sea for up to 8 months, most of them using waters off the Antarctic coast. They spent 90 percent of this time submerged, averaging 2.2 dives per hour and usually staying on the surface only 2–4 minutes between dives. The mean depth per dive ranged from 269 to 589 meters for each seal. The maximum depth was 1,430 meters and the longest, a record for all marine mammals, was 120 minutes. The animals spent 10–20 minutes at the bottom of each foraging dive but also evidently dived for other purposes that did not involve such prolonged stays at maximum depth.

Adult males arrive at the breeding beaches from early August to September. They remain there up to 9 weeks, during which time they do not feed. Adult females begin coming ashore a few weeks later and also do not feed until their departure, about 4–5 weeks after arrival. The females initially form small, peaceful groups, but eventually thousands of females may crowd onto a favorable beach. They become more aggressive and more widely spaced after giving birth. Social activity is much the same as in *M. angustirostris.* Bulls are not territorial but form a dominance hierarchy, with the highest-ranking individuals being able to mate with the most females. A single dominant male, or "beachmaster," may be able to control a small group of females, but when more than about 60 females are present in the cluster, a second bull gains some access. When the group grows to around 130 females, a third male wins a place, and so on. Some "harems" of this kind contain up to 1,350 females. Males hoping to enter the scene challenge the established males with a bubbling roar that may carry for several kilometers. If the call is answered and the challenger is not scared off, a confrontation develops as described above for *M. angustirostris.* Serious fighting is not common; when it does occur, the opponents lift the anterior two-thirds of their bodies off the ground, move their heads from side to side, and lunge with the canine teeth. The loser deflates his proboscis and retreats to the water or to the periphery of the breeding area.

Within 2–7 days of arriving on land females give birth to the young conceived in the previous breeding season. After another 17–22 days they enter estrus and mate. Gestation lasts just over 11 months and begins with a delayed-implantation period of about 4 months. The young are born in September or October, but there is a synchronous peak throughout the range of the species in mid-October. The newborn is 130 cm long, weighs 35–50 kg, and has black woolly hair. At the time of weaning, 20–25 days later, it weighs 140–80 kg and its coat is silvery gray. The mother loses about a third of her body weight during lactation and then promptly returns to sea to forage, permanently abandoning her pup. The young then form groups of their own, learn to swim by themselves, and move out to sea when they are 8–10 weeks old. Females become sexually fertile at about 2–3 years, and nearly all females more than 4–6 years old become pregnant annually. Males are sexually mature at 6 years but do not begin to compete for mating privileges until they are 10 years old and usually cannot become beachmasters until 12–14 years. In the formerly exploited population of South Georgia, where large bulls were killed regularly, males began to breed 2–3 years earlier than in other populations. Hindell and Little (1988) reported that two females were observed suckling pups on Macquarie Island within 1 km of where they had been born and marked 23 years earlier.

The southern elephant seal formerly bred in Tasmania but was eliminated there by aboriginal peoples. Since 1977 there have been a few isolated records of individuals visiting and even giving birth on Tasmania and nearby islands (Pemberton and Skira 1989). A colony on King Island, in Bass Strait, between Tasmania and Victoria, survived until the early nineteenth century. The species also once bred in the Juan Fernandez Islands and occurred on St. Helena and possibly the Seychelles. Commercial exploitation began around 1800 and intensified a few decades later as fur seals became scarce. Elephant sealing often was done in association with whaling, the main objective in each case being oil. A large bull *M. leonina* could produce about 750 liters of oil (Busch 1985). All of the major populations were greatly reduced, but most survived because some breeding beaches were not accessible to boats loading oil and because the demand for whale oil declined. The last major sealing expeditions were undertaken in the 1870s, and the species subsequently recovered to some extent. In 1910 commercial exploitation resumed on South Georgia, but under regulations that limited the annual kill to a maximum of 6,000 bulls. Mistakes in management, including an inappropriate increase in the yearly quota, had led to new declines by the 1940s. Improved controls were implemented in 1952, but hunting ceased in 1964, again in connection with a decline in whaling (Bonner 1982*a*).

According to Laws (1994) and McCann (1985), except for the group in Argentina, southern elephant seal populations are no longer increasing, and some are declining. Overall numbers had approached 1 million in the 1950s but are now about 664,000, not including the approximately 190,000 pups born each year. South Georgia has a relatively stable population of about 357,000 individuals; Kerguelen, 143,500; Heard Island, a sharply declining population of 40,500; and the Valdez Peninsula of Argentina, about 34,000. Hindell (1991) and Hindell and Burton (1988) calculated the size of the original population on Macquarie Island at 93,000–110,000 seals. This group was reduced to about 30,000 in the early nineteenth century but had recovered to perhaps 150,000 by 1959. That number apparently was excessive for the habitat, and numbers since have fallen to around 80,000. *M. leonina* is on appendix 2 of the CITES.

PINNIPEDIA; PHOCIDAE; **Genus ERIGNATHUS**
Gill, 1866

Bearded Seal

The single species, *E. barbatus*, is found along coasts and ice floes in the Arctic Ocean and adjoining seas and regularly occurs as far south as the Sea of Okhotsk, Hokkaido, the Alaska Peninsula, Hudson Bay, and Labrador (Burns 1981*a*; J. E. King 1983; Rice 1977). Individuals have reached eastern China, Tokyo Bay, the Gulf of St. Lawrence, Scotland, Normandy, and northern Spain (J. E. King 1983; McLaren 1984; Zhou 1986) but probably not Cape Cod Bay as reported previously (Ray and Spiess 1981). Except as noted, the information for the remainder of this account was taken from Burns (1981*a*) and J. E. King (1983).

Length is about 200–260 cm and weight is about 200–360 kg. The sexes are approximately the same size, though the heaviest individuals are pregnant females. Coloration is variable, most animals being light to dark gray, others being brownish, and all having a dark area on the back. The face and foreflippers are often rust or reddish brown. The hair is generally short and straight, but there is a great profusion of long, very sensitive, glistening white mustachial whiskers, the basis for the name of the species. The body is robust, but the head is disproportionately small. Norwegian sailors used the name "square flipper" for *Erignathus* because the third digit on the forelimb is slightly longer than the others, so that the appendage has a blunt, squared appearance. The nails on the foreflippers are strong, whereas those on the hind flippers are slender and pointed. Females, like those of *Monachus*, have four mammae (females of all other phocids have only two mammae).

The bearded seal generally is restricted to shallow waters over the continental shelf and near coastlines and often is associated with sea ice. It avoids regions of continuous, thick, shorefast or drifting ice but is common where ice is in constant motion and has many openings. Favorable ice conditions are especially prevalent in the Bering and Chukchi seas. There *Erignathus* moves as the ice seasonally advances and recedes. The population migrates south through the Bering Strait in late autumn and early winter and spends the winter amidst the drifting ice of the Bering Sea. In the spring there is a northward movement into the Chukchi Sea. Certain other populations, such as that of the Sea of Okhotsk, have no ice available during the summer and so come ashore, preferably on gravel beaches, to rest or molt. *Erignathus* attempts to avoid the polar bear by resting very near the water and diving immediately when disturbed. It evidently has good vision and hearing and a fair sense of smell. It is able to make breathing holes through thin ice by ramming with its head. Dives to depths of 200 meters have been reported, but most feeding is done in shallower waters. The diet varies by area but consists almost entirely of bottom-dwelling animals. Important items are crabs, shrimps, gastropods, clams, whelks, arctic cod, and flounder. The whiskers may be used to help sort out smaller food organisms.

Erignathus tends to be solitary, though loose aggregations sometimes form at favorable hauling-out sites. It is highly vocal and produces a distinctive underwater song that Ray, Watkins, and Burns (1969) described as a long, oscillating warble that may last for more than one minute, followed by a short, low-frequency moan. This sound is known to be made only by mature males during the spring and may be a proclamation of territory or breeding condition. Mating occurs mainly in the first three weeks of May in Alaska and the Canadian Arctic and from the end of March to late May in the Barents Sea. Implantation of the blastocyst is delayed about 2 months, until July or early August. The total period of pregnancy is approximately 11 months. The pups are born in the open on ice floes from mid-March to early May. The peak birth period is late April in most regions and early to mid-April in the Sea of Okhotsk and the southern Bering Sea. Most females become pregnant each year and enter estrus just before or after they stop nursing the young conceived in the previous breeding season. The single pup is about 130 cm long, weighs 30–40 kg, and has a short, woolly, grayish brown coat with a light-colored face and four broad, transverse, light bands on the crown and the back. It is weaned after 12–18 days, by which time it has molted and become adult-like in color and has grown to about 85 kg. It is promptly abandoned by its mother at this time, but it is already actively swimming and feeding on its own. Sexual maturity

Bearded seal *(Erignathus barbatus)*, photo from Zoological Garden Berlin-West.

usually is attained at 5–6 years by females and 6–7 years by males. Maximum known longevity is 31 years.

Erignathus formerly occurred in substantial numbers along the Norwegian coast as far south as Trondheim, but now it only rarely wanders into that area (Smit and Wijngaarden 1981). Otherwise there appears to have been little reduction in distribution, and commercial exploitation has been limited because of the nongregarious habits of the genus. Its meat and skin are of importance to local subsistence economies. The strong, durable, and elastic hide is used to make boot soles, heavy ropes, dog harness, and kayak covering. Formerly the intestines were used to make waterproof clothing and the blubber was used to obtain oil for lamps. About 10,000–13,000 individuals are taken annually by subsistence and commercial hunters in Norwegian, Russian, U.S., and Greenland waters (Reijnders et al. 1993; U.S. National Marine Fisheries Service 1987). Population estimates are around 150,000 in the Sea of Okhotsk, 300,000 in the Bering and Chukchi seas, and 300,000 along the rest of northern Eurasia. There are no recent estimates of numbers or harvests for the eastern Canadian Arctic, but the population there was thought to be about 185,000 in the late 1950s.

PINNIPEDIA; PHOCIDAE; **Genus CYSTOPHORA**
Nilsson, 1820

Hooded Seal

The single species, *C. cristata,* occurs mainly along ice floes in the North Atlantic region, from Baffin Island and Newfoundland, through the waters around Greenland, to Svalbard and Iceland. There also are records of wandering individuals from as far west as northern Alaska, as far north as eastern Ellesmere Island, as far east as the Kara Sea and the Yenisey River, and as far south as southwestern Hudson Bay, Montreal, Florida, the British Isles, and Portugal. Recently there were reports of the most southerly known occurrences on both sides of the Atlantic: a young individual at Fort Lauderdale, Florida (Fletemeyer 1985) and a pregnant female at the mouth of the Guadalquivir River, in southern Spain (Ibáñez et al. 1988). An apparently wild individual also was captured in 1990 at San Diego, California (Dudley 1992). The information for this account was taken mostly from J. E. King (1983), Kovacs and Lavigne (1986), and Reeves and Ling (1981); major exceptions are noted.

Adult males, which on average are slightly larger than females, are about 250–70 cm long and weigh 200–400 kg; adult females are about 200–220 cm long and weigh 145–300 kg. The background color is silvery or bluish gray, and there are numerous black or brownish black patches of irregular size and shape ranging from 50 to 80 sq cm on the back and sides to much smaller on the neck and abdomen. The face is black to behind the eyes. There are large claws on both the hind and foreflippers. The cranium is comparatively short, and the snout is long and wide. In association with the development of the prominent fleshy proboscis, the frontonasal area of the skull is elevated and the nasal openings are wider than in any other phocid.

The most striking feature of this seal is the enlargement of the nasal cavity of the adult male to form an inflatable crest, or hood, on top of the head. The hood begins to develop when a male is about 4 years old and continues to grow with age and increasing body size. It is divided into two by the nasal septum and is lined by a continuation of the mucous membrane of the nose. Its skin is very elastic,

Hooded seal *(Cystophora cristata):* A. Adult male, photo by Bernhard Grzimek; B. Bladder nose fully inflated while swimming; C. Adult male with hood inflated; D. Young in "blueback" stage. Photos by Erna Mohr.

and there are muscle fibers, particularly around the nostrils. When not inflated, the hood, which starts just behind the level of the eyes, hangs down in front of the mouth. When blown up, however, it forms a high cushion on top of the head; it then is large enough to hold about 6.3 liters of water.

The presence of an inflatable proboscis in males of both *Cystophora* and *Mirounga* sometimes has been considered indicative of close phylogenetic relationship but now generally is thought to be the result of convergent evolution. In *Cystophora* the hood is inflated with air, and the nostrils are closed before inflation. In *Mirounga* the proboscis is erected partly by muscular action and blood pressure, assisted by inflation, and the nostrils remain open. In addition, *Cystophora* has the ability to blow a bright red balloon-shaped structure from one nostril (usually the left). This balloon, or bladder, is an inflation of the highly elastic mucous nasal septum and may reach at least the size of an ostrich egg. It is produced when an animal closes one nostril, fills the hood on that side with air, and then uses that air to force the nasal septum out of the other nostril. The hood and balloon may be used as a threat, as they have been observed during the mating season and when *Cystophora* is disturbed, but they also sometimes are inflated when an animal is lying quietly on the ice. Females also have a hood, but it is not well developed and not inflatable.

The hooded seal is found in the same general region as the harp seal *(Phoca groenlandica)* but tends to stay farther from shore and to occupy deeper water and thicker, drifting ice. It rarely hauls out on land or on shorefast ice except in the Gulf of St. Lawrence. It follows a regular annual cycle of migration and dispersal. In February the adults concentrate in widely scattered groups on the thick ice of three areas: (1) off the east coast of Labrador and Newfoundland, and to a lesser extent in the Gulf of St. Lawrence; (2) in Davis Strait, between Canada and Greenland; and (3) around Jan Mayen Island in the Norwegian Sea east of Greenland. Following the breeding season in these areas there is a period of dispersal for foraging, and then the adults from the first two concentrations move into the Denmark Strait, between Greenland and Iceland, for purposes of molting. The animals from the Jan Mayen area apparently molt at a point farther north along the east coast of Greenland. This molting phase occurs from June to early August, and the seals then disperse to little-known feeding waters until the next breeding season. Some individuals move to the south and then up the west coast of Greenland. Others spread to the vicinity of Svalbard and Bear Island. Available evidence indicates that there is much mixing of the breeding stocks, and Wiig and Lie (1984) found no significant differences between the skulls of the three groups.

Cystophora evidently does not feed while on the ice for breeding and molting. At other times it may move well out to sea to forage and is capable of diving to great depths. A one-month-old individual descended to 75 meters on its first recorded dive, and adults probably can reach depths greater than 180 meters. The diet consists mainly of cephalopods, shrimp, mussels, and fish such as halibut, cod, redfish, capelin, herring, and flounder.

Cystophora is solitary for most of the year, but during the breeding season females come together in favorable habitat to give birth, with individuals spaced about 50 meters apart on the ice. Based on aerial observations of sexual distribution at this time, Boness, Bowen, and Oftedal (1988) suggested a polygynous mating system, with one adult male usually associated with a cluster of two to five females. Prior to such association, however, up to seven males at a time compete fiercely for access to each female or cluster. Eventually one large old male will chase the others away, either to the periphery of the ice floe or into the water. He then stays near the female while she nurses her newborn and is highly aggressive toward intruders, both people and other seals. Probably, however, he merely is defending his right to mate with the female when she enters estrus and is not primarily interested in protecting her or her newborn. After mating he may depart to look for other available females. During this period the males frequently inflate their hood and nasal balloon, but known vocalizations are limited to a few low-frequency pulses.

Mating takes place in the water after lactation ceases. Nearly all adult females become pregnant each year. Implantation of the blastocyst is delayed about 4 months, and the total gestation period is almost a full year. The young are born mainly from mid- to late March, occasionally in early April, and usually in the middle of a large ice floe. Births occur at about the same time in all three breeding areas. The newborn averages about 100 cm in length and 20 kg in weight and can swim and move about almost immediately. Its beautiful furry coat is silvery blue-gray above, darker on the face, and creamy white ventrally. The period of lactation usually has been reported as 5–12 days, but Bowen, Oftedal, and Boness (1985) found that pups were weaned after an average of only 4 days. This period, during which the pup grows to around 47 kg, is the shortest known for any mammal. Its significance may be related to a need to take rapid advantage of the interval between the time of heavy ice, when polar bears might have easy access to the young, and the final breakup of the ice. Bowen, Boness, and Oftedal (1987) found that females lose approximately 16 percent of their own weight during this period. They then abandon their young and return to the sea to forage. The pups remain on the ice until mid- to late April, not feeding, and lose about 13 kg. They then enter the water and are capable of fending for themselves. Most females reach sexual maturity at around 3 years. Males are sexually mature at 4–6 years but probably cannot compete successfully for mating until they are 13 years old. Maximum known longevity is 35 years.

Native people of the Greenland coast long have taken the hooded seal for its meat and skin. Commercial exploitation of the breeding aggregation around Jan Mayen Island started in the eighteenth century. Regular annual hunts on the ice off Newfoundland began at least 150 years ago and were carried out in conjunction with the killing of the more abundant harp seal. Harvests of the two species were not differentiated in the statistics until 1895, but the total take of harp and hooded seals was around 500,000 animals annually from 1820 to 1860 and subsequently declined. Before 1930 *Cystophora* was sought mainly to obtain oil and leather, and so both adults and pups were taken. Since then the beautiful pelt of the newborn has been the primary objective.

According to Wiig and Lie (1984), the annual kill of *Cystophora* averaged about 40,000 east of Greenland and 20,000 off Newfoundland around the turn of the century and 24,000 at Jan Mayen and 12,000 off Newfoundland in the 1970s. Since 1971 there has been increasing regulation of the kill through international agreements and quotas. There has been concern about the hunt both for humanitarian reasons and because of its possible adverse effects on overall populations. Numbers apparently declined during the first part of the twentieth century, partially recovered when hunting ceased during World War II, and then declined again in response to heavy kills in the 1950s and 1960s. J. E. King (1983) cited the following numerical estimates for total population size and the number of pups born annually, respectively: Newfoundland, 100,000 and 30,000; Davis Strait, 40,000 and 11,300; Jan Mayen, 250,000 and 50,000. Studies by Bowen, Myers, and Hay (1987) indicate larger population sizes, however, based on an estimated annual pup production of 62,400 for Newfoundland and 19,000 for Davis Strait. They suggested that numbers may have increased recently and that the commercial kill dropped sharply because of poor markets for pelts. Hammill, Stenson, and Myers (1992) reported a likely recent increase in the Gulf of St. Lawrence and that pup production there in 1991 was at least 2,000.

Campbell (1987) provided total population estimates of about 300,000 for Newfoundland and 90,000 for Davis Strait. He thought that these populations were at or above historical levels and that they would not be adversely affected by properly managed hunting. However, the market for seal pelts was largely eliminated in the late 1980s through an import ban by the European Community (the United States already had banned such trade through the Marine Mammal Protection Act of 1972). A kill of about 5,000 hooded seals annually, supposedly for subsistence purposes, continues off eastern Greenland. Reijnders et al. (1993) compiled a worldwide estimate of 500,000–600,000 for *Cystophora* and did not consider current harvest levels to be a threat but cautioned that there is potential for overexploitation and loss of prey to commercial fisheries.

PINNIPEDIA; PHOCIDAE; **Genus HALICHOERUS**
Nilsson, 1820

Gray Seal

The single species, *H. grypus*, occurs in temperate and subarctic waters of three regions of the North Atlantic: (1) along the coasts of Labrador, Newfoundland, Nova Scotia, and New England and in the Gulf of St. Lawrence and on associated islands as far south as Nantucket; (2) around Iceland, the British Isles and many associated islands, and Brittany and along the coasts of Norway and the Kola Peninsula; and (3) in the Baltic Sea, the Gulf of Bothnia, and the Gulf of Finland. Individuals occasionally reach New Jersey, Svalbard, the continental coast of the North Sea, and Portugal. Recent mitochondrial DNA analyses suggest that *Halichoerus* may be congeneric with *Phoca* (see account thereof). Except as noted, the information for this account was taken from Bonner (1981*a*) and J. E. King (1983).

Halichoerus shows more sexual dimorphism than any other phocid except *Mirounga*. In the eastern Atlantic adult males, which may be up to three times the size of females, are 195–230 cm long and weigh 170–310 kg; adult females are 165–95 cm long and weigh 105–86 kg. Canadian seals are larger, with males sometimes weighing more than 400 kg and females up to 256 kg at the beginning of the breeding season (Reijnders et al. 1993). Coloration varies widely, and all shades of gray, brown, black, and silver may be found. Both sexes are darker dorsally and lighter ventrally with varying degrees of spotting, but in males the darker tone forms a continuous background upon which there may be irregular spotting, while in females the lighter tone forms the background, which is marked with spots and patches of a darker color. Some males are almost completely black, and some females are predominantly creamy white with only a few scattered dark markings on the back.

Adult males have massive shoulders, and the skin there

Gray seals *(Halichoerus grypus)*, male and female, photo by Lothar Schlawe.

and on the chest is heavily scarred and thrown into heavy folds and wrinkles. They also have an elongated snout with a convex outline above the broad muzzle, which results in an equine appearance or "Roman nose." Adult females have a straight profile to the dorsal surface of the head. Miller and Boness (1979) suggested that the prominent snout of the males is used as a visual signal to communicate status; it is displayed prominently during short-range agonistic encounters in the breeding season. The massive skull of *Halichoerus* shows the elevated frontonasal area and large narial openings also found in *Cystophora*, but the cheek teeth are larger and stronger.

The gray seal generally breeds on rocky or cliff-backed coasts or on small offshore islands and is most often observed foraging in waters adjacent to such areas. It is also seen frequently in estuaries, particularly when salmon are present. Certain breeding colonies are present on open sandy beaches, most notably on Sable Island, a sand bar about 40 km long and 1.5 km wide east of Nova Scotia. Some breeding also occurs in sea caves. Bonner (1972) noted that North Rona Island, off northern Scotland, offers no beaches sheltered from oceanic swells, and so *Halichoerus* moved above the rocky shore to vegetated slopes to breed. In the Baltic Sea and the Gulf of St. Lawrence breeding usually takes place on drifting or shorefast ice.

There are no regular migrations, but young animals disperse widely. A pup tagged on Sable Island was found 25 days later on the coast of New Jersey, 1,280 km away. Pups from the Farne Islands sometimes appear in Norway, Denmark, and the Netherlands. Such individuals probably return to their natal beaches during the breeding season. Adults may forage in coastal waters after breeding but then come ashore again to molt, generally not at the same sites used for breeding. Molting occurs from around January to March in the eastern Atlantic, in April and May on the Baltic ice, and in May in the western Atlantic. The seals eat little or nothing during molting and generally fast for several weeks while ashore to breed. When they do seek prey, they forage at various depths, down to at least 70 meters. The diet includes octopus, squid, and crustaceans but consists largely of whatever fish are most abundant in coastal waters. Salmon, cod, herring, skates, mackeral, and flounder are all of importance.

Aggregations of *Halichoerus* form ashore during the molting and breeding seasons. About a month before the young are born in the British Isles large numbers of pregnant females and males of all ages assemble on or near the breeding beaches. The largest bulls tend to keep apart from the others. The birth of the first pups seems to signal the start of male territorial behavior. They compete for locations, generally inland, where there is access to the most females. Serious fighting is minimal, but through challenges and ritualized displays subordinate individuals are forced to the periphery of the breeding grounds or back into the water. If females are few in number and widely spaced, each male may defend only a single female. On crowded beaches, however, dominant bulls may be only 3–4 meters apart, and each may attempt to control 6–7 females. The area dominated varies from day to day; there is no fixed territory. The males remain ashore at least 18 days, often much longer, and attempt to approach all available estrous females. Mating occurs on land or in the water. A female usually mates with several males, and several times with each.

The timing of the breeding season varies widely between the three major populations and between individuals within the colonies. In the British Isles the young are born generally from September to December but occasionally from March to May. The birth peak is in September in Wales, in October in the Orkney Islands, in November in the Farne Islands, and from late December to early January in the Scroby Islands. In Iceland and Norway the peak also occurs in October, but in the Baltic breeding takes place on the ice in late February and early March. In the western Atlantic most births are from mid-January to mid-February.

Soon after coming ashore females give birth to the young conceived during the previous breeding season. They enter estrus and mate within 3 weeks, at the end of the nursing period. The fertilized egg develops to the blastocyst stage after 8–10 days, implantation is then delayed 100 days, and active gestation lasts 240 days; the total period of pregnancy is thus about 11.5 months. There normally is a single pup, though Spotte (1982) recorded a

Gray seals *(Halichoerus grypus)*, mother suckling young, photo from North American Newspaper Alliance, Inc. Inset: young adult *(H. grypus)*, photo by Erna Mohr.

number of twin births. The newborn averages 76 cm in length and 14 kg in weight and is covered with creamy white and rather silky fur. Lactation stops after 16–21 days, by which time the pup weighs around 50 kg and has acquired adultlike pelage. It is capable of swimming at birth but usually does not enter the water until its first molt is complete. The female may remain constantly with the pup until weaning or may go off to forage and then reestablish contact upon her return through vocalization and olfaction. She loses about a quarter of her own weight during lactation, then mates and quickly abandons her young. The latter stays ashore for another 2–4 weeks but then goes to sea and can provide for itself. Females usually have their first young at 4–5 years but continue to grow until they are nearly 15 years old. Males reach sexual maturity at about 6 years but do not usually appear on the breeding grounds until 8 years and continue to grow until age 11. Most breeding bulls are 12–18 years old. According to McLaren (1984), maximum known longevity is 46 years.

Halichoerus is of little commercial value but is persecuted extensively because of its alleged competition with valuable fisheries. There have been bounties and government control operations in Canada, but the population there, especially the colony on Sable Island, is increasing. According to Bonner (1982*a*), gray seal numbers had been reduced greatly in the British Isles by 1900. Subsequently it received protection during part of the breeding season, and by the 1920s it had increased substantially in numbers and reportedly was causing problems for fishermen. Controversy has continued to the present, pitting economic interests, especially in Scotland, against seal preservationists. The closed season now has been extended generally to cover the entire breeding period, but provision has been made for some culling. Numbers more than doubled between the 1960s and the 1980s. There now are about 87,000 animals in the British Isles, most of them in island groups off northern Scotland; there are another 20,000 throughout the remainder of the eastern Atlantic and 80,000–110,000 in Canada (Reijnders et al. 1993). The only U.S. breeding colony, that at Nantucket and Muskeget islands, Massachusetts, consists of about 20 seals. The Baltic subspecies, *H. g. macrorhynchus*, has been reduced drastically through bounties, water pollution, erosion of rookeries, and drowning of young in the wakes of ships. Numbers have fallen from over 100,000 a century ago to fewer than 3,000 (Reijnders et al. 1993; Wachtel 1986). The IUCN now classifies the Baltic population as endangered.

PINNIPEDIA; PHOCIDAE; **Genus PHOCA**
Linnaeus, 1758

Harbor, Ringed, Harp, and Ribbon Seals

There are four subgenera and seven species (Bigg 1981; Burns and Fay 1970; G. S. Jones 1984; J. E. King 1983; Prestrud 1990; Rice 1977; Shaughnessy and Fay 1977; Zhou 1986):

subgenus *Pagophilus* Gray, 1844

P. groenlandica (harp seal), primarily in coastal waters and pack ice from Hudson Bay and the Gulf of St. Lawrence across the Arctic to northwestern Siberia, occasional wanderers as far as the Mackenzie River Delta, Virginia, Iceland, the British Isles, and the Elbe River;

subgenus *Pusa* Scopoli, 1777

P. hispida (ringed seal), primarily in Arctic Ocean and adjoining seas, Sea of Japan, Sea of Okhotsk, Bering Sea,

Hudson and James bays, Baltic Sea, Gulf of Bothnia, and several inland lakes in southern Finland and adjacent parts of Russia, occasional wanderers as far as northeastern China, the Pribilof Islands, Newfoundland, the British Isles, France, Germany, and Portugal;

P. sibirica (Baikal seal), Lake Baikal, a body of fresh water in south-central Siberia;

P. caspica (Caspian seal), Caspian Sea;

subgenus *Phoca* Linnaeus, 1758

P. largha (spotted seal, or larga seal), regularly in coastal waters and ice floes from northeastern Siberia and northern Yukon to northeastern China,occasional wanderers have reached southeastern China and possibly Taiwan;

P. vitulina (harbor seal, or common seal), coastal waters from Hokkaido around the North Pacific to Baja California, from northwestern Greenland and Hudson Bay to South Carolina, and from Iceland and Svalbard to the Baltic Sea and Portugal, and in several inland lakes near Hudson Bay;

subgenus *Histriophoca* Gill, 1873

P. fasciata (ribbon seal), mainly in pack ice from Hokkaido and the Sea of Okhotsk to northern Alaska and the Aleutian Islands.

Seals of the genus *Phoca* are the most abundant pinnipeds of the Northern Hemisphere. They are mostly small and show little sexual dimorphism in size but vary greatly in coloration. From *Halichoerus* they are distinguished by a much smaller snout, smaller nasal openings, and more complex cheek teeth. From *Erignathus* they are distinguished by a relatively longer and narrower jugal, the first and second digits of the foreflipper longer (rather than shorter) than the third, and only two mammae in the female (E. R. Hall 1981). Data compiled by Burns and Fay (1970) suggest that *Phoca* may still be in the process of differentiation and that the various subgenera—*Pagophilus, Pusa, Phoca,* and *Histriophoca*—may eventually reach generic level but that there is no consistent cranial difference between them. *P. groenlandica* appears to be the most primitive species in the group, and *P. fasciata,* the most divergent from the ancestral stock. This arrangement has been commonly followed (Corbet and Hill 1991; Reijnders et al. 1993; Wozencraft *in* Wilson and Reeder 1993), but some authorities (including Jones et al. 1992) have treated the above subgenera as full genera (see also Wyss 1988*b*). Recent analyses of mitochondrial DNA (Mouchaty, Cook, and Shields 1995; Perry et al. 1995) indicate that *Pagophilus* (probably including *Histriophoca*) is a separate genus and is more closely related to *Cystophora* than to *Phoca* and that *Phoca* (in the above subgeneric sense) and *Pusa* are congeneric with *Halichoerus.* Additional information is provided separately for each species.

Phoca groenlandica (harp seal)

Except as noted, the information for the account of this species was taken from J. E. King (1983) and Ronald and Healey (1981). Adult males are 171–90 cm long and average 135 kg in weight; adult females are 168–83 cm long and average 120 kg. Coloration changes a number of times with age and also varies considerably in adults. Generally, adult males are silvery gray, with the head black to just behind the eyes and a black band running up each flank, the bands joining across the front of the back to form the shape of a horseshoe or harp. In adult females the facial and dorsal markings are paler and may be broken up into spots.

This species is divided into three major populations, one breeding in the Gulf of St. Lawrence and on the ice off eastern Newfoundland, one breeding to the north of Jan Mayen Island in the Greenland Sea, and one breeding in the White Sea. All three undergo an annual north-south migration of 6,000–8,000 km. The northwest Atlantic population spends the summer feeding in Hudson Bay and the waters off Baffin Island, northwestern Greenland, and northern Labrador. It starts south in September and by January has entered the breeding areas. After breeding, the adults molt in early May and then head north again. The Jan Mayen population summers between Svalbard and Greenland, moves south to the breeding area in late winter, molts on the ice north of Jan Mayen in April, and returns north by mid-May. The White Sea group summers mainly in the Kara and Barents seas close to the ice edge, moves south in the autumn, and rapidly enters the White Sea in January and early February. The adults molt from mid-April to late May and then migrate north out of the White Sea. The pups of this population are carried by drifting ice into the Barents Sea, where they begin to feed independently. The young of all populations generally do not move south until after the breeding season, and they join the adults to molt in the spring. Some young remain in the north throughout the year.

The harp seal is dependent on ice for breeding and molting and also may be found near the pack ice while foraging. It prefers rough, hummocky ice at least 0.25 meter thick and penetrates deep into large floes by following leads or channels. It maintains natural holes 60–90 cm in diameter for purposes of access to the water and breathing. It is capable of moving quickly on the ice and also is a powerful, high-speed swimmer. While migrating it sometimes leaps from the water like a dolphin. Its vision and hearing are thought to be acute, especially underwater, but not its sense of smell. Dives may reach a depth of 150–200 meters. The diet consists mainly of pelagic fish, especially capelin, and also pelagic and benthic crustaceans.

This species is gregarious, and a number of individuals may join in pursuing schooling fish. Tens of thousands come together in the spring molting aggregations, and as many as 40 seals may share a breathing hole. The varied vocalizations include trills, clicks, and numerous birdlike sounds and are heard largely during the breeding season. At that time the females form groups on the ice, spacing themselves about 1.5–2.5 meters apart. The males fight for access to the females, using teeth and flippers. Seasonal monogamy is thought to be the practice, and mating occurs on the ice.

A few days after hauling out on the ice, the females give birth to the young conceived the year before. They enter estrus and mate just before weaning the young. Implantation of the blastocyst is delayed about 4 months, and active gestation lasts another 7.5 months. Births occur from late January to early April, mostly from late February to mid-March. At birth the single pup is about 100 cm long and weighs about 12 kg; within 2–3 days it develops a beautiful coat of silky white fur. It is weaned after 10–12 days and then is abandoned by the mother, who has fed little, if at all, during lactation. The pup's weight increases to about 33 kg at weaning but drops to 27 kg over the next 2 weeks as it remains on the ice and molts into a grayish pelage. It then enters the water and begins to forage, initially staying near the surface and taking small crustaceans. Both sexes become fertile at around 5.5 years, but females may not begin to produce young on an annual basis for several more years, and males are not able to actively compete for mat-

Harp seal *(Phoca groenlandica):* A. Adult female; B. Two-year-old young. Photos by A. Pedersen.

ing privileges until about 8 years. Both sexes remain reproductively active through their twenties and may live until the early thirties.

Each of the three populations is thought originally to have numbered about 3 million individuals; thus *P. groenlandica* would have been the second most abundant pinniped after *Lobodon carcinophagus.* Unlike the latter, however, the harp seal forms large seasonal concentrations in accessible areas and thus can be practically and intensively exploited by humans. It long has been taken by native peoples of the North as a source of food and fiber, and about 10,000 individuals per year still are killed for this purpose (Ronald, Selley, and Healey 1982). Commercial hunting was begun in the sixteenth century by Basque whalers off Newfoundland. The first recorded expedition specifically seeking the harp seal was to Jan Mayen in 1720. There have been large-scale hunts on both sides of the Atlantic in most years since the early nineteenth century (Bonner 1982*a;* Busch 1985; Sergeant 1976). At first, seals of all ages were sought for skins and oil, but in recent years the main target has been the valuable white pelt of pups aged 2–10 days. Perhaps 50 million harp seals have been taken commercially in the nineteenth and twentieth centuries, far more than any other pinniped. The cost in human life also has been high, with about 1,000 men and 400 sailing vessels lost in the ice from 1800 to 1865, prior to the regular use of stronger, steam-powered craft to reach the seal concentrations. In 1857 nearly 400 vessels and 13,000 men participated in the hunt in the western Atlantic. The industry has been dominated by Norway since the 1880s, though many Canadians have been hired to do the hunting and processing. Russia began pelagic sealing with icebreakers in the White Sea in the 1920s and has been responsible for nearly the entire take there since the 1940s.

The Jan Mayen population was greatly reduced by hunting in the nineteenth century, and the kill there has been well under 50,000 in most years since 1890. As a result, the White Sea population came under increasing pressure, with the annual harvest there peaking at over 450,000 seals in 1924. The maximum recorded kill in the western Atlantic was around 687,000 in 1831. The annual take in each region was about 100,000–200,000 in the 1930s, but populations evidently were declining then. A cessation of hunting during World War II allowed some recovery, perhaps to near original numbers in the western Atlantic. After the war, sealing resumed, with the western kill alone reaching 456,000 in 1951 and averaging 316,000 per year from then until 1960. Subsequent evidence of declining populations, along with public protests in Europe and America on both conservation and humanitarian grounds, led to governmental restrictions on harp sealing. A regulatory and quota system was established in 1971 through an international agreement between Canada, Norway, and Russia. The annual kill quota for the western Atlantic has fallen steadily and was set between 170,000 and 186,000, including 100,000 pups, from 1977 to 1983. In response to public concern, the European Community in October 1983 placed a two-year ban on imports of products of harp and hooded seals (U.S. National Marine Fisheries Service 1984). This ban subsequently was extended. The United States, through the Marine Mammal Protection Act of 1972, already had prohibited importation. Therefore, commercial marketing and killing of the harp seal has been curtailed. An annual quota of 186,000 remains in effect for the western Atlantic, but the actual harvest in recent years has averaged about 60,000. About 40,000–80,000 are still being taken each year in the White Sea and about 7,000 at Jan Mayen (Reijnders et al. 1993).

Recent numerical estimates place the Jan Mayen population at 100,000–400,000 individuals, the White Sea population at around 500,000, and the western Atlantic herds at 2–4 million (Reijnders et al. 1993). There is controversy regarding the future outlook. On the one hand, Roff and Bowen (1986) reported that the annual pup production in the western Atlantic, at around 480,000, is far in excess of what would be necessary to sustain the population even if commercial kill quotas were being met and thus a rapid increase is likely. On the other hand, Bonner (1982*a*) pointed out that recent intensification of the capelin fishery had depleted that important food source of the harp seal, with possibly ominous consequences. Reijnders et al. (1993) also considered reduced food availability to be a major threat and expressed concern about pollution and accidental killing of seals by fishing gear.

Phoca hispida (ringed seal)

Except as noted, the information for the account of this species was taken from Frost and Lowry (1981) and J. E. King (1983). Adults in Alaskan waters average 115 cm in length and 49 kg in weight, with reported maximums of 168 cm and 113 kg. Seals of the Baltic population tend to be larger than those elsewhere, with males more than 10 years old averaging 141 cm in length and weighing 71–128 kg, and females 137 cm and 61–142 kg (Helle 1980). Coloration is variable, but the upper parts are commonly gray, heavily spotted with black. The spots may merge to form large dark areas, but many of them are surrounded by a ring of lighter color.

The ringed seal is found in seasonally or permanently ice-covered waters and makes annual movements in relation to the advance and retreat of the ice. Highest densities of breeding adults are on and under stable landfast ice. Nonbreeders occur mainly in the flow zone and in association with moving pack ice. In the summer, all age and sex classes concentrate at the edge of the permanent arctic pack ice or on inshore ice remnants. Individuals, however, may be found wherever there is open water in the ice, even as far as the North Pole. In the Sea of Okhotsk spring breeding occurs on the pack ice, but the seals must haul out on shore during the summer. Molting occurs from March to July, with a peak in June. Foraging in the water then slows but does not cease altogether. During the early spring, late summer, and autumn the seals spend most of their time swimming among ice floes and feeding.

As the ice thickens during the late autumn and winter the seals maintain openings for breathing and access by abrading the ice with the claws of their foreflippers. Breathing holes, which can be kept open through ice more than 2 meters thick, are cone-shaped, with the small end up. Rather than constructing their own openings, the seals maintain and expand natural ones, which tend to occur along cracks. Freezing pressure may then force the ice up into ridges and heaps in the vicinity of the holes, and snow may accumulate around the uplifts and over the holes. The seals take advantage of the situation by coming up through the holes and hollowing out lairs in the snow. There may be several such dens in a single snowdrift. Some are used simply for resting by a single individual and usually have one oval chamber about 30 cm high and 120 cm in diameter. Others are larger, have several chambers, and are used by a number of seals (Smith and Stirling 1975). Pregnant females construct more elaborate subnivean lairs where they can give birth and nurse their young. The pup then sometimes digs additional tunnels into the surrounding snow. Since *P. hispida* has a longer lactation period than most phocines, the snow-covered lair may provide necessary insulation while the pup builds up fat reserves. The lair also serves as a resting site for mother and young and may provide some protection against predators. Smith, Hammill, and Taugbol (1991) reported that bears and foxes frequently break into the lairs and force the occupants to swim out into the frigid waters but that the seals may then move to an alternate lair, where the pup can regain thermoneutrality. On Lake Saimaa, Finland, birth lairs are constructed right along the shoreline and usually are partly over ground and immediately above flat, rather than ridged, ice (Helle, Hyvärinen, and Sipilä 1984; Sipilä 1990). In the Sea of Okhotsk the young are born on shorefast ice but not in lairs.

P. hispida will not use a breathing hole that has been at all disturbed and is very cautious when coming up onto open ice. It first uses its apparently excellent senses of sight,

Ringed seal *(Phoca hispida)*, photo by Lothar Schlawe.

hearing, and smell to inspect the vicinity of the access hole. It evidently can detect a human at least 200 meters away by scent alone (Smith and Hammill 1981). It may spend months in darkness, within its lair or under ice-covered waters, and even in summer may be unable to use the sense of sight during deep dives. Echolocation may be used to some extent for orientation, but the seal's extremely well-developed vibrissae, with innervation more than 10 times greater than normally found in mammals, are probably the most important means for underwater environmental assessment (Hyvärinen 1989). While feeding in the water *P. hispida* generally alternates periods of 12–130 seconds at the surface with dives of 18–720 seconds. However, it has been known to remain underwater for as long as 43 minutes under natural conditions, and a forced dive in captivity lasted 68 minutes. It may reach depths as great as 91 meters while searching for food. The bulk of the diet is made up of fishes of the cod family (usually less than 20 cm long but occasionally up to 127 cm), pelagic amphipods, euphausiids (krill), shrimp, and other crustaceans.

P. hispida is abundant but not gregarious and is widely dispersed throughout the Arctic. Recorded population densities average around 1/sq km in favorable areas but are as high as 15/sq km in June on the fast ice off Baffin Island. During the nonbreeding season, seals are usually seen alone, occasionally in small groups, though herds of thousands once were reported in the Sea of Okhotsk. Several animals sometimes share a breathing or access hole and the associated ice surface or lair. A dominance hierarchy probably is established in such aggregations. According to Smith and Hammill (1981), the seals are then aggressive toward one another, slapping with their foreflippers, splashing water, lunging, and biting. Overall social structure appears comparable to that of *Leptonychotes*. Adult males probably exhibit limited polygamy, establishing large underwater territories that comprise the areas and lairs used by several females. Four underwater vocalizations have been identified: high- and low-pitched barks, a yelp, and a chirp; they probably are involved in reproductive activity and in maintaining social structure around access holes (Calvert and Stirling 1985; Stirling 1973).

Mating takes place in the water, mostly in late April and early May, about a month after parturition. Implantation of the blastocyst is delayed for about 3.5 months, and the total gestation period is 11 months. Most births occur from mid-March to mid-April, but the peak is in early May on Lakes Saimaa and Ladoga. Twins have been reported, though normally there is a single pup. It averages 65 cm in length and 4.5 kg in weight and is covered with woolly white fur considerably longer and finer than that of young ribbon and spotted seals. It begins to shed this coat at 2–3 weeks and attains an adultlike pelage by 6–8 weeks. Nursing lasts 5–7 weeks in most areas but only 3 weeks in the Sea of Okhotsk, where births usually occur on drifting ice. The pups are abandoned after weaning, when they weigh 9–12 kg and the ice begins to break up. Growth continues for 8–10 years. Sexual maturity is reached at about 5–7 years, but the maximum annual pregnancy rate of around 78 percent is not achieved until females are more than 10 years old. Maximum known longevity is 43 years. The major natural sources of mortality are predation on pups by the arctic fox, which evidently has little difficulty in locating and penetrating the subnivean lairs, and predation on older seals by the polar bear.

Many thousands of ringed seals are killed annually by native peoples for skins, furs, oil, and meat. Some commercial hunting takes place, but since *P. hispida* does not form large breeding aggregations, there have been no massive harvests and numerical declines comparable to those suffered by *P. groenlandica*. There are no accurate population estimates, but *P. hispida* is generally considered to be the most common northern pinniped, with perhaps 6–7 million individuals present in the Arctic Ocean and major adjoining bodies of water. The subspecies *P. hispida ladogensis*, in Lake Ladoga in Russia, now numbers about 5,000 individuals and is protected. However, the subspecies *P. h. saimensis*, in Lake Saimaa, Finland, has declined sharply because of the entanglement and drowning of young seals in fishing gear; only 160–80 individuals survive (Helle, Hyvärinen, and Sipilä 1984; Reijnders et al. 1993). Recent development of hydroelectric power plants has resulted in periodic lowering of water levels and consequent collapse of some of the lairs that *saimensis* constructs along the shoreline, just above the surface of the ice (Marine Mammal Commission 1994; Sipilä 1990). The IUCN classifies *P. h. ladogensis* as vulnerable and *P. h. saimensis* as endangered; the latter subspecies also is listed as endangered by the USDI.

The Baltic subspecies, *P. h. botnica*, formerly was common, with up to 20,000 taken annually by hunters in the early twentieth century. Excessive harvests and bounties subsequently led to greatly reduced numbers, and the total population now is only 5,000–8,000 (Reijnders et al. 1993). The decline is continuing, evidently in connection with a reduction in the annual pregnancy rate to only 28 percent in mature females, which in turn is probably the result of water pollution (Helle 1980). *P. h. botnica* is classified as vulnerable by the IUCN.

Phoca sibirica (Baikal seal)

Except as noted, the information for the account of this species was taken from J. E. King (1983) and Thomas et al. (1982). Adults are 110–42 cm long and weigh 50–130 kg. The color is dark brownish gray, shading to a lighter yellowish gray ventrally. The fur is dense and unspotted. The forelimbs and foreclaws are larger and stronger than those of *P. hispida* and *P. caspica*.

This seal is the only pinniped restricted to fresh water. It is confined almost entirely to Lake Baikal, though on occasion individuals are seen in connecting rivers. Seasonal movements and activities are determined primarily by ice conditions. During the winter the lake is completely covered by ice averaging 80–90 cm thick. At this time the seals may be found throughout the lake, mainly in the water and at breathing and access holes through the ice. They probably keep these holes open by abrading the ice with their strong foreclaws, but they may also use their heads, teeth, and rear flippers. Immature seals generally utilize one hole, whereas adults have a primary hole and several auxiliary openings. In the late winter or early spring pregnant females move onto the ice and make a subnivean lair for giving birth (see account of *P. hispida*). Around 1 April all seals begin to concentrate to feed along new natural openings in the ice, and in May they move toward the north end of the lake as the ice breaks up. The annual molt occurs from late May to early June. In the summer the seals concentrate to the southeast, using the shore and offshore rocks for hauling out. In the autumn they again move to areas where the ice is forming.

Observations suggest that surface resting peaks in the winter between 1300 and 1700 hours and during the summer at 1100 and 1800 hours and that most foraging year-round is done at twilight and at night. Based on the seal's anatomy, the maximum natural diving time has been estimated at 20–25 minutes, but frightened captives have remained underwater for 2–3 times as long. The diet consists

Baikal seal *(Phoca sibirica)*, photo by Eugene Maliniak.

largely of noncommercially valuable fish, such as gomyanka, which reportedly move to depths of 20–180 meters at night, and also includes some invertebrates.

Although *P. sibirica* is basically solitary, a number of individuals may share access holes, and large aggregations form in areas of favorable habitat during certain times of the year. The spring feeding assemblies of 200–500 animals are begun by juveniles, which are then joined in turn by the adult males, pups of the year, and adult females. Other large groups form on shore during the summer. During the breeding season adult males are apparently polygynous. It is not known whether they are territorial, and available evidence suggests that they do not physically fight one another.

Mating is thought to occur underwater in May, at about the time of weaning. There apparently is a short period of delayed implantation, and total gestation lasts just over 9 months. The young are born on the ice from mid-February through March. There commonly is a single pup, but the twinning rate, about 4 percent, is unusually high for pinnipeds. Moreover, both twins often survive to weaning and then remain together for a while. The newborn is about 70 cm long and 3 kg in weight and is covered with a long, white, woolly coat. This pelage persists for about 6 weeks and then is replaced during the next 2 weeks with an adult-like coat. Lactation is unusually long, about 2.0–2.5 months. Males reach sexual maturity by 7 years. Most females are breeding by age 6 and can continue to age 30, with about 88 percent producing pups each spring. Maximum reported longevity is 56 years.

The Baikal seal is thought to have recovered from excessive hunting that seriously reduced its numbers in the 1930s. The total population now is estimated at 60,000–70,000 individuals. There is a regulated annual commercial harvest of 5,000–6,000 for pelts, oil, and meat. However, the actual kill may be much higher since many seals are shot and not recovered, and there is concern that population viability is being disrupted (Reijnders et al. 1993). The IUCN classifies the species as near threatened.

Phoca caspica (Caspian seal)

Except as noted, the information for the account of this species was taken from J. E. King (1983). Adult males may reach a length of 150 cm and a weight of 86 kg. Females are slightly smaller, reaching about 140 cm. Both sexes are deep gray dorsally and grayish white ventrally. Males also generally have dark spots all over, whereas in females the spots are lighter and mainly on the back.

Annual movements are associated with the availability of ice in the Caspian Sea. In late autumn the seals migrate into the northeastern part of the sea, where the water is shallower and freezes. The ice forms large floes that then smash against one another in the wind, creating upthrusts and rough hills. Here the seals maintain small breathing holes and larger exit holes; females give birth to their young on sheltered stretches of ice. Molting occurs in the spring, also in the north, but the ice then melts, the water becomes warm, and most of the seals migrate south to spend the summer in the deeper, cooler part of the sea. This shift also is associated with the abundance of prey, which includes mainly small fish and crustaceans.

P. caspica usually is found in pairs and may be monogamous. Mating occurs from late February to mid-March, about a month after the young conceived the previous year are born. The gestation period, which probably includes several months of delayed implantation, is about 11 months. Births take place from late January to early February. The newborn are about 64–79 cm long, weigh about 5 kg, and are covered with long, white fur that is replaced in about 3 weeks with short, dark gray hair. Lactation lasts

Caspian seal *(Phoca caspica)*, photo by Lothar Schlawe.

about 1 month. Sexual maturity comes at 5–7 years. McLaren (1984) indicated that maximum known longevity is 50 years.

Total numbers are estimated at 500,000–600,000 animals and are considered reasonably stable. About 60,000 pups are taken each year for their skins. Nonetheless, the species is threatened by the overall degradation of the Caspian ecosystem and by loss of food to commercial fisheries (Reijnders et al. 1993). The IUCN now classifies the species as vulnerable.

Phoca largha (spotted seal, or larga seal)

Except as noted, the information for the account of this species was taken from Bigg (1981), J. E. King (1983), and McLaren (1984). Adult males are 150–70 cm long and weigh about 90 kg; females are 140–60 cm long and weigh about 80 kg. The coat is generally pale gray, darker dorsally, and sprinkled with small brown or black spots. Compared with that of *P. vitulina,* the skull of *P. largha* is small and fragile, and the teeth are set straight in the jaws.

This species is closely related to *P. hispida* but, unlike the latter, is strongly associated with ice. There is an annual migration in the autumn and winter to the edge of the pack ice, where the seals haul out on floes. Breeding occurs during late winter and spring in eight areas of ice distributed from the Bering Sea to the northwestern Yellow Sea. After the breeding season the seals remain on the ice to molt and then move back to the shoreline for the summer. They may then ascend rivers, possibly including the Yangtze. Adults are thought to be capable of diving to 300 meters and are known to take a wide variety of fish, cephalopods, and crustaceans. Newly weaned pups feed on small amphipods around ice floes.

Monogamy is evident during the breeding season. Pairs form about 10 days before the female gives birth to the young conceived the previous year and stay together for at least a month afterwards. Such family groups are separated by at least 0.25 km and may defend a territory around the natal ice floe. The reproductive cycle is thought to be similar to that of *P. vitulina.* Where the two species overlap in distribution, *P. largha* gives birth first, before the ice melts. The young are born as early as February in the southern parts of the range and from late March to mid-May in the Bering Sea. The pup is about 85 cm long and has long white fur that is replaced by adultlike pelage after about a month. It cannot swim at birth, but by the time of weaning, a month later, its weight has tripled and it can dive to a depth of 80 meters. Sexual maturity is reached by 4–5 years, and maximum known longevity is 32 years.

P. largha has been reported to do some damage to commercial fisheries in the Sea of Okhotsk but apparently has not been extensively persecuted. There is a regulated annual kill of about 13,000 individuals in Russian waters. Total numbers are estimated at about 130,000 in the Sea of Okhotsk and fewer than 100,000 in the Bering Sea (Reijnders et al. 1993).

Phoca vitulina (harbor seal, or common seal)

Except as noted, the information for the account of this species was taken from Bigg (1981) and J. E. King (1983). Adult males are about 150–200 cm long and weigh about 70–170 kg; females are about 120–50 cm long and weigh 50–150 kg. Coloration is variable, but generally there is a gray or brownish gray background, liberally sprinkled with small spots. In some populations the spots fuse, so that the entire dorsal region is dark, and in some cases there are pale rings around the spots. Compared with that of *P. groenlandica* and *P. hispida,* the skull of *P. vitulina* is more thickly boned and has a broader facial area.

The harbor seal is nonmigratory and makes only limit-

Harbor seals, or common seals *(Phoca vitulina):* A. Adult male, photo by Erna Mohr; B. Two females, photo from Miami Seaquarium; C. Immature seals, photo by A. Pedersen; D. Head, showing nostrils closed, photo by Erwin Kulzer.

ed movements in association with foraging and breeding, though young have been known to disperse up to 300 km. This species, unlike most northern phocids, is not associated with ice. It lives mainly along shorelines and in estuaries, commonly resting on sandbanks, easily accessible beaches, reefs, and protected tidal rocks. It frequently ascends rivers for many kilometers and may remain year-round in freshwater lakes. It occasionally visits Loch Ness and other lakes in Scotland (Williamson 1988). Northerly populations tend to stay in areas where swift currents and tides maintain ice-free conditions in winter. Breeding occurs as far north as Svalbard (Gjertz and Borset 1992). *P. vitulina* is a strong swimmer and leaps completely out of the water (porpoising). Captive individuals have made dives as deep as 206 meters and lasting as long as 30 minutes, but the normal foraging time underwater is only 3 minutes. Newly weaned young feed mainly on shrimp and other small, bottom-dwelling crustaceans. Older animals take herring, anchovies, trout, smelt, cod, flounder, salmon, other fish, and octopus.

Although this species usually is solitary, aggregations of several hundred may form ashore during the breeding season, and a number of individuals may also come together at a favorable hauling-out site. There then is much agonistic behavior involving biting, head butting, flipper waving, snorting, and growling. Sullivan (1982) reported that adults and juveniles form strong linear hierarchies based on size and sex, the adult males dominating all other animals. In a study of the largest breeding colony in eastern Canada, containing about 2,000 individuals, Godsell (1988) found no evidence of social cohesion other than that between mother and pup. Although *P. vitulina* generally is considered one of the least vocal of pinnipeds, Ralls, Fiorelli, and Gish (1985) found that an adult male made a variety of sounds and learned to mimic a number of human words and phrases.

Males are thought to be polygynous and may fight for mating privileges. Females enter estrus and mate shortly after weaning the young conceived during the previous breeding season. The fertilized egg develops to the blastocyst stage, and implantation then is delayed for 1.5–3.0 months. The total period of pregnancy is 10.5–11.0 months (Reijnders et al. 1993). The peak of births varies widely by area, being in February in Baja California, March–April in California, May–June in Newfoundland and Nova Scotia, and June–July in Europe, the North Pacific, and the arctic North Atlantic. The single pup is 75–100 cm long, weighs 10–12 kg, usually has adultlike pelage, and is able to swim immediately. The mother is solicitous and will even take the pup in her mouth and dive if there is danger, but she promptly abandons it after weaning. Lactation lasts 3–6 weeks. Sexual maturity is attained at 3–5 years in females and about 6 years in males. About 90 percent of adult females bear a pup each year. Ohtaishi and Yoneda (1981) reported a 34-year-old individual.

P. vitulina is of little commercial value, though it is sometimes killed for its fur. It does come into conflict with commercial fisheries, especially in northwestern North America, the Sea of Okhotsk, and the Netherlands, and therefore has been persecuted. The U.S. National Marine Fisheries Service (1987) estimated numbers at 48,000–51,500 in Europe, 10,000–15,000 on the coast of eastern Asia, 260,000 in Alaska, 42,000 on the rest of the west coast of North America, and 30,000–45,000 in the western Atlantic. As a result of excessive hunting, pollution, disturbance, and other problems, populations have declined in much of the range, and all five recognized subspecies have been affected.

Data compiled by Reijnders et al. (1993) suggest that numbers in Asian waters have fallen to fewer than 4,000. The subspecies there, *P. v. stejnegeri*, had been classified as vulnerable by the IUCN but in 1996 was removed from that category. Recent information from the Marine Mammal Commission (1994) and the U.S. National Marine Fisheries Service (1994) indicates that there has been a major decline of the subspecies in Alaskan waters, *P. v. richardsi*, with fewer than 50,000 seals surviving, possibly because of loss of food to intensifying commercial fisheries, entanglement in debris and fishing gear, and emigration. Surveys by Teilmann and Dietz (1994) found that the subspecies of Greenland, *P. v. concolor*, also has suffered a substantial decline. Heide-Jorgensen and Härkönen (1988) reported that populations in the waters between Norway, Sweden, and Denmark were recovering from past human persecution. Shortly thereafter a severe epidemic, a form of distemper, wiped out more than half of the seals in that region and the southern North Sea—about 18,000 animals—and also spread to British waters (Harwood 1990; Heide-Jorgensen et al. 1992). Numbers subsequently increased again, but European populations remain threatened by pollution and other problems (Reijnders et al. 1993; Simmonds 1991). The subspecies of Europe, *P. v. vitulina*, is classified as endangered by Russia. *P. v. mellonae*, recently confirmed to be a distinct subspecies, is restricted to several hundred individuals in a few freshwater lakes on the Ungava Peninsula of Quebec, to the east of Hudson Bay; it is jeopardized by hydroelectric development and pollution (Reijnders et al. 1993; Smith and Lavigne 1994).

Phoca fasciata (ribbon seal)

Except as noted, the information for the account of this species was taken from Burns (1981*b*) and J. E. King (1983). Adults of both sexes are about 150 cm long and weigh about 90 kg. The largest individual on record, a pregnant female, was 180 cm long and weighed 148 kg. Adult males are reddish brown before the molt and dark chocolate brown afterwards, with white or yellowish white bands encircling the neck, the hind end of the body, and each fore-

Ribbon seal *(Phoca fasciata)*, photo from San Diego Zoological Society.

flipper. Females are paler with less distinct markings. Compared with other northern phocids, *P. fasciata* has a relatively longer and more flexible neck, a shorter rostrum, and more widely spaced teeth.

The ribbon seal is associated with sea ice during the late winter, spring, and early summer. At that time the species is concentrated in ice-covered parts of the Chukchi, Bering, and Okhotsk seas and adjacent straits and bays. It prefers to haul out on moderately thick, clean ice and depends on such for giving birth and molting in the spring. It seldom hauls out on land and generally is found on ice floes well out to sea. The Bering Sea population evidently utilizes a remnant zone of ice in the middle of the sea that does not melt until mid-June. This population, like that of the Sea of Okhotsk, then apparently becomes completely pelagic for the summer and autumn. Some individuals may wander a considerable distance, there being one record from central California and another from the central North Pacific south of the Aleutians (Stewart and Everett 1983). Possibly as an evolutionary consequence of its disassociation with the land, *P. fasciata* shows little caution when resting on ice and can easily be approached by humans or vessels. It can, however, move faster on ice than a person can sprint, using its foreflippers and body in a manner similar to that described for *Lobodon.* Its diet consists mainly of crustaceans, fish, and cephalopods.

The ribbon seal is usually solitary. Aggregations sometimes form on favorable ice floes, but individuals are spaced well apart. Males are apparently polygynous and do not stay long with any one female. Mating takes place at about the same time as weaning, and gestation, which probably includes a period of delayed implantation, lasts around 11 months. Births occur from 3 April to 10 May, mostly from 5 to 15 April. The single pup is about 86 cm long, weighs about 10.5 kg, and has a coat of long, white hair that is shed within 5 weeks. The mother is not especially solicitous, often goes off to forage, and promptly abandons her offspring after 3–4 weeks of lactation. The pup then is not a capable swimmer and remains mostly on the ice for a few more weeks, its weight falling from around 30 kg to 22 kg before it finally becomes proficient at foraging. Nearly all females reach sexual maturity and begin producing a pup annually by the time they are 4 years old. Males are sexually mature at 4–5 years. Potential longevity is probably about 30 years.

Because it is rarely found near land, *P. fasciata* is not regularly hunted by native peoples, but its lack of caution while on the ice makes it vulnerable to commercial pelagic sealers. In the 1960s the annual Soviet catch averaged about 13,000 animals, and overall populations began to fall. In 1969 the yearly quota was reduced to 3,000, and numbers rebounded. There now are estimated to be about 140,000 individuals in the Sea of Okhotsk and 100,000 in the Bering Sea.

Cetaceans

Whales, Dolphins, and Porpoises: Cetacea

This order of wholly aquatic mammals occurs in all the oceans and adjoining seas of the world, as well as in certain lakes and river systems. Living cetaceans are traditionally divided into two suborders: the Odontoceti, individuals of which have teeth (generally all of one kind) and an asymmetrical skull, and the Mysticeti, individuals of which have plates of baleen, instead of teeth, and a symmetrical skull. In accordance with the classification adopted by Klinowska and Cooke (1991) and Reeves and Leatherwood (1994), the Odontoceti comprise four superfamilies: the Platanistoidea, with the families Platanistidae, Lipotidae, Pontoporiidae, and Iniidae; the Delphinoidea, with Monodontidae, Phocoenidae, and Delphinidae; the Ziphioidea, with Ziphiidae; and the Physeteroidea, with Physeteridae. The Mysticeti comprise the families Eschrichtiidae, Neobalaenidae, Balaenidae, and Balaenopteridae. Rice (1977, 1984) considered the Odontoceti and the Mysticeti to be distinct orders because it is questionable whether the two groups had a common origin and because the differences between them are as great as those between some of the universally recognized orders of mammals. Certain authorities, such as Ellis (1980), have followed Rice's procedure. Gaskin (1976), however, summarized data supporting a monophyletic origin for cetaceans and treated the Odontoceti and the Mysticeti as suborders. The latter procedure also is supported by new cytological data (Arnason, Höglund, and Widegren 1984) and now has returned to general acceptance (Barnes, Domning, and Ray 1985; Corbet and Hill 1991; Fordyce and Barnes 1994; Heyning 1989*a;* Jones et al. 1992; Mead and Brownell *in* Wilson and Reeder 1993). Recent analysis of mitochondrial DNA suggests even that the above subordinal division is not warranted and that it may be more appropriate to divide the Cetacea only into three superfamiles: the Delphinoidea, with the families Platanistidae, Lipotidae, Pontoporiidae, Iniidae, Monodontidae, Phocoenidae, and Delphinidae; the Ziphioidea, with the family Ziphiidae; and the Physeteroidea, with Physeteridae, Eschrichtiidae, Neobalaenidae, Balaenidae, and Balaenopteridae (Milinkovitch, Meyer, and Powell 1994).

There are several problems regarding the division of the Cetacea at the family level, but perhaps the major one involves the freshwater or river dolphins of the genera *Inia, Lipotes, Pontoporia,* and *Platanista.* These genera resemble one another superficially as well as in some internal characters and often have been united in a single family, the Platanistidae. However, some authorities have long recognized that the resemblances may result from convergent evolution and that at least some of the genera belonged in different families. Rice (1977) placed all of these genera in the family Platanistidae but recognized three subfamilies: the Iniinae, with the genera *Inia* and *Lipotes;* the Pontoporiinae, with *Pontoporia;* and the Platanistinae, with *Platanista.* Kasuya (1973) gave familial rank to all three groups, and Zhou, Qian, and Li (1979) thought that *Lipotes* should also be in a monotypic family, the Lipotidae. Although the new general mammal lists by Corbet and Hill (1991) and Mead and Brownell (*in* Wilson and Reeder 1993) have continued to place all four genera in the Platanistidae, specialists in cetacean systematics now generally recognize at least three full families, the Iniidae, the Pontoporiidae, and the Platanistidae; the Lipotidae now are treated either as a separate family or as a subfamily of the Pontoporiidae (Barnes, Domning, and Ray 1985; Fordyce and Barnes 1994; Leatherwood and Reeves 1994; Pilleri and Gihr 1981*b;* Rice 1984; Zhou 1982). Pilleri and Gihr used the name Stenodelphidae in place of Pontoporiidae.

Considering the above, as well as the increasing recognition of the Phocoenidae as a family distinct from the Delphinidae and of the Neobalaenidae as distinct from the Balaenidae, 13 Recent cetacean families are accepted here, along with 41 genera and 78 species. The sequence of these taxa presented here generally follows that adopted by Klinowska and Cooke (1991) and Reeves and Leatherwood (1994). However, a cladistic analysis of morphological data by Heyning (1989*a*) suggests the following phylogenetic sequence for the odontocete families: Physeteridae, Ziphiidae, Platanistidae, Iniidae (including Lipotidae and Pontoporiidae), Monodontidae, Phocoenidae, and Delphinidae. In contrast, recent analyses of mitochondrial DNA suggest that the family Physeteridae is more closely related to the suborder Mysticeti than to the rest of the Odontoceti, that the Mysticeti diverged from the Physeteridae comparatively recently on the evolutionary scale (about 25 million years ago), that such divergence occurred after the origin of other odontocete families, and thus that a revision of cetacean subordinal classification may be required (Milinkovitch, Ortí, and Meyer 1993, 1995).

The length of the cetacean head and body is taken in a straight line from the tip of the snout to the notch between the tail flukes. Head and body length varies from about 1.2 to 31.0 meters, and weight ranges from 23 to 160,000 kg. The tail flukes are set in a horizontal plane, which immediately distinguishes cetaceans from fish, whose tail fins are in a vertical position. Other external features conspicuous in cetaceans are the torpedo-shaped body, front limbs that are modified into flippers (pectoral fins) and ensheathed in a covering, the absence of hind limbs, and the usual presence of a dorsal fin. There is no covering of fur, but all whales have some hairs in the embryonic stage, and in

The back of a humpback whale *(Megaptera novaeangliae)* that has surfaced to breathe. The two openings on top of the head are the "blowholes," which the whale closes before submerging. These are comparable to the nostrils of other mammals. Photo by Vincent Serventy.

adult mysticetes a few bristles persist around the mouth. Cetaceans lack sweat glands and sebaceous glands. They have a fibrous layer (blubber) filled with fat and oil just beneath the skin, which assists in heat regulation. There are no external ears or ear muscles and no scales or gills.

The nostrils open externally, usually at the highest point on the head. The odontocetes have a single blowhole, and the mysticetes have a double one. The asymmetry of the skull of the toothed whales is in correlation with the reduction of one of the nasal passages; both features seem to have evolved in connection with modifications needed for sound production (Yurick and Gaskin 1988). There is a direct connection between the blowhole and the lungs, so that a suckling calf cannot get milk into its lungs. The blowhole is closed by valves when the animals are submerged. Cetaceans do not blow liquid water out of the lungs. When the animals exhale, the visible spout is from the condensation of water vapor entering the air from the lungs, and possibly from the discharge of the mucous oil foam that fills the air sinuses.

The bones are spongy in texture, and the cavities are filled with oil. In some genera the vertebrae in the neck are fused, but all lack the complex articulation of vertebrae of land mammals and present a graded series from head to tail. There are no bony supports for the dorsal fin and the tail flukes. The pelvic girdle, represented by two small bones embedded in the body wall and free from the backbone, serves only as the place of attachment for the muscles of the external reproductive organs.

Propulsion is obtained by means of up-and-down movements of the tail such that the flukes present an inclined surface to the water at all times. The force generated at right angles to the surface of the flukes is resolvable into two components, one raising and lowering the body and the other driving the animal forward. The fins serve as balancing and steering organs. Cetaceans are the swiftest animals in the sea; some dolphins can maintain a sustained speed of 26–33 km/hr (Rice 1967).

Certain questions regarding the physiological adapta-

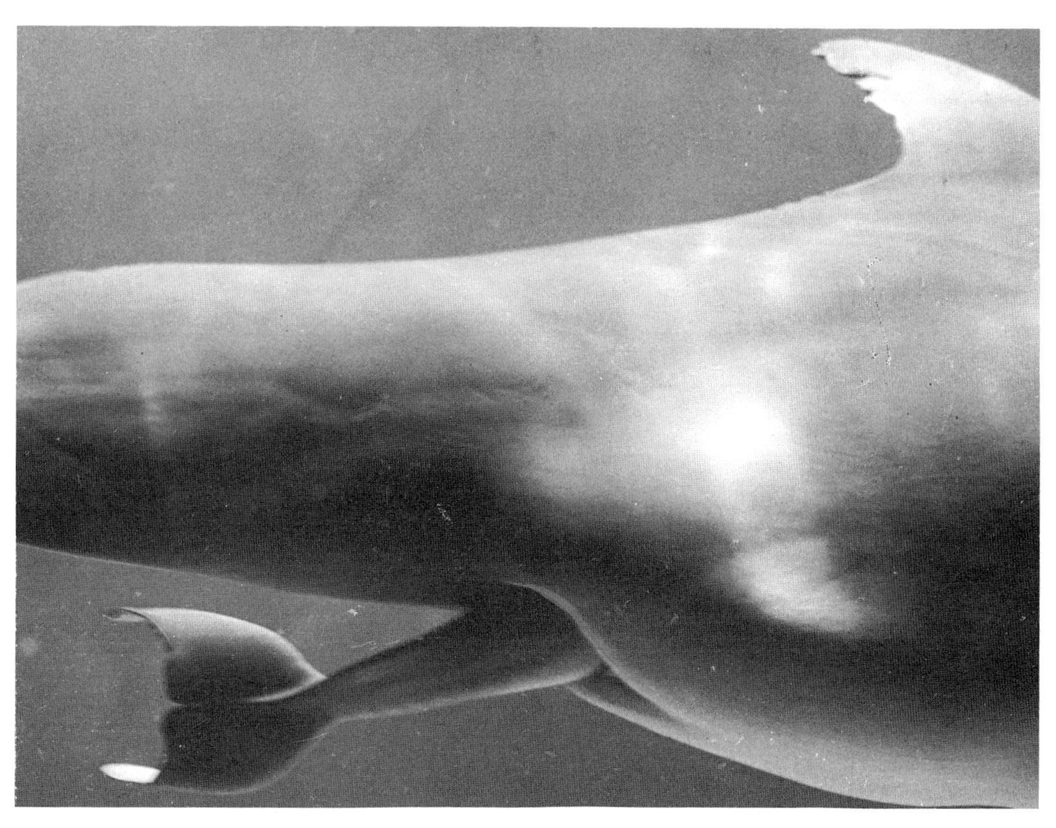

Bottle-nosed dolphin *(Tursiops truncatus)*, showing a baby being born tail first, photo from Miami Seaquarium.

tions of cetaceans in diving and temperature tolerance remain unanswered, but some of the general mechanisms are known. Before diving, a cetacean expels the air from its lungs. The following are some of the adaptations that make long dives possible: (1) the oxygen combined with the hemoglobin of the blood and with the myoglobin of the muscles accounts for 80–90 percent of the oxygen supply utilized during prolonged diving; (2) arterial networks seem to act as shunts, maintaining the normal blood supply to the brain but effecting a reduced supply to the muscles and an oxygen debt that the animal can repay when it surfaces; (3) a decreased heartbeat further economizes the available oxygen; and (4) the respiratory center in the brain is relatively insensitive to an accumulation of carbon dioxide in the blood and tissues. The hydrostatic pressures encountered at great depths are alleviated by not breathing air under pressure and by the permeation of the body tissues with noncompressible fluids. The only substances in the body of a cetacean that can be compressed appreciably by the pressure of great depths are the free gases, found mainly in the lungs. The collapse of these gases drives them into the more rigid, thick-walled parts of the respiratory system. The body temperature is regulated by the insulation of the blubber, which retains body heat when the animals are in cold water, and by the thin-walled veins associated with arteries in the fins and flukes.

Some cetaceans are the only animals other than elephants that have a brain larger than that of humans; brain weight varies from about 0.2 kg in *Pontoporia blainvillei* to 9.2 kg in *Physeter catodon* (Kamiya and Yamasaki 1974; Rice 1967). In adult *Homo sapiens* the brain weighs about 1.3 kg. Like that of humans, the brain of cetaceans is highly convoluted. Moreover, cetaceans evolved a brain the size of humans' 30 million years ago, whereas the human brain has been its present size for only about 100,000 years (Lilly 1977). These factors, along with the ability of captive odontocetes to learn rapidly and to form social bonds with humans, suggest that some cetaceans are highly intelligent.

Most cetaceans have eyes well adapted for underwater vision and can also see well above water. A greasy secretion of the tear glands protects the eyes against the irritation of salt water. The sense of smell is vestigial in mysticetes and absent in odontocetes. Cetaceans have good directional hearing underwater (Rice 1967).

Hydrophonic studies have shown that cetaceans produce numerous underwater sounds and that some genera depend largely on echolocation for orientation and securing food (Caldwell and Caldwell 1977, 1979; McNally 1977; Rice 1984; Thompson, Winn, and Perkins 1979). Odontocetes commonly produce two kinds of sounds: clicklike pulses, which last 0.001–0.010 second, have a frequency band of 100 to more than 200,000 cps and are uttered in series of five to several hundred per second; and whistles, which last about 0.5 second and have a frequency band of 4,000–20,000 cps. Clicks are used in echolocation and sometimes in communication; whistles may be highly individualized and are used primarily for communication. There is still some question about where sounds are produced, but there is increasing evidence that both clicks and whistles originate in a series of air sacs that lie above the bony cranium in the soft tissues around the blowhole on top of the head. Sounds are reflected off the concave dorsal surface of the skull and are then focused and directed by the melon, a large pocket of fat on the forehead of most odontocetes. This pocket is especially well developed in genera that regularly feed in the lightless bathypelagic zone (e.g., *Hyperoodon*) and those that live in turbid rivers (e.g., *Platanista*). The enormous square head of the sperm whale *(Physeter)* is filled in large part by the spermaceti organ, which probably functions much like the melon in smaller odontocetes. Mysticetes seem to employ only a crude form of echolocation but do produce a variety of moans and squeals, which are sometimes combined into elaborate "songs" and are primarily for communication. It is probable that the large size of the cetacean brain is associated with the development of a precise sense of hearing and the analysis of complex echolocative and communicative signals.

Odontocetes generally feed on fish, cephalopods (such as squid and octopus), and crustaceans. The conical teeth seize slippery prey but are not adapted for chewing. The killer whale *(Orcinus)* regularly takes warm-blooded prey such as penguins, pinnipeds, and other cetaceans. Mysticetes feed on many different kinds of small animals, mostly crustaceans, which are collectively referred to as zooplankton. Mysticetes are often referred to as filter feeders or grazers. Their baleen is composed of modified mucous membrane and arranged in thin plates, one behind the other and suspended from the palate. There are two rows of plates, one on each side of the mouth. The plates hang at right angles to the longitudinal axis of the head. The outer borders of the plates are smooth, whereas the inner parts are frayed into brushlike fibers. These plates act as sieves or strainers. Most mysticetes feed by swimming with their mouth open through swarms of plankton. They then use their huge tongue to force water out through the lowered baleen, within which food organisms are trapped.

Most cetaceans are gregarious to some extent, and most have a relatively long period of parental care and maturation. The gestation period ranges from 9.5 to 17 months. There is almost always a single offspring, which at birth is usually one-fourth to one-third the length of the mother. Immediately after being born in the water, baby cetaceans must reach the surface for a supply of air. Most, if not all, cetacean mothers probably push their offspring to the surface. When nursing, the mother floats on her side so that the calf can breathe. Later the calf can suckle underwater. The teats of the mammary glands lie within paired slits on either side of the reproductive opening. The mammary glands have large reservoirs in which the milk collects, and the contraction of the body muscles forces the milk, by way of the teats, into the mouth of the young. The rapid growth rate of most cetaceans is at least partly related to the high calcium and phosphorus content of the milk. Because cetaceans live in an aquatic environment and need not support their own weight, they can attain great size. Maximum size sometimes is not reached until many years after sexual maturity. Most species have a potentially long life span, and some individuals are thought to have lived more than 100 years.

Cetaceans were spared from the wave of extinction of large mammals that swept the earth toward the close of the Pleistocene (about 10,000 years ago) probably in association with the spread of advanced human hunters (Martin and Wright 1967). Some whales and dolphins were killed by people in ancient times, but generally in small numbers and near the shore. By about the year A.D. 1000 the Basques of coastal France and Spain had developed a fishery involving the pursuit and harpooning of the right whale *(Eubalaena glacialis)* from small boats. As the whales became fewer, the Basques went farther and farther out to sea, perhaps reaching waters off Newfoundland just before Columbus's arrival in the West Indies. Over the next 500 years ships became bigger and faster, hunting and processing techniques steadily improved, human population and demand for cetacean products increased, and whale populations declined (Allen 1942; Brownell, Ralls, and Perrin 1989; Committee for Whaling Statistics 1980, 1984; Ellis

1980; Grzimek 1975; D. O. Hill 1975; McHugh 1974; Small 1971; U.S. National Marine Fisheries Service 1981, 1984, 1987, 1989).

For centuries one of the most valuable cetacean products was baleen (sometimes erroneously called "bone"), thin strips of which were used to stiffen various articles of clothing and for other purposes requiring a combination of strength and flexibility. Changes in fashion and the development of spring steel and plastics greatly reduced the demand for baleen. The most consistently important objective of whaling was the oil derived from blubber, which was a major fuel for lamps. Some approximate average yields of barrels of oil (1 barrel = about 105 liters) for individuals of certain species of whales are: sperm, 35; fin, 40; blue, 75; humpback, 30; and bowhead, 90. The invention of kerosene and eventually of the electric light helped to alleviate the pressure on whale populations in the latter half of the nineteenth century. In 1905, however, a new process was developed for hardening fat, and this led to the use of baleen whale oil in the production of margarine. Whale oil also came to be used in the manufacture of soaps, lubricants, waxes, explosives, and numerous other products. The true bones of whales were ground up and used to make glue, gelatin, and fertilizer. Whale meat was eaten by the people of some nations and was used as dog food and (when ground up) as cattle feed.

For many years the right whale, rich in oil and baleen, remained the prime target of the industry. It was hunted from bases on both sides of the Atlantic and in Iceland. As it became rare the fishery slackened, finally terminating around 1700 in Europe and 1800 in America. Meanwhile, however, whalers had been turning increasingly toward arctic waters in search of the bowhead *(Balaena mysticetus)*. From the early seventeenth century to the late nineteenth century, first at Spitsbergen and then in the waters off northern North America and Russia, one population after another was devastated. Both the right whale and the bowhead, as well as the gray *(Eschrichtius robustus)* and the humpback *(Megaptera novaeangliae)*, which also were sometimes taken in the early days of whaling, tended to remain near the coast and so could be easily killed and processed. With the decline of these species in the Northern Hemisphere, deep-sea whaling intensified, lengthy voyages became the rule, and the sperm whale *(Physeter catodon)* became the main target. The pursuit of this species was led by the United States from the early eighteenth century to the middle of the nineteenth. Populations of the right whale in the North Pacific and the Southern Hemisphere also came under intensive exploitation, starting about 1800. When the industry reached its peak in 1847 nearly 700 American whaling ships were operating throughout the world. Over the next 20 years the industry declined, with some suggested causes being a reduction in whale numbers, increasing expenses, destruction of much of the U.S. whaling fleet in the Civil War, and the development of kerosene, which could be used in place of whale oil for illumination.

American whaling never again attained the importance it had in the early nineteenth century, but developments elsewhere allowed the industry to recover and even expand. During the 1860s in Norway a gun was perfected that fired a harpoon with an explosive head. Subsequently, people learned how to inflate the carcass of a dead whale with air so that it would not sink. These techniques encouraged large-scale exploitation of the blue *(Balaenoptera musculus)*, fin *(B. physalus)*, and sei *(B. borealis)* whales, which previously had been too big or too fast to approach, harpoon by hand, and keep afloat. In the early twentieth century enormous stocks of these species, as well as the humpback, were discovered in the waters around Antarctica. Starting in 1904, whaling stations were established on islands in this region. The humpback, often found in the vicinity of land, was quickly and drastically reduced. In the 1920s came the first floating factories, huge ships with slip sternways for drawing up dead whales, no matter how large, and with all the facilities for processing them. Accompanied by a fleet of swift, diesel-powered catcher boats, the factories could remain for lengthy periods in the midst of whale populations in remote waters. Eventually the whalers began to use such sophisticated methods as sonar and aircraft observation to locate their objective.

The value of cetacean products, together with improvements in human technology, resulted in the twentieth century's becoming the most destructive period in whaling history. From 1904 to 1939, 580,931 blue, fin, and humpback whales were recorded killed in the Southern Hemisphere. The peak annual tonnage taken was in the 1930–31 season (a season refers to the austral summer, December–March), when 29,410 blue, 10,017 fin, and 576 humpback whales were killed in antarctic waters. The rate of kill fell sharply during World War II but had rapidly built up by 1947. With the decline of the blue whale, pressure increased on the fin and sei whales and, once again, on the sperm whale. There also was renewed interest in whaling in the Northern Hemisphere, especially in the Pacific Ocean. The peak annual catch of individual whales was reached during the 1961–62 season, when the worldwide kill was 65,966 blue, fin, sei, humpback, and sperm whales. During that season 21 factory ships and 269 catcher boats operated in the Antarctic. The major whaling nations were Japan, Norway, the Soviet Union, Great Britain, and the Netherlands. As the larger species became rare and international regulations became more effective, the harvest fell. The blue and humpback whales received complete protection in 1966. The recorded kill of fin, sei, and sperm whales was 41,640 in the 1968–69 season and 9,429 in 1978–79. There was, however, a corresponding rise in the take of the smaller minke *(Balaenoptera acutorostrata)* and Bryde's *(B. edeni)* whales, the total catch of which was 4,238 in 1968–69 and 10,777 in 1978–79.

International control of whaling was first seriously discussed in 1927 at a meeting of the League of Nations. In 1935 the United States, Norway, Great Britain, and several other nations entered into an agreement under which the right and bowhead whales were protected (effective in 1937) and certain other whaling regulations were set forth. In 1946 the International Whaling Commission, a regulatory body now consisting of 38 member nations, was formed. Although the commission established various limits and refuge areas, its objectives were long thwarted by lack of cooperation by those members most involved in the whaling industry. Finally, as all the large species of whales approached or passed commercial extinction, regulations became more meaningful. The total kill quota fell from 45,673 whales in the 1973–74 season to 14,523 in 1980–81, only 2,121 of which were from the larger species (fin, sei, sperm). By the latter season only Japan and the Soviet Union still conducted major whaling operations, each employing one or two floating factories. Several other countries had small, land-based whale fisheries, and illegal "pirate" whaling ships were sometimes active.

At its 1982 meeting the International Whaling Commission voted to end commercial whaling altogether by the end of the 1984–85 season; the last annual quota was 6,623, of which none were sperm whales. The U.S. government, reacting to immense public concern, long had advocated a total ban on commercial whaling. In 1972 the U.S. Congress passed the Marine Mammal Protection Act, which essentially ended the taking and importation of cetaceans and their products by persons subject to U.S. jurisdiction. Even after 1985 several nations continued to conduct whaling operations, first under a formal "objection," which legally removed their obligation to comply with the International Whaling Commission's moratorium, and later on the grounds that the whales were being taken for legal research purposes. Prospects for enforcement of U.S. laws providing for trade restrictions against these countries, especially Japan, helped to end the strictly commercial operations by 1987, but "research" whaling has continued, as has limited subsistence hunting by native peoples. The latter two activities were responsible for a kill of fewer than 700 whales during 1988.

Harvests at such levels continued through 1993 (Cooke and Thomsen 1993), but Japan subsequently increased its take of minke whales by about 200 animals. In December 1995 the United States determined that trade sanctions were warranted against Japan, but those measures were not actually imposed (*Traffic Bull.* 16[1] [1996]: 1). A similar sequence of events involving Norway had transpired in 1993 (*Oryx* 28 [1994]: 86). Such inaction, together with an increasing tendency by some countries to act independently of, or even to withdraw from, the International Whaling Commission, is causing concern about a possible return of uncontrolled commercial exploitation (Stoett 1993). All cetaceans, however, are on either appendix 1 or appendix 2 of the CITES.

One of the immediate goals of the United States Marine Mammal Protection Act was to reduce the number of small cetaceans (notably *Stenella* and *Delphinus*) being killed or injured by commercial fishing operations. The most critical problem was the incidental catch of these mammals in seines intended for tuna, especially in the eastern Pacific. Although many kinds of small cetaceans had been intentionally hunted by people and some populations had declined, there had been no worldwide pursuit and devastation comparable to what befell the larger whales (Mitchell 1975*a*, 1975*b*). In the 1960s, however, improved types of tuna seines began trapping large numbers of dolphins commonly found in association with the fish. The total kill from 1959 to 1972 has been estimated at 4.8 million (Lo and Smith 1986). The number of dolphins killed or seriously injured in 1972 was estimated at 368,600 by U.S. fishing vessels and 55,078 by vessels of other countries. Subsequent regulations, modifications of fishing gear, and cooperation by fishermen reportedly reduced the respective figures for 1979 to 17,938 and 6,837. Unfortunately, the kill by foreign fishing fleets rose drastically in the 1980s, sometimes approaching or exceeding 100,000 per year. Protest by the American public, including a threatened boycott of imported tuna, led to international agreements and new U.S. legislation that intensified fishing restrictions and fostered conservation measures. The total known incidental kill fell sharply and was estimated at only 4,000 dolphins in 1993 (for additional details see account of *Stenella*).

The known geological range of the Cetacea is early Eocene to Recent. In an extinct Eocene suborder, the Archaeoceti, the skull is symmetrical and the teeth are clearly differentiated into incisors, canines, premolars, and molars. Remains of the earliest archaeocete, *Pakicetus*, were found in sediments in Pakistan dated to about 54 million years ago; it is known only by cranial and dental material, but its dense auditory bullae suggest that it was at least partly aquatic. *Ambulocetus*, which lived in the same area around 2 million years later, is known by more complete skeletal remains. In superficial appearance and habits the latter genus may have resembled a crocodile. It had limbs and feet and apparently could both walk on land and swim in the water. Terrestrial movement may have been like that of a modern sea lion or fur seal. Swimming would have been by foot propulsion but also involved undulation of the vertebral column, such as occurs in modern cetaceans. The remains of these and other early archaeocetes provide evidence of a phylogenetic association of the Cetacea and a primitive group of ungulates, the mesonychid condylarths (Berta 1994; Maas and Thewissen 1995; Thewissen 1994; Thewissen, Hussain, and Arif 1994). *Rodhocetus*, about 47 million years old and also from Pakistan, is the first whale discovered in deposits representing a deep-water marine environment and the first that evidently swam primarily by dorsoventral oscillation of a heavily muscled tail (Gingerich et al. 1994). According to Fordyce and Barnes (1994), the oldest definitely known Odontoceti are from the late Oligocene, but their diversity indicates a significant earlier radiation. The oldest reported fossil members of the Mysticeti are from late Eocene or early Oligocene strata; they had teeth but may have been filter feeders. Recent analyses of mitochondrial DNA suggest that the latter specimens may have been misidentified and that the Mysticeti did not actually originate until the end of the Oligocene, about 25 million years ago (Milinkovitch, 1995; Milinkovitch, Meyer, and Powell 1994).

CETACEA; **Family PLATANISTIDAE; Genus PLATANISTA**
Wagler, 1830

Ganges and Indus Dolphins, or Susus

The single Recent genus, *Platanista*, contains two species (Pilleri and Gihr 1971; Reeves and Brownell 1989; Rice 1977):

P. gangetica, Ganges-Brahmaputra-Meghna river system and the Karnaphuli River of India, Bangladesh, and Nepal;
P. minor, Indus River system of Pakistan and India.

Mead and Brownell (*in* Wilson and Reeder 1993) included within the Platanistidae what are here treated as the separate families Iniidae, Lipotidae, and Pontoporiidae (see accounts thereof and also account of order Cetacea). Pilleri and Gihr (1977*a*) supported use of the name *P. indi* in place of *P. minor*, but Van Bree (1976) argued that the latter designation has priority and most recent authorities have agreed. Corbet and Hill (1992) included *P. minor* in *P. gangetica*, noting that it does not seem possible to provide clear differential diagnoses that would justify even subspecific separation.

Head and body length is variable, but adults are usually 200–300 cm long. The female is the larger sex, there being a questionable report of one that measured about 400 cm. The expanse of the tail flukes is about 46 cm. Reeves and Brownell (1989) referred to weights of 51–89 kg. The back is dark lead gray to lead black, and the belly is somewhat lighter. The snout is slender, curved slightly upward, well differentiated from the steeply rising forehead, and about 18–21 cm long. The neck is distinctly constricted, the dorsal fin is low and ridgelike, and the pectoral fin is cut off squarely at the end (Kellogg 1940). There are 26–39 teeth on each side of each jaw. The upper toothrows are merged together and almost in contact.

Platanista resembles *Inia* in external features of the head, beak, neck, and dorsal fin but differs in the truncate

Susus (*Platanista* sp.), photos by G. Pilleri.

pectoral fin. Other distinguishing characters, as listed by Zhou (1982), include the lack of hairs on the snout, a longitudinal and slitlike blowhole, the presence of a fore-stomach and a two-compartment main stomach, the presence of a caecum, the presence of a maxillary crest, contact between the palatal portions of the maxillae, and teeth with a simple crown. The maxillary crest is a unique feature and involves the development of thin plates of bone, one on each side of the skull, that project outward from the base of the rostrum and nearly meet in front of the blowhole.

Susus occur only in fresh water but are found from the tidal limits of rivers to the foothills of the Himalayas. They may undergo local migrations, moving into small tributaries during the monsoons but returning to large rivers in the dry winter (Kasuya and Aminul Haque 1972). *Platanista* rises to breathe about every 30–120 seconds, usually leaping out of the water in an upward or forward direction but occasionally exposing only the blowhole. Captives normally swim on their side, apparently because the tiny, deeply set eye can receive light only at such an angle (Herald et al. 1969; Purves and Pilleri 1974–75). Although the eye of *Platanista* lacks a lens and usually is not visible externally, and the genus is sometimes referred to as being blind, the eye does seem to function as a direction-finding device. Waller (1983) suggested that the eye may form a visual image in air when the dolphin lifts its head above water. Captives swim about continuously over a 24-hour period, perhaps because the streams they naturally inhabit flow swiftly during the monsoons and constant activity is necessary to avoid injury (Pilleri et al. 1976). There is also a continuous transmission of sound over a 24-hour period. Mizue, Nishiwaki, and Takemura (1971) found 87 percent of the sounds to be clicks for echolocation and 5 percent to be communicative. To find food, *Platanista* probably uses echolocation and also probes with its sensitive snout for fish, shrimp, and other organisms in the bottom mud.

There are reports that *Platanista* travels and feeds in schools of 3 to 10 or more individuals. In 90 percent of the sightings of *P. gangetica* by Kasuya and Aminul Haque (1972), however, only a single animal was seen. Births of *P. gangetica* may occur at any time of the year but are mostly from October to March, with a peak in December and January, at the beginning of the dry season (Kasuya 1972*b*). Although the gestation period once was estimated at only 8–9 months, it probably is not less than 11–12 months (Reeves and Brownell 1989). The single young is 70 cm long at birth and is weaned within a year. Sexual maturity is attained at about 10 years. Two males were reported to be still growing at 16 and 28 years of age.

Both species of *Platanista* are now classified as endangered by the IUCN and are on appendix 1 of the CITES. According to the U.S. National Marine Fisheries Service (1978), *P. gangetica* was still fairly numerous and not endangered, but Brownell, Ralls, and Perrin (1989) reported that it seemed to be declining. Klinowska and Cooke (1991) summarized numerous problems, including pollution and siltation of rivers, boat traffic, fragmentation of populations by dam construction, accidental capture in fishing gear, and deliberate hunting to obtain meat, oil, and medicinal products. Overall population estimates of about 4,000–5,000 dolphins have been issued but are not considered reliable. The species has apparently disappeared in the Karnaphuli River, which empties into the Bay of Bengal, to the east of the Ganges.

P. minor has declined drastically in range and numbers and is now also listed as endangered by the USDI. The main threat seems to be construction of numerous barrages—barriers to impound water for subsequent use in irrigation. Therefore, dolphin populations are split up, seasonal movements are restricted, and water quality deteriorates. The animals have also been hunted for their meat and oil (Klinowska and Cooke 1991). In 1974 the total population was estimated to contain only 450–600 individuals, most of which were in a 130-km stretch of the Indus between Sukkur and Guddu barrages in Sind Province, Pakistan (Kasuya and Nishiwaki 1975). Although protected by law, *P. minor* still was being taken regularly by fishermen (Pilleri and Pilleri 1979*b*). Current overall estimates are about the same, as the tendency to increase with improving protection and education has been balanced by deteriorating habitat conditions. About 400 individuals remain between Sukkur and Guddu barrages, there are a few more downstream, and perhaps 150, split into two or three highly vulnerable subpopulations, survive farther north in Punjab Province (Khan and Niazi 1989; Klinowska and Cooke 1991; Reeves and Leatherwood 1994).

The geological range of the Platanistidae is middle Miocene in eastern North America, early and middle Miocene in western North America, and Recent in south-central Asia (Rice 1984).

CETACEA; **Family LIPOTIDAE; Genus LIPOTES**
Miller, 1918

Baiji, or Chinese River Dolphin

The single known genus and species, *Lipotes vexillifer*, has been recorded from the mouth of the Yangtze (Chang Jiang) to a point about 1,900 km up that river, from Dongting and Poyang lakes, and from the Quiantang River just south of the mouth of the Yangtze (Zhou, Qian, and Li 1977). Early reports indicating that the species was restricted to Dongting Lake and some nearby waters were incorrect. As discussed above in the account of the order Cetacea, some authorities (including Mead and Brownell *in* Wilson and Reeder 1993) place *Lipotes* in the family Platanistidae and others (including Klinowska and Cooke 1991) place it in the Pontoporiidae.

Males generally are larger than females (Gao and Zhou 1992). According to Chen (1989), 12 recently collected adult males had head and body lengths of 141–216 cm and weights of 42–125 kg, and 10 adult females had head and body lengths of 185–253 cm and weights of 64–167 kg. Coloration is pale blue gray above and whitish below. Although the lower surfaces of the flippers and tail flukes are white, the name "whitefin" or "white flag" dolphin, sometimes applied to this genus, is based on an incorrect translation. The long, beaklike snout is curved upward, and there are 30–36 teeth in each side of each jaw. The greatly reduced eyes are functional. Externally *Lipotes* resembles *Inia* but differs in having an upwardly curved rostrum, a relatively smaller and more rounded pectoral fin, and a larger and more triangular dorsal fin (Brownell and Herald 1972). Other distinguishing characters of *Lipotes*, as listed by Zhou (1982), include the lack of hairs on the snout, a blowhole that is longitudinal and elliptic, no fore-stomach but a three-compartment main stomach, the absence of a caecum, the absence of a maxillary crest, contact between the palatal portion of the maxillae, and teeth that have crowns with reticulate enamel rugosity.

Zhou, Qian, and Li (1977) considered *Lipotes* to be a fluvatile and estuarine odontocete, contrary to early reports that the genus was primarily an inhabitant of shallow lakes. During periods of high water, as in the late spring and summer, *Lipotes* also goes upstream into lakes and small rivers. In extensive surveys along the middle and lower Yangtze, Hua, Zhao, and Zhang (1989) found the baiji to be associated with large eddies in the vicinity of sandbars, islands, and beaches. Individuals commonly made seasonal movements of 30–200 km to remain near the eddies. Feeding activity was primarily diurnal; at night the dolphins often rested in areas of very slow current. The long beak of *Lipotes* is used to probe muddy bottoms for food. Dives are short, usually lasting only 10–20 seconds (Zhou, Pilleri, and Li 1979). Several underwater acoustic signals are emitted, including a whistle, which apparently function in communication (Jing, Xiao, and Jing 1981). There also are series of clicks that probably are used for echolocation (Wang, Lu, and Wang 1989). The diet consists of fish, and the stomach of the type specimen contained 1.9 liters of an eel-like catfish (Brownell and Herald 1972).

Zhou, Pilleri, and Li (1979, 1980) found an overall average of one *Lipotes* every four km along part of the Yangtze. The genus usually occurred in pairs, which in turn made up a larger social unit of about 10 individuals. Births were thought to take place in March and April. Based on surveys along the lower Yangtze from 1979 to 1986, Zhou and Li (1989) reported groups usually to number 3–4 but sometimes 8–10. Calves thought to be less than three months old were seen from February through May, and lactating females were captured in July and September. Chen (1989) recorded births in February and March and indicated that mating could occur in April and May. Brownell and Herald (1972) stated that a lactating female was captured on 21 December. Data cited by Hayssen, Van Tienhoven, and Van Tienhoven (1993) suggest that gestation lasts about 10–11 months, the interbirth interval is 2 years, there is a single young about 80 cm long at birth, and females reach sexual maturity at 8 years. A young male rescued from a fishing net in 1980 was still living in captivity in 1994 (Leatherwood and Reeves 1994). Based on dentition, a wild individual was estimated to have lived 24 years (Gao and Zhou 1992).

Lipotes long was protected by custom, though if an individual was killed accidentally, its meat would be eaten and its fat used for medicinal purposes. In recent decades its numbers and distribution have declined seriously through hunting, accidental catching by fishermen, collision with motorized vessels, development of irrigation facilities, dynamiting for channel maintenance, and reduction of prey species through dam construction (Chen and Hua 1989; Chen et al. 1979; Zhou, Pilleri, and Li 1979; Zhou, Qian, and Li 1977). Because of sedimentation, there are no longer permanent populations in Dongting or Poyang lakes, though individuals may enter those lakes during high-water periods. Legal protection has been provided in China since 1949 (Klinowska and Cooke 1991). Brownell, Ralls, and Perrin (1989) stated that the baiji probably is the most endangered of all cetaceans, with only a few hundred surviving. In 1986 total numbers were estimated at 300 animals and were thought to be declining (Perrin and Brownell 1989). In 1993 the estimate was 200 or fewer, and subsequent surveys covering about half of the range located only 9 individuals (Reeves and Leatherwood 1994). Many conser-

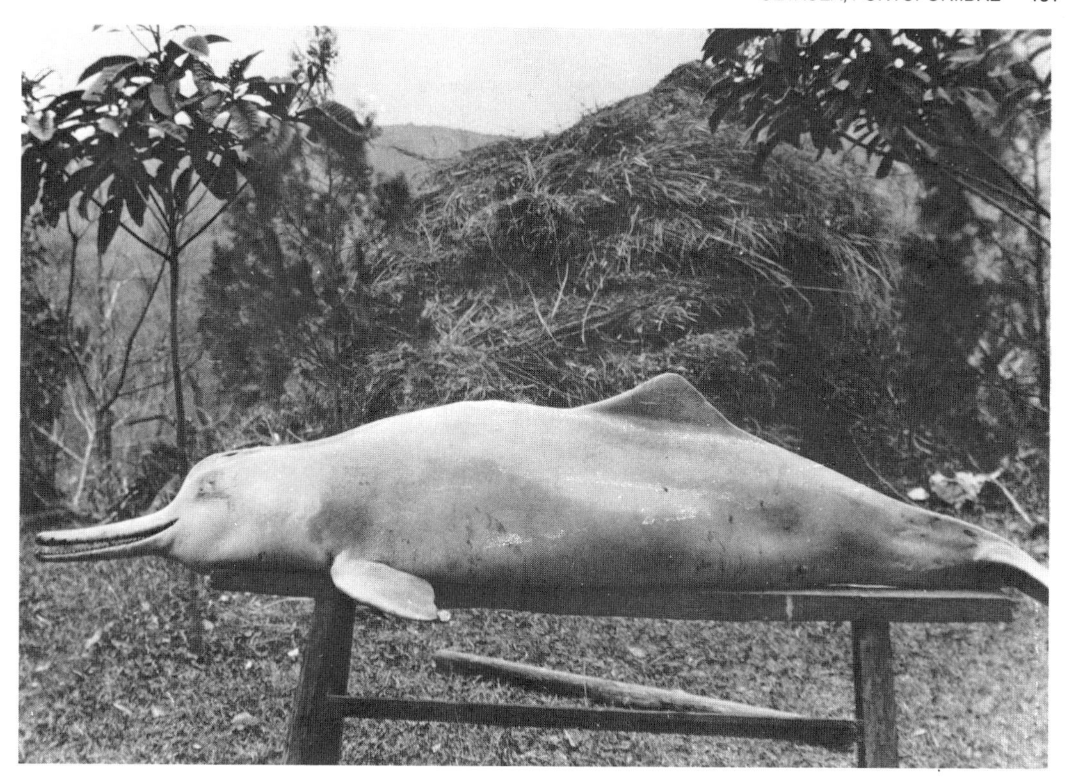

Baiji *(Lipotes vexillifer)*, photo by Clifford Pope through Robert L. Brownell, Jr.

vation measures have been discussed, including captive breeding and reintroduction at protected sites, but little has been accomplished (Leatherwood and Reeves 1994). *Lipotes* now is classified as critically endangered by the IUCN, is listed as endangered by the USDI, and is on appendix 1 of the CITES.

The only fossil that may be referable to the Lipotidae is a fragment from the late Miocene or Pliocene of China (Fordyce and Barnes 1994).

CETACEA; **Family PONTOPORIIDAE; Genus PONTOPORIA**
Gray, 1846

Franciscana, or La Plata Dolphin

The single Recent genus and species, *Pontoporia blainvillei*, occurs in the coastal waters and estuaries off southeastern South America, from the Doce River at 19°37′ S in Brazil to the Valdez Peninsula at 42°30′ S in Argentina (Brownell 1975*b*, 1989). As discussed above in the account of the order Cetacea, some authorities (such as Mead and Brownell *in* Wilson and Reeder 1993) include the Pontoporiidae in the family Platanistidae. In addition, as noted by Rice (1984), some earlier workers had regarded the Pontoporiidae as a subfamily of the Delphinidae. Some authorities, such as Pilleri and Gihr (1981*b*), use the name Stenodelphidae in place of Pontoporiidae.

Head and body length is 125–75 cm and weight is 20–61 kg. Females are generally larger than males. The color is grayish above and paler below. The young are usually brownish. The beak is extremely long, slender, slightly downwardly curved, and moderately demarcated from the bulging forehead (Rice 1984). There are 48–61 teeth on each side of each jaw (Kasuya and Brownell 1979), and the total number of teeth per individual has been counted at 210–42.

Pontoporia bears some resemblance to *Inia* in external features of the head, neck, and pectoral fin but differs in that the beak is more slender and the dorsal fin is prominent and triangular in shape. The neck, while distinct and having all the vertebrae free, is not so obvious as in *Inia*. Other distinguishing characters, as listed by Zhou (1982), include the lack of hairs on the snout, a blowhole that is transverse and crescentic, no fore-stomach and a single main stomach, the absence of a caecum, the absence of a maxillary crest, contact between the palatal portion of the maxillae, and teeth that have a simple crown.

Although commonly included among the river dolphins for purposes of classification and study, *Pontoporia* is the only member of that group that occurs in salt water. It is found in the estuary of the Rio de la Plata but has not been recorded from the adjoining Parana and Uruguay rivers (Brownell 1975*b*). According to Kellogg (1940), *Pontoporia* is rarely seen in the Rio de la Plata during winter, perhaps because most schools then migrate out to sea or northward along the Brazilian coast. *Pontoporia* presumably locates its prey by echolocation and by probing the bottom with its long snout. Examination of stomach contents suggests that most food is taken at or near the bottom. The diet includes fish, squid, and shrimp (Fitch and Brownell 1971).

Examination of specimens taken off the coast of Uruguay (Brownell 1975*b*, 1989; Harrison and Brownell 1971; Harrison et al. 1981; Kasuya and Brownell 1979) indicates that most females have a 2-year reproductive cycle, with mating from December to February, births from Septem-

La Plata dolphin *(Pontoporia blainvillei)*, photos by Cory T. de Carvalho.

ber to December, lactation until the following August or September, and then a rest of several months. However, some females may undergo a postpartum estrus and breed once a year. The gestation period is about 10.5–11.1 months. The single young is generally 70–80 cm long and weighs 7.3–8.5 kg at birth. It starts to take solid food at 3 months, but lactation may last at least 9 months. Both males and females become sexually mature at 2–3 years and physically mature 1–2 years later. One male specimen was 16 years old at the time of death.

Pontoporia is taken regularly along the coast of South America in the nets of fishermen who are primarily seeking sharks (Brownell, Ralls, and Perrin 1989; Kasuya and Brownell 1979; Klinowska and Cooke 1991; Mitchell 1975*b;* Pilleri 1971; Praderi, Pinedo, and Crespo 1989; Reeves and Leatherwood 1994). The dolphins are used for pig feed and as a source of oil. About 1,500–2,000 individuals were killed annually in this manner off Uruguay during the late 1960s and early 1970s. The rate of kill there subsequently declined together with the shark fishery, but there also are dangerously high takes off Argentina and Brazil, and this problem may be adversely affecting overall population size. *Pontoporia* is on appendix 2 of the CITES.

The geological range of the Pontoporiidae is late Miocene or early Pliocene to Pleistocene in western North America and late Miocene or Pliocene to Recent in southeastern South America (Fordyce and Barnes 1994; Rice 1984).

CETACEA; **Family INIIDAE; Genus INIA**
D'Orbigny, 1834

Boto, or Amazon Dolphin

The single Recent genus and species, *Inia geoffrensis,* occurs in the Amazon and Orinoco basins of South America. The range includes the Tocantins drainage of eastern Brazil and extends as far inland as the eastern parts of Colombia, Ecuador, and Peru. *I. boliviensis* of the upper Madeira River system in Bolivia was considered a distinct species by Pilleri and Gihr (1977*b,* 1981*a*) but has been at least tentatively treated as a subspecies of *I. geoffrensis* by most other recent authorities (Best and Da Silva 1989*a,* 1989*b,* 1993; Corbet and Hill 1991; Klinowska and Cooke 1991; Mead and Brownell *in* Wilson and Reeder 1993; Perrin and Brownell 1989). A 400-km series of rapids along the Madeira River near the border between Brazil and Bolivia may be a barrier between *geoffrensis* and *boliviensis,* but Best and Da Silva (1989*a*) suggested that the dolphins may be able to pass the rapids at times of high water, as do the

Amazon dolphins *(Inia geoffrensis):* A. Photo from Fort Worth Zoological Park through Lawrence Curtis; B. Photo by James N. Layne.

large migratory catfish in the area. The population of the Orinoco Basin has been assigned to a third subspecies, *humboldtiana*, and reportedly is separated from *geoffrensis* by two sets of rapids just below the waterway in southern Venezuela that connects the upper Orinoco with the Amazon system. However, Meade and Koehnken (1991) observed the dolphins above, below, and between these rapids and doubted that there was an effective barrier. As discussed above in the account of the order Cetacea, some authorities (including Mead and Brownell *in* Wilson and Reeder 1993) do not consider the Iniidae to be a family separate from the Platanistidae.

Head and body length is about 170–300 cm, and the expanse of the tail flukes is about 51 cm. Males weigh as much as 160 kg, females as much as 96.5 kg (Best and Da Silva 1989*b*). The variable coloration of this genus is apparently associated with age. All of the younger animals kept by Trebbau (1975) had dark bluish metallic gray upper parts and paled to silvery gray on the lateral and ventral parts. Older and larger individuals were much lighter and generally pink in color. The lateral and ventral parts were clear pinkish gray, the flukes and most of the snout were pinkish, and the melon (forehead) was bluish and pink. The darkest parts of the older animals were the regions around the blowhole, eyes, neck, and middorsum. There were no distinct borders to the colors. With respect to wild individuals, Best and Da Silva (1989*a*) observed that those resident in opaque waters tend to be predominantly pink as adults, whereas those in rivers where light penetration is greater are dark, showing at most a pink flush on the underside and flanks.

The skull is nearly symmetrical. The rounded head bears the blowhole on its summit. The eyes are small. The forehead is blunt, and all the vertebrae of the distinct external neck are free. The beak is long, slender, curved slightly downward, and clearly distinguishable from the rest of the head. For most of their length the lower jaws are fused or at least closely appressed. The teeth number about 24–35 on each side of each jaw and usually total around 100–130 (Best and Da Silva 1989*b*); the back 8–9 teeth have a distinct keel. The dorsal fin is low, long, and ridgelike and the pectoral fin is short and broad. The upper arm bone is longer than the lower ones.

Inia has a sparse covering of short, stiff bristles on its snout, whereas *Lipotes, Pontoporia,* and *Platanista* lack such hairs. Other distinguishing characters, as listed by Zhou (1982), include a transverse and crescentic blowhole, the presence of a fore-stomach and a single main stomach, the absence of a caecum, the absence of a maxillary crest, the separation of the palatal portion of the maxillae by a vomer, and teeth that have crowns with nodular enamel rugosity.

This dolphin is restricted to fresh water. According to Trebbau and Van Bree (1974), it generally is found in

Amazon dolphin *(Inia geoffrensis)*, photo by Bob Noble through Marineland of the Pacific.

brownish, turbid, and slow-moving or temporarily stagnant streams. It may migrate into flooded forests, as well as small streams and lakes, during periods of high water. It then sometimes becomes trapped in lakes during the dry season but is able to survive by preying on fish that also are trapped. If fish are abundant in an area, the boto can be seen there for weeks at a time. It surfaces to breathe every 30–60 seconds. Sometimes only the blowhole and top of the head emerge, but frequently the dorsal fin and ridge of the back are also exposed. When swimming rapidly, and apparently when feeding, *Inia* rolls to breathe. This genus seems to be less active than the Delphinidae but will occasionally leap out of the water to heights of 125 cm.

The senses of touch and hearing are probably acute; *Inia* probes for food on the bottom with its sensitive snout and may use echolocation to detect underwater obstacles and prey. The eyes are small, but Caldwell, Caldwell, and Evans (1966) found vision to be acute and apparently to be the preferred method of environmental investigation. They also recorded 12 types of sounds, which generally were less varied, lower in intensity, and of slightly lower frequencies than those heard in most other odontocetes. The diet consists of fish, usually less than 30 cm long. Captive adults ate 4–5 kg of fish per day (Trebbau and Van Bree 1974). Observations by Defler (1983) suggest that *Inia* may sometimes associate with the giant otter *(Pteronura)* in the hope of securing fish driven by the latter from shallow water.

Best and Da Silva (1989*a*, 1989*b*) reported that a boat survey indicated an average population density of 0.22/sq km, that most animals are seen alone, and that there is no evidence of social hierarchies or territorial defense. Caldwell, Caldwell, and Brill (1989) stated that captives may be very aggressive toward one another, with large males attacking both females and smaller males. In contrast, Caldwell and Caldwell (1969) had observed wild *I. g. geoffrensis* in schools of 12–15, occasionally as many as 20, and noted that captives tended to cluster together. And Trebbau (1975) found *Inia* mostly in small groups that had a tendency to occupy a defined territory. When an individual was taken captive, others would come to its assistance. As many as 8 captives were held together, evidently without strife. Johnson (1982) observed a large captive male sharing its food with a female and a smaller male held in the same tank. Pilleri and Gihr (1977*b*) reported that *I. g. boliviensis* usually occurs alone or in pairs. It is likely that the reports of groups involve temporary association during the breeding season.

Harrison and Brownell (1971) collected a pregnant female *I. g. geoffrensis* in February and pregnant and lactating females and a calf in April. They suggested that implantation can occur in October and November, and probably earlier, and that births take place from July to September in the upper Amazon. Based mostly on observations of captive and dead individuals, Best and Da Silva (1989*b*) reported that the calving season is May–July, the gestation period is 10–11 months, length at birth averages about 79 cm, lactation may last more than a year, and females become sexually mature when they are 160–75 cm long. According to Brownell (1984), one specimen was still alive after 18 years in captivity and longevity for the genus probably is about 30 years.

This curious and inoffensive dolphin often swims around fishing boats. Pilleri (1979) stated that *Inia* seems common in the Orinoco Basin and is not hunted there but is threatened by motorboats, pollution, and dam construction. Best and Da Silva (1989*a*) recorded no major exploitation or downward trends in range or numbers but expressed concern about the likely future effects of deforestation, pollution, hydroelectric development, and expanding commercial fisheries. Klinowska and Cooke (1991) pointed out that dams, several of which already have been built across major tributaries of the Amazon and many more of which are projected, can fragment dolphin populations, degrade their habitat, and greatly reduce their food supply. Pilleri and Gihr (1977*b*) reported that *I. g. boliviensis* has been severely reduced in numbers, mainly through hunting by people for its hide and fat. *I. geoffrensis* (including *boliviensis* and *humboldtiana*) is classified as vulnerable by the IUCN and is on appendix 2 of the CITES.

The geological range of the Iniidae is late Miocene to Recent in South America (Fordyce and Barnes 1994). Grabert (1984) theorized that during the Miocene the Iniidae crossed from the Pacific through former gaps in the Andes into central South America. There *I. g. boliviensis* evolved, and in the early Pleistocene some populations expanded into the Amazon Basin, where they gave rise to *I. g. geoffrensis*. About 10,000 years ago a barrier of nonturbid, "blackwater" formed, which separated the subspecies *I. g. geoffrensis* in the Amazon and *I. g. humboldtiana* in the Orinoco system.

CETACEA; **Family MONODONTIDAE**

Beluga and Narwhal

This family of two known genera, each with one species, occurs in the Arctic Ocean and nearby seas. Heide-Jorgensen and Reeves (1993) reported a skull and a number of observations suggesting the occurrence of hybridization between the two genera, *Delphinapterus* and *Monodon*. The family has sometimes been considered part of the Delphinidae. In contrast, Barnes, Domning, and Ray (1985) transferred the genus *Orcaella* from the Delphinidae to the Monodontidae. The latter arrangement has not been generally accepted (Arnold and Heinsohn 1996; Heyning 1989*a*; Lint et al. 1990; Mead and Brownell *in* Wilson and Reeder 1993; Reeves and Leatherwood 1994).

Head and body length (not including the tusk of the narwhal) is usually 280–490 cm. The body form resembles that of the Delphinidae. The forehead is high and globose, the snout is blunt, and there is no beak. The blowhole is located well back from the tip of the snout. There are no external grooves on the throat. The pectoral fin is short and rounded, and there is no dorsal fin. The skull lacks crests. There are 50–51 vertebrae. The cervical vertebrae are not fused, so that the neck is more flexible than in most cetaceans.

These animals live in cold waters. They migrate in response to shifting pack ice and hard winters. Individuals that become trapped in ice fields can sometimes break through the ice by ramming it from the underside; the cushion on top of the head lessens the shock to the animal. The small spout is not well defined. Both genera sometimes ascend rivers. They emit various sounds and are sensitive to sounds in the water but apparently disregard noises originating on land. They seem to feed mainly on the bottom. The diet includes fish, cephalopods, and crustaceans. Groups sometimes comprise more than 100 individuals. Both genera are extensively hunted by natives of the Arctic.

Geologically the Monodontidae now are known to have occupied temperate waters as far south as Baja California during the late Miocene and Pliocene (Barnes, Domning, and Ray 1985).

CETACEA; MONODONTIDAE; **Genus DELPHINAPTERUS**
Lacépède 1804

Beluga, or White Whale

The single species, *D. leucas*, occurs primarily in the Arctic Ocean and adjoining seas, the Sea of Okhotsk, the Bering Sea, the Gulf of Alaska, Hudson Bay, and the Gulf of St. Lawrence (Rice 1977). It also ascends large rivers, such as the Amur, Anadyr, Ob, Yenesei, Yukon, Churchill, and St. Lawrence. Individuals have appeared as far south as Japan, Washington, Connecticut, New Jersey, Ireland, Scotland, the Rhine River, and the Baltic Sea (Banfield 1974; Ellerman and Morrison-Scott 1966; Ellis 1980; E. R. Hall 1981). The use of the name "beluga" has sometimes caused confusion, since the name is also applied to the great white sturgeon, one of the principal sources of caviar.

According to Fay (1978), head and body length is usually about 340–460 cm in males and 300–400 cm in females, and average weight is about 1,500 kg in males and 1,360 kg in females. Kleinenberg et al. (1969) listed maximum size as 700 cm and 2,000 kg. Pectoral fin length is about 20–45 cm. The general adult color is creamy white. The young are dark gray, black, or bluish in the first year of life; then become yellowish, mottled brown, or pale gray; and finally attain the white coloration at around five years. The lightening in color is caused by a reduction in the melanin of the skin.

The body is fusiform, the tail is strongly forked, there is a constriction at the neck, and the snout is blunt. The upper jaw protrudes slightly ahead of the bulbous melon. There is no dorsal fin, but a low dorsal ridge is present on the back. There are 8–10 teeth on each side of each jaw.

The beluga occurs both along coasts and in deep offshore waters. Some populations, such as that of the Gulf of Alaska, seem to reside permanently in relatively small areas, while others undertake extensive migrations. Wintering sites may be either north or south of summering sites and may be either nearer to or farther from shore. The environmental conditions that prevail in the range of a particular population, such as prey distribution and extent of the pack ice, apparently affect the choice of seasonal movements. One population of about 11,500 whales concentrates in the Mackenzie River estuary from late June to early August, spreads through the deep waters of the Beaufort Sea during the late summer, and then migrates about 5,000 km to the southwest and is distributed mainly in shallow coastal areas, bays, and river mouths around the Bering Sea in winter. Another population concentrates in river estuaries adjoining southern Hudson Bay in the summer but moves northward and into the open waters of the bay in winter. In the waters off northern Eurasia there seems to be a general tendency to move away from shore in the autumn and toward shore in the spring, but there are exceptions, and times of migration vary from year to year. An advantage of being close to shore in the summer is that rivers are unfrozen and may be entered for purposes of finding food and rearing young. Some individuals have ascended rivers in Siberia for distances of up to 2,000 km (Banfield 1974; Fay 1978; Finley, Hickie, and Davis 1987; Harrison and Hall 1978; Kleinenberg et al. 1969; Sergeant 1973).

Delphinapterus, a very supple animal, can scull with its tail and thus swim backward. Normally it swims about 3–9 km/hr, but when pursued, it can attain a speed of 22 km/hr. It usually surfaces to breathe every 30–40 seconds, but radio-tracking studies by Martin and Smith (1992) show that it also routinely dives for periods of 9.3–13.7 minutes and

Beluga whale *(Delphinapterus leucas)*, photos by Warren J. Houck.

to depths of 20–350 meters, presumably for feeding. One trained individual is known to have remained underwater for 20 minutes on a dive to a depth of 647 meters (Ridgway 1986*a*). The beluga is capable of covering a distance of 2–3 km underwater, and therefore the presence of surface ice is not necessarily a problem (Kleinenberg et al. 1969). The beluga cannot remain in waters extensively covered by thick ice, however, because it is unable to break through more than a very thin layer to make breathing holes (Harrison and Hall 1978).

Delphinapterus is known to produce a variety of sounds. One, "a low liquid trill, like the cries of curlews in the spring," has given rise to the vernacular name "sea canary." Many other animals produce underwater noises, but few such sounds can be heard so readily above water as those of the beluga. Also emitted are whistles comparable to those used by certain other cetaceans for communication and clicks for echolocation (Brodie 1989; Fay 1978; Kleinenberg et al. 1969). That *Delphinapterus* employs echolocation extensively is also indicated by its huge frontal melon, an organ thought to be used in the directing of sonar signals, and by the fact that this genus lives in regions that lack sunlight for months at a time, as well as in waters that may be partly covered by ice. Echolocation presumably is used to avoid obstacles and to search for prey on the bottom. The diet includes a great variety of fish, such as cod

Beluga whales *(Delphinapterus leucas)*, photo by Warren J. Houck.

and herring, as well as octopus, squid, crabs, and snails (Fay 1978; Kleinenberg et al. 1969).

Schools of as many as 10,000 individuals have been reported, but such aggregations are formed only temporarily for migration or in the presence of abundant food sources; permanent social units are much smaller (Kleinenberg et al. 1969). The latter groups are of two kinds: nursery herds of as many as 200 individuals, which in turn are divided into family units of several adult females, their newborn, and one or more older calves; and separate but associated groups of 8–16 mature males (Brodie 1989).

There is conflicting information on reproduction, but studies in arctic Canada (Braham 1984*b;* Brodie 1971, 1989; Sergeant 1973) indicate that the calving season lasts from April to September and peaks from late June to August in different areas, that each adult female normally gives birth every 3 years, and that the gestation period is 14–14.5 months. The newborn is about 160 cm long and weighs about 80 kg. Lactation lasts for 20–24 months. Sexual maturity is attained at 4–7 years in females and 8–9 years in males. Potential longevity is 25–30 years.

The beluga has been exploited by the native people of the Arctic since ancient times and has been subject to commercial fishing operations in the twentieth century. It is often caught in nets set across migratory routes. The meat is used for human and domestic animal consumption. The 10- to 25-cm-thick layer of blubber yields 100–300 liters of oil, which is used in the production of soap, lubricants, and margarine. The fat of the head is rendered into a high-quality lubricant. The bones are ground up for fertilizer, and the skin is used to make boots and laces. The tanned hide of the beluga is sometimes called "porpoise leather." In recent decades the demand for these products has declined. From the 1950s to the 1980s annual catches fell from about 4,000 to 1,200 in Russia and now may be even lower. There also are annual kills of about 1,000 in Canada, 1,000 in Greenland, and 200 in Alaska. The current number of *Delphinapterus* in the world is estimated at close to 100,000, of which about half are in North American waters (Braham 1984*b;* International Whaling Commission 1992; Kleinenberg et al. 1969; Klinowska and Cooke 1991; Mitchell 1975*a,* 1975*b;* Reeves and Leatherwood 1994; Sergeant and Brodie 1975; U.S. National Marine Fisheries Service 1989). The beluga now is classified as vulnerable by the IUCN and is on appendix 2 of the CITES.

Some North American populations, such as that in Ungava Bay, have been almost eliminated (Reeves and Mitchell 1987*a,* 1989), and others are being overexploited. Originally about 7,000 white whales summered in eastern Hudson and James bays and associated rivers, but that number was reduced by about 75 percent through hunting, and recovery may now be impossible because of river diversion, harbor construction, and other human activities in the habitat (Reeves and Mitchell 1987*c,* 1989; Smith and Hammill 1986). A much larger population, about 23,000 animals, survives in the western part of Hudson Bay (Richard 1993), but a separate population off southeast Baffin Island was estimated at only 500 in 1986 and has been greatly reduced by hunting since then (Richard 1991*b*). The population off western Greenland and in Lancaster Sound has been reduced substantially but still contains at least 10,000 belugas (Reeves and Mitchell 1987*b;* Richard and Pike 1993), and the population of 11,500 that summers in the Beaufort Sea is considered to be relatively safe, though there is concern about the potential effects of human development of the Mackenzie River estuary (Finley, Hickie, and Davis 1987; International Whaling Commission 1992). The isolated population in the St. Lawrence estuary numbered at least 5,000 individuals in 1885 but declined drastically through commercial, sport, and bounty hunting. It now contains about 350 animals and is legally protected, but its reproductive rate is low, evidently because of chemical pollution of its habitat (Prescott 1991; Reeves and Mitchell 1984; Sergeant and Hoek 1988).

CETACEA; MONODONTIDAE; **Genus MONODON**
Linnaeus, 1758

Narwhal

The single species, *M. monoceros,* occurs in the Arctic Ocean and nearby seas, from 65° to 85° N, mainly between 70° and 80° N. There have been occasional records as far south as the Alaska Peninsula, Newfoundland, Great Britain, and Germany (Reeves and Tracey 1980).

Head and body length, exclusive of the tusk, is 360–620 cm, pectoral fin length is 30–40 cm, and the expanse of the tail flukes is 100–120 cm. According to Reeves and Tracey (1980), average head and body length is about 470 cm in males and 400 cm in females, and average weight is 1,600 kg in males and 900 kg in females. About one-third of the weight is blubber. Coloration becomes paler with age. Adults have brownish or dark grayish upper parts and whitish underparts, with a mottled pattern of spots throughout. The head is relatively small, the snout is blunt, and the flipper is short and rounded. There is no dorsal fin, but there is an irregular ridge about 5 cm high and 60–90 cm long on the posterior half of the back. The posterior margins of the tail flukes are strongly convex rather than concave or straight as in most cetaceans.

There are only two teeth, both in the upper jaw. In females the teeth usually are not functional and remain embedded in the bone. In males the right tooth remains embedded, but the left tooth erupts, protrudes through the upper lip, and grows forward in a spiral pattern to form a straight tusk. The tusk is about one-third to one-half as long as the head and body and sometimes reaches a length of 300 cm and a weight of 10 kg. In rare cases the right tooth also forms a tusk, but both tusks are always twisted in the same direction. Occasionally one or even two tusks develop in a female. The distal end of the tusk has a polished appearance, and the remainder is usually covered by a reddish or greenish growth of algae. There is an outer layer of cement, an inner layer of dentine, and a pulp cavity that is rich in blood. Broken tusks are common, but the damaged end is filled by a growth of reparative dentine. The form of this plug sometimes gives the erroneous impression that another narwhal has jammed its own tusk directly into the damaged one (Newman 1978; Reeves and Tracey 1980). The spiral mode of growth ensures that the tusk will grow straight and thus that there will be no interference with swimming, as would occur with a curved shape (Kingsley and Ramsay 1988).

It traditionally has been suggested that the tusk of *Monodon* is used to break through surface ice, to probe the bottom for food, or to skewer prey (Ellis 1980). More recent hypotheses are that the tusk serves to radiate heat or to enhance echolocation (Reeves and Mitchell 1981). New evidence, however, supports an old view that the tusk is exactly what it appears to be, namely, a weapon. According to Silverman and Dunbar (1980), the presence of many scars on the heads of adult males, the large proportion of broken tusks seen, the discovery of a tusk tip embedded in the jaw of a male, and actual observations of narwhals crossing and striking their tusks strongly indicate that the tusk is used in intraspecific aggression, most probably during the mat-

Narwhals *(Monodon monoceros)*, painting by Richard Ellis.

ing season. It also may communicate dominance rank or serve in ritualized displays, much like the horns and antlers of some ungulates. Best (1981) emphasized the role of tusk length as a means of nonviolent assessment of hierarchical status and doubted that the tusk actually was used to physically damage an opponent, but Gerson and Hickie (1985) rejected this view in favor of the weapon hypothesis. They found the degree of head scarring in adult males to be correlated positively with body size and tusk girth and weight (but not tusk length). Based on knowledge of other mammals, the amount of scarring resulting from tusk-inflicted injuries would be expected to be greatest in dominant males, who must fight more often and more intensively in order to defend a harem. Females and immature males were found to have significantly fewer head scars.

Monodon has the most northerly distribution of any mammal; its normal range is almost entirely above the Arctic Circle. It is generally found in deeper waters than those used by *Delphinapterus* and tends to avoid shallow seas like those around Alaska. It seems to be most common in the eastern Canadian Arctic and in waters off Greenland. One or more major populations winter in the open waters of Baffin Bay and Davis Strait. When the ice breaks up in the summer, there are large-scale migrations to the north, east, and west, into the fjords and inlets of Greenland and the islands of the eastern Canadian Arctic. Another group winters in the Greenland Sea and summers along the east coast of Greenland and around Svalbard (Newman 1978; Reeves and Tracey 1980). There apparently is a third stock that winters in Hudson Strait and summers in northern Hudson Bay (Richard 1991*a*).

The narwhal swims rapidly near the surface when migrating but moves erratically underwater while foraging (Hay and Mansfield 1989). It maintains breathing holes in pack ice by upward thrusts of the large melon on the forehead (not the tusk) and can break through a layer of ice 18 cm thick. Sometimes several individuals cooperate to break the ice. *Monodon* emits a wide variety of sounds. The deep groans, roars, and gurgles that have been reported are probably associated with respiration. There are also screeches and whistles, probably for communication, and series of clicks, much like those known to be used by some other cetaceans for echolocation (Ford and Fisher 1978; Reeves and Tracey 1980). There has been no conclusive demonstration that *Monodon* actually does echolocate, but such an ability would seem useful in finding obstacles and prey in ice-covered waters or on long arctic nights. Radiotracking studies have revealed that it commonly dives to the seabed, where it apparently feeds, and confirm that it reaches depths of 257 meters and remains underwater for up to 15.1 minutes; it probably is capable of going more than twice as deep (Martin, Kingsley, and Ramsay 1994). The diet consists of fish, cephalopods, and crustaceans.

As many as 2,000 individuals may join in a migratory school, but permanent groups contain 3–20. Most of the smaller groups are apparently families, and each seems to have only a single large male. Some observations also suggest that during the summer there is a segregation into groups of large adult males, immature males, and mature females and calves. Births occur from June to August, the period when the breakup of pack ice allows entrance to protected bays and fjords. Each female apparently produces a single calf every 3 years. Gestation is thought to last about 15.3 months. The newborn is around 160 cm long and weighs just over 80 kg. It is weaned at 1–2 years. Sexual maturity may be reached at about 8 years in males and 6 years in females. Maximum longevity evidently approaches 50 years (Best and Fisher 1974; Ellis 1980; Hay and Mansfield 1989; Mansfield, Smith, and Beck 1975; Reeves and Tracey 1980; Strong 1988).

The narwhal long has been hunted for subsistence by the native peoples of the Arctic. The animals are traditionally harpooned from kayaks when they enter inlets during the summer. Occasionally a large number can be easily killed when they become trapped by the rapid formation of pack ice at the end of summer. The sinews are used for thread, the meat is sometimes eaten by humans and sled dogs, and the skin, high in vitamin C, is probably what prevented scurvy in the arctic peoples (Reeves 1977). The skin, called *maktaq* in Canada, is still an esteemed delicacy and is a major factor in the conservation of the narwhal (Reeves 1993).

Commercial hunting of *Monodon* seems to have begun in the tenth century A.D., soon after the establishment of Viking colonies in Greenland. The tusk was long sold at high prices, as it was said to be the horn of the mythical unicorn and to have magical properties, such as rendering poison harmless. By the seventeenth century the true origin and general availability of the tusk were recognized, and prices fell accordingly. Some demand continued, especially in the Orient, where the tusk was used in decorative work and medicinal preparations. Skins were imported to Europe to make boots, laces, and gloves. In the 1960s, as in the case of other kinds of ivory, the value of the narwhal's tusk again began to increase. The price went from about U.S. $2.75 per kg in 1965 to nearly $100 per kg in 1979. Most tusks, however, go whole to collectors and decorators, who have paid up to $4,500 for a choice specimen. Such a market caused an intensification of hunting, which is now often done with motorboats, explosive harpoons, and high-powered rifles. In 1976 the Canadian government established regulations that set quotas for various areas and required full utilization of the carcass as well as the tusk, but these rules are nearly impossible to enforce. Hundreds of tusks continue to enter the market each year, having been purchased from native hunters at an average cost of over $1,000 (Braham 1984*b*; Broad, Luxmoore, and Jenkins 1988; Davis et al. 1978; Ellis 1980; Klinowska and Cooke 1991; Mansfield, Smith, and Beck 1975; Newman 1978; Reeves 1977, 1993; Reeves and Mitchell 1981; Reeves and Tracey 1980; Smith et al. 1985; Strong 1988).

The annual kill of *Monodon* in Canada and Greenland is now thought to be around 1,000. It is not known whether overall populations in those countries are being reduced by human hunting, but there apparently have been severe declines in Russian waters. There are an estimated 34,000 individuals in the major stock of *Monodon* in Davis Strait and Baffin Bay to the west of Greenland, several thousand more in the stock to the east of Greenland, and about 1,300 in northern Hudson Bay. *Monodon* is on appendix 2 of the CITES; a 1985 effort to move it to appendix 1 was stopped by the efforts of Canada and Greenland. The genus has no definitive IUCN classification but probably warrants designation as vulnerable. Its tusk is one of the few cetacean products still in international commerce. Prices for both the tusk and *maktaq* have continued to increase in recent years, and the resulting harvest may not be sustainable (Klinowska and Cooke 1991; Reeves and Leatherwood 1994).

CETACEA; **Family PHOCOENIDAE**

Porpoises

This family of four Recent genera and six species is found in the North Pacific and adjoining seas, in coastal waters on both sides of the North Atlantic and around South America, in the Black Sea, off southern Asia and the East Indies, and around certain subantarctic islands. The Phocoenidae sometimes are included in the Delphinidae but were given familial rank by Barnes, Domning, and Ray (1985), Pilleri and Gihr (1981*b*), and Rice (1984). Barnes (1985) recognized two phocoenid subfamilies: the Phocoeninae, with the genera *Phocoena* and *Neophocaena;* and the Phocoenoidinae, with *Australophocaena* and *Phocoenoides.*

These are relatively small cetaceans with a head and body length of about 120–220 cm and a normal weight range of 25–125 kg. They share many characters with the Delphinidae, but as noted in the account of that group, porpoises have a blunt snout and a rather stocky body form. The genus *Neophocaena* lacks a dorsal fin. The Phocoenidae have 15–30 teeth on each side of each jaw. These teeth are spade-shaped (though greatly reduced in *Phocoenoides*) with laterally compressed, weakly two- or three-lobed crowns, whereas the teeth of the Delphinidae are unlobed (Rice 1984).

According to Barnes (1985), all phocoenids share a unique suite of cranial characters: an eminence or boss on each premaxilla anterior to the narial opening, a small (atrophied) posterior termination of the premaxilla that projects posteriorly adjacent to each naris and does not reach the nasal bone, palatine bones that are relatively widely exposed on the palate and thereby separate the hamular processes of the pterygoids, a branch of the preorbital lobe of the air sinus system that extends from the orbit dorsally into a recess that lies between the frontal bone and the facial portion of the maxillary bone on either side of the skull, and an asymmetrical cranial vertex that is only slightly offset to the left side. Each of these characters is found also in at least one species of some other odontocete family, but the combination is found only in the Phocoenidae.

According to Rice (1984), porpoises of the genera *Phocoena* and *Neophocaena* inhabit coastal waters, bays, estuaries, swamps, and large rivers. They normally are slow swimmers and almost never accompany vessels. *Phocoenoides* usually inhabits oceanic waters, though it also enters fjords and straits with deep, clear water. It is among the swiftest of cetaceans and habitually races ahead of vessels. It feeds on deep-dwelling squid and fish. The biology of *Australophocaena* is not well known, but Barnes (1985) noted that it also appears to be an offshore animal. Porpoises usually travel in small groups. Females have a gestation period of about 11 months and give birth during the warm months.

The geological range of the Phocoenidae is late Miocene to early Pliocene in western North America, early Pliocene in western South America, Pleistocene in eastern Asia, and Recent in the present range (Barnes 1985; Rice 1984).

CETACEA; PHOCOENIDAE; **Genus PHOCOENA**
G. Cuvier, 1817

Common Porpoises, or Harbor Porpoises

There are three species (Brownell and Praderi 1984; Gaskin 1984; Klinowska and Cooke 1991; Rice 1977; Villa 1976):

P. phocoena, coastal waters from Davis Strait to North Carolina, from Iceland and Novaya Zemlya to the Cape Verde Islands and Gambia, from northern Alaska to Japan and Baja California, and in the Mediterranean and Black seas;
P. sinus, Gulf of California;
P. spinipinnis, coastal waters from northern Peru and southern Brazil to Tierra del Fuego.

Guiler, Burton, and Gales (1987) identified a skull from Heard Island in the southern Indian Ocean as *P. spinipinnis,* but Brownell, Heyning, and Perrin (1989) referred the specimen to *Australophocaena.*

Head and body length is about 120–200 cm, pectoral fin length is 15–30 cm, dorsal fin height is 15–20 cm, and the expanse of the tail flukes is 30–65 cm. In *P. phocoena* the mean head and body length is about 150–60 cm and the usual weight is 45–60 kg. The respective maximums are 186 cm and 90 kg (Gaskin, Arnold, and Blair 1974). Four physically mature specimens of *P. sinus* were about 135–44 cm long and weighed 43–47 kg (Brownell et al. 1987). The sexes are approximately equal in size. Coloration is variable

Common porpoise *(Phocoena phocoena)*, photo by Stephen Spotte, Mystic Marinelife Aquarium.

but usually is either dark gray or black above grading to whitish below or entirely dark. The conical head is not beaked, and the dorsal fin is usually triangular and located just behind the middle of the back. There are 16–28 teeth on each side of each jaw. The spade-shaped teeth are entirely crowned or bear crowns with two or three lobes.

Common porpoises frequent coastal waters, bays, estuaries, and the mouths of large rivers. They sometimes ascend the rivers. In the western Atlantic, at least, *P. phocoena* apparently moves well out to sea at the end of summer and reappears in the spring. In other regions *P. phocoena* also moves away from shore and/or toward the south during the autumn and winter in order to avoid the buildup of ice (Gaskin 1984). This species generally swims quietly near the surface, rising about four times per minute to breathe. It seems to be less frolicsome than most members of the family Delphinidae, seldom jumps out of the water, ignores boats, and rarely rides bow waves. When pursued it can attain a speed of about 22 km/hr. When diving for food it stays underwater for an average of about four minutes. One individual was caught in a fish trap at a depth of 79 meters. *P. phocoena* produces clicklike sounds such as those known to be used by some members of the Delphinidae for echolocation. Its diet consists mainly of smooth, nonspiny fish about 10–25 cm long, such as herring, pollack, mackerel, sardines, and cod (Gaskin, Arnold, and Blair 1974). Captives do not seem to differ much from *Tursiops* in trainability (Andersen 1976).

In a study of *P. phocoena* off New Brunswick, Gaskin and Watson (1984) learned that when known individuals returned each spring and summer they reestablished specific ranges of about 1.0–1.5 sq km and patrolled these areas in a territorial fashion. Although *P. phocoena* may sometimes travel in schools of nearly 100 individuals, it is usually seen in pairs or in groups of 5–10 (Leatherwood, Caldwell, and Winn 1976). The reproductive season of this species could be extensive, but mating seems to occur mainly from June to September and calving from May to August. Gestation lasts 10–11 months and includes 6–7 weeks of preimplantation pregnancy. The single newborn weighs 6–8 kg and is 70–100 cm long. The young is brought to a sheltered cove by the mother and nurses for about 8 months. Sexual maturity is attained at 3–4 years. The life span is short for a cetacean, usually about 6–10 years and rarely more than 13 years (Gaskin 1977; Gaskin, Arnold, and Blair 1974; Gaskin and Blair 1977; Gaskin et al. 1984; Read 1990; Simons 1984).

Little is known about the life history of the other species of *Phocoena*. A pregnant female *P. spinipinnis* with a near-term fetus was collected off Uruguay in late February or early March (Mitchell 1975*b*). Additional specimens suggest that calving occurs during about that same period throughout the range of the species (Klinowska and Cooke 1991).

Phocoena phocoena has been taken heavily by people in various areas, its meat being used for human and domestic animal consumption and its oil for lamps and lubricants. In the Baltic Sea this species now has declined to a critically low level following centuries of exploitation, during which catches reached up to 3,000 per year off Denmark alone. Numbers also have fallen sharply in the Netherlands, probably mainly because of water pollution, as well as excessive harvest of food fishes. The isolated population in the Black Sea and adjoining Sea of Azov, once estimated to contain 25,000–30,000 individuals, sustained an annual harvest of up to 2,500. That fishery was terminated by the Soviet, Rumanian, and Bulgarian governments in 1966 and by Turkey in 1983, though there is some concern that the latter country will reopen a season. Only about 10,000 porpoises survive in the Black Sea, and the population of the Mediterranean Sea, never very numerous, may now be extinct. Until the late nineteenth century, Indian tribes in eastern Canada caught several thousand porpoises yearly and sold the oil. About 50,000 individuals are thought to survive in the Bay of Fundy and the Gulf of Maine, and there are substantial populations to the north in the Gulf of St. Lawrence and off Newfoundland. About 700–1,500 *P. phocoena* are still deliberately or accidentally taken off western Greenland each year, and this kill may be adversely affecting the overall resident population of 10,000–15,000 in that region. Other major populations, including about 100,000 animals off Norway and in the northern North Sea and perhaps 50,000 off Alaska and the western conterminous United States, are not currently undergoing substantial and intentional exploitation (Buckland, Smith, and Cattanach 1992; Gaskin 1977, 1984; Jefferson and Curry 1994;

Klinowska and Cooke 1991; Mitchell 1975*a*, 1975*b*; Read and Gaskin 1988; Smeenk 1987; U.S. National Marine Fisheries Service 1978).

However, throughout its range *P. phocoena* is subject to accidental entanglement in modern synthetic fishing nets, particularly gill nets, which are visually and acoustically almost undetectable to porpoises. These nets are probably the single greatest threat to the species, which is concentrated in the same nearshore waters where fishing activity is greatest, and several populations are in decline as a result. Especially severe killing has been reported all along the western and northeastern coasts of North America, off Greenland, and in the North Sea (Jefferson and Curry 1994). An estimated 7,000 are taken annually by Danish fisheries, and the species apparently is becoming rare in the southern North Sea and parts of the Baltic (Simmonds 1994). At least another 2,000 are being caught each year by the Gulf of Maine groundfish gill-net fishery, and the population there is being considered for classification as threatened by the USDI. Efforts are underway in the United States to develop remedial measures such as education, modified fishing gear, and acoustical devices to make the animals avoid nets (Marine Mammal Commission 1994). The entire species *P. phocoena* now is classified as vulnerable by the IUCN and is on appendix 2 of the CITES.

Phocoena spinipinnis is subject to a direct commercial kill, especially off Peru, where its meat is used for human consumption. At least 1,500–2,500 individuals were taken in Peru in 1988. It also is taken incidentally in gill nets along the coasts of Peru, Chile, Argentina, and Uruguay and is sometimes killed for use as crab bait (Klinowska and Cooke 1991). It is on appendix 2 of the CITES.

The species *P. sinus* is classified as critically endangered by the IUCN and as endangered by the USDI and is on appendix 1 of the CITES. Villa (1976) considered it to be nearly extinct. It formerly may have occurred throughout the Gulf of California, but Brownell (1986) reported that all precise and all recent records are from the upper part of the gulf and stated that the species has the most limited natural distribution of any marine cetacean. *P. sinus* was considered abundant in the early twentieth century but declined in conjunction with the intensification and modernization of commercial fisheries starting in the 1940s. A main problem is accidental capture in gill nets set for fish and shrimp. Several hundred individuals may have died annually in this manner during the early 1970s (Brownell 1983), and the current estimated annual kill is 30–40 (Jefferson and Curry 1994). In addition, excessive fishing is reducing food supplies, and dams in the southwestern United States and western Mexico may be cutting off the flow of nutrients to the Gulf of California. Silber (1988) reported several new sightings but stated that the species is exceedingly rare. Reeves and Leatherwood (1994) noted that the results of four independent surveys indicate that not more than a few hundred animals survive.

CETACEA; PHOCOENIDAE; **Genus NEOPHOCAENA**
Palmer, 1899

Finless Porpoise

Klinowska and Cooke (1991) and Rice (1977) recognized a single species, *N. phocaenoides*, occurring in warm coastal waters and certain rivers from the Persian Gulf to Korea, Japan, Borneo, and Java. Pilleri and Gihr (1975, 1981*b*) thought that there were actually three species: *N. phocaenoides*, of southern Asia and the East Indies; *N. asiaeorientalis*, of the Yangtze (Chang Jiang) River Valley in China; and *N. sunameri*, of Japan and Korea. Recent morphological and biochemical analyses support designation of those three taxa as subspecies of a single species (Amano, Miyazaki, and Kureha 1992; Zhou, Gao, and Sun 1993).

This is the smallest cetacean. Head and body length is 120–206 cm, pectoral fin length is about 28 cm, and the expanse of the tail flukes is about 55 cm. Weight is 25–40 kg (Lekagul and McNeely 1977). Although there are numerous literary references indicating that the color is black or dark plumbeous gray, observations of live animals by Pilleri, Zbinden, and Gihr (1976) show that the upper parts are actually pale gray with a bluish tinge on the back and sides and the underparts are whitish. Apparently, however, the skin quickly darkens after death. Distinctive features are the small size, the abruptly rising forehead, and the absence of a dorsal fin. There are 15–21 spade-shaped teeth on each side of each jaw.

In southern Asia and the East Indies *Neophocaena* seems to be closely associated with mangroves and salt marshes (Pilleri and Gihr 1975). In China the genus is found in shallow coastal waters, in the middle and lower Yangtze (Chang Jiang), and in Dongting and Boyang lakes, often in association with *Lipotes* (Chen et al. 1979; Zhao, Gao, and Sun 1993; Zhou, Pilleri, and Li 1979). Part of the Japanese population is migratory, concentrating in the Inland Sea in the spring and moving out to the Pacific coast from late summer to mid-winter (Kasuya and Kureha 1979). *Neophocaena* moves rather slowly, rolls to the surface to breathe, and does not jump out of the water. The diet includes fish, shrimp, and small squid.

About half of all sightings made in Japan by Kasuya and Kureha (1979) involved solitary animals, and most of the others were of two animals, generally a cow and calf or a mated pair. There is usually a 2-year breeding cycle in Japan, with gestation lasting approximately 11 months. Available specimens (Shirakihara, Takemura, and Shirakihara 1993) indicate that births occur from August to April off western Kyushu and from March to August, with a peak in April, in the Inland Sea. Average neonatal body length is 78 cm. Sexual maturity probably is attained by males at 4–6 years and by females at 6–9 years. One pregnant female was 21 years old and a lactating female was 23.

In China *Neophocaena* is usually seen in groups of 2–5 individuals and occasionally in schools of as many as 50. Births occur in April and May in the Yangtze River and the Yellow Sea and from June through March, with a peak from August to December, in the South China Sea. Gestation may be about 10–11 months, the single newborn is about 68–84 cm long and weighs about 7 kg, and weaning occurs at 6–15 months. The calf clings with its flippers and is carried on the back of the female in a spot where her skin is roughened. The extent of this roughened area, however, is considerably less in the Chinese population than in that of southern Asia. Sometimes a group of as many as 6 females is seen, each carrying its calf in this manner. In Chinese waters sexual maturity is attained at 4–10 years and maximum known longevity is 30 years (Chen et al. 1979; Chen, Liu, and Harrison 1982; Gao and Zhou 1993; Hayssen, Van Tienhoven, and Van Tienhoven 1993; Pilleri and Chen 1979; Zhou, Gao, and Sun 1993).

Neophocaena has been taken regularly in nets set for shrimp and fish and is also deliberately hunted for its meat, skin, and oil (Mitchell 1975*a*). Gill nets probably are a severe problem in many coastal and estuarine areas (Jefferson and Curry 1994). For these reasons, and also perhaps because of pollution and collision with motorboats, *Neophocaena* has declined or completely disappeared in some areas (Reeves and Leatherwood 1994). Modification of fish-

Finless porpoise *(Neophocaena phocaenoides)*, photos by T. Kasuya.

ing methods in Japan has reduced the accidental catch there. The population using the Inland Sea, perhaps the largest in Japanese waters, contains about 4,900 individuals (Kasuya and Kureha 1979). *Neophocaena* is on appendix 1 of the CITES, and the population in the Yangtze River has been designated as endangered by the IUCN.

CETACEA: PHOCOENIDAE; **Genus AUSTRALOPHOCAENA**
Barnes, 1985

Spectacled Porpoise

The single species, *A. dioptrica,* is known from the coastal waters of Uruguay, Argentina, the Falkland Islands, South Georgia, Heard Island, the Kerguelen Islands, New Zealand, the Auckland Islands, and Macquarie Island (Baker 1977; Brownell 1975*a;* Brownell, Heyning, and Perrin 1989; Fordyce, Mattlin, and Dixon 1984). This species long was assigned to *Phocoena* but was shown by Barnes (1985) to represent a separate genus more closely related to *Phocoenoides* and perhaps the Southern Hemisphere counterpart of the latter.

According to Klinowska and Cooke (1991), head and body length ranged up to 224 cm in nine males and was 125–204 cm in seven females. Pagnoni and Saba (1989) reported that an adult male from Argentina was 202 cm long and weighed 115 kg. According to Brownell (1975*a*), an adult male and an adult female measured 204 and 186 cm in length and 255 and 150 mm in dorsal fin height. The dorsal and upper lateral surfaces to just above the eyes are black, the sharply demarcated lower lateral and ventral surfaces are white, the pectoral flippers are white, and the eyes are surrounded by a black patch. There is some external resemblance to *Phocoena,* but *Australophocaena* is distinguished by its large, triangular dorsal fin and its relatively smaller and more rounded pectoral fins. There are 18–23 upper and 16–19 lower teeth on each side.

Barnes (1985) observed that the skull of *Australophocaena* resembles that of *Phocoenoides* in having a high, anteroposteriorly elongated and broad cranial vertex; a relatively small temporal fossa; an inflated zygomatic process of the squamosal; a short and anteroposteriorly expanded postorbital process of the frontal; and a flat rostrum that does not curve downward distally. In *Phocoena* the skull has a lower and generally more transversely compressed and knoblike cranial vertex, a larger and more circular temporal fossa, a more slender zygomatic process, a longer and pointed postorbital process, and a rostrum on which the premaxillary bones arch upward and the anterior tip and lateral margins curve more ventrally. *Australophocaena* has more vertebrae than *Phocoena* but fewer than *Phocoenoides.* The teeth of *Australophocaena* have spade-shaped crowns, as do those of *Phocoena,* and are not as reduced as those of *Phocoenoides*. From *Phocoenoides*, *Australophocaena* also differs in having larger nasal bones, a much shorter rostrum, and a higher and more compressed braincase.

The biology of the spectacled porpoise is not well known, but it appears to be an offshore animal, like *Phocoenoides* but in contrast to *Phocoena* and *Neophocaena* (Barnes 1985). Pregnant females were collected in July and August (Brownell 1975*a*). *Australophocaena* is on appendix 2 of the CITES.

CETACEA; PHOCOENIDAE; **Genus PHOCOENOIDES**
Andrews, 1911

Dall Porpoise

The single species, *P. dalli*, occurs from Japan, around the rim of the North Pacific, to Baja California (Kasuya 1978; Mitchell 1975*b*; Morejohn 1979; Rice 1977). The name *truei* has sometimes been used in a specific, subspecific, or other categorical sense for a stock that breeds during the summer in the central Sea of Okhotsk and then summers off the Pacific coast of Japan (International Whaling Commission 1992). This name, however, seems to be based on a color variant of *P. dalli*. Szczepaniak, Webber, and Jefferson (1992) reported an individual with this coloration from California. Miyazaki, Jones, and Beach (1984) observed both forms in the same school during the breeding season and also cited records of females of one form giving birth to young of the other form. In addition to the population of *truei*, there are recognized breeding stocks of *P. dalli* in the southern Sea of Okhotsk, in the northern Sea of Okhotsk, south of the Kamchatka Peninsula, in the central Bering Sea, south of the Aleutian Islands, in the central Gulf of Alaska, and in the eastern North Pacific (International Whaling Commission 1992).

Head and body length is about 170–236 cm, pectoral fin length is 18–25 cm, dorsal fin height is 13–20 cm, and the expanse of the tail flukes is 40–56 cm. Weight is usually 80–125 kg, but the U.S. National Marine Fisheries Service (1978) reported a maximum of 218 kg. Both sexes are about the same size (Morejohn 1979). The upper parts of *Phocoenoides* are black. In most populations there is a large white area on each side of the body, but it does not extend far in front of the anterior margin of the dorsal fin. In the population to which the name *truei* is often applied (see above) the white area usually extends laterally to above the flipper and there may also be a white throat patch. Some individuals have intermediate color patterns, a few are entirely black, and a few are mostly brownish or gray.

The head of *Phocoenoides* is sloping, and the lower jaw projects slightly beyond the upper. The dorsal fin is low and triangular. There are 97–98 vertebrae, compared with only 64–65 in *Phocoena* (E. R. Hall 1981). The spade-shaped teeth, the smallest for any phocoenid or delphinid (Morejohn 1979), number 19–27 on each side of each jaw. An unusual feature is the presence of horny protuberances of the gums between the teeth. These "gum teeth" function as gripping organs and probably wear down with use to expose the teeth; the true teeth thus seem to function mainly in older animals.

Phocoenoides occurs mainly in cool waters sometimes near land but generally well offshore in seas more than 180 meters deep. In the western Pacific there is a well-defined

Dall porpoise *(Phocoenoides dalli)*: A. Color pattern of most populations; B. Color pattern commonly found off Pacific coast of Japan. Photos by Warren J. Houck.

annual migration in which most Japanese animals shift northward for the summer to the Sea of Okhotsk and the Kuril Islands. In the eastern Pacific there are apparently no large-scale migrations, and large numbers of *Phocoenoides* remain throughout the year from Alaska to California. There is, however, a tendency to concentrate near the shore and to the south during the autumn and winter and offshore and to the north in the spring and summer. Such seasonal movements are probably related to distributional changes in prey organisms (Kasuya 1978; Leatherwood and Reeves 1978; Mitchell 1975*b;* Morejohn 1979; U.S. National Marine Fisheries Service 1978).

Phocoenoides, unlike *Phocoena,* often plays about ships and leaps out of the water and also emerges farther from the water when rolling. It is reputedly the fastest cetacean, sometimes briefly attaining speeds of 55 km/hr. It does not bother to ride bow waves unless the vessel is moving at least 26 km/hr. It commonly preys on animals that live at depths in excess of 180 meters. It produces clicklike sounds that presumably are used for echolocation. The diet includes fish, mainly small unarmed fish such as herring and anchovies, and squid (Jefferson 1988; Leatherwood and Reeves 1978; Morejohn 1979).

Groups of 2–12 individuals are usual, but aggregations of several thousand have been seen (Jefferson 1988). Available evidence indicates that births occur between early spring and early autumn in the eastern Pacific, with a strong peak in June, July, and August (Jefferson 1989). More detailed studies in the western Pacific (Amano and Kuramochi 1992; Jefferson 1988; Kasuya 1978) have shown that births take place from May to August, that females commonly give birth every 1 or 2 years, and that the gestation period averages 11.4 months. Estimates of the length of lactation vary widely, from as short as 1.6–4.0 months to 2 years. The single newborn is about 100 cm long. The age of sexual maturity is 5–8 years in males and 3–7 years in females. Maximum known age attained in the western Pacific is 22 years.

Phocoenoides is designated as conservation dependent by the IUCN and is on appendix 2 of the CITES. Its meat has been used for human consumption, its blubber is a source of oil, and its bones have been ground up and used for fertilizer (Banfield 1974). Direct exploitation has been especially heavy in Japan, and the population there appears to have declined as a result. This kill intensified during the 1980s, reportedly to compensate for the shortage of meat resulting from the whaling moratorium. Approximately 111,500 individuals were taken from 1986 to 1989, though governmental controls have since reduced the rate of harvest. Another major problem in this region has been the accidental kill by fishermen using gill nets to catch salmon. This kill has approached 20,000 porpoises annually and seems to involve mainly immature animals (International Whaling Commission 1992; Kasuya 1978; Mitchell 1975*a,* 1975*b;* Reeves and Leatherwood 1994; U.S. National Marine Fisheries Service 1978). The current estimate of the total number of *Phocoenoides* in the Pacific is 1.2 million (U.S. National Marine Fisheries Service 1994).

CETACEA; **Family DELPHINIDAE**

Dolphins

This family of 17 Recent genera and 33 species inhabits all the oceans and adjoining seas of the world as well as the estuaries of many large rivers. Some species occasionally ascend rivers. The porpoises of the family Phocoenidae (see account thereof) sometimes have been included in the Delphinidae. In accordance with the classification adopted by Klinowska and Cooke (1991) and Reeves and Leatherwood (1994), the Delphinidae are divided into six subfamilies: the Steninae, with the genera *Steno, Sousa,* and *Sotalia;* the Delphininae, with *Lagenorhynchus, Grampus, Tursiops, Stenella, Delphinus,* and *Lagenodelphis;* the Lissodelphinae, with *Lissodelphis;* the Orcaellinae, with *Orcaella;* the Cephalorhynchinae, with *Cephalorhynchus;* and the Globicephalinae, with *Peponocephala, Feresa, Pseudorca, Orcinus,* and *Globicephala.* Barnes, Domning, and Ray (1985) accepted the same subfamilies but transferred the Orcaellinae to the family Monodontidae. The latter arrangement was rejected by Arnold and Heinsohn (1996) following an extensive morphological analysis. Rice (1977, 1984) used the name Stenoninae for the subfamily Steninae, and Pilleri and Gihr (1981*b*) considered that group to be a full family with the name Stenidae.

Head and body length ranges from as little as 139 cm in some specimens of *Sotalia* to as much as 980 cm in *Orcinus,* and weight in full-grown individuals ranges from about 50 to 9,000 kg. The common name "dolphin" is generally applied to small cetaceans having a beaklike snout and a slender, streamlined body form, whereas "porpoise" refers to those small cetaceans with a blunt snout and a rather stocky body form. The blowhole is located well back from the tip of the beak or the front of the head. The pectoral and dorsal fins are sickle-shaped, triangular, or broadly rounded, and the dorsal fin is located near the middle of the back. The genus *Lissodelphis* lacks a dorsal fin.

In the Delphinidae the vertebrae number 50–100 and the first 2 neck vertebrae are fused, whereas in the Iniidae, Lipotidae, Pontoporiidae, and Platanistidae all the neck vertebrae are free. Delphinids also have a lesser number of double-headed ribs than do members of these other four families. The fusion of the lower jaws in delphinids does not exceed one-third of their length, but this fusion is more than half the length in the other four families. In the Delphinidae, unlike in the Platanistidae, the skull lacks a crest. There are usually many functional teeth in both the upper and lower jaws of delphinids, the maximum number being about 260; in the genus *Grampus,* however, there are only 4–14 teeth, and these are confined to the lower jaw.

The spout of dolphins usually is not well defined, but that of *Globicephala* extends about 150 cm. Breathing is often accompanied by a low hissing or puffing noise. The respiration rate of bottle-nosed dolphins *(Tursiops)* taken in an aquarium where they could swim freely was 1.5–4.0 respirations per minute. This genus has a heart rate of 81–137 beats per minute (the average is 100 per minute). Bottle-nosed dolphins have been observed sleeping in calm water about 30 cm below the surface; slight movements of the tail brought the head above water so that the animals could breathe.

The Delphinidae include the most agile and some of the speediest cetaceans. They commonly surface several times a minute and frequently leap clear of the water. Many species follow ships and seem to frolic about the bow. They have remarkable group precision and regularity of movement. Migration is known to occur in some species.

The ability to perceive objects by means of reflected sound (echolocation) was demonstrated in *Tursiops* in 1958, the first time for a cetacean. The sense of vision is not as well developed as that of hearing, but *Tursiops* can see moving objects in the air at least 15 meters away. Observations of captive dolphins indicate that these animals are very intelligent. Adult delphinids will engage in complex play with various objects and can be trained to perform tricks.

Pacific white-sided dolphins *(Lagenorhynchus obliquidens)* trained to leap unusually high. Many of the smaller cetaceans regularly leap out of the water but rarely go more than a foot or two above the surface and reenter the water at a distance of only a few feet. Photo from Marineland of the Pacific.

Dolphins and porpoises usually associate in schools of five to several hundred individuals, though they are sometimes seen alone or in pairs. Groups may assemble when frightened and will attack intruders. They sometimes kill large sharks by ramming them. They utter a wide variety of underwater sounds that appear to function in communication. Cooperative behavior has often been observed; one or more individuals will come to the aid of another that is injured, sick, or giving birth, pushing it to the surface so that it can breathe.

The earliest documented delphinid is from the late Miocene (Barnes, Domning, and Ray 1985). The known geological range of the family is late Miocene to Recent in North America and Europe, upper Pliocene to Recent in Japan, Pleistocene to Recent in New Zealand, and Recent in all oceans.

CETACEA; DELPHINIDAE; **Genus STENO**
Gray, 1846

Rough-toothed Dolphin

The single species, *S. bredanensis,* occurs in tropical and warm temperate waters of all oceans and adjoining seas (Rice 1977). On the basis of reported hybridization between *Steno* and *Tursiops* and between *Tursiops* and *Grampus,* Van Gelder (1977) recommended making *Steno* a synonym of *Grampus* (the earliest available name of the three).

Head and body length of adults is 209–65 mm, pectoral fin length is 36–49 mm, dorsal fin height is 18–28 mm, and weight is 90–155 kg (Miyazaki and Perrin 1994). The upper parts are slate-colored or purplish black, with scattered spots and markings. In life, according to Miyazaki and Perrin (1994), the ventral margin of a dark cape passes low over the eye, arches high toward the dorsal fin, and then sweeps downward at midlength. The belly is pinkish white or rose-colored, with slaty spots. The pectoral fin, dorsal fin, and tail

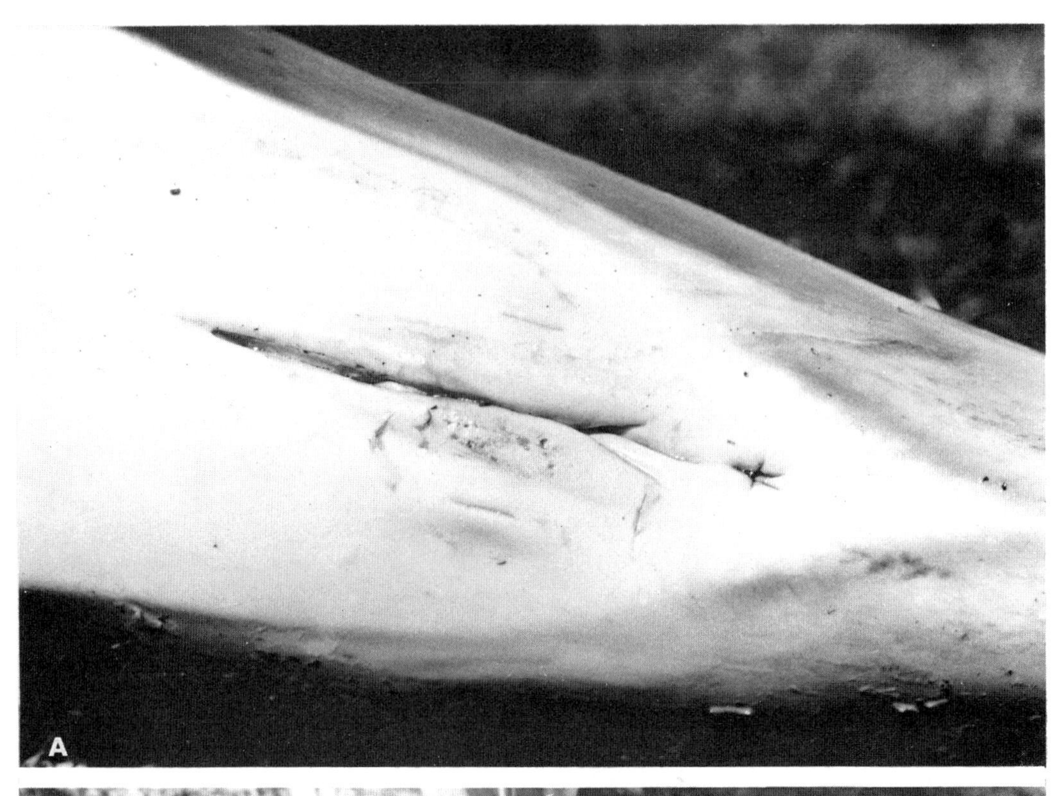

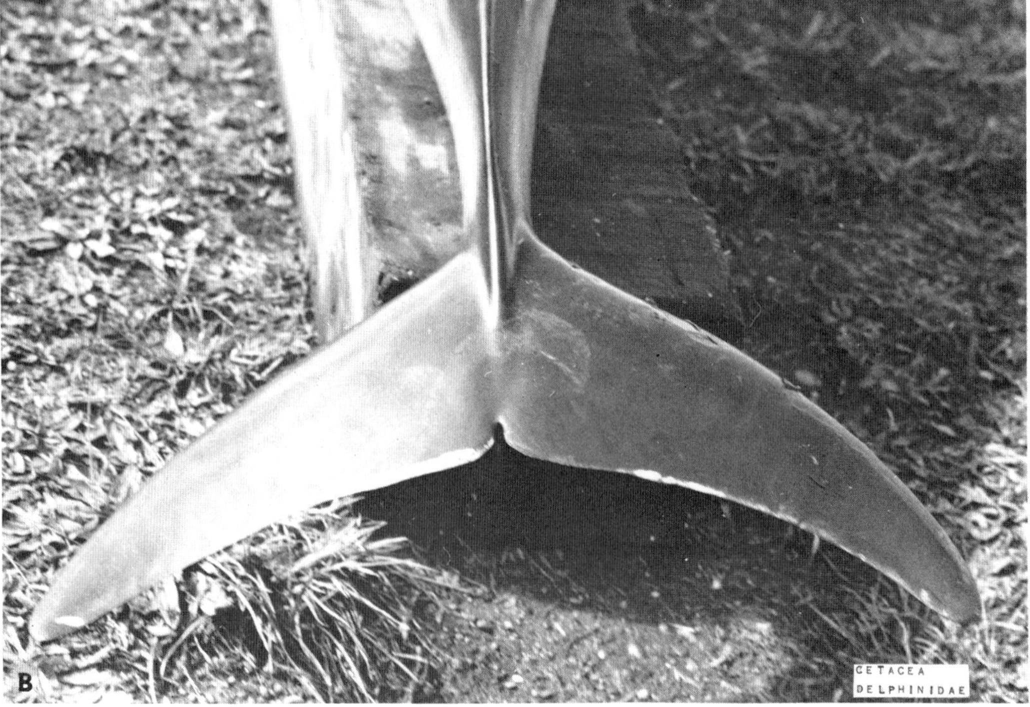

Pacific white-sided dolphin *(Lagenorhynchus obliquidens):* A. Urogenital region of a young female, showing the mammary slits on each side of the urogenital opening, the anus to the right; B. Dorsal view of caudal flukes, showing the deep tail notch typical of the Delphinidae and the prominent ridge on the top of the tail. Photos by Warren J. Houck.

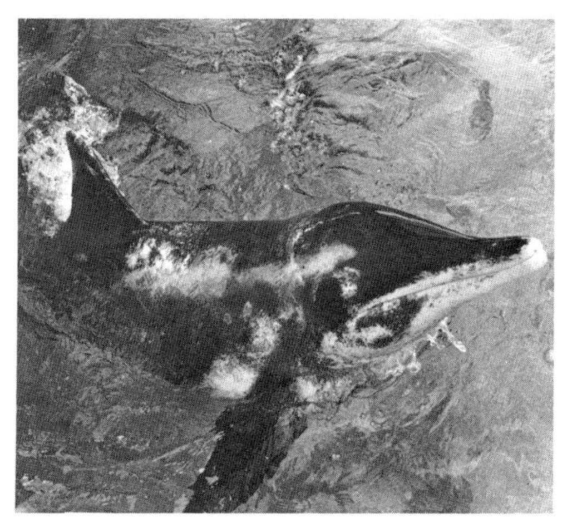

Rough-toothed dolphin *(Steno bredanensis):* Top, photo by Gary L. Friedrichsen through National Marine Fisheries Service; Bottom, photos by Michael S. Sinclair.

flukes are dark. The beak is generally dark above and white below; it is slender and compressed from side to side. There are 19–27 teeth on each side of each jaw. The surfaces of the teeth are roughened and furrowed by vertical ridges and wrinkles.

Steno is found mainly in tropical and subtropical waters, but a few individuals have stranded in colder areas outside the normal range. It dives extensively, staying submerged for up to 15 minutes, and utilizes echolocation. Stomach contents have included fish and squid. Groups usually are made up of 50 individuals or fewer, but one school of more than 100 animals reportedly stranded. The smaller groups may be components of larger aggregations of as many as 300 individuals (Miyazaki and Perrin 1994). *Steno* has been found in schools together with *Tursiops* and *Stenella*. Captives are easily trained and are reportedly even more intelligent than *Tursiops*. Four females taken off West Africa in May contained fetuses 60–87 cm long (Leatherwood, Caldwell, and Winn 1976; Mitchell 1975*b;* Perrin and Walker 1975; Rice 1967). Sexual maturity is attained at about 14 years in males and 10 years in females (Miyazaki and Perrin 1994). Individuals estimated to be as old as 32 years have been found (Perrin and Reilly 1984).

In the Mediterranean Sea near Sicily, Watkins et al. (1987) observed an aggregation of approximately 160 rough-toothed dolphins divided into eight groups of about 20 individuals each. The animals varied in size and included many mother-calf pairs. A few broke away from their groups and rode the bow wave of the ship, traveling at 16 km/hr. Some leaped clear of the water, and some dived to a depth of 70 meters, where they vocalized into a hydrophone. Both broad-spectrum clicks and whistles were recorded.

Steno is on appendix 2 of the CITES. There is little human effort to hunt this genus, but some individuals are caught and used for food in Japan, Africa, and the Caribbean, and some are captured accidentally in the tropical Pacific tuna fishery (Mitchell 1975*a*). The number of *Steno* in the eastern tropical Pacific has been estimated at 145,900 individuals (U.S. National Marine Fisheries Service 1994).

CETACEA; DELPHINIDAE; **Genus SOUSA**
Gray, 1866

Humpback Dolphins

Two species usually are recognized (Beaubrun 1990; Mitchell 1975*b*; Rice 1977):

S. chinensis, coastal waters of eastern and southern Africa, southern Asia, the East Indies, and Australia;
S. teuszii, coastal waters from Western Sahara to Angola.

The above arrangement was accepted by Corbet and Hill (1991, 1992), Klinowska and Cooke (1991), Mead and Brownell (*in* Wilson and Reeder 1993), and Reeves and Leatherwood (1994). However, some authorities, including Lekagul and McNeely (1977) and Medway (1977), consider *Sousa* to be part of *Sotalia*. Pilleri and Gihr (1981*b*) accepted *Sousa* as a full genus and divided it into the following distinct species: *S. plumbea*, coastal waters from eastern Africa to Thailand; *S. lentiginosa*, coastal waters from eastern Africa to Thailand; *S. chinensis*, coastal waters of southern China; *S. teuszii*, coastal waters of western Africa; and *S. borneensis*, coastal waters of Borneo and Australia. Meester et al. (1986), Ross (1984), and Ross, Heinsohn, and Cockcroft (1994) regarded *S. plumbea*, but not *S. lentiginosa*, as a species distinct from *S. chinensis*. Rice (1977) and Ross (1984) suggested that *S. teuszii* might be only a subspecies of *S. chinensis* (including *S. plumbea*).

Head and body length is 120–280 cm, pectoral fin length is 30 cm or less, dorsal fin height is about 15 cm, and the expanse of the tail flukes is about 45 cm. A specimen of *S. teuszii* 230 cm long weighed 139 kg. Coloration is variable; most forms are brown, gray, or black above and paler beneath, but some populations are whitish, speckled with gray, or freckled with brown spots.

The skull of *Sousa* differs from that of other dolphins in the rather long symphysis of the jaws and in the widely separated pterygoid bones that do not close together behind the palate. There are 23–50 teeth on each side of each jaw. *Sousa* resembles *Tursiops* and *Steno* but is distinguished by having more teeth and 10–15 fewer vertebrae. The rostrum of *Sousa* is always narrower than that of *Tursiops*.

These dolphins are found in both salt and fresh water, inhabiting seas, estuaries, and the mouths of rivers. They sometimes ascend rivers and have been reported 1,200 km up the Yangtze (Chang Jiang) in China (Lekagul and McNeely 1977). One population of about 500 individuals seems to remain all year in muddy mangrove creeks in the delta of the Indus River (Pilleri and Pilleri 1979*a*). A population off South Africa also does not appear to migrate, but its habitat consists of deep waters over sand and reefs (Saayman and Tayler 1979).

Sousa sometimes leaps out of the water to a height of 120 cm but seems to be slower than most dolphins. It rolls to breathe and appears to roll faster when feeding. Saayman and Tayler (1979) found *Sousa* to ride waves rarely, to remain submerged for up to three minutes, and to hunt individually, generally in the vicinity of reefs. The diet evidently consists of fish and cephalopods (Barros and Cockcroft 1991). Purves and Pilleri (1983) reported the animals in the Indus Delta to produce clicks, whistles, and screams.

Sousa often is seen alone or in pairs, but Ross (1984) and Saayman and Tayler (1979) reported group size off South Africa to average about 7 and range up to about 30 individuals. In that area births occurred all year, with a peak in summer (December–February). The single young is about 100 cm long at birth (Ross, Heinsohn, and Cockcroft 1994). Lactation apparently lasts at least two years (Barros and Cockcroft 1991).

The bioconservation status of humpback dolphins is not well understood, but Reeves and Leatherwood (1994) indicated that populations are generally low and some continue to decline in part because of their dependence on restricted coastal habitat. Small numbers are caught in the Arabian Sea, the Red Sea, and the Persian Gulf and are used for human consumption (Mitchell 1975*b*). The population in the Indus Delta is not deliberately fished but is threatened by accidental catching and industrial pollution (Pilleri and Pilleri 1979*a*). Villagers on the coast of Mauritania are said to protect the dolphins, which can be induced to drive mullet into nets, but foreign fishing fleets in the same region may be reducing supplies of fish (Klinowska and Cooke 1991). Antishark nets set to protect bathing beaches in South Africa and Australia apparently kill many of these dolphins (Ross, Heinsohn, and Cockcroft 1994). *Sousa* is on appendix 1 of the CITES.

CETACEA; DELPHINIDAE; **Genus SOTALIA**
Gray, 1866

Tucuxi, or River Dolphin

Da Silva and Best (1994, 1996), Reeves and Leatherwood (1994), and Rice (1977) recognized a single species, *S. fluviatilis*, with two populations or subspecies: the riverine *S. f. fluviatilis*, in the Amazon River and its tributaries, as far west as Ecuador and Peru; and the marine *S. f. guianensis*, in coastal waters and the lower reaches of rivers from Honduras to southern Brazil. Borobia et al. (1991) tabulated a record from the upper Tocantins drainage in eastern Brazil and also discussed occurrences in the middle Orinoco that could involve either inland movement of the marine form or crossing from the Amazon system by the riverine form. The latter reports were questioned by Da Silva and Best (1996), who suggested confusion with *Inia*. Husson (1978) treated *guianensis* as a full species.

Head and body length is 139–206 cm (Da Silva and Best 1994; Perrin and Reilly 1984), dorsal fin height is 11–13 cm, and the greatest girth is 70–98 cm. An adult male 160 cm long weighed about 47 kg. Best and Da Silva (1984) listed weights of 40–53 kg for sexually mature animals. In *S. f. guianensis* the upper parts range from pale bluish gray to brown or blackish. In at least some populations the color of the back extends to a circle around the eye, onto the pectoral fin, and to the sides of the tail. In some populations the sides are yellowish orange and there is a bright yellow patch on each side of the dorsal fin near the top. The un-

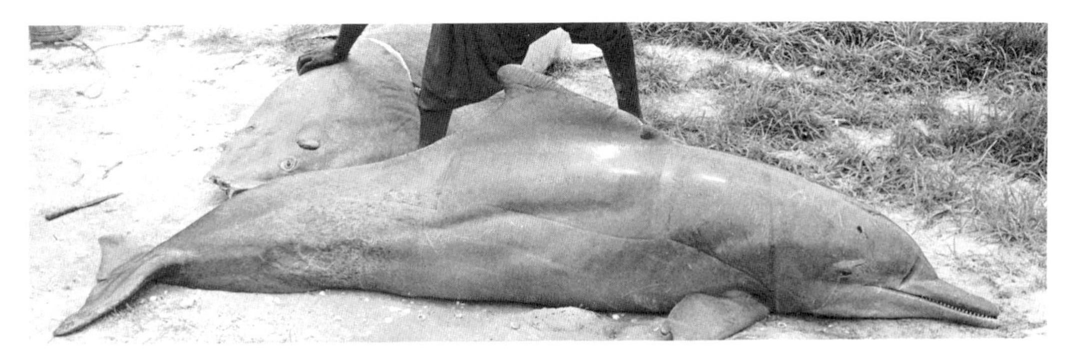

Humpback dolphin: Top, *Sousa teuszii,* photo by J. Cadenat. Middle and bottom, *S. chinensis,* photos by Michael M. Bryden.

River dolphins *(Sotalia fluviatilis guianensis)*, photos by W. Gewalt.

derparts of *S. f. guianensis* are white, pinkish, or grayish. In *S. f. fluviatilis* the upper parts are bluish or pearl gray, the color being darker anteriorly. Larger animals are noticeably paler above. The pectoral fin, both above and below, is the same color as the back. The underparts are pinkish white to white. A prominent band of the ventral coloration extends upward on the sides of the body to slightly above the level of the eye. The dorsal color, however, extends down to a distinct line from the corner of the mouth to the base of the pectoral fin and includes the eye. The tip of the beak and the apex of the dorsal fin are conspicuously white.

Sotalia is often found in the same area as *Inia,* but the latter has a more prominent dorsal fin, a longer beak, and a more prominently bulging forehead. From *Steno* and *Tursiops, Sotalia* is distinguished by the separation of its pterygoid bones and by having fewer caudal vertebrae and more teeth. There are 26–35 teeth on each side of each jaw.

Sotalia is found in both salt and fresh water. *S. f. guianensis* occurs along the coast but often enters large rivers. *S. f. fluviatilis* occurs in much of the Amazon Basin, as far west as the Andes (Mitchell 1975*b*). The latter subspecies has been reported to be more active but less inquisitive than *Inia. Sotalia* swims rather slowly and rarely leaps clear of the water but has been seen to jump 120 cm above the surface. It always rolls out of the water to breathe, the head and trunk usually appearing in smooth sequence and a short puff being the only sound heard. The reported interval between breaths is 5–85 seconds, with a mean of 33 seconds. *Sotalia* is most active in the early morning and late afternoon. Captives have been found to emit echolocation-type clicks (Caldwell and Caldwell 1970) and have been trained to perform various acrobatics (Gewalt 1979*b*). They are reportedly more difficult to train than *Tursiops* but more reliable in day-to-day performance (Terry 1986). The diet consists mostly of fish, but shrimp are also eaten.

The tucuxi has regular movements within a home range and is gregarious. Groups often swim and roll in tight formation and in nearly perfect synchrony, the individuals almost touching sides when they appear at the surface. No social interaction between *Sotalia* and *Inia* has been observed, though the two genera are often in close proximity. Data compiled by Da Silva and Best (1994) suggest that *Sotalia* is seen regularly within its range, there being an average occurrence of about 1 individual per kilometer of river, and that groups of as many as 30 have been observed in coastal areas. Husson (1978) reported *Sotalia* to be rather common in the mouths of large rivers in Surinam, where it occurs in groups of about 10 individuals. Da Silva and Best (1996) indicated that *S. f. guianensis* has an estimated gestation period of 11–12 months and gives birth during the winter along the coast of Guyana, the newborn being 60–65 cm long. Best and Da Silva (1984) found that a birth peak of *S. f. fluviatilis* apparently occurs in October and November, during the low-water period of the Amazon River system. The gestation period there is estimated to be 10.0–10.3 months, and the single newborn is 71–83 cm long.

Some South American natives regard *Sotalia* as a sacred animal, consider it a good friend and protector, and believe that it will bring the bodies of drowned persons to shore. However, it is often taken accidentally in the nets of fishermen (Mitchell 1975*b*) and is losing its habitat to rapid human development in the Amazon (Brownell, Ralls, and Perrin 1989). Da Silva and Best (1994) reported that the greatest direct threat to *Sotalia* in both marine and freshwater habitats is incidental capture in the nets of intensifying modern commercial fisheries. Other serious problems include pollution of waterways and planned hydroelectric dams, which could disrupt dolphin populations and eliminate their food supplies. *Sotalia* is on appendix 1 of the CITES.

CETACEA; DELPHINIDAE; **Genus LAGENORHYNCHUS**
Gray, 1846

White-sided and White-beaked Dolphins

There are six species (Klinowska and Cooke 1991; Leatherwood, Grove, and Zuckerman 1991; Rice 1977; Testaverde and Mead 1980):

L. albirostris (white-beaked dolphin), North Atlantic and adjacent waters from Davis Strait and Cape Cod to Barents Sea, the Baltic Sea, Portugal, and possibly Turkey;
L. acutus (Atlantic white-sided dolphin), North Atlantic from southern Greenland and Virginia to western Norway and the British Isles, with reports from as far as the southern Barents Sea, the Baltic Sea, the Azores, and the Adriatic Sea;
L. obliquidens (Pacific white-sided dolphin), waters from southeastern Alaska to Baja California and from the Commander and Kuril islands to Japan and Taiwan;
L. obscurus (dusky dolphin), temperate waters near the coasts of South America, South Africa, Kerguelen Island, southern Australia, and New Zealand;
L. australis (Peale's dolphin, or blackchin dolphin), temperate waters off southern South America and the Falkland Islands, one record from the Cook Islands in the South Pacific;
L. cruciger (hourglass dolphin), Antarctic and cold temperate waters of the Southern Hemisphere.

Head and body length is 150–310 cm, pectoral fin length is about 30 cm, dorsal fin height is up to 50 cm, and the expanse of the tail flukes is 30–60 cm. Sergeant, St. Aubin, and Geraci (1980) stated that adult weight of *L. acutus* is as much as 234 kg in males and 182 kg in females. Banfield (1974) gave the weight of *L. obliquidens* as 82–124 kg. Most species have a gray or black back, a paler belly, and various bands and stripes on the sides. In *L. albirostris* the beak is white, and in *L. obliquidens* there are conspicuous yellowish brown and grayish streaks on the sides. The dor-

Pacific white-sided dolphin *(Lagenorhynchus obliquidens)*, photo from New York Zoological Society.

Lagenorhynchus sp., photo by R. L. Pitman.

sal and pectoral fins are pointed. The beak is poorly defined and only about 5 cm long. There are 73–92 vertebrae, and there are 22–45 teeth on each side of each jaw (E. R. Hall 1981).

These dolphins generally occur in cool waters. *L. acutus* and *L. cruciger* are primarily pelagic species, whereas *L. obscurus* and *L. australis* are usually found near coasts (Mitchell 1975*b;* Rice 1977). In the western part of its range *L. albirostris* moves north into Davis Strait during the spring and summer, then turns back in the autumn and spends the winter as far south as Cape Cod (Leatherwood, Caldwell, and Winn 1976). *L. obliquidens* may move close to the southern California shore in the winter and spring but then moves north and farther out to sea in the summer and autumn (Leatherwood and Reeves 1978). Such movements are probably related to the availability of food.

Neither *L. acutus* nor *L. albirostris* commonly rides the bow waves of ships (Leatherwood, Caldwell, and Winn 1976). *L. obliquidens* often leaps clear of the water and is the only dolphin of the eastern Pacific known to turn complete somersaults under natural conditions (Leatherwood and Reeves 1978). *L. obscurus* also is highly acrobatic, its displays apparently serving to communicate social information (Würsig and Würsig 1980). In radio-tracking studies off the coast of southern Argentina, Würsig (1982) found the minimum mean daily movement of *L. obscurus* to be 19.2 km, and Würsig and Bastida (1986) determined several individuals to have ranges extending about 50–160 km from the site of capture. The diet of *Lagenorhynchus* includes herring, mackerel, capelin, anchovies, hake, squid, crustaceans, and whelks (gastropods).

These dolphins are gregarious. Groups may number as many as 1,000 in *L. acutus* and 1,500 in *L. albirostris.* Groups are usually much smaller, however, especially in the western Atlantic (Leatherwood, Caldwell, and Winn 1976; Mitchell 1975*b*); off Canada *L. acutus* generally is seen in groups of only 6–8 (Banfield 1974). Investigations based on strandings (Sergeant, St. Aubin, and Geraci 1980) determined that large groups contained many more adult females than adult males and no immature animals (2.5–6.0 years old) but that immature animals predominated in individual strandings and occasionally formed schools of their own. *L. obliquidens* is found in groups of up to several thousand individuals (Leatherwood and Reeves 1978) but segregates by age and sex during the breeding season (Banfield 1974). In a study of *L. obscurus* off Argentina,

[Dusky dolphin *(Lagenorhynchus obscurus)*, photo by Stephen Leatherwood.

Würsig and Würsig (1980) observed group size to range from 6 to 300 but noted that the larger schools were feeding aggregations, that nonfeeding groups usually contained only 6–15 animals, and that the latter individuals included both sexes and sometimes young. Aggregation apparently is facilitated through signals involving leaping and vocalization, and the different groups seem to cooperate in herding schools of fish and keeping them from escaping. After coming together and feeding in such a manner, there may be considerable social interaction and sexual activity (Würsig 1986). Two individuals marked off the coast of Argentina on 3 January 1975 were still together on 1 December 1982 (Würsig and Bastida 1986).

A study of many specimens of *L. obscurus* killed in Peruvian coastal fisheries showed most births to occur in late winter (August–October), a gestation period of 12.9 months, a lactation period of 12.0 months, and an interbirth interval of 28.6 months (Van Waerebeek and Read 1994). Other investigated populations seem to give birth from late spring to early autumn, with gestation estimated to last 10–12 months. There is normally a single young per birth, but there is at least one record of a pregnant female with two large embryos. Birth size is about 90–125 cm. In *L. acutus* nursing lasts about 18 months, the young leave the mother at 2 years, and maximum longevity is estimated to be 22 years for males and 27 years for females. Captive *L. obliquidens* have been maintained for approximately 20 years, and specimens taken in the wild are thought to have lived as long as 46 years (Banfield 1974; Klinowska and Cooke 1991; Mitchell 1975*b;* Sergeant, St. Aubin, and Geraci 1980; U.S. National Marine Fisheries Service 1978; Würsig and Würsig 1980).

Lagenorhynchus generally is abundant, and none of the species is known to be in severe decline, but overall population data for most species are not available (Klinowska and Cooke 1991). All are on appendix 2 of the CITES. There have been recent total estimates of 931,000 individuals of *L. obliquidens* (Buckland, Cattanach, and Hobbs 1993), 144,000 of *L. cruciger,* and at least tens of thousands of *L. albirostris* (Reeves and Leatherwood 1994). There is a large directed fishery for *L. obscurus* along the coast of Peru, with the fresh meat being sold in the markets of Lima for U.S. $1.00–$1.25 per kg; nearly 10,000 individuals are killed there each year (Read et al. 1988). Fishermen also deliberately have taken small numbers of *L. acutus* and *L. albirostris* off Newfoundland, Norway, and the British Isles. *L. obliquidens* has been taken both intentionally and incidental to other fisheries off Japan (Mitchell 1975*a,* 1975*b*). *L. cruciger* and *L. obscurus* are taken in large numbers by Chilean fishermen for use as crab bait (Klinowska and Cooke 1991). That practice, as well as incidental capture, also may be jeopardizing *L. australis,* which is the most restricted and perhaps least abundant species of the genus (Reeves and Leatherwood 1994).

CETACEA; DELPHINIDAE; **Genus GRAMPUS**
Gray, 1828

Risso's Dolphin, or Gray Grampus

The single species, *G. griseus,* occurs in all temperate and tropical oceans and adjoining seas (Rice 1977).

Head and body length is 360–400 cm, pectoral fin length is about 60 cm, dorsal fin height is about 40 cm, and the expanse of the tail flukes is about 76 cm. An adult male stranded in Florida was estimated to weigh 400–450 (Paul 1968). Coloration varies with age, but adults are usually slaty or black tinged with blue or purple above and paler beneath. The fins and tail are black. There are usually whitish streaks on the body, probably healed scars from attacks by other *Grampus* and perhaps by squid.

Grampus is distinguished from other dolphins in having only three to seven pairs of teeth. These pairs are in the front end of the lower jaw, but occasionally one or two vestigial teeth are present in the upper jaw. It has no beak, and the front of the head rises almost vertically from the tip of the upper jaw. Field marks include the blunt snout and the pale gleam of the back in front of the high, pointed, recurved dorsal fin.

Grampus generally occurs well out to sea in waters deeper than 180 meters (Mitchell 1975*b*). It apparently migrates northward during the warmer months (U.S. National Marine Fisheries Service 1978), and it has been observed as far north as the Gulf of Alaska (Braham 1983). It is frolicsome, sometimes riding the bow waves of vessels and leaping clear of the water. It may fall back into the water head up and occasionally waves the back third of its body in the air. The diet includes cephalopods and fish. *Grampus* usually is seen in groups of fewer than 12, but several such schools may associate in one area (Leatherwood, Caldwell, and Winn 1976). An aggregation of more than 2,000 was observed in an area 2–4 km long and 4 km wide off the coast of Washington in August (Braham 1983). Births are thought to occur during the warmer months, which are December–April off South Africa (Skinner and Smithers 1990). A pregnant female with a near-term embryo was taken in December. The newborn is approximately 150 cm long (Mitchell 1975*b*).

Although it is locally abundant, the gray grampus is intensively exploited in Japan and the Philippines and may be undergoing an unsustainable rate of harvest in Sri Lanka (Reeves and Leatherwood 1994). It also has been taken sporadically by small whale fisheries in Newfoundland, the Lesser Antilles, Japan, and Indonesia (Mitchell 1975*a,* 1975*b*). It is on appendix 2 of the CITES. The number of *Grampus* in the eastern tropical Pacific Ocean has been estimated at 175,800 individuals (U.S. National Marine Fisheries Service 1994). The famous dolphin "Pelorus Jack," which received lifelong protection from the government of New Zealand, is believed to have been a *Grampus.* This individual had the habit of playing about ships and seemed to guide them into Pelorus Sound. It was observed from about 1896 to 1916, which gives some idea of the longevity of *Grampus.*

CETACEA; DELPHINIDAE; **Genus TURSIOPS**
Gervais, 1855

Bottle-nosed Dolphins

Most authorities now follow Rice (1977) in recognizing a single worldwide species, *T. truncatus,* found primarily in temperate and tropical waters of the Atlantic, Pacific, and Indian oceans and adjoining seas (Corbet and Hill 1991, 1992; Klinowska and Cooke 1991; Mead and Brownell *in* Wilson and Reeder 1993; Reeves and Leatherwood 1994; Ross and Cockcroft 1990). However, assessments by E. R. Hall (1981), Ross (1977), and some other authorities indicated the presence of at least three species: *T. truncatus* in the Atlantic; *T. aduncus* in the Indian, South Pacific, and western and southern North Pacific oceans; and *T. gillii* in the eastern North Pacific (including the Gulf of California).

Gray grampus *(Grampus griseus)*, photos by Warren J. Houck.

E. R. Hall (1981) used the name *T. nesarnack* in place of *T. truncatus*. Zhou and Qian (1985) reported the presence of both *T. truncatus* and *T. aduncus* off the coast of China. Van Gelder (1977) recommended making *Tursiops* a synonym of *Grampus* on the basis of reported hybridization between the two genera. Such hybridization now is known to have occurred both in the wild and in captivity, and hybrids also have been produced in captivity by interbreeding *Tursiops* with the genera *Pseudorca, Steno,* and *Globicephala* (Sylvestre and Tasaka 1985).

Irrespective of the above taxonomic issues, in all areas where the systematics of *Tursiops* have been studied two main ecotypes have been recognized: a generally smaller form staying in shallow waters near the mainland and a larger form occurring farther offshore (Klinowska and Cooke 1991; Leatherwood, Caldwell, and Winn 1976; Leatherwood and Reeves 1978, 1983). Recent studies of specimens from Florida and the east coast of the United States have indicated complete distinction between the two ecotypes on the basis of skull morphology and hemoglobin analysis (Hersh and Duffield 1990; Mead and Potter 1995). Although some authorities have suggested the application of different specific names to the two forms in various local areas (Mead and Potter 1990, 1995), there has been no effort to apply such division to *Tursiops* on a worldwide scale.

Head and body length is 175–400 cm, pectoral fin length is 30–50 cm, dorsal fin height is about 23 cm, and the expanse of the tail flukes is about 60 cm. Adults usually weigh 150–200 kg, but weights in excess of 650 kg have been reported (Leatherwood, Caldwell, and Winn 1976). The upper parts are usually dark gray or slaty blue, the flippers and flukes are darker, and the underparts are paler. The genus is distinguished by the short, well-defined snout or beak, about 8 cm long and supposedly resembling the top of an old-fashioned gin bottle. There are 20–28 teeth on each side of each jaw. Each tooth is about 1 cm in diameter.

Available data (Klinowska and Cooke 1991; Leatherwood, Caldwell, and Winn 1976; Leatherwood and Reeves 1978; Mead and Potter 1990; Mitchell 1975*a*) indicate that *Tursiops* is found in coastal waters, in offshore areas as far as the edge of the continental shelf, and around some oceanic islands and atolls. The coastal ecotype frequents bays, lagoons, and estuarine complexes and sometimes ascends large rivers. Some populations remain in a restricted area of favorable habitat throughout the year. Others, es-

Bottle-nosed dolphin *(Tursiops truncatus)*, photo from Marineland of the Pacific. Inset: *T. gillii*, photo by Warren J. Houck.

pecially of the offshore ecotype, appear to be migratory in some regions, and movements may relate to seasonal changes in temperature or food distribution. Along the east coast of the United States both ecotypes concentrate south of Cape Hatteras in the winter, and then some segments of the populations spread northward in the spring and summer. *Tursiops* sometimes rides the surf or the bow wave of vessels and may jump to heights of six meters above the surface. In 26 separate observations along a tidal creek on the Georgia coast Hoese (1971) saw *Tursiops* rush its entire body, except the tail, up onto a mudbank to capture small fish that had been driven in front of it. The genus also has been observed to "throw" fish onto a beach and then retrieve them and to disable fish by "kicking" them into the air with its tail (Shane 1990).

Tursiops is commonly seen along the coasts of the United States and is also familiar to people through its displays of agility and sagacity under captive conditions. Compared with most wild animals, it is easily trained to perform acrobatics, locate hidden objects, and play with balls. It is also used widely in research work involving cetacean physiology, psychology, and sociology. Such studies have demonstrated that *Tursiops* has a high degree of what humans commonly refer to as intelligence. Some of these dolphins were even able to learn complex procedures quickly merely by watching other individuals perform (Adler and Adler 1978).

Tursiops is also noted for having a brain larger than that of *Homo* and for the variety of its vocalizations. Certain authorities have argued that these dolphins, and probably other cetaceans as well, have a complex language and may eventually be able to communicate meaningfully with people. The preponderance of current evidence, however, suggests that the large cetacean brain is associated mainly with development of the sense of hearing and the processing of echolocation and communicative data. Each individual *Tursiops* appears to have its own "signature" whistle, which it can vary according to the situation and by which it can communicate to others of its kind a limited amount of information about its identity, location, and condition (Caldwell and Caldwell 1977, 1979; Caldwell, Caldwell, and Tyack 1990; Hickman and Grigsby 1978; Jerison 1986; Ridgway 1986*b*). Other sounds and visual signals are also used in communication. Both the whistles and the high-frequency, clicklike pulses used for echolocation are thought to be produced not in the larynx but in the nasal sacs within the forehead (Hollien et al. 1976). Echolocation—the detection of objects by bouncing sound waves off of them—is utilized by *Tursiops* to find obstacles and food underwater. Hult (1982) found that series of high-intensity clicks are used also to herd schools of fish. The diet consists mainly of fish, especially bottom-dwelling species in coastal areas, and also includes cephalopods. Captive dolphins eat 6–7 kg of food per day (Mitchell 1975*b*).

According to Shane, Wells, and Würsig (1986), feeding peaks in the morning and afternoon, but activity may occur throughout the day and night. Individuals seem to make regular use of a particular area, but some animals move to different areas seasonally. Off the west coast of Florida a population of about 105 dolphins appears to maintain an overall home range of approximately 85 sq km. Within this group, mother-calf pairs and subadult males have an average range of 40 sq km, and other adult females, subadult females, and adult males have a range of 15–20 sq km. The density in this region is about average for the species, but other reported densities range from 0.06 to 4.80 per sq km. Territorial defense has not been observed in *Tursiops*, though individuals sometimes appear to recognize range boundaries. At least some groups in the eastern Pacific appear to have a limited home range associated with particular islands (Leatherwood and Reeves 1978).

Tursiops may form large aggregations, which usually comprise smaller groups of 2–15 individuals (Leatherwood, Caldwell, and Winn 1976; Shane, Wells, and Würsig 1986). From place to place, there is considerable variation in population stability and group size, but in most areas

Bottle-nosed dolphins *(Tursiops truncatus)*, photo by Bob Noble through Marineland of the Pacific.

there are subunits of individuals that remain together for long periods of time, as well as other resident animals that associate with many different subgroups (Ballance 1990). The average size of 688 observed subunits within a population off the west coast of Florida was 4.8 (Irvine et al. 1981). In the eastern Pacific, group size is usually 15 or fewer near the shore and 25–50 far offshore (Leatherwood and Reeves 1978). However, some aggregations of *Tursiops* in that region have been estimated to contain as many as 10,000 dolphins (Scott and Chivers 1990). Schools of up to 600 were reported off North Carolina in the late nineteenth century, and sexes were about equally represented there (Mead 1975). In 93 sightings of *Tursiops* in the Indian Ocean, Saayman and Tayler (1973) found the number of individuals per group to range from 3 to 1,000 and to average about 140.

The members of a group may cooperate in hunting fish, with some individuals crowding the fish toward shore and others patrolling offshore to prevent the fish from escaping, or with different units carrying out a synchronized attack at opposite ends of a school of fish (Würsig 1986). Social hierarchies are evident in captivity, with the largest adult male in a tank being dominant to all other individuals and the larger females ranking above smaller females. Dominance is expressed by biting, ramming, and tail slapping. In the wild, however, some groups may consist solely of females and young, with these units being joined temporarily by small bands of adult males (Shane, Wells, and Würsig 1986). A permanently resident "community" of dolphins in the area of Sarasota Bay, Florida, comprises four main structural units: bands of females with their most recent offspring, mother-calf pairs within the female bands, separate groups of subadults containing one or both sexes, and adult males (Irvine et al. 1981; Scott, Wells, and Irvine 1990; Wells 1991). The average female band has seven individuals, which share a common home range and maintain a close association for many years. Adult males in the same vicinity move over a larger area, encompassing the ranges of several female bands, and sometimes associate with other communities. On separation from their mothers, young dolphins join the subadult groups. Males seem to predominate in the subadult groups apparently because females tend to remain for a shorter period and then move into a female band, where they begin to participate in reproduction.

In European waters the young of *Tursiops* are born in midsummer. Births off Florida occur throughout the year, mostly from about March to October, but peak periods of populations differ (Urian et al. 1996). There is a prolonged reproductive season off South Africa, with most births occurring in late spring and summer (Ross 1977). The normal interval between calves is two years, but if the young dies at birth, another offspring may be produced a year later. The gestation period has been estimated to last 11.5–14.0 months. The single newborn weighs about 10–20 kg and is 100–135 cm long. Lactation usually lasts 18–20 months, but the young may begin to take some solid food at less than 6 months. Mother-calf bonds usually last 3–6 years. Sexual maturity reportedly is attained at some point between 5 and 14 years in females and usually at 9 to 13 years in males (Baird, Walters, and Stacey 1993; Mead and Potter 1990; Mitchell 1975*b*; Odell 1975; Ross 1977; Sergeant, Caldwell, and Caldwell 1973; Shane, Wells, and Würsig 1986). The average age of adult animals found in the wild is 19 years for males and 26 years for females; the oldest individuals known are a 39-year-old male and a 49-year-old female in the Sarasota Bay population (Klinowska and Cooke 1991).

There are many documented cases of fully wild dolphins, mostly *Tursiops*, approaching boats and people and apparently seeking friendly association, even to the point of allowing human swimmers to hold and ride them (Lockyer 1990). *Tursiops* also sometimes "cooperates" with fishermen, indicating the location of fish and then driving them into nets (Pryor et al. 1990). Nonetheless, bottle-nosed dolphins have been deliberately hunted by people in many parts of the world (Mead 1975; Mitchell 1975*a*, 1975*b*). Products include meat for human consumption, fertilizer, body oil for cooking and illumination, and jaw oil for use as a lubricant in watches and precision instruments. Several commercial fisheries existed in the eastern United States until the early twentieth century. The largest was that at Cape Hatteras, North Carolina, where 2,000 or more dolphins were taken annually during the peak years 1885–90. *Tursiops* also is occasionally taken accidentally in nets set for tuna and other fish and sometimes is shot by fishermen, who consider it a competitor. However, this genus is not among those that have been devastated through incidental killing by commercial fisheries in the tropical Pacific. Many bottle-nosed dolphins have been taken alive in waters off the United States, Japan, and Italy for purposes of display and human entertainment. Under the United States Marine Mammal Protection Act of 1972 such collecting requires a permit, and all other taking and importation of cetaceans and their products is prohibited. *Tursiops* also is on appendix 2 of the CITES.

The population of *Tursiops* in the eastern tropical Pacific has been estimated recently to comprise 243,500 individuals (U.S. National Marine Fisheries Service 1994). Otherwise there are no general indications of population size and trends over large regions. Information compiled by Baird, Walters, and Stacey (1993), Klinowska and Cooke (1991), and Reeves and Leatherwood (1994) indicates little immediate concern for the genus as a whole but suggests that some local populations are declining or already seriously depleted and that long-term prospects for *Tursiops* are problematical. The inshore populations are particularly vulnerable to human encroachment, hunting, disturbance, and pollution. There are still small directed fisheries in West Africa, Sri Lanka, Indonesia, Taiwan, Japan, Peru, and much of the Caribbean region. There also has been intensive hunting to protect fisheries in Japanese waters, which in some years has killed more than 1,000 dolphins out of an estimated total population of only 37,000. Anti-shark nets protecting beaches along the east coast of South Africa may be killing *Tursiops* at a greater rate than can be sustained by the local population. In 1987–88 at least half of the bottle-nosed dolphins off the northeastern United States perished, reducing the population there to about 10,000–13,000. There may be several thousand more off Florida and about 40,000 in the Gulf of Mexico. In the Mediterranean Sea there are now fewer than 10,000, and the Black Sea population was sharply reduced by an intensive fishery that persisted until 1966 in most of the surrounding countries and until 1983 in Turkey.

CETACEA; DELPHINIDAE; **Genus STENELLA**
Gray, 1866

Spinner, Spotted, and Striped Dolphins

There seem to be five species (Brownell and Praderi 1976; E. R. Hall 1981; Hubbs, Perrin, and Balcomb 1973; Mullen 1977; Perrin 1975*a*, 1975*b*; Perrin et al. 1981, 1987; Rice 1977):

S. longirostris (pantropical spinner dolphin), tropical and warm temperate waters of the Atlantic, Indian, and Pacific oceans and adjoining seas;
S. clymene (Atlantic spinner dolphin), tropical and subtropical waters of the Atlantic Ocean and adjoining seas;
S. coeruleoalba (striped dolphin), tropical and temperate waters of the Atlantic, Indian, and Pacific oceans and adjoining seas;
S. attenuata (bridled or pantropical spotted dolphin), tropical and warm temperate waters of the Atlantic, Indian, and Pacific oceans and adjoining seas;
S. frontalis (Atlantic spotted dolphin), tropical and warm temperate waters of the Atlantic Ocean and adjoining seas.

Although E. R. Hall (1981) suspected that there are only two species of spotted dolphins, he, along with Schmidly, Beleau, and Hildebran (1972) and Leatherwood, Caldwell, and Winn (1976), recognized a third species to take the place of *S. attenuata* in the Atlantic and adjoining seas. Hall used the name *S. frontalis* for that species and *S. pernettensis* in place of what is listed above as *S. frontalis.* Hall also listed a fourth specific name, *S. dubia,* that might apply to certain spotted dolphin populations. In a detailed revision of the spotted dolphins, Perrin et al. (1987) recog-

Spotted dolphin *(Stenella attenuata)*, photo by R. L. Pitman through National Marine Fisheries Service.

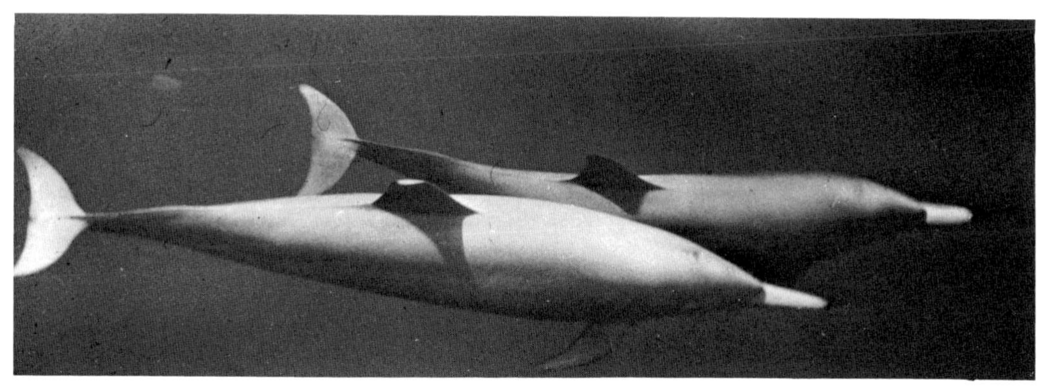

Spinner dolphins *(Stenella longirostris)*, photo by Gary L. Friedrichsen through National Marine Fisheries Service.

nized only two species, found *S. frontalis* to be the correct name of the Atlantic spotted dolphin, placed *S. plagiodon* and *S. pernettensis* in the synonymy thereof, and considered *S. dubia* a nomen nudum.

Perrin (1990) divided *S. longirostris* into three highly distinctive subspecies: *S. l. longirostris*, found in most of the range given above; *S. l. orientalis* (eastern spinner), mainly in offshore waters within 1,000 km of the coast of Mexico, Central America, and northern South America; and *S. l. centroamericana*, in coastal waters of the Pacific from southern Mexico to Panama. A fourth form, the whitebelly spinner, occurs mainly to the south and east of *S. l. orientalis* and is thought to be a hybrid between the latter and *S. l. longirostris*. In addition, Perrin, Miyazaki, and Kasuya (1989) reported the presence of a distinct and very small form of *S. longirostris* in the Gulf of Thailand.

Head and body length is 150–350 cm, pectoral fin length is 15–38 cm, dorsal fin height is 12–30 cm, the expanse of the tail flukes is about 35–66 cm, and weight is about 60–165 kg. Males average larger than females, but females have relatively longer rostra (Perrin 1975*b*; Schnell, Douglas, and Hough 1985). There is much variation in color pattern. In general, spinner dolphins are dark gray to black above, light gray or tan on the sides, and paler below; mature spotted dolphins have a sprinkling of pale spots on a dark background above and a sprinkling of dark spots on a pale background below; and the striped dolphin is grayish above with mostly pale underparts and sides and one dark stripe extending from each eye to the anus and another stripe extending from the eye to the pectoral fin.

There is a distinct beak, the rostrum is long and narrow, the palate is not grooved on the inner side of the toothrow, and the union of the two branches of the lower jaw is relatively short. There are 29–65 teeth on each side of each jaw and 68–81 vertebrae (E. R. Hall 1981; Leatherwood, Caldwell, and Winn 1976).

These dolphins seem to prefer the deeper, clearer offshore waters, mainly in tropical and subtropical parts of the world. *S. frontalis* is generally found more than 8 km from the shore but in the Gulf of Mexico may move closer during spring and summer (Leatherwood, Caldwell, and Winn 1976). There may be two forms of *S. frontalis*, the larger inhabiting the continental shelf and the smaller being oceanic, though the size difference could reflect differential feeding habits (Perrin et al. 1987). There also are several morphologically separate stocks of *S. attenuata*, *S. coeruleoalba*, and *S. longirostris* in the eastern Pacific (Douglas et al. 1992; Perrin et al. 1985; Perrin, Akin, and Kashiwada 1991; Schnell, Douglas, and Hough 1986). With respect to *S. attenuata*, analysis of skulls indicates a sharp demarcation between a smaller pelagic and a larger coastal form; the masticatory structure of the latter suggests adaptation for feeding on larger prey (Douglas, Schnell, and Hough 1984). In this same region *S. attenuata* and *S. longirostris* have been found to occur primarily in tropical waters north of the equator and south of the Galapagos, where there is relatively little annual variation in surface temperatures; *S. coeruleoalba* appears to prefer equatorial and subtropical waters with relatively large seasonal changes in surface temperature (Au and Perryman 1985). *S. longirostris* is often found far out to sea, where its distribution fluctuates in accordance with changing oceanographic conditions (Mitchell 1975*b*). *S. coeruleoalba* apparently is migratory in the western Pacific, with some populations approaching Japan in the autumn, swimming along the coast, and then moving back out to sea the following spring (Nishiwaki 1975).

These dolphins can swim at speeds of 22–28 km/hr. They often seem to frolic about ships and ride the bow waves. They sometimes jump clear of the water, and the spinner dolphins derive their name from their habit of leaping above the surface and turning two or more times on their longitudinal axis (Leatherwood, Caldwell, and Winn 1976). Observations by Norris, Würsig, and Wells (1994) indicate that the spinning behavior of *S. longirostris* helps to define the shape of a moving school and allows the members to keep track of one another as they move along. In Hawaiian waters that species spends most of the daylight hours resting and moving very slowly in relatively shallow places. Spinning, diving, swimming in a zigzag pattern, and other intense activity starts up abruptly in the afternoon, and around sunset the schools move offshore to feed (Würsig et al. 1994). *S. attenuata* feeds near the surface, but *S. longirostris* may sometimes feed at depths of 250 meters or more (Fitch and Brownell 1968; Mitchell 1975*b*). The diet of *Stenella* consists mainly of squid and small fish.

These dolphins have been found to produce two specialized kinds of sounds: whistles, probably highly individualized, for communication; and regular trains of clicks for exploring the environment and finding food through echolocation (Caldwell and Caldwell 1971*b;* Caldwell, Caldwell, and Miller 1973). Brownlee and Norris (1994) reported that whenever *S. longirostris* is active it is very noisy. In addition to clicks for echolocation and pure-tone whistles with harmonic structure, it emits various burst-pulsed signals that may sound like quacks, blatts, banjo twangs, barks, chuckles, and even the lowing of cattle. The whistles, which can vary greatly, are thought to define the limits and movements of a school and to convey information about food, danger, and other factors to members. Each dolphin apparently has its own "signature" whistle, by which it can be recognized by others in the group. The burst-pulsed signals seem to be associated mainly with social activity and close contact.

According to Perrin et al. (1987), schools of *S. attenuata* may number as many as several thousand dolphins, and oceanic populations may have home ranges several hundred kilometers or more in diameter. Based on observations during fishing operations and in captivity, Pryor and Shallenberger (1991) reported that schools of *S. attenuata* are divided into relatively stable subgroups of around 20 individuals, which may be adult females and their calves, adult males, young adults, or juveniles. Of these four categories, the first three, especially females, tend to be concentrated in breeding schools of fewer than 300 dolphins. Juveniles, mostly males, tend to concentrate in separate, smaller schools. Groups of *S. longirostris* may contain up to several hundred individuals (Leatherwood, Caldwell, and Winn 1976). Except for mother-calf bonds, social groups of *S. longirostris* are very fluid, with large schools continually forming and breaking down and individuals moving freely among several sets of companions over periods of minutes, hours, days, or weeks (Perrin and Gilpatrick 1994). *S. frontalis* also appears to have a complex social organization, with groups numbering up to 100 individuals but usually fewer than 50 and typically 1–15 in coastal waters (Perrin, Caldwell, and Caldwell 1994).

Schools of as many as 3,000 *S. coeruleoalba* are found off Japan (Nishiwaki 1975), though most contain fewer than 500 individuals. Schools in that area consist of adults, juveniles, or animals of all ages. Some schools, termed breeding schools, contain subgroupings of estrous females and adult males. The latter may leave after mating, and the remaining animals are then said to form a nonbreeding school. Calves evidently leave their mothers at about 2–3 years and then join juvenile schools until they reach sexual maturity. Most subadults then rejoin an adult school, the young females usually entering a nonbreeding school (Perrin, Wilson, and Archer 1994). Much the same pattern has been reported in Japanese populations of *S. attenuata* (Kasuya, Miyazaki, and Dawbin 1974).

In the western Mediterranean, *S. coeruleoalba* has a single annual calving peak, in the autumn, and at birth its young are about 92.5 cm long and have an average weight of 11.3 kg (Perrin, Wilson, and Archer 1994). Strandings of *S. longirostris* in Florida indicate a calving season from May to July, and newborn *S. frontalis* are 88–120 cm long; otherwise little is known about the biology of *Stenella* in Atlantic waters (Mead et al. 1980; Perrin, Caldwell, and Caldwell 1994; Perrin et al. 1987). In the last two decades, however, studies of many individuals killed through direct exploitation off Japan and incidental catching by the tuna fishery of the eastern Pacific have greatly increased our knowledge of the reproduction and life history of *Stenella* (Kasuya 1972*a*, 1976, 1985; Kasuya, Miyazaki, and Dawbin 1974; Miyazaki 1984; Myrick et al. 1986; Nishiwaki 1975; Perrin, Coe, and Zweifel 1976; Perrin and Gilpatrick 1994; Perrin and Henderson 1984; Perrin and Hohn 1994; Perrin, Holts, and Miller 1977; Perrin, Miller, and Sloan 1977; Perrin, Wilson, and Archer 1994; Perrin et al. 1987). Births may occur at any time, but there are pronounced spring and autumn peaks on both sides of the Pacific. The interval between births is about 36 months in *S. longirostris* in the eastern Pacific. Off Japan the birth interval in an intensely exploited population of *S. coeruleoalba* declined from about 4 years in 1955 to 2.8 years in 1977, and the age of female sexual maturity declined from 9.7 to 7.2 years. The birth interval of *S. attenuata* is about 48 months off Japan, where there is little disturbance, but only 26–36 months in the eastern Pacific, where there have been heavy losses to the tuna fishery. Such developments apparently reflect a greater compensatory reproductive effort by the exploited populations. However, such a compensation factor has not been determined to differentiate between the especially hard-hit population of *S. longirostris* found within several hundred kilometers of the Pacific coast of Mexico and Central America and the less heavily exploited population of that species found farther out to sea.

Reported gestation periods are about 10.5 months in *S. longirostris*, 11.2–11.5 months in *S. attenuata*, and 11–12 months in *S. coeruleoalba*. There is normally a single young, but twins have been reported to occur on rare occasion. The mean length of the newborn is about 77 cm in *S. longirostris*, 83 cm in *S. attenuata*, and 100 cm in *S. coeruleoalba*. Lactation continues for about 12–24 months in *S. longirostris*, 11–25 months in *S. attenuata*, and 18 months in *S. coeruleoalba*. A small percentage of females have been found to be simultaneously lactating and pregnant. In *S. longirostris* males reach sexual maturity at 7–10 years, and females at 4–7 years. In *S. attenuata* the average age at sexual maturity is 12–15 years for males and 9–11 years for females, depending on the population. In *S. coeruleoalba* both sexes become sexually mature at 5–9 years and physically mature at 14–17 years. Pregnant female *S. attenuata* up to 35 years old have been found. Natural annual mortality is less than 14 percent for both adults and juveniles, and the life span is relatively long. The maximum age has been estimated at 46 years for *S. attenuata* and 57 years for *S. coeruleoalba*.

There are fisheries directed against *Stenella* in various parts of the world, the most important being that of Japan (Kasuya 1976; Mitchell 1975*a*, 1975*b*; Nishiwaki 1975). Fishermen of that country formerly took about 20,000 *S. coeruleoalba* and 500–2,000 *S. attenuata* each year by driving and harpooning the animals. The dolphins are used locally for human consumption. The population of *S. coeruleoalba* off Japan is estimated to have contained 400,000–600,000 individuals originally, but excessive exploitation has reduced this number by about half. According to Perrin, Wilson, and Archer (1994), Japanese fishermen voluntarily lowered their catch of *S. coeruleoalba* to only about 1,000 annually in 1989–93. Small numbers also are taken in the western Mediterranean, where the total population has been estimated at 225,000.

Stenella is the genus that has been most seriously affected by the incidental killing of dolphins by tuna fisheries (Boreman 1992; International Whaling Commission 1992; Joseph 1994; Klinowska and Cooke 1991; Marine Mammal Commission 1994; Mitchell 1975*a*, 1975*b*; U.S. National Marine Fisheries Service 1978, 1981, 1987, 1989, 1994). Fishermen know that *Stenella* often associates with schools of tuna, and they therefore set their nets around groups of dolphins. Deployment of a new type of purse seine in the 1960s led to the accidental death of great numbers of dolphins, which became entangled as the nets were closed. The species *S. attenuata* and *S. longirostris* in the eastern tropical Pacific suffered the most serious losses, caused by American vessels. In that region the primary range of both species extends approximately from Baja California and Peru west to Hawaii and the Marquesas Islands. About 6 million individuals of those two species and more than 1 million of other species were killed there from 1959 to 1989. In 1972, one of the worst years on record, the estimated number of dolphins taken by U.S. fishermen was 368,600, of which 178,000 were *S. attenuata* and 42,000 were *S. longirostris*. That same year, however, the United

States Marine Mammal Protection Act was passed, prohibiting the killing and importation of cetaceans by persons subject to U.S. jurisdiction except under permit. To avoid economic hardship, the U.S. government issued permits allowing a certain number of dolphins to be taken accidentally, but the set quota was reduced each year, dropping from 78,000 in 1976 to 31,150 in 1980 and 20,500 in 1990. Regular inspection, development of improved fishing gear, cooperation by fishermen, and a considerable decline in the U.S. fishing fleet led to a marked reduction in mortality. In 1980 American vessels were estimated to have accidentally taken only 15,000 dolphins of all kinds. For the next decade the kill by U.S. tuna fishermen remained about the same or slightly higher, but reduced quotas and more intensive enforcement thereof, in response to public concern, resulted in a kill of only 115 dolphins in 1993.

Unfortunately, concomitant with this falling accidental take and the decline of U.S. fishing activity there was an enormous rise in incidental taking by foreign fishing fleets in the eastern Pacific. To some extent this situation resulted merely from a change in registration by vessels unwilling to comply with the new U.S. regulations. The total annual kill was near or above 100,000 from 1986 to 1989, and once again there was concern for the survival of the affected species. Protest by the environmental community and a threatened boycott of imported tuna led to voluntary restrictions by American canners and to congressional action requiring that foreign countries seeking to export tuna to the United States demonstrate that they have conservation programs comparable to the U.S. program. In 1992, 10 nations involved in the tuna fishery established the International Dolphin Conservation Program, aimed at achieving ecologically sound and commercially viable means of taking tuna while totally eliminating dolphin mortality. The pursuant International Dolphin Conservation Act of 1992 in the United States established the basis for regulatory, research, and conservation efforts needed to attain the latter objective. Although there are questions about how well these measures are being implemented, and even though much disagreement continues in the United States and between the involved nations, there has been a steady reduction in the incidental kill since 1990. The take by foreign fishing fleets in 1993 was estimated at only 3,900 dolphins.

All species of *Stenella* are on appendix 2 of the CITES, and the IUCN now designates the three Pacific species, *S. longirostris, S. attenuata,* and *S. coeruleoalba,* as conservation dependent. Current population estimates for the eastern tropical Pacific are: *S. longirostris* (eastern and whitebelly spinners), 1.7 million; *S. attenuata,* 2.1 million; and *S. coeruleoalba,* 1.9 million (U.S. National Marine Fisheries Service 1994; Wade and Gerrodette 1993). These figures may seem large, but in some cases they represent severe declines that have occurred since the 1960s as a consequence of the tuna purse-seine fishery. Some estimates place the eastern spinner population alone at as many as 5.6 million in 1959 (Klinowska and Cooke 1991), though Wade (1993) calculated that current numbers, 632,000, are 44 percent as great as the historical norm. The eastern spinner has been described as a highly distinctive subspecies (*S. longirostris orientalis*) that warrants special conservation attention (Douglas et al. 1992; Perrin, Akin, and Kashiwada 1991). The U.S. National Marine Fisheries Service (1992, 1993*a*) classified the eastern spinner as "depleted" in accordance with the Marine Mammals Protection Act but rejected a petition to list the subspecies as threatened pursuant to the Endangered Species Act.

Another morphologically distinct subspecies, *S. longirostris centroamericana,* is found within several hundred kilometers of the coast of Mexico and Central America. As early as 1979 its numbers were calculated to be only about 20 percent as great as originally (Perrin and Henderson 1984), and overall numerical estimates for the species are now less than half what they were then. The northern offshore stock of *S. attenuata,* found mainly within 1,000 km of the coast from Baja California to northern South America, now contains about 731,000 individuals, fewer than a quarter of the number thought to have been present originally (Marine Mammal Commission 1994). This stock has been listed as "depleted" by the U.S. National Marine Fisheries Service (1993*b*). The stock of *S. attenuata* along the coast of Mexico and Central America has been described as a separate subspecies, *S. a. grafmani,* and is thought to contain only about 30,000 individuals of the total given above for the species (Wade and Gerrodette 1993).

CETACEA; DELPHINIDAE; **Genus DELPHINUS**
Linnaeus, 1758

Common Dolphins, or Saddleback Dolphins

Heyning and Perrin (1994) recognized two species:

D. delphis, temperate and tropical waters throughout the Pacific Ocean and adjoining seas, the North Atlantic Ocean, and the Mediterranean and Black seas;
D. capensis, nearshore tropical and temperate waters from Japan to Taiwan, from southern California and the Gulf of California to Peru, from Venezuela to northeastern Argentina, along the western and southern coasts of Africa, around Madagascar, and in the northern Indian Ocean and adjoining waters.

Heyning and Perrin indicated that additional study was needed to determine whether the population of *D. capensis* in the northern Indian Ocean and adjoining waters actually represents a third species, *D. tropicalis,* as suggested earlier by Van Bree and Gallagher (1978). Prior to the work of Heyning and Perrin most authorities treated *D. capensis* as part of *D. delphis* (E. R. Hall 1981; Klinowska and Cooke 1991; Mead and Brownell *in* Wilson and Reeder 1993; Rice 1977; Van Bree and Purves 1972). However, Banks and Brownell (1969) had recognized *D. capensis* as a distinct species (using the name *D. bairdii*) based on specimens from waters off California and Baja California. This distinction is supported by recent analysis of mitochondrial DNA from the same region (Rosel, Dizon, and Heyning 1994).

Head and body length is 150–250 (rarely 260) cm, pectoral fin length is about 30 cm, dorsal fin height is about 40 cm, and the expanse of the tail flukes is about 50 cm. Average adult weight is about 80 kg, and the maximum is 136 kg (Evans 1994). *Delphinus* is among the most colorful of cetaceans. Over most of its range the back is brown or black, the belly is whitish, and the sides have bands and stripes of gray, yellow, and white. Some populations, however, lack the side markings. The well-defined beak is narrow and sharply set off from the forehead by a deep V-shaped groove. The beak of *D. capensis* is longer than that of *D. delphis* (Heyning and Perrin 1994). There are 40–50 teeth on each side of each jaw.

Delphinus is a pelagic genus, generally occurring well out to sea, but it has been reported to enter fresh water occasionally. Populations off southern California evidently shift northward and farther offshore during the spring and

Common dolphin *(Delphinus delphis)*, photo by M. Scott Sinclair through National Marine Fisheries Service.

summer (U.S. National Marine Fisheries Service 1978). Intensive radio-tracking studies have shown that the common dolphin tends to occur in places with significant bottom relief—canyons, escarpments, seamounts—where currents are interrupted and the resultant upswell supports high levels of oceanic productivity. Schools tend to divide into small parties to feed in the afternoon and at night and then regroup at dawn. Dives generally last two to three minutes and reach depths of up to 280 meters (Leatherwood and Reeves 1978; Leatherwood et al. 1982). *Delphinus* is among the swiftest of cetaceans; although it usually travels at about 10 km/hr, it is capable of reaching more than 40 km/hr. It often rides the bow waves of ships and seems to frolic about them. It sometimes feeds in the company of *Lagenorhynchus.* The diet includes cephalopods and small fish, including flying fish. *Delphinus* produces clicks for echolocation, whistles for group coordination, and various other sounds common to most dolphins (Evans 1994), such as described in the accounts of *Tursiops* and *Stenella.*

Delphinus is among the most gregarious of mammals, often being seen in groups of 1,000 or more (Leatherwood and Reeves 1978). Schools of as many as 300,000 individuals have sometimes formed over concentrations of fish in the Black Sea (Mitchell 1975*a*). During the spring and summer the large schools may break up into groups of 50–200 animals, and the sexes may segregate between mating seasons (U.S. National Marine Fisheries Service 1978). Some observations suggest that permanent social units contain fewer than 30 dolphins (Evans 1994). Individuals have been seen aiding wounded members of their group, supporting them in the water and pushing them to the surface to breathe (Lekagul and McNeely 1977).

In the northeastern Pacific there are apparently two periods of mating, January–April and August–November, and two periods of calving, March–May and August–October. Each adult female gives birth every 2 or 3 years. Gestation lasts 9–11 months. The offspring weighs about 7 kg and is 80–90 cm long at birth and is weaned after about 6 months. Estimates of the age at sexual maturity range from 2 to 7 years in males and 3 to 12 years in females depending on the region. The potential longevity is more than 20 years (Evans 1994; Leatherwood and Reeves 1978; Perrin and Reilly 1984; U.S. National Marine Fisheries Service 1978).

Delphinus has long been hunted by people in various parts of the world. Perhaps the largest fishery was in the Black Sea, where the population is estimated to have been 1 million individuals and where up to 200,000 were taken annually in the 1930s. This population subsequently underwent a precipitous decline, and its hunting was banned by the Soviet government in 1966 (Mitchell 1975*a*). Turkish fishermen, however, continued to catch large numbers, perhaps 88,000 in 1971 alone (Berkes 1977). The Turkish fishery was legally banned in 1983, but only about 50,000 dolphins were estimated to survive in the Black Sea, and there are reports of continued killing (Evans 1994). *Delphinus* has been accidentally taken by the Pacific tuna fishery in the same manner as described in the account of *Stenella.* The kill was around 20,000–25,000 individuals annually in the 1970s and 1980s and appears to have contributed to severe reductions in some of the subpopulations in the region (Klinowska and Cooke 1991; Reeves and Leatherwood 1994) but has subsequently declined in response to regulation. The total current population of the common dolphin in the eastern tropical Pacific is estimated at 3.1 million (U.S. National Marine Fisheries Service 1989, 1994). *Delphinus* is on appendix 2 of the CITES.

CETACEA; DELPHINIDAE; **Genus LAGENODELPHIS**
Fraser, 1956

Short-snouted White-belly Dolphin, or Fraser's Dolphin

The single species, *L. hosei,* occurs primarily in tropical parts of the Atlantic, Indian, and Pacific oceans and adjoining seas, with some records from temperate waters (Caldwell, Caldwell, and Walker 1976; Hersh and Odell 1986; Perrin, Leatherwood, and Collet 1994; Rice 1977). Until 1971 *Lagenodelphis* was known to science only by a single

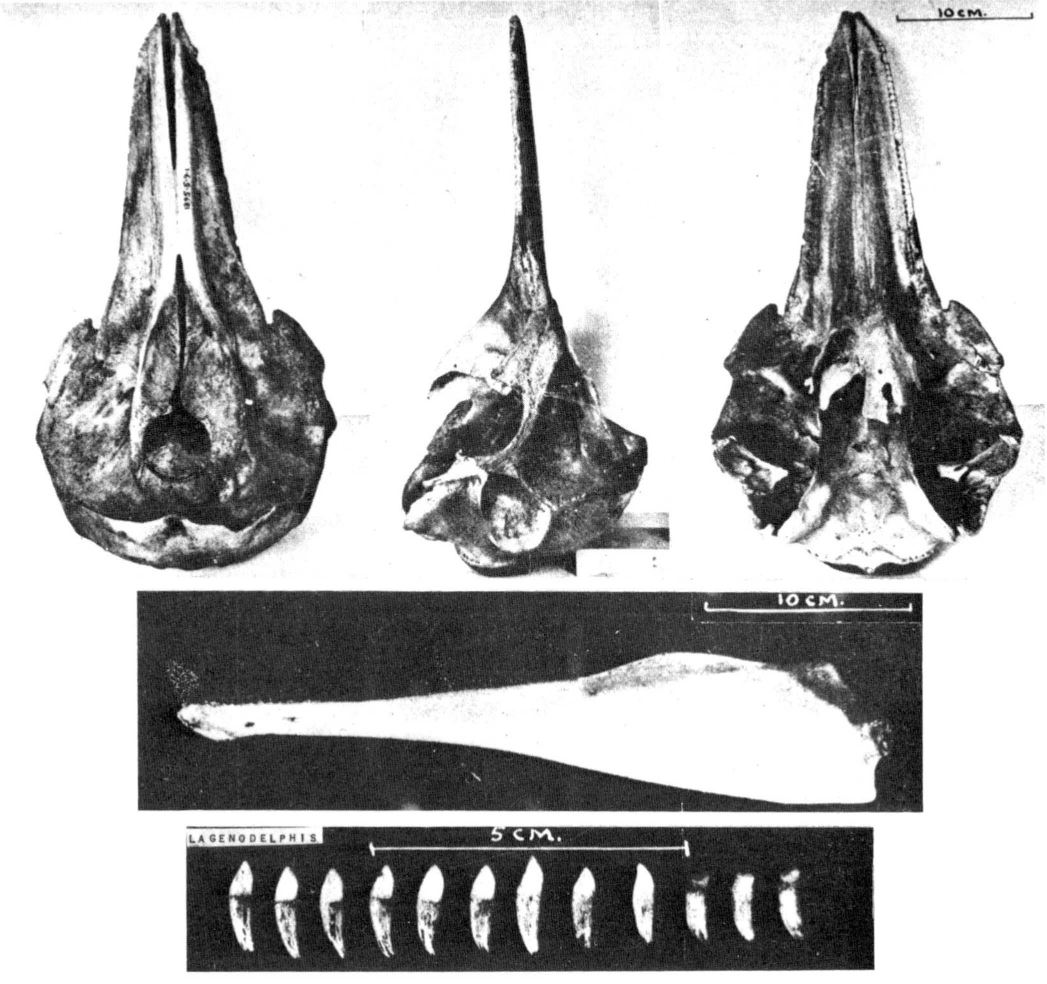

Fraser's dolphin *(Lagenodelphis hosei)*, photos from *Sarawak Museum Journal*, F. C. Fraser.

skeleton collected before 1895 at the mouth of the Lutong River in Borneo.

Head and body length is 206–70 cm, pectoral fin length is 22–29 cm, dorsal fin height is 13–22 cm, and the expanse of the tail flukes is 41–59 cm; two adult males weighed 129 kg and 209 kg and an adult female weighed 164 kg (Perrin, Leatherwood, and Collet 1994). The back is gray and the belly is white. Along each side, extending from the region of the eye to the anus, are three stripes, a creamy white one above, a blackish one in the middle, and another white one below. The body is robust, the pectoral and dorsal fins are relatively small, and the beak is short and indistinct. There are 36–44 teeth on each side of each jaw (Leatherwood, Caldwell, and Winn 1976). The skull is similar to that of *Lagenorhynchus,* but the facial part has a pair of deep palatal grooves and the premaxillary bones are fused dorsally in the midline; this fusion resembles that seen in *Delphinus* but is not as extensive.

Most records are from well offshore. *Lagenodelphis* appears to be a deep diver, and when it surfaces to breathe it often charges up, creating a spray from its head. It has been reported to leap clear of the water. Recorded stomach contents include deep-sea fish, squid, and shrimp (Caldwell, Caldwell, and Walker 1976; Mitchell 1975*b*). In the Caribbean, Watkins et al. (1994) observed two "pods" (family groups) numbering about 60 and 80 individuals herding and capturing fish near the surface. These individuals emitted broadband clicks for apparent echolocation and communicative whistles similar to those of *Tursiops. Lagenodelphis* also commonly occurs in groups of as many as 1,000, and occasionally 2,500, and sometimes is seen in the company of *Stenella attenuata* (Jefferson and Leatherwood 1994; Leatherwood, Caldwell, and Winn 1976; Leatherwood et al. 1982). A pregnant female was collected off South Africa on 17 February 1971 (Perrin et al. 1973). Available data do not suggest strong seasonality in breeding in the eastern Pacific (Perrin, Leatherwood, and Collet 1994). The gestation period probably is 10–12 months, the single newborn is approximately 100 cm long, sexual maturity evidently occurs at 7 years in both sexes, and the maximum known longevity is 16 years (Jefferson and Leatherwood 1994). Many of the known specimens have been taken accidentally in the tuna fishery in the same manner as described in the account of *Stenella* (U.S. National Marine Fisheries Service 1978). An estimated 289,300 individuals are present in the eastern tropical Pacific (Wade and Gerrodette 1993). *Lagenodelphis* is on appendix 2 of the CITES.

Right whale dolphins *(Lissodelphis borealis):* Top and middle, photos by M. Scott Sinclair through National Marine Fisheries Service; Bottom, photo by K. C. Balcomb.

CETACEA; DELPHINIDAE; **Genus LISSODELPHIS**
Gloger, 1841

Right Whale Dolphins

There are two species (Jefferson et al. 1994; Rice 1977):

L. borealis, temperate waters of the North Pacific, from Japan and the Kuril Islands to British Columbia and California, with extralimital records as far north as the Aleutians;

L. peronii, temperate waters of the oceans of the Southern Hemisphere, the southernmost record being from about 64° S.

In *L. borealis* head and body length is up to 307 cm in males and 230 cm in females, pectoral fin length is about 28 cm, and the expanse of the tail flukes is about 35 cm. A mature male 282 cm long weighed 113 kg. A mature female 226 cm long weighed 81 kg (Leatherwood and Walker 1979). The largest measured *L. peronii* was 297 cm long, and an individual 251 cm long weighed 116 kg (Jefferson et al. 1994). *L. borealis* is mostly black or dark brown, but there is a broad white patch on the chest and a narrow midventral white band from the chest to the tail. In *L. peronii* the top of the head and back are black or bluish black, and the beak, flippers, tail, sides, and underparts are white. The beak is short but distinct. There are 36–47 sharp, pointed teeth on each side of each jaw. The vernacular name refers to the lack of a dorsal fin in both *Lissodelphis* and the right whales (*Balaena*).

The information in the remainder of this account was compiled from Fitch and Brownell (1968), the International Whaling Commission (1992), Jefferson et al. (1994), Leatherwood et al. (1982), Leatherwood and Walker (1979), Mitchell (1975*b*), and the U.S. National Marine Fisheries Service (1978). Right whale dolphins are primarily oceanic, usually remaining well offshore in cool, deep waters. Numbers of *L. borealis* apparently increase over the continental shelf of California during the autumn and winter, a period corresponding with the maximum abundance of squid in the region. In the late spring and summer that species moves farther north and out to sea. *L. peronii* generally occurs between the subtropical and Antarctic convergences but in some years extends to the south of the latter. *Lissodelphis* is quick and active and sometimes travels by smooth, low leaps above the surface, each of which may cover up to 7 meters. Speeds as great as 45 km/hr may be briefly attained; a boat traveling at 33 km/hr was easily outdistanced during a 10-minute chase. Bow waves of vessels are usually avoided. Entire schools have been observed to dive for as long as 6 minutes and 15 seconds. Otoliths found in the stomach of an adult *L. borealis* indicated that the animal had been feeding on fish that dwell at depths of at least 200 meters. The diet consists of various kinds of fish, especially lanternfish, and cephalopods, such as squid. *L. borealis* produces series of clicks much like those used by other genera for echolocation. Whistles, apparently for communication, are also emitted but are not common.

Groups contain as many as 3,000 individuals but usually number 100–200. Newborn *L. borealis* are seen most frequently in winter and early spring and are about 60–100 cm long. Female *L. peronii* with near-term fetuses have been found in November and April.

Fishermen deliberately take *L. peronii* off Chile and *L. borealis* off Japan. A more substantial problem is incidental taking by the squid driftnet fleets of Japan, Taiwan, and South Korea. An estimated 15,000–24,000 *L. borealis* were being killed annually in such a manner during the late 1980s. As a result, populations in the northwestern Pacific have been reduced by 24–73 percent. Populations off California, estimated to number around 79,000 individuals, are subject to a small incidental kill. Both species are on appendix 2 of the CITES.

CETACEA; DELPHINIDAE; **Genus ORCAELLA**
Gray, 1866

Irrawaddy River Dolphin

The single species, *O. brevirostris*, occurs in coastal waters from the Bay of Bengal to New Guinea and northern Australia and ascends far up the Ganges, Irrawaddy, Mekong, and other rivers (Rice 1977).

Head and body length is 180–275 cm (Lekagul and McNeely 1977). A mature female weighed 190 kg (Marsh et al. 1989). The coloration is slaty blue or slaty gray throughout, or the underparts may be slightly paler. *Orcaella* has a bulging forehead, a short and shelflike beak, and 12–20 teeth on each side of each jaw. The pectoral fin is broadly triangular. The dorsal fin is small, sickle-shaped, and located on the posterior half of the back. The neck and tail are very flexible (Mitchell 1975*b*).

Orcaella lives in warm, tropical, and often silty waters. It regularly enters rivers, has been recorded from 1,440 km up the Irrawaddy, and can live permanently in fresh water. It often accompanied river steamboats. It breathes at intervals of 70–150 seconds; the head appears first and then disappears, and then the back emerges, but the tail is rarely seen. The diet consists of fish and crustaceans, often found on the bottom (Lekagul and McNeely 1977; Mitchell 1975*b*). Although groups usually consist of no more than 10 animals, solitary individuals are rarely seen (Leatherwood and Reeves 1983). Mating has been reported to occur from March to June, and the gestation period has been estimated at 14 months; a captive newborn calf was 96 cm long and weighed 12.3 kg, began eating solid food at 6 months, and was fully weaned at 2 years. Physical maturity may be attained at 4–6 years, and longevity may be about 30 years (Marsh et al. 1989).

This dolphin occasionally is taken accidentally, and its oil reportedly has been used as a remedy for rheumatism in parts of India, but it is generally unexploited. According to Thein (1977), the fishermen of Burma attract *Orcaella* by tapping the sides of their boats with oars. The dolphin swims around the boat in ever-diminishing circles, thereby forcing the fish into nets. The fishermen share their catch with *Orcaella* and consider it a friend, not to be harmed. Some fishermen in Cambodia and Viet Nam regard *Orcaella* as sacred and release it if it becomes entangled in their nets, but others kill it for use as food (Marsh et al. 1989). Logging and other human disruption of vulnerable coastal and riverine habitat has reduced the size of some populations of *Orcaella* (Klinowska and Cooke 1991). A relatively large population still exists in the coastal waters of the Northern Territory of Australia (Reeves and Leatherwood 1994). *Orcaella* is on appendix 2 of the CITES.

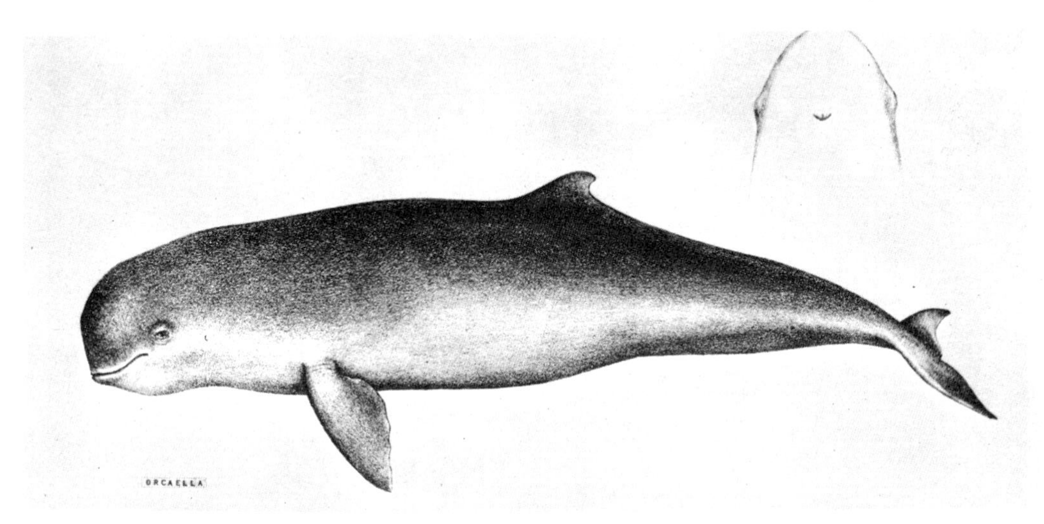

Irrawaddy River dolphin *(Orcaella brevirostris)*, photo from *Anatomical and Zoological Researches: Zoological Results of the Two Expeditions to Western Yunnan in 1868 and 1875*, John Anderson.

CETACEA; DELPHINIDAE; **Genus CEPHALORHYNCHUS**
Gray, 1846

Southern Dolphins, or Piebald Dolphins

There are four species (Brownell and Praderi 1985; Goodall, Galeazzi, et al. 1988; Goodall, Norris, et al. 1988; Rice 1977):

C. commersonii, coastal waters of Argentina, Tierra del Fuego, Falkland Islands, Kerguelen Islands, and possibly South Georgia and South Shetland Islands;
C. eutropia (black dolphin), coastal waters of Chile and Tierra del Fuego;
C. heavisidii, coastal waters of southwestern Africa;
C. hectori, coastal waters of New Zealand.

Head and body length is 110–80 cm, pectoral fin length is 15–30 cm, dorsal fin height is about 7–15 cm, and the expanse of the tail flukes is about 22–41 cm. Females generally are slightly larger than males. The various species do not differ greatly in weight, and the overall range is about 26–86 kg (P. B. Best 1988; Goodall 1994; Goodall, Galeazzi, et al. 1988; Goodall, Norris et al. 1988; Slooten and Dawson 1988). The color pattern comprises areas of sharply contrasting black and white. Generally the head, pectoral and dorsal fins, tail region, and posterior part of the back are black and the chin, sides, belly, and anterior part of the back are white; sometimes the upper parts are entirely black. There is no beak, and there are 24–35 teeth on each side of each jaw.

These dolphins are found in coastal waters (Rice 1977). According to Baker (1978), *C. hectori* usually occurs within 8 km of the coast of New Zealand, often in shallow, muddy waters. It often is attracted to boats and briefly rides the bow waves and follows in the wake. It sometimes feeds on

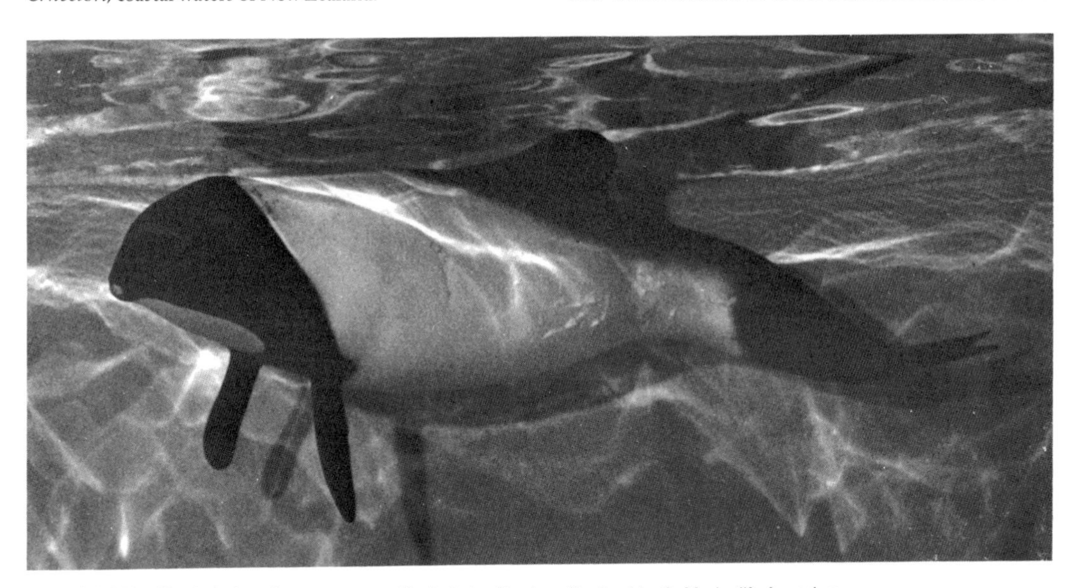

Piebald dolphin *(Cephalorhynchus commersonii)*, photo by Stephen Spotte, Mystic Marinelife Aquarium.

Piebald dolphin *(Cephalorhynchus commersonii)*, photo by Stephen Spotte, Mystic Marinelife Aquarium.

the sea floor, taking various kinds of fish and perhaps such invertebrates as squid and crustaceans, but Slooten and Dawson (1994) noted that it also feeds at various depths and has been observed chasing fish at the surface. Groups usually contain 2–8 individuals, but occasionally more than 20 are seen together, and there is one record of an aggregation of 200–300. Slooten and Dawson (1988) indicated that social interaction takes place within such aggregations. They reported also that mating peaks early in the winter, that *C. hectori* gives birth during the austral spring and early summer (November–February), that the calving interval probably is at least 2 years, and that the young evidently begin taking solid food by the age of 6 months. Slooten and Dawson (1994) added that individuals associate freely with many others, courtship is characterized by frequent vertical leaps out of the water, calves usually remain with the mother until they are 2 years old, sexual maturity is attained at 6–9 years, and the maximum observed longevity is 20 years.

The behavior and ecology of *C. commersonii* and *C. eutropia* is basically similar to that of *C. hectori* (Goodall, Galeazzi, et al. 1988; Goodall, Norris, et al. 1988). Both are strong swimmers and often ride bow waves. *C. eutropia* is found in areas of strong tidal flow above a steeply dropping continental shelf and is known to ascend rivers. *C. commersonii* occurs along the coast, where the continental shelf is wide and flat; its diet consists largely of bottom-dwelling invertebrates. Such habitat, which is characterized by turbid water, probably makes echolocation important for navigation and locating prey. The narrow-band, high-frequency pulses emitted by *Cephalorhynchus* are very much like those of *Phocoena*, which occupies the same kind of habitat. Watkins, Schevill, and Best (1977) recorded three kinds of sounds made by all four species of *Cephalorhynchus:* clicks at slow rates, short bursts of clicks, and a cry.

Limited data indicate that *C. heavisidii* generally is found in groups of 2 or 3, that the young are born in the summer, and that the neonate is about 85 cm long (Best and Abernethy 1994). Groups of *C. eutropia* usually consist of 2–5 individuals, but aggregations of 20–50 frequently are seen (Goodall, Norris, et al. 1988). Groups of more than 110 *C. commersonii* have been reported, but the usual number is about 2–4 (Goodall 1994; Goodall, Galeazzi, et al. 1988; Leatherwood, Kastelein, and Miller 1988). A captive group of 6 *C. commersonii* was kept together without strife, each animal consuming 3–4 kg of food per day (Gewalt 1979*a;* Spotte, Radcliffe, and Dunn 1979). In the wild *C. commersonii* gives birth during the spring and summer (October–March). Gestation lasts about 12 months, the newborn is 55–75 cm long and weighs 4.5–7.5 kg, sexual maturity is reached at 5–8 years, and individuals up to 18 years old have been found (Goodall 1994; Goodall, Galeazzi, et al. 1988; Lockyer, Goodall, and Galeazzi 1988). *C. hectori* also gives birth during the spring and summer off New Zealand (November–February), there is an interbirth interval of 2–3 years, sexual maturity is attained at 6–9 years, and maximum known longevity is 20 years (Slooten 1991).

Cephalorhynchus is taken deliberately off the coast of South America and is subject to accidental killing by fishermen and crabbers throughout its range; the genus is considered to be especially vulnerable to coastal purse seining (Goodall, Galeazzi, et al. 1988; Goodall, Norris, et al. 1988; Mitchell 1975*a*, 1975*b*). An intensifying deliberate kill for use as crab bait appears to be adversely affecting populations of *C. commersonii* and *C. eutropia* off Chile and Argentina (Klinowska and Cooke 1991). *C. hectori*, of New Zealand, appears particularly susceptible because of the trawl fishing and pollution in its limited habitat, but Baker (1978) did not consider reports of a decline to be substantiated. Cawthorn (1988) estimated this species to number 5,000–6,000 and indicated that entanglement in gill nets posed a threat. Dawson and Slooten (1988) concluded that the species was declining because of this problem and estimated that only 3,000–4,000 survived. This information led the government of New Zealand to prohibit commercial use of gill nets within a sanctuary of 1,170 sq km off the Banks Peninsula, on the east coast of South Island, where there is a high concentration of *C. hectori* (Slooten and Dawson 1994). *C. hectori* now is classified as vulnerable by the IUCN, and all species of *Cephalorhynchus* are on appendix 2 of the CITES.

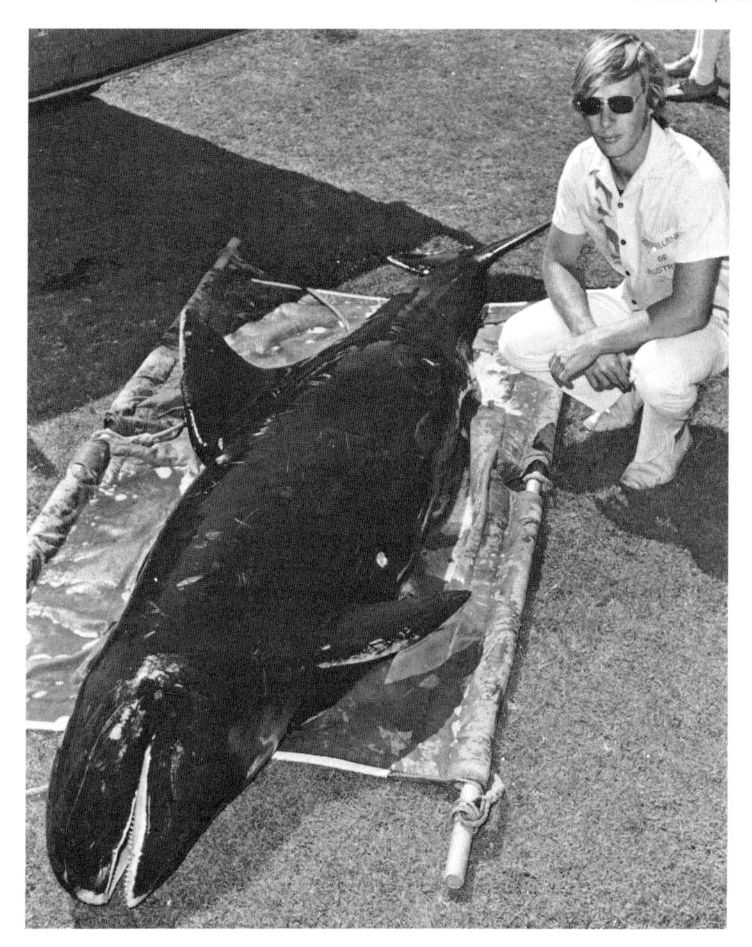

Melon-headed whale *(Peponocephala electra)*, photo by Michael M. Bryden.

CETACEA; DELPHINIDAE; **Genus PEPONOCEPHALA**
Nishiwaki and Norris, 1966

Many-toothed Blackfish, or Melon-headed Whale

The single species, *P. electra,* occurs in tropical waters of the Atlantic, Indian, and Pacific oceans and also is known from Maryland, England, South Africa, Japan, and eastern Australia (Best and Shaughnessy 1981; Caldwell, Caldwell, and Walker 1976; Perrin 1976; Perryman et al. 1994; Rice 1977). This species long was placed in the genus *Lagenorhynchus,* but Nishiwaki and Norris (1966) considered it to resemble *Feresa* and *Pseudorca* more closely and placed it in a distinct genus.

Head and body length at physical maturity is about 250–80 cm, pectoral fin length is about 50 cm, dorsal fin height is about 25 cm, and the expanse of the tail flukes is about 60 cm. Best and Shaughnessy (1981) reported an adult male to weigh 206 kg. The upper parts are black and the belly is slightly paler. The lips and an oval patch on the abdomen are often white or unpigmented. The body is rather long and slim and the tail stock is long. Unlike in *Lagenorhynchus,* there is no beak, the forehead being rounded, curving smoothly from the anterior tip of the rostrum to the blowhole, and overhanging the lower jaw to some extent. *Peponocephala* resembles *Pseudorca* but is smaller, has a sharper appearance to the snout, and has more teeth. There are 20–26 teeth on each side of each jaw. The anterior three cervical vertebrae are fused (Bryden, Harrison, and Lear 1977; Caldwell, Caldwell, and Walker 1976; Nishiwaki and Norris 1966).

Most of what is known about the natural history of this genus was summarized by Bryden, Harrison, and Lear (1977), Perrin and Reilly (1984), and Perryman et al. (1994). *Peponocephala* seems normally to stay well out to sea but may be seen near land at any time of the year. One group was observed to swim at 20 km/hr, and many individuals leaped almost clear of the water. The diet consists of fish, squid, and possibly pelagic mollusks. *Peponocephala* is probably a social animal. Most schools contain 15–1,500 individuals. One stranding of 53 specimens in Australia included 16 males, 35 females, and 2 individuals of undetermined sex. Limited data suggest that in the Southern Hemisphere births occur from August to December and that gestation lasts 12 months. Newborn individuals have been found in Hawaii in June and in the Philippines in April and June. Length at birth is about 100 cm. Males seem to attain sexual maturity by the time they are 7 years old, and females by 12 years. Maximum longevity has been estimated to be 47 years.

Peponocephala has been taken in moderate numbers by

fisheries in Japan, Indonesia, and other countries, but there is no evidence of any substantive effect on overall populations (Klinowska and Cooke 1991; Perryman et al. 1994). An estimated 45,400 individuals of this genus inhabit the eastern tropical Pacific Ocean (Wade and Gerrodette 1993). It is on appendix 2 of the CITES.

CETACEA; DELPHINIDAE; **Genus FERESA**
Gray, 1871

Pygmy Killer Whale

The single species, *F. attenuata,* occurs in tropical and warm temperate waters of the Atlantic, Indian, and Pacific oceans and adjoining seas (Leatherwood, Caldwell, and Winn 1976; Rice 1977).

Head and body length is 207–87 cm, pectoral fin length is about 40–50 cm, dorsal fin height is 20–30 cm, the expanse of the tail flukes is 50–66, and recorded weights of five individuals were 110–56 kg (Klinowska and Cooke 1991; Leatherwood, Caldwell, and Winn 1976; Ross and Leatherwood 1994). The back is black or dark gray, the sides are often paler, the chin is often white, and there are usually white areas around the lips and on the belly. The forehead is rounded. There are 8–13 teeth on each side of each jaw. *Feresa* bears some resemblance to *Orcinus,* but the latter is much larger.

Until 1952 *Feresa* was known only by two skulls, but subsequently the genus was found to be widely distributed though perhaps rare. Most records are from places with little or no continental shelf. *Feresa* apparently resides year-round off Japan, Hawaii, and South Africa, but short, seasonal migrations may occur (Caldwell and Caldwell 1971*a*). A specimen recently found stranded near Buenos Aires, Argentina, represents the southernmost record (Lichter, Fraga, and Castello 1990). The diet is not well

Pygmy killer whales *(Feresa attenuata),* photo by Kenneth S. Norris (top) and Gary L. Friedrichsen (bottom) through National Marine Fisheries Service.

known, but a captive in Japan ate sardines, horse mackerel, sauries, and squid (Nishiwaki 1966). According to the U.S. National Marine Fisheries Service (1978, 1981), *Feresa* feeds on other kinds of dolphins, especially the young. Captives are aggressive, and other kinds of dolphins kept with them show fright reactions. *Feresa* usually travels in groups of 5–10, but aggregations of several hundred have been seen. Ross and Leatherwood (1994) reported that 16 groups seen in the eastern tropical Pacific contained an average of 25 individuals, and that available data suggest that most calves are born during the summer. Small numbers of *Feresa* are taken for human utilization, but there may be a more substantial loss to incidental gill-net fisheries, especially in Sri Lanka. The genus is on appendix 2 of the CITES.

CETACEA; DELPHINIDAE; **Genus PSEUDORCA**
Reinhardt, 1862

False Killer Whale

The single species, *P. crassidens,* occurs in all the temperate and tropical oceans and seas of the world (Rice 1977). The species was described in 1846 on the basis of a fossilized skull found in the Lincolnshire Fens in England. Stranded animals collected in 1862 in the Bay of Kiel allowed assessment of the external morphology (Allen 1942).

Head and body length is as much as 610 cm in males and 490 cm in females. Dorsal fin height is about 40 cm, and weight reaches 1,360 kg (Scheffer 1978*a;* U.S. National Marine Fisheries Service 1978). The coloration is black throughout. *Pseudorca* bears some resemblance to *Orcinus* but can be distinguished by its uniformly dark color, more slender build, more tapering head, smaller and more backwardly curving dorsal fin, and tapering flippers, which average about one-tenth of the head and body length. There are usually 8–11 teeth on each side of each jaw.

Pseudorca usually is found well out to sea. It often rides the bow waves of ships and jumps completely out of the water (Leatherwood, Caldwell, and Winn 1976). The diet is known to include squid, medium to large fish, and young dolphins (U.S. National Marine Fisheries Service 1981). A captive, however, was not aggressive toward other kinds of dolphins in the same pool and formed a close social relationship with them. This captive also was friendly to people and learned tricks rapidly. The wild school from which it was taken contained about 300 individuals that comprised small subgroups of 2–6 animals each. They emitted

False killer whales *(Pseudorca crassidens):* A. Photo by K. C. Balcomb; B. Photo from Marineland of the Pacific.

a variety of vocalizations, including a piercing whistle (Brown, Caldwell, and Caldwell 1966). Schools of as many as 835 individuals have stranded, and these have included animals of all ages and sexes (U.S. National Marine Fisheries Service 1978). Average group size has been reported to be 55 off Japan and 180 in South Africa (Stacey, Leatherwood, and Baird 1994). Breeding may occur throughout the year (Scheffer 1978*a*), but in Japanese waters there is a mating peak in December–January and a parturition peak around March. The average interval between births is 6.9 years, gestation lasts about 15.5 months, the newborn are usually 160–200 cm long, and lactation is estimated to continue for 18–24 months. Sexual maturity is attained by females at 8–11 years but not for another 8–10 years by males. Females more than 45 years old may be postreproductive, but some are estimated to have lived for 62.5 years (Perrin and Reilly 1984; Purves and Pilleri 1978; Stacey, Leatherwood, and Baird 1994).

Pseudorca is notorious to fishermen in various parts of the world because of its habit of stealing fish from lines (Leatherwood, Caldwell, and Winn 1976; Scheffer 1978*a*). *Pseudorca* is taken in small numbers for human consumption and incidental to the tuna fishery of the eastern tropical Pacific (Mitchell 1975*a*, 1975*b*). Except perhaps for Japanese waters, there do not appear to be any acute bioconservation problems for the genus (Reeves and Leatherwood 1994). An estimated 39,800 individuals are present in the eastern tropical Pacific (Wade and Gerrodette 1993). *Pseudorca* is on appendix 2 of the CITES.

CETACEA; DELPHINIDAE; **Genus ORCINUS**
Fitzinger, 1860

Killer Whale

Rice (1977) and most other authorities have recognized a single species, *O. orca*, which occurs in all the oceans and adjoining seas of the world, chiefly in coastal waters and cooler regions. The species *O. nanus* and *O. glacialis*, described from antarctic waters by Soviet authorities, were not accepted by Heyning and Dahlheim (1988).

Orcinus is the largest member of the dolphin family. Sexual maturity is attained by males when they are about 550 cm long and by females when they reach 490 cm (Mitchell 1975*b*). Subsequently, as in all cetaceans, growth continues for some years. Maximum head and body length is about 980 cm in males and 850 cm in females. Maximum weight is about 9,000 kg in males and 5,500 kg in females (Scheffer 1978*b*). In males the body becomes stocky with age, and there is a disproportionate increase in the size of certain organs: the broad, round pectoral fin reaches a length of 200 cm; the erect, triangular dorsal fin reaches a height of 180 cm; and the tail flukes reach an expanse of 275 cm. In females the body remains less stocky and the dorsal fin is smaller and backwardly curved. There is a sharply contrasting color pattern. The upper parts are black except, usually, for a light gray area behind the dorsal fin; the underparts are white; and there are white patches on the sides of the head and on the posterior flanks. The snout is bluntly rounded. There are 10–14 large teeth on each side of each jaw.

The killer whale seems to prefer coastal waters and often enters shallow bays, estuaries, and the mouths of rivers. Some populations appear to reside permanently in relatively small areas, but others migrate, probably in association with food supplies. In the North Atlantic during the spring and summer there appear to be movements both closer to the southeastern Canadian coast and northward into arctic waters. *Orcinus* usually swims at about 10–13 km/hr but can exceed 45 km/hr. Its dives usually last 1–4 minutes, but one individual is known to have remained underwater for 21 minutes; another was found entangled in a submarine cable at a depth of about 1,000 meters.

Orcinus produces clicklike sounds such as those known to be used by other delphinids for echolocation, and its ability to echolocate has been demonstrated experimentally. Also emitted are a variety of other sounds apparently used in communication, including screams, distinct tonal whistles, and pulsed calls. Whistles usually are associated with low levels of whale activity. Most sounds heard in stable social groups engaged in active behavior are repeated pulsed calls. Each such group has a repertoire of discrete pulsed calls that is unique to the group and that may persist for 10 years or more (Dahlheim and Awbrey 1982; Ford and Fisher 1983; Heyning and Dahlheim 1988).

Orcinus adjusts to captivity and learns quickly. Within

Killer whale *(Orcinus orca)*, photo by Scanlan. Skull: photo by Warren J. Houck.

Killer whale *(Orcinus orca)*, photo by Dr. Bubenik through Bernhard Grzimek.

two months of its first training session one individual was performing before the public, and after five months its trainer could safely place his head into its mouth (Banfield 1974; Leatherwood, Caldwell, and Winn 1976; Scheffer 1978*b*).

Orcinus received its common name because it is the world's largest predator of warm-blooded animals, though in relative terms it is no more of a "killer" than an insectivorous frog or a beef-eating human. And although marine mammals are evidently the preferred prey in certain areas at certain times, an examination of the stomach contents of 364 *Orcinus* taken off Japan revealed more fish and cephalopods than cetaceans and pinnipeds (Scheffer 1978*b*). *Orcinus* does have a throat large enough to swallow seals, young walrus, and the smaller cetaceans. Steltner, Steltner, and Sergeant (1984) described an attack by 30–40 killer whales on a herd of several hundred narwhals *(Monodon)*. A group of killer whales sometimes joins in an attack on a large baleen whale, literally tearing it to pieces. Penguins and other aquatic birds are also seized. When *Orcinus* detects a seal or bird near the edge of the ice in arctic waters, it may dive deeply and then rush to the surface, breaking ice up to a meter thick and dislodging its prey into the water. The approach of a group of killer whales usually seems to cause other marine vertebrates to panic. The U.S. National Marine Fisheries Service (1981) stated that in Argentina *Orcinus* has been seen flopping well up onto the land to drive pinnipeds into the water. Of 568 observations made by Lopez and Lopez (1985) of *Orcinus* hunting pinnipeds along the Argentine coast, 203 involved two or more whales corraling prey in the water, and 365 involved one or more whales rushing up onto the surf zone of the beach and then moving back to deep water; 122 of these beachings ended in the capture of a sea lion or an elephant seal.

Orcinus is a highly social genus, and individuals have been seen to assist others that were trapped or injured. Schools of up to 250 individuals have been reported (Hoyt 1977), but groups usually consist of 2–40 animals (Scheffer 1978*b*). Reports of very large groups may refer to temporary joining of a number of "pods" (family groups), possibly in relation to seasonal peaks in prey availability or to social activity (Heyning and Dahlheim 1988). Based on investigations from 1973 to 1987, Bigg et al. (1990) reported that there is a population of about 261 killer whales in the partly enclosed waters off British Columbia and Washington. They are divided into two "communities," one with 16 pods to the north of Vancouver Island and the other with 3 pods around Vancouver Island, in Puget Sound, and southward along the Washington coast. Individuals within each community share a common range, associate with one another, and only rarely enter the range of the other community. Within the communities are four "clans," each of which comprises a number of pods that have a common acoustical dialect and appear to be distantly related but do not form a particular social unit. In contrast, the members of a pod travel together most of the time. Within each pod are 1–3 cohesive "subpods," which temporarily fragment from the pod to travel about separately. Subpods, in turn, usually contain 2 or more "intrapod groups," each with 2–9 individuals that are always in close proximity. The intrapod groups are matrilines generally comprising a grandmother, her adult son, her adult daughter, and the offspring of her daughter. Subpods and pods appear to be composed of related matrilineal groups.

Pods generally are well organized and may be led by a large male. A pod observed by Cousteau and Diole (1972) contained 1 large alpha male, 1 large female, 7 or 8 medium-sized females, and 6–8 calves. Some of the larger pods in the vicinity of Vancouver Island contain several adult males; from 1974 to 1981 the largest pod there was observed to have 7–10 adult males, 22 adult females, and 17–22 immatures (Balcomb and Bigg 1986). It has been suggested that intrapod groups often consist of a mated pair, a barren female, and a calf (Heimlich-Boran 1986). However, as indicated above, the adult pair are more likely the offspring of the elder female and the calf is probably the result of mating between the younger female and a male from another group. Movements of an intrapod group are led by the mother even when her fully adult sons are present (Bigg et al. 1990).

Breeding can occur at any time of the year, though in the Northern Hemisphere mating may peak from May to July and births take place mainly in the autumn. The single newborn weighs about 180 kg and is about 240 cm long (U.S. National Marine Fisheries Service 1978, 1981). Observations in captivity indicate a gestation period of 517 days and a weaning age of 14–18 months (Asper, Young, and Walsh 1988). Various reported calculations of the calving interval range from 3.0 to 8.3 years (Perrin and Reilly 1984). Studies of the population along the coast of British Columbia and Washington indicate that females give birth to their first viable calf at 12–16 years of age, produce about 5 calves during an average reproductive life of 25 years, and have a maximum longevity of 80–90 years. Males in that area typically attain sexual maturity at 15 years and physical maturity at 21 years and have a maximum longevity of 50–60 years (Olesiuk, Bigg, and Ellis 1990).

There long has been controversy about whether the

killer whale is a threat to human swimmers and boaters. Documented records of attacks are rare and usually involve provocation by the persons involved (Leatherwood, Caldwell, and Winn 1976; Scheffer 1978*b*). In contrast, people often kill or harass *Orcinus* because it is considered to be a competitor to fisheries. In the 1950s the U.S. Navy reportedly machine-gunned hundreds of killer whales at the request of the government of Iceland (Mitchell 1975*a*). *Orcinus* has been reported to take a substantial amount of commercially valuable fish off Alaska and to have been killed by fishermen in retaliation (Marine Mammal Commission 1994). The genus also has been killed by people in such places as Norway, Greenland, and Japan to obtain meat and oil. These problems continue, though overall populations do not seem to be declining (Klinowska and Cooke 1991). During the 1979–80 season the whaling fleet of the Soviet Union took 916 killer whales (Committee for Whaling Statistics 1980). Only 16 were reported taken in the 1982–83 season, however, and subsequently during the 1980s all commercial hunting ceased. *Orcinus* has been live-captured regularly off the coasts of British Columbia and Washington, mainly for eventual use in human entertainment (Bigg and Wolman 1975). Since 1976, however, most individuals taken for public display have come from waters off Iceland (Marine Mammal Commission 1994). The total number of killer whales in the wild is probably at least 100,000 (Reeves and Leatherwood 1994). *Orcinus* now is designated as conservation dependent by the IUCN and is on appendix 2 of the CITES.

CETACEA; DELPHINIDAE; **Genus GLOBICEPHALA**
Lesson, 1828

Pilot Whales, or Blackfish

There appear to be two species (E. R. Hall 1981; Kasuya 1975; Mitchell 1975*b*; Reilly 1978; Rice 1977; Van Bree 1971):

G. melas, cool temperate waters of the Southern Hemisphere and the North Atlantic, also occurred in the North Pacific off Japan until at least the tenth century A.D.;
G. macrorhynchus, tropical and temperate waters of the Atlantic, Indian, and Pacific oceans and adjoining seas.

The name *G. melaena* often has been used in place of *G. melas*, and the name *G. sieboldii* sometimes has been used in place of *G. macrorhynchus*, but the designations listed above are now generally accepted (Klinowska and Cooke 1991; Mead and Brownell *in* Wilson and Reeder 1993; Payne and Heinemann 1993; Reeves and Leatherwood 1994). *G. melas* has a divided distribution. In the North Atlantic region it occurs from Greenland and the Barents Sea to Cape Hatteras and the Mediterranean. *G. macrorhynchus* has been recorded as far north as Delaware Bay and the Bay of Biscay but generally is found no farther north than Cape Hatteras and northwestern Africa.

Head and body length is 360–850 cm, pectoral fin length is about one-fifth the head and body length, dorsal fin height is about 30 cm, and the expanse of the tail flukes is about 130 cm. Kasuya, Marsh, and Amino (1993) noted that in *G. macrorhynchus* mature males are twice as heavy as mature females. Banfield (1974) stated that the average weight of *G. melas* is 800 kg and that the largest recorded male weighed 2,750 kg. The coloration is black throughout, except for a white area often present below the chin. The head is swollen, so the forehead bulges above the upper jaw. The pectoral fin is narrow and tapering, and the dorsal fin is located just in front of the middle of the back. There are 7–11 teeth on each side of each jaw.

According to Banfield (1974), *G. melas* may occur either near the shore or well out to sea. During the colder months a population of this species remains in the Gulf Stream to the south of the Grand Banks of Newfoundland. In the summer this population shifts shoreward and northward to the Gulf of St. Lawrence, the Labrador Sea, and Greenland. Such movements are evidently in response to the migrations of the squid *Illex*, the major source of food. There is an apparently separate population of *G. melas* off northwestern Europe (Mitchell 1975*a*). In the eastern Pacific *G. macrorhynchus* seems to become more abundant near the shore during the winter (Reilly 1978).

Pilot whales sleep by day and feed by night. They normally swim at about 8 km/hr but can attain speeds of 40 km/hr (Banfield 1974; U.S. National Marine Fisheries Service 1978). A radio-tagged individual was tracked to a depth of 610 meters, and a captive was used in a U.S. Navy experiment to recover objects about 500 meters below the surface. Pilot whales are among the most intelligent and affable of cetaceans, adapting well to captivity and being easily trained. They have been demonstrated to use echolocation efficiently and to have a variety of sounds for communication, including an individualized "signature" whistle (Banfield 1974; Reilly 1978; Taruski 1979). The preferred diet is squid, up to 27 kg of which may be consumed daily. Other cephalopods and small fish are also eaten.

Group size varies. The U.S. National Marine Fisheries Service (1978) stated that *G. melas* occurs in schools of hundreds and thousands, but Banfield (1974) gave average group size as only 20 individuals. Observations in the Faeroe Islands from 1709 to 1990 indicate an average school size of 149 and a range of 1–1,200 (Bloch et al. 1993). Each of five stranded groups of *G. melas* in Great Britain contained 23–40 animals and consisted mostly of females but also included more than 1 mature male (Martin, Reynolds, and Richardson 1987). Analysis of specimens taken off Japan (Kasuya, Marsh, and Amino 1993) indicates that *G. macrorhynchus* lives in cohesive schools of 15–20 whales. These groups are breeding units comprising both sexes and all age classes. All schools contain several mature males, which produce spermatozoa continuously, and on average there are 8 mature females per mature male. Reilly (1978) referred to the same kind of school in the eastern Pacific as a traveling-hunting group, a large and well-organized unit that may be divided into harems of females and young led by 1 or a few large males, as well as other subgroups based on age and sex. He also identified two other kinds of herds: the feeding group, a loose aggregation of animals that come together only to exploit a source of food, and the loafing group, a cluster of 12–30 animals that float in one area for such purposes as mating and nursing young. Polygamy seems to be the rule in both species of *Globicephala*. Adult males are often scarred, presumably from battles over control of a harem. Fighting involves biting, butting with the large head, and slapping with the tail.

Social organization seems to be highly developed in *Globicephala*, and this factor sometimes works to the disadvantage of the animals. Individuals harpooned by people usually rush forward in panic and can be driven toward the shore. The rest of the group seems to respond to the movements and cries of the wounded animals and to follow them into shallow water, where all may be easily killed. The same tendency may be associated with mass stranding, an especially common occurrence in *Globicephala*. It is possible that if a leader's echolocation mechanism fails to function

Pacific pilot whale *(Globicephala macrorhynchus)*, photos from Marineland of the Pacific.

properly, because of pathology or unusual environmental conditions, it will guide the entire group onto the shore. The cries of the first animals to become stranded attract others to the same fate (Banfield 1974; Reilly 1978).

Analysis of specimens of *G. macrorhynchus* taken in fisheries off the Pacific coast of Japan (Kasuya and Marsh 1984; Kasuya and Tai 1993) indicates diffusely seasonal reproductive seasons that differ between northern and southern stocks. In the north matings peak in October–November, births in December–January. Respective periods for the southern stock are April–May and July–August. The gestation period averages about 15 months. The mean length of the single newborn is 185 cm in the north and 140 cm in the south. The young begin to take some solid food at 6–12 months, but lactation continues for at least 2 years. Many calves suckle until they are 6 years old, and a few, particularly those of older mothers, do so until past 10 years. The mean calving interval is about 7 years. Young males may move to another school after weaning, but females remain in their maternal group for life, thus creating a matrilineal society. Males reach sexual maturity at 15–22 years and have a maximum known longevity of 46 years. Females attain sexual maturity at 7–12 years and subsequently produce up to 4 or 5 calves. Their reproductive activity slows after age 28 and stops altogether by age 40, but they may live up to 63 years.

Strandings of *G. melas* on the British coast indicate no evident reproductive seasonality (Martin, Reynolds, and Richardson 1987). However, examination of many specimens caught in the Faeroe Islands (Desportes et al. 1992; Martin and Rothery 1993) suggests diffusely seasonal reproduction. The gestation period in the latter area evidently is only about 12 months, with both matings and births peaking from April to June and to a lesser extent in September. The single newborn averages 177 cm in length and 74 kg in weight. Lactation lasts an average of 3.4 years, and the mean interbirth interval is 5.1 years. Females usually attain sexual maturity at 6–9 years of age, males at 11–18 years. A 55-year-old female was found to be pregnant.

Pilot whales have been among the more heavily exploited of the small cetaceans (Banfield 1974; Buckland et al. 1993; Kasuya and Marsh 1984; Klinowska and Cooke 1991; Mitchell 1975*a*, 1975*b*; Reilly 1978; U.S. National Marine Fisheries Service 1978; Zachariassen 1993). There have been active fisheries in Japan, the Caribbean, the northeastern United States, Newfoundland, Ireland, Norway, and several islands north of Great Britain. The main products are meat for human and domestic animal consumption, blubber oil, and oil from the bulbous head. Records of the Faeroe Islands fishery extend back nearly 400 years. Two accountings of cumulative kill there are 117,546 animals from 1584 to 1883 and 240,721 from 1709 to 1992. Despite this large take, the annual average has not declined, and the current population of *G. melas* in the entire region between Greenland and Europe has been estimated at 778,000 individuals. In contrast, the western Atlantic population, estimated to have once contained 60,000 animals, evidently declined sharply following commercial organization of the Newfoundland fishery in 1947. About 40,000 pilot whales were killed there from 1951 to 1959, but only 6,902 were taken from 1962 to 1973. Recent estimates put the number off Canada and the northeastern United States at about 24,000. The catch of *G. macrorhynchus* on the Pacific coast of Japan was about 1,000 annually in the late 1940s and early 1950s and has been 500–700 per year more recently. This kill may be more than can be sustained by the estimated 58,000 individuals in the populations off Japan. Wade and Gerrodette (1993) estimated that another 160,000 pilot whales occur in the eastern tropical Pacific. *G. macrorhynchus* now is designated as conservation dependent by the IUCN, and both species are on appendix 2 of the CITES.

CETACEA; **Family ZIPHIIDAE**

Beaked Whales

This family of 6 Recent genera and 19 species occurs in all the oceans and adjoining seas of the world.

Head and body length is 3.3–12.8 meters and weight ranges from 1,000 to more than 11,000 kg. The vernacular name is derived from the long, narrow snout, which is sharply demarked from the high, bulging forehead in *Berardius* and *Hyperoodon* and forms a continuous smooth profile with the head in *Ziphius, Tasmacetus,* and *Mesoplodon* (the external appearance of the sixth genus, *Indopacetus,* is not known). The pectoral fin is rather small and ovate. The dorsal fin is small, usually sickle-shaped, and located on the posterior half of the back. The genus *Ziphius* has a low median keel from the dorsal fin to the tail. The tail flukes of beaked whales usually are not so strongly notched in the center as in other cetaceans. A pair of grooves on the throat converge anteriorly to form a V pattern at the chin. The stomach has 4–14 chambers but no esophageal chamber (Rice 1967).

The genus *Tasmacetus* has one pair of large functional teeth in the lower jaw and 17–28 small functional teeth on each side of both the upper and lower jaws. In the other genera there are only one or two pairs of large functional teeth, and these are in the lower jaw. These teeth push through the gums sooner in males than in females and often never erupt in the latter sex. There also are frequently series of small nonfunctional teeth in the upper and lower jaws. Except in *Berardius,* the bones of the skull are asymmetrical. Certain cranial bones are crested or elevated, forming large ridges in *Hyperoodon* and lesser ridges in some of the other genera. There are 43–49 vertebrae, and those of the neck tend to fuse.

Beaked whales are the least known of cetacean families. They usually remain well out to sea, avoid ships, and dive to great depths to secure cephalopods and fish. The presence of a well-developed melon on the forehead suggests that ziphiids are echolocators (U.S. National Marine Fisheries Service 1981). Some appear to be solitary, some travel in groups of 2–12, and some, particularly *Ziphius,* associate in schools of 40 or more. They generally swim and dive in unison. The hides of many individuals are scratched from intraspecific fighting. The newborn are about one-third as long as the mother.

The geological range of this family is early Miocene to late Pliocene in Europe, early Miocene in Africa, middle to late Miocene and Pliocene in North America, early Miocene in South America, early Miocene and Pliocene in Australia, Miocene in New Zealand, and Recent in all oceans (Rice 1984).

CETACEA; ZIPHIIDAE; **Genus BERARDIUS**
Duvernoy, 1851

Giant Bottle-nosed Whales

There are two species (Goodall 1978; Klinowska and Cooke 1991; Rice 1977):

B. arnuxii, known from waters near the Antarctic pack ice, in the southern South Pacific, and off South Africa, Australia, New Zealand, Argentina, Tierra del Fuego, the Falkland Islands, South Georgia, the South Shetlands, and the Antarctic Peninsula;

B. bairdii, the North Pacific from the Bering Sea to Japan and Baja California.

Balcomb (1989) summarized evidence suggesting that *B. arnuxii* is not specifically distinct from *B. bairdii.*

These are the largest ziphiids. *B. bairdii* attains sexual maturity when it attains a head and body length of about 10 meters, and growth continues to a maximum of 12.8 meters in females and 12.0 meters in the slightly smaller males. A female *B. bairdii* 11.1 meters long weighed about 11,380 kg (Rice 1984). Pectoral fin length is about 100–135 cm, height of the dorsal fin is about 30 cm, and the expanse of the tail flukes is 250–300 cm. *B. arnuxii* is smaller than *B. bairdii,* not being known to exceed 9.9 meters in head and body length (McCann 1975). Coloration is uniformly blackish brown or dark gray, sometimes with white blotches on the underparts. The skin is always extensively covered with pairs of parallel scratches, probably made by the teeth of conspecifics.

The skull is more nearly symmetrical than in the other beaked whales. The snout is tapered, and the forehead is well defined. Both sexes have two pairs of large teeth in the lower jaw. The anterior pair is completely visible, even when the mouth is closed, because the lower jaw protrudes well beyond the upper.

Giant bottle-nosed whales usually stay well offshore in waters more than 1,000 meters deep. Substantive natural history data are available only for *B. bairdii.* This species is alleged to have dived to depths as great as 2,400 meters after being harpooned, and feeding dives to 1,000 meters or more are routine. It may raise its flukes in the air before diving. It is said to be alert and hard to capture. It normally stays underwater for 15–25 minutes at a time but has remained submerged for just over an hour. Its diet consists mostly of squid and also includes octopus, crustaceans, and deepwater fish (Ellis 1980; Rice 1978*a*, 1984). In both the eastern and western Pacific *B. bairdii* is migratory, arriving at the continental slope during summer and autumn, when surface water temperatures are the highest (Balcomb 1989). One population appears off the Pacific coast of southern Japan in the summer and moves northward to the waters east of Hokkaido in the autumn. During this period it concentrates over the continental shelf in waters 1,000–3,000 meters deep and feeds on organisms living on the ocean floor. Schools spend four or five times as long in dives as at the surface (Kasuya 1986).

Groups of *B. bairdii* are tightly organized and commonly contain 3–30 individuals of all ages and both sexes (Ellis 1980; Rice 1978*a*). Studies of this species in Japan (Kasuya 1977) indicate that mating peaks in October and November and births occur mainly in March and April. The gestation period is about 17 months, apparently the longest among cetaceans. The newborn is around 460 cm long, and physical maturity is not attained until after 20 years. According to Klinowska and Cooke (1991), sexual maturity is reached by males at 6–10 years and by females at 10–14 years, and maximum estimated ages are 84 years for males and 54 years for females. Balcomb (1989) reported that a female thought to be 44 years old was reproductively active.

Hunting of *B. bairdii* has been undertaken in Japan since at least 1612. The dried meat is considered a delicacy in some parts of the country. For many years the annual kill amounted to only a few individuals, but following World War II modern fishing methods resulted in a sharply increased harvest. The peak kill was 382 whales in 1952. A subsequent decline in the take, which averaged 69 per year from 1969 to 1977, may be associated in part with a reduction in the whale population (Ellis 1980; Mitchell 1975*a*,

Giant bottle-nosed whale *(Berardius bairdii):* A & B. Photos from Fisheries Research Board of Canada through I. B. MacAskie; C. Head showing teeth in lower jaw, photo from Tokyo Whales Research Institute through Hideo Omura; D & E. Photos by Warren J. Houck.

1975*b*; Nishiwaki and Oguro 1971; Rice 1978*a*). In recent years the Japanese government has maintained an annual quota of about 60 individuals, which amounts to 1 percent of the estimated 6,000 *B. bairdii* in the western North Pacific (Reeves and Leatherwood 1994). Both species of *Berardius* are designated as conservation dependent by the IUCN and are on appendix 1 of the CITES.

CETACEA; ZIPHIIDAE; **Genus ZIPHIUS**
G. Cuvier, 1823

Goose-beaked Whale

The single species, *Z. cavirostris,* occurs in the temperate and tropical waters of all oceans and adjoining seas (Rice 1977).

Data compiled by Heyning (1989*b*) indicate no clear sexual dimorphism in size. However, females have been reported to reach sexual maturity when they attain a head and body length of 6.1 meters and to then continue growing to a maximum of 7 meters. Males have been reported to reach sexual maturity when they attain 5.4 meters and to then continue growing to a maximum of 6.7 meters. A frequently cited report of an individual 8.5 meters long is erroneous (Mitchell 1975*b*). Pectoral fin length is about 50 cm, dorsal fin height is about 40 cm, and the expanse of the tail flukes is about 150 cm. An adult male 5.7 meters long weighed 2,450 kg (Ross 1984), and an adult female 6.6 meters long weighed 2,952.5 kg. Coloration is variable, but two frequently observed patterns are: face and upper back creamy white, remainder of body black; and entire body grayish fawn, with some small blotches of slightly darker gray below.

The beak is short and blends into the sloping forehead. The opening of the mouth is relatively small. Males have two functional teeth, which protrude from the tip of the lower jaw; these teeth usually are not visible in females. Rows of small rudimentary teeth usually are present in both jaws. As in other ziphiids, the tail flukes generally lack a median notch, but some specimens of *Ziphius* do have such a notch.

According to Leatherwood, Caldwell, and Winn (1976),

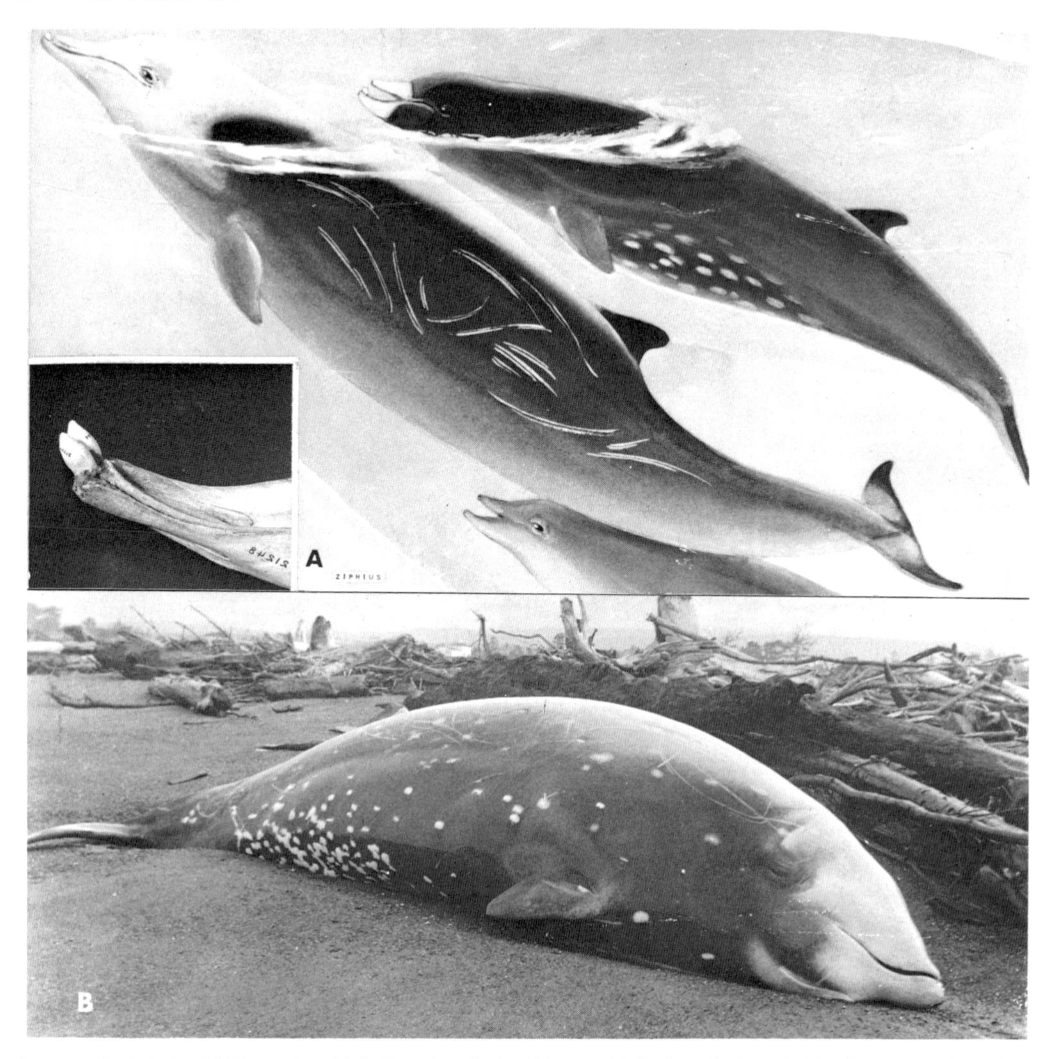

Goose-beaked whales *(Ziphius cavirostris):* A. Photo from National Geographic Society of painting by Else Bostelmann; Inset: lower jaw showing the two teeth, photo by P. F. Wright of specimen in U.S. National Museum of Natural History; B. Photo by Warren J. Houck.

Ziphius seems to be primarily tropical in distribution but moves northward into temperate waters during the summer. There are also records from as far north as the Aleutian Islands and the Gulf of Alaska (C. S. Harrison 1979; Rice 1978*a*). *Ziphius* is generally found well offshore and often dives to great depths, remaining underwater for 30 minutes or more. Before diving, it raises its tail flukes straight above the surface. It has been observed to leap clear of the water. Its diet consists primarily of squid and also includes deep-water fish.

Groups of up to 40 individuals have been reported, but schools usually contain fewer than half that number. Some white-headed adults, possibly old males, are solitary (Rice 1978*a*). The members of a group often travel, dive, and feed together in fairly close association. Banfield (1974) stated that births occur in late summer or early autumn, but data compiled by Heyning (1989*b*) indicate no distinct calving season. A gestation period of about 1 year has been reported. The newborn is about 200–300 cm long (Mitchell 1975*b*). According to Mead (1984), newborn average 270 cm in length, and maximum known longevity is 36 years. There are estimates suggesting that some animals have lived as long as 62 years (Heyning 1989*b;* Klinowska and Cooke 1991).

Ziphius is on appendix 2 of the CITES. It has been exploited to a modest extent for human utilization in Japan, the Caribbean, and other areas but is not known to be declining (Klinowska and Cooke 1991). There is some concern about incidental taking off California (Reeves and Leatherwood 1994). About 20,000 individuals are estimated to occur in the eastern tropical Pacific (Wade and Gerodette 1993).

CETACEA; ZIPHIIDAE; **Genus TASMACETUS**
Oliver, 1937

Shepherd's Beaked Whale

The single species, *T. shepherdi,* is known by 13 specimens stranded in Australia, New Zealand, Stewart and Chatham

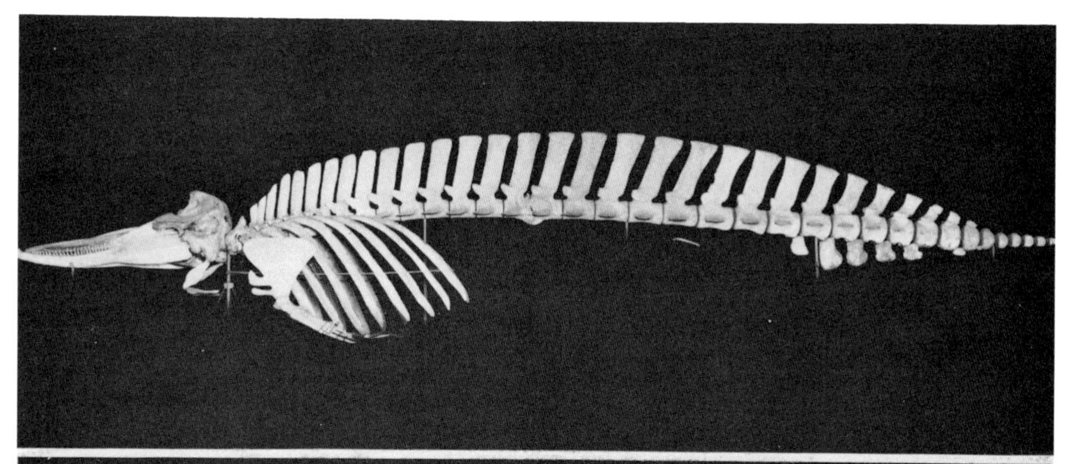

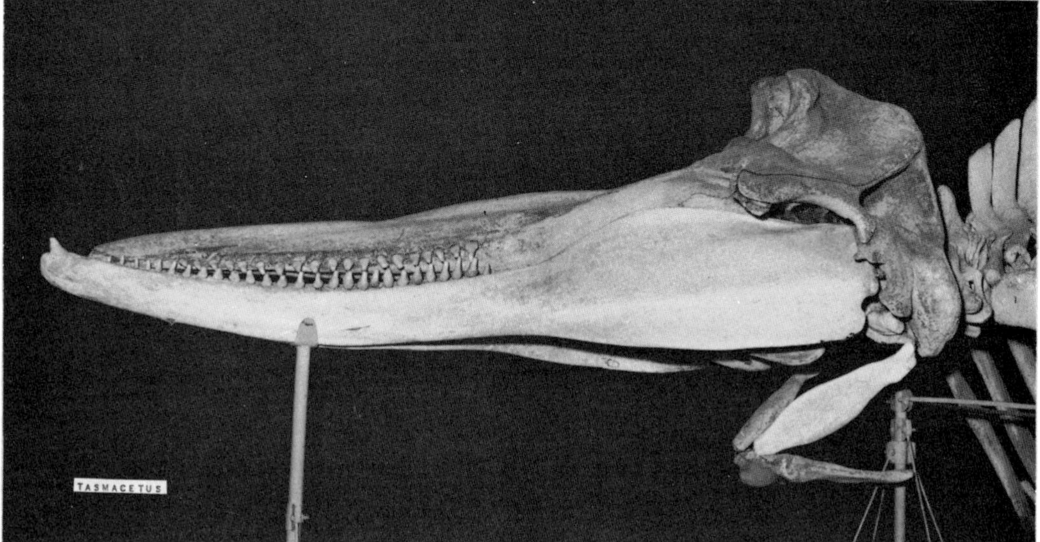

Shepherd's beaked whale *(Tasmacetus shepherdi)*, photos by Eldon V. Burkett through Wanganui Public Museum, New Zealand.

islands near New Zealand, the Juan Fernandez Islands off central Chile, the Valdez Peninsula of east-central Argentina, Tierra del Fuego, and Isla Gable east of Tierra del Fuego and by a probable sighting of a live individual off New Zealand (Brownell, Aguayo, and Torres N. 1976; Goodall 1978; Lichter 1986; Mead and Payne 1975; Watkins 1976). It also has been reported in waters off Tristan da Cunha in the South Atlantic (Mead and Brownell *in* Wilson and Reeder 1993).

On the basis of four specimens, Mead and Payne (1975) provided the following data. Head and body length is 6.1–7.0 meters, pectoral fin length is 69 cm, dorsal fin height is 34 cm, and the expanse of the tail flukes is 135–52 cm. The general coloration is thought to be dark, but the ventral surface is pale, there is a light area anterior to the pectoral fin, and there are two light stripes along part of the side. In males there are two large teeth at the tip of the lower jaw, but in the one known female specimen these teeth did not erupt. In contrast to all other ziphiids, these ziphiids also have numerous smaller functional teeth. The total tooth count is 17–21 on each side of the upper jaw and 18–28 on each side of the lower jaw. The stomach of one specimen contained a number of fish, as well as a small crab and a small squid beak that may have been eaten by the fish. The stomach contents suggest that *Tasmacetus* had been feeding on the bottom in fairly deep water.

According to Klinowska and Cooke (1991), records of this genus have been increasing and it may be less rare than suggested by the number of known occurrences. There are no reports of direct or incidental taking. *Tasmacetus* is on appendix 2 of the CITES.

CETACEA; ZIPHIIDAE; **Genus HYPEROODON**
Lacépède, 1804

Bottle-nosed Whales

There are two species (Ellis 1980; Klinowska and Cooke 1991; Mead 1989*b*; Rice 1977):

H. ampullatus, the North Atlantic from Davis Strait and the White Sea to Rhode Island and the English Channel, occasionally in the Mediterranean;
H. planifrons, known from waters off Australia, New

Bottle-nosed whale *(Hyperoodon ampullatus)*, photo from *Endeavour.*

Zealand, Chile, Brazil, Argentina, Tierra del Fuego, the Falkland Islands, South Georgia, the South Orkney Islands, South Africa, and much of Antarctica.

Leatherwood et al. (1982) reported a sighting by K. C. Balcomb of a group of *Hyperoodon,* its species unidentified, near the equator in the central Pacific. Klinowska and Cooke (1991) cited several other reports from that region and suggested that a third species of *Hyperoodon* may be involved, but Mead (1989*b*) thought identification as *H. planifrons* to be plausible. Moore (1968) placed *H. planifrons* in its own subgenus, *Frasercetus.*

Maximum head and body length is 9.8 meters in males and 8.7 meters in females (Mead 1989*b*). The dorsal fin is at least 30 cm high and distinctly hooked (Leatherwood, Caldwell, and Winn 1976). A female 6 meters long weighed 2,500 kg. Coloration becomes lighter with age. Calves are grayish brown to black, immature animals are often spotted yellowish brown and white, and old individuals may be completely yellowish white. There are usually two large teeth in the lower jaw of young males, but in older males one or both may be lost. In females the functional teeth are smaller or do not emerge at all. Rows of vestigial teeth are often present in the lower and upper jaws. In females and young males the forehead slopes rather smoothly into the beak (Ellis 1980). In older males the forehead rises abruptly from the beak and may become so bulbous that when viewed in profile it appears to protrude forward at the top. This development results in part from the enlargement of crests on the maxillary bones. The forehead encloses a huge melon filled with oil that is much the same in appearance, and probably in function, as that of the spermaceti of *Physeter.*

Bottle-nosed whales generally stay well out to sea. They seem to occur mainly in cooler waters and may approach the polar ice packs during the summer. Limited observations suggest that *H. planifrons* appears off southern Africa during the summer months, November–January (Ross 1984). Substantive natural history data are available only for *H. ampullatus.* This species is usually found in waters more than 1,000 meters deep. During the spring and early summer it migrates to the northern parts of its geographic range, and in the late summer and autumn it returns south. It dives suddenly and with great speed. Although dives usually last about 30–45 minutes, there are reports that this whale has remained underwater for 2 hours, which, if accurate, would be a record for the Cetacea. One harpooned individual is said to have dived vertically and pulled out 1,000 meters of line. Feeding probably takes place at great depths. The diet consists mainly of squid and also includes fish and bottom-dwelling echinoderms (Benjaminsen and Christensen 1979; Leatherwood, Caldwell, and Winn 1976; Rice 1967).

A variety of sounds has been reported for *H. ampullatus,* most notably series of clicks and whistles such as are known to be used by certain other odontocetes for, respectively, echolocation and communication (Winn, Perkins, and Winn 1970). Groups usually comprise 2–4 individuals, generally of the same age and sex. Larger groups are sometimes seen and usually contain mature animals of both sexes. Solitary whales, often young individuals, are common. Groups are said not to abandon an injured member (Benjaminsen and Christensen 1979). The group reported by Leatherwood et al. (1982) comprised about 25 individuals.

Based on observations off South Africa, the calving season of *H. planifrons* appears to be in spring or summer (Ross 1984). One individual of that species may have lived for more than 50 years (Klinowska and Cooke 1991). Both the mating and calving seasons of *H. ampullatus* peak in April. Females probably give birth every 2 years. The gestation period is thought to last 12 months. The single newborn is about 300–330 cm long, and lactation may last about a year. Sexual maturity is attained at 8–12 years in females and 7–11 years in males. Maximum longevity is at least 37 years (Benjaminsen and Christensen 1979).

A large bottle-nosed whale can yield up to 200 kg of spermaceti oil, the uses of which are given in the account of *Physeter*, as well as about 2,000 kg of blubber oil (Ellis 1980). Following the decline of the bowhead whale *(Balaena mysticetus)* in the late nineteenth century, commercial interest in *H. ampullatus* increased, especially in Norway. From 1882 to 1920 approximately 50,000 individuals were taken off northwestern Europe. The annual kill fell from 2,000–3,000 in the 1890s to 20–100 in the 1920s. After some years of reduced hunting, and perhaps of recovery by the whale population, the fishery again intensified: the annual kill peaked at around 700 individuals in 1965 but then declined to near zero. The population in the eastern Atlantic is estimated to have originally contained 40,000–100,000 whales (Benjaminsen and Christensen 1979; Ellis 1980; U.S. National Marine Fisheries Service 1978). The current number is much lower, perhaps around 12,000, and the western Atlantic population is also small, but the exact status is unknown (Klinowska and Cooke 1991). An apparently nonmigratory population near Sable Island, off Nova Scotia, contains only a few hundred individuals and may be threatened by development of offshore oil and gas fields (Reeves, Mitchell, and Whitehead 1993). Both species of *Hyperoodon* are designated as conservation dependent by the IUCN and are on appendix 1 of the CITES.

CETACEA; ZIPHIIDAE; **Genus INDOPACETUS**
Moore, 1968

Indo-Pacific Beaked Whale

The single species, *I. pacificus*, is known with certainty only by two skulls, one found at Mackay on the east coast of Queensland and the other at Danane on the east coast of Somalia (Moore 1968). *Indopacetus* was tentatively included within *Mesoplodon* by Klinowska and Cooke (1991) and Mead (1989*a*) but was treated as a distinct genus by Bannister (*in* Bannister et al. 1988), Corbet and Hill (1991), Mead and Brownell (*in* Wilson and Reeder 1993), and Reeves and Leatherwood (1994).

The first skull to be found was nearly 122 cm long and was thought to have come from a fully mature animal about 7.6 meters long (Ellis 1980). Moore (1968, 1972) listed the following cranial characters as distinguishing the genus: (1) the alveoli of the developed teeth are a single pair, apical on the mandible, and in an old adult male become progressively at least as shallow as 30 mm; (2) the frontal bones occupy an area of the synvertex of the skull approximating or exceeding that occupied by the nasal bones; (3) there is almost no posterior process of the premaxillary crest extending posteriorly on the synvertex between the nasal and maxillary bones or between the frontal and maxillary bones; (4) in the lateral extension of the maxillary bone over the orbit there is a deep groove about half as long as the orbit; (5) at about the midlength of the beak there is a swelling caused by the lateral margins proceeding forward a short distance without convergence, or even with a little divergence and then convergence again; (6) there is fusion of a considerable length of the mesethmoid bone to both premaxillary rims of the mesorostral canal; and (7) proliferation of bone from the vomer (and distally the premaxillae) into the mesorostral canal at the onset of adulthood is absent or minimal in the one adult specimen.

Pitman, Aguayo L., and Urbán R. (1987) reported that an unidentified beaked whale observed many times in the eastern Pacific off Mexico, Central America, and northern South America might represent *I. pacificus*. The unidentified whale has an estimated length of 5.0–5.5 meters, a relatively flat head, a moderately long beak, and a distinctive dorsal fin that is low, wide-based, and triangular. There are two morphs: the larger, evidently representing adult males, is black with a broad white or cream-colored swathe originating immediately posterior to the dorsal surface of the head and running posteroventrally on either side of the animal and is marked with scratches and scars; and the smaller, probably being females and young, is uniform gray-brown or bronze-colored and is unscarred. The two kinds have been seen together in groups of as many as eight individuals. Although some of these observations may represent the newly described *Mesoplodon peruvianus*, which occurs in the same general region, Ralls and Brownell (1991) and Urbán-Ramírez and Aurioles-Gamboa (1992) suggested that an unidentified species is involved.

Klinowska and Cooke (1991) discussed several records from the Seychelles that could refer to *Indopacetus*, including a sighting of a group of three adults and a juvenile. They noted, however, that the estimated length of the adults, 4.6 meters, would not correspond to the postulated size of the animal associated with the skull of *Indopacetus* (see above). Although there are no reports of exploitation, the genus apparently is rare; it is on appendix 2 of the CITES.

CETACEA; ZIPHIIDAE; **Genus MESOPLODON**
Gervais, 1850

Beaked Whales

There are 3 subgenera and 12 species (Ellis 1980; Goodall 1978; Klinowska and Cooke 1991; Lichter 1986; Loughlin and Perez 1985; Mead 1989*a;* Mead and Baker 1987; Mead and Brownell *in* Wilson and Reeder 1993; Moore 1968; Reiner 1986; Reyes 1990; Reyes, Mead, and Van Waerebeek 1991; Rice 1977, 1978*a;* Robineau and Vely 1993; Urbán-Ramírez and Aurioles-Gamboa 1992):

subgenus *Mesoplodon* Gervais, 1850

M. peruvianus, known by 12 specimens from Baja California and Peru;
M. hectori, known by 20 specimens from southern California, Argentina, Tierra del Fuego, the Falkland Islands, South Africa, Tasmania, and New Zealand;
M. mirus, temperate waters from Nova Scotia and the Bahamas to the British Isles and the Canary Islands, also known from South Africa, Australia, and New Zealand;
M. europaeus, western North Atlantic from New York to the West Indies, also records from the English Channel, Ireland, the Canary Islands, Mauritania, Guinea-Bissau, and Ascension Island;
M. ginkgodens, known from Sri Lanka, Indonesia, Taiwan, China, Japan, Australia, the Chatham Islands, the Galapagos Islands, southern California, and Baja California;
M. grayi, known from the Netherlands, South Africa, the southern Indian Ocean, southern Australia, Tasmania, New Zealand, the Chatham Islands, Peru, Chile, Tierra del Fuego, Argentina, and the Falkland Islands;
M. carlhubbsi, temperate waters of the North Pacific from Japan to British Columbia and California;
M. bowdoini, known from Western Australia, Victoria, Tasmania, New Zealand, and Campbell Island;

True's beaked whales *(Mesoplodon mirus)*, photo from National Geographic Society of painting by Else Bostelmann. Inset: *M.* sp., photo from American Museum of Natural History.

M. stejnegeri, cool temperate waters from the Bering Sea to Japan and southern California;
M. bidens, cool temperate waters of the North Atlantic from Newfoundland and Massachusetts to southern Norway and the Azores, one specimen from the Gulf Coast of Florida, one possible record from Italy;

subgenus *Dolichodon* Gray, 1866

M. layardii, known from southern Africa, Heard Island, Australia, Tasmania, New Zealand, Chatham Island, Chile, Tierra del Fuego, Argentina, Uruguay, and the Falkland Islands;

subgenus *Dioplodon* Gervais, 1850

M. densirostris, tropical and warm temperate waters of all oceans.

E. R. Hall (1981) used the generic name *Micropteron* Eschricht, 1849, in place of *Mesoplodon.* Mead (1989*a*) questioned the validity of the above subgenera and considered a thorough revision necessary. He and some other authorities have included *Indopacetus* (see account thereof) within *Mesoplodon.*

Head and body length is 3.3–6.2 meters, pectoral fin length is 20–70 cm, dorsal fin height is about 15–20 cm, and the expanse of the tail flukes is about 100 cm. Males are generally larger than females. *M. carlhubbsi* reaches a maximum length of about 530 cm and a weight of about 1,500 kg, with no discernible size difference between the sexes (Mead, Walker, and Houck 1982). Color is variable but is usually slaty black to bluish black or dark gray above and paler below. Only two teeth become well developed, one on each side of the lower jaw, and these may be lost in old age. In most species they are exposed when the mouth is closed and are used in intraspecific combat. In females the functional teeth are much smaller than those of males and often do not erupt above the gums. There may be small nonfunctional teeth in both jaws. A row of such vestigial teeth on each side of the upper jaw seems to be characteristic of *M. grayi.*

The structure of the pair of functional teeth in males varies remarkably (Ellis 1980; Rice 1978*a*). They can be briefly described as follows: *M. hectori,* shaped like laterally flattened triangles, located at tip of jaw; *M. mirus,* small and angled slightly forward, located at tip of jaw; *M. europaeus,* shaped like laterally flattened triangles, located one-third of the way from tip to apex of mouth; *M. ginkgodens,* shaped in profile like the leaf of the ginkgo tree, located one-third of the way from tip to apex of mouth; *M. grayi,* shaped in profile like a flattened onion, located in middle of mouth; *M. carlhubbsi,* large and straight-sided, located near middle of mouth and exposed when mouth closed; *M. bowdoini,* large and flattened, located in raised sockets near front of mouth; *M. stejnegeri,* large and tusklike, exposed, located in back of mouth; *M. bidens,* small and pointed slightly to rear, located near middle of mouth; *M. layardii,* shaped somewhat like the tusks of a boar (Suidae), extend out of mouth and curve over upper jaw; and *M. densirostris,* set at a point where the lower jaw becomes extremely high and broad, so the teeth are well exposed and extend above the upper jaw.

Mesoplodon is apparently a pelagic genus, staying mainly in deep waters far from shore. It is relatively sluggish when at the surface (Leatherwood et al. 1982). The diet includes squid, other cephalopods, and fish. Rice (1978*a*) observed a group of 10–12 *M. densirostris* dive; after 45 minutes they still had not resurfaced. Most of his other sightings of the genus have been of schools of 2–6 individuals. There is one record of a stranding of 28 *M. grayi* (Ellis 1980). Usually, however, *Mesoplodon* is encountered alone or in small groups (Leatherwood, Caldwell, and Winn 1976). *M. stejnegeri* has been observed in tight groups of 5–15 individuals of various sizes swimming and diving in unison (Loughlin et al. 1982). Adult male *Mesoplodon* often are heavily scarred from fighting one another. The scars evidently are inflicted by the two functional teeth. Heyning (1984) showed that in *M. carlhubbsi* the scars are often parallel and probably result from attacks delivered with the mouth closed.

Reproductive data are very limited (Ellis 1980; Mead, Walker, and Houck 1982; Rice 1978*a*). In *M. bidens* mating usually takes place in late winter and spring, gestation lasts about 1 year, the newborn is about 213 cm long, and weaning occurs after about 1 year. *M. carlhubbsi* evidently gives birth during the summer after a gestation period of 12 months, the newborn averaging 250 cm. A female *M. hectori* and a young calf stranded near San Diego in May. A female *M. europaeus* and a young calf stranded in Florida in October. Female *M. densirostris* attain sexual maturity at about 9 years and are known to have lived for 27 years (Mead 1984). A specimen of *M. europaeus* may have been more than 48 years old (Mead 1989*a*).

Very small numbers of *Mesoplodon* are taken by commercial fisheries (Mitchell 1975*a*). There is some concern about incidental taking in fishing nets, but there are no indications of major conservation problems (Reeves and Leatherwood 1994). All species are on appendix 2 of the CITES.

CETACEA; **Family PHYSETERIDAE**

Sperm Whales

This family of two Recent genera and three species occurs in all the oceans and adjoining seas of the world. The sequence of genera presented here follows that of E. R. Hall (1981) and not that of Rice (1977). The genus *Kogia* sometimes is placed in a subfamily or family, the Kogiidae, distinct from that containing the genus *Physeter.* The Kogiidae were recognized as a full family by Fordyce and Barnes (1994), Jones et al. (1992), Klinowska and Cooke (1991), and Reeves and Leatherwood (1994) but not by Caldwell and Caldwell (1989), Corbet and Hill (1991, 1992), Heyning (1989*a*), or Mead and Brownell (*in* Wilson and Reeder 1993).

The characters common to both genera include: a broad, flat rostrum; a great depression in the facial part of the skull to accommodate a highly developed spermaceti organ, which is bounded posteriorly by a high occipital crest; an S-shaped blowhole located on the left side of the snout; no preorbital or postorbital processes on the skull; a simple air sinus system; a left nasal passage for respiration and a right one apparently modified for sound production; numerous short grooves on the throat; a small, narrow lower jaw that ends well short of the anterior end of the snout; functional teeth only in the lower jaw; and relatively small pectoral appendages (E. R. Hall 1981; Rice 1967). The two genera differ markedly in size and certain other morphological characters. There are also major differences in known aspects of natural history.

The geological range of the Physeteridae is middle Miocene to Pleistocene in North America, early Miocene in South America, early Miocene to Pleistocene in Europe, late Pliocene in Japan, early Pliocene in Australia, and Recent in all oceans (Rice 1984). Certain Oligocene fossils from the Caucasus also may be referable to the Physeteridae (Fordyce and Barnes 1994).

CETACEA; PHYSETERIDAE; **Genus KOGIA**
Gray, 1846

Pygmy and Dwarf Sperm Whales

There are two species (Caldwell and Caldwell 1989; E. R. Hall 1981; Nagorsen 1985; Pinedo 1987; Rice 1977):

K. breviceps (pygmy sperm whale), worldwide in tropical and temperate waters;
K. simus (dwarf sperm whale), worldwide in tropical and temperate waters.

Head and body length is 210–70 cm in *K. simus* and 270–340 cm in the larger *K. breviceps.* Pectoral fin length is about 40 cm, and the expanse of the tail flukes is about 61 cm. The usual weight has been estimated at 136–272 kg for *K. simus* and 318–408 kg for *K. breviceps,* but the U.S. National Marine Fisheries Service (1981) indicated that the latter species sometimes exceeds 772 kg. Males are generally larger and heavier than females. Both species are dark steel gray on the back, shade to a lighter gray on the sides, and gradually fade to a dull white on the belly (Leatherwood, Caldwell, and Winn 1976).

The head is only about one-sixth the total length and resembles that of some delphinid genera in external outline, but the blunt snout projects beyond the lower jaw. Among cetaceans, the facial part of the skull of *Kogia* is among the shortest. The blowhole of *Kogia,* like that of *Physeter,* is on the left side of the head, but it is located on the forehead rather than toward the tip of the snout. *Kogia* resembles *Physeter* in having a spermaceti organ in the head and in having the functional teeth confined to the lower jaw. The teeth are sharp and curved; the number on each side of the lower jaw is 8–11 in *K. simus* and 12–16 in *K. breviceps* (E. R. Hall 1981). A few smaller, nonfunctional teeth may be present in the upper jaw of *K. simus.* The dorsal fin is hooked and curved like a sickle; it is relatively high and located near the middle of the back in *K. simus* but much lower and located posterior to the middle of the back in *K. breviceps.* As for *Physeter* (see account thereof), there has been debate regarding the function of the spermaceti organ, with the most convincing evidence in the case of *Kogia* being that it serves as an acoustical transducer for the production of sound pulses for echolocation (Nagorsen 1985).

A few observations of live animals indicate that *Kogia* is rather sluggish and sometimes floats motionless on the surface. Otherwise the genus is known largely on the basis of strandings. Some of these records suggest that *K. breviceps* moves away from the shores of the eastern United States and South Australia during the summer and that *K. simus* migrates toward Japan in the summer. South African records, however, indicate that both species are present throughout the year and that neither makes major seasonal movements (Fitch and Brownell 1968; G. J. B. Ross 1979). The diet includes mostly cephalopods but also a variety of fish and some crustaceans. The stomachs of two specimens of *K. simus* from the northeastern Atlantic contained crustaceans known to inhabit depths of 500–1,300 meters (Nagorsen 1985). Analysis of diet by Ross (1984) showed that 93 percent of the cephalopods consumed by immature *K. simus,* as well as by calves and accompanying females, consisted of species that dwell over the continental shelf, whereas 71 percent of the food of adults consisted of oceanic species. Respective figures for *K. breviceps* were 45 percent and 85 percent. These figures suggest that juvenile and immature *Kogia,* especially those of *K. simus,* live closer to shore than do adults.

According to Rice (1978*d*), *K. breviceps* usually is seen in pairs, perhaps a female and a calf, or in groups of 3–5. Data summarized by G. J. B. Ross (1979) suggest that the school size of *K. simus* is 10 animals or fewer, that females with calves may group together, that immature animals may form their own groups, and that sexually mature males and females can be found in the same school. Intraspecific fighting has been witnessed. Ross also analyzed records of small and near-term fetuses in *K. breviceps*, which indicate that in both the Northern and Southern hemispheres the mating and calving seasons extend over a period of approximately 7 months, from autumn to spring. There were too few records of *K. simus* to determine whether there is a distinct breeding season in that species, though there does appear to be an extended calving season of at least 5–6 months. In *K. breviceps* the gestation period may be more than 11 months. A number of females have been found to be pregnant and lactating at the same time, and it seems likely that many females give birth every year. The single newborn is about 100 cm long in *K. simus* and 120 cm long in *K. breviceps*.

Pygmy and dwarf sperm whales seem to be rare, but occasionally they are taken and utilized by fishermen, and Sylvestre (1983) reported data on 33 stranded specimens that had been maintained alive for brief periods in captivity. The rarity may reflect an early, unrecorded exploitation. *Kogia* is lethargic and easy to approach, and a few records indicate that this genus was sometimes harpooned by whalers (Ellis 1980; Mitchell 1975*a*). The number of individuals of *K. simus* in the eastern tropical Pacific Ocean has been estimated at 11,200 (Wade and Gerrodette 1993). Both species are on appendix 2 of the CITES.

Pygmy sperm whale *(Kogia breviceps)*, photos from Marineland of Florida.

CETACEA; PHYSETERIDAE; **Genus PHYSETER**
Linnaeus, 1758

Sperm Whale

The single species, *P. catodon*, occurs in all oceans and adjoining seas of the world except in polar ice fields (Rice 1977). Many authorities have followed Husson and Holthuis (1974) in using the name *P. macrocephalus* for this species, but E. R. Hall (1981) considered *P. catodon* to be the proper designation. In recent debate detailed arguments

Sperm whale *(Physeter catodon)*, photos from *British Mammals*, Archibald Thorburn (top) and R. L. Pitman (bottom).

were presented by Schevill (1986) in support of *P. catodon* and by Holthuis (1987) in support of *P. macrocephalus.*

This is the largest toothed mammal in the world and the most sexually dimorphic of all cetaceans. Sexual maturity is attained by males when they reach a length of about 12.2 meters and by females at about 8.5 meters; growth subsequently continues for a number of years. Maximum head and body length is nearly 20 meters in males, though individuals more than 15.2 meters long are rare. Head and body length in females seldom exceeds 12 meters. Physically mature males usually weigh 35,000–50,000 kg; females weigh only about one-third as much. The pectoral fin is about 200 cm long, and the expanse of the tail flukes is usually 400–450 cm. The color is gray to dark bluish gray or black. With increasing age males may become paler and sometimes piebald. Most specimens have some white in the genital and anal regions and on the lower jaw (Best 1974, 1979; Ellis 1980; Leatherwood, Caldwell, and Winn 1976). Several completely white individuals are on record (Hain and Leatherwood 1982).

The head is enormous and squarish, especially in males. The skull is the most asymmetrical of any mammal's. The blowhole is located toward the tip of the snout and on the left side. The lower jaw is extraordinarily slender and has 16–30 (usually 20–26) strong, conical teeth on each side. These teeth, which may be more than 20 cm long, fit into sockets in the palate when the mouth is closed. Smaller, nonfunctional teeth are present in the upper jaw. *Physeter* is the only cetacean with a gullet large enough to swallow a man. There is no dorsal fin, but there is a longitudinal row of bumps on the posterior half of the back. The vertebrae number 50–51, and most of the neck vertebrae are fused. The blubber is up to 35 cm thick. A specimen of *Physeter* 13 meters long and weighing 20,000 kg had a heart that

Sperm whale *(Physeter catodon)*, photo from American Museum of Natural History.

weighed 116 kg. The brain of *Physeter*, the largest of any mammal's, weighs about 9.2 kg (Rice 1967).

The most striking morphological feature, and the one that gives *Physeter* its vernacular name, is the huge spermaceti organ in the head, filled with up to 1,900 liters of waxy oil. At the anterior and posterior ends of the organ are air sacs. The left nasal passage goes along the left side of the spermaceti organ, leads directly to the blowhole, and also connects to the anterior air sac. The much narrower right nasal passage goes through the lower part of the spermaceti organ, opens into the anterior air sac through large internal lips, and also connects to the posterior air sac. This general structure has been known for centuries, but its exact functions are still being debated (Best 1979; D. W. Rice 1989). It may either assist in evacuating the lungs prior to a dive and in absorbing nitrogen at extreme pressures, regulate buoyancy during deep diving, or reverberate and focus sounds. Clarke (1979), noting that *Physeter* apparently lies motionless at great depths to await and snap up squid, suggested that buoyancy is then controlled by varying the temperature and thus the density of the spermaceti oil. Cooling may be accomplished by drawing water into the nasal passages and around the oil. Norris and Harvey (1972) opposed this theory and suggested that the spermaceti organ is involved mainly in the production of burst pulses for long-range echolocation. According to this view, air goes through the right nasal passage, produces sounds at the lips leading into the anterior air sac, goes back up the left nasal passage, and is then pumped around again. The two air sacs may serve as sound mirrors.

Physeter produces a variety of sounds, including groans, whistles, chirps, pings, squeaks, yelps, and wheezes (Ellis 1980). The voice is very loud and can be heard many kilometers away by persons with proper underwater listening equipment. The most common sounds are series of clicklike pulses generally much lower in frequency than those of the Delphinidae. The spermaceti organ may be used in the directional beaming as well as the production of these pulses, which apparently are involved in both echolocation and communication. In addition to other sounds, each whale has its own individualized "coda," a stereotyped, repetitive sequence of 3–40 or more clicks that is heard only when one animal meets another (Rice 1978*d;* Norris and Harvey 1972).

The sperm whale generally is found in waters conducive to the production of squid—at least 1,000 meters deep and with cold-water upswellings (Ellis 1980). The best areas are off the coasts of South America and Africa, in the North Atlantic and the Arabian Sea, between Australia and New Zealand, in the western North Pacific, and all along the Equator. Most animals stay between 40° N and 40° S, but during the summer the bachelor males of medium size move to between 40° and 50°, and at least some of the older males venture beyond 50° into or near arctic and antarctic waters. Although the other animals do not wander as far, there are northward shifts of concentration in the boreal summer and southward shifts in the austral summer as the movements of squid are followed (Berzin 1972; Best 1974, 1979; Rice 1978*d*). Certain major populations or stocks seem to move together on a seasonal basis, but migrations are not as predictable as those of the baleen whales. *Physeter* normally swims at a maximum of around 10 km/hr, but when pursued it can attain speeds of up to 30 km/hr (Berzin 1972). It sometimes lifts its head vertically out of the water, apparently to look or listen. Before a dive it gives a spout, characteristic in being directed obliquely forward to the left. It then lifts its tail flukes high in the air and descends almost vertically. There is commonly a prolonged initial submergence at around 360 meters for 20–75 minutes, followed by a series of shallow dives. Females do not generally go as deep as the larger males (Ellis 1980). There are at least 14 instances on record of sperm whales becoming entangled in underwater communication cables, one at a depth of 1,135 meters. *Physeter* has been tracked by sonar to a depth of 2,500 meters. An individual killed after an 82-minute dive in waters 3,200 meters deep was found to have just consumed a kind of shark known to dwell on the bottom (Rice 1978*d*). Several species of sharks and other fish are included in the diet, but the predominant food is squid. Most of the squid taken are less than 1 meter long, but the stomach of one sperm whale contained a squid about 10.5 meters long. It has been estimated that each whale eats about 3 percent of its weight in squid per day (Ellis 1980). *Physeter* often bears the scars of combat with large squid.

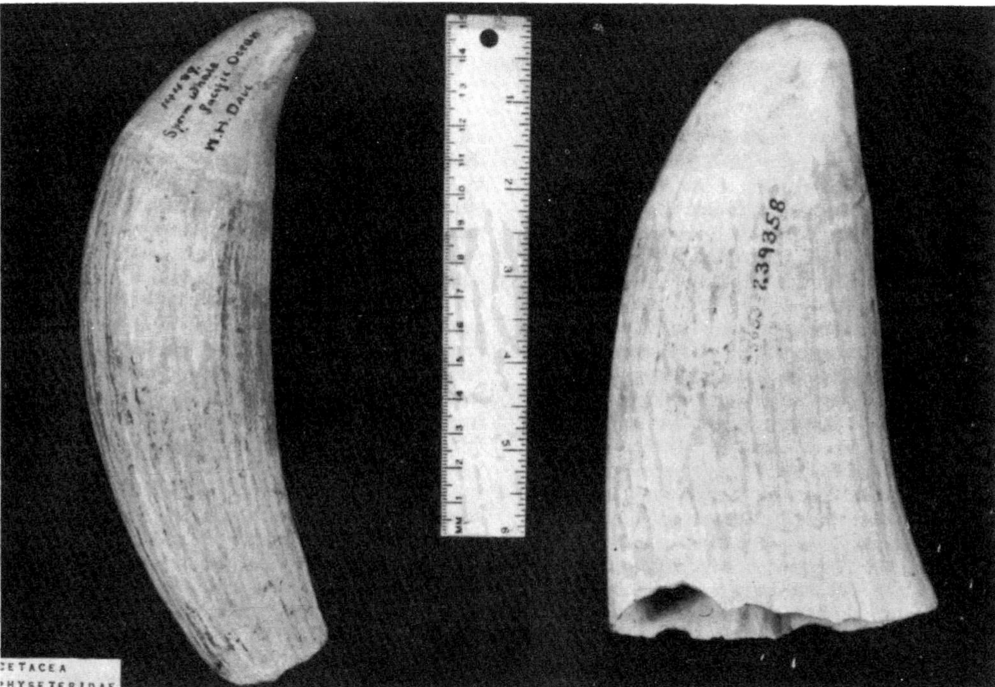

Top, dead sperm whale *(Physeter catodon)*, photo by Warren J. Houck. Bottom, teeth of *P. catodon*, photo by P. F. Wright of specimen in U.S. National Museum.

Hundreds of individuals sometimes join in migratory schools, and there is one record of 3,000–4,000 being seen together off Patagonia. The basic social unit, however, is the mixed school, which has a year-round membership of adult females, calves, and some juveniles, usually 20–40 individuals in all. These units are stable, and there is evidence that the bonds between females last many years. If nursing calves are present, the group may be referred to as a nursery school. Shortly after weaning, young males and some young females begin to leave and to join in juvenile schools, usually made up of 6–10 individuals. The females eventually return to a mixed school before reaching puberty. Males from the juvenile schools and some coming directly from mixed schools form bachelor schools. When the bachelors are small these schools contain 12–15 whales, though several groups may aggregate. As the bachelors grow larger they divide into smaller groups, and then often into pairs and lone animals. Even if an association is maintained by older males the individuals may stay some distance from each other. During the breeding season each mixed school is commonly joined by 1–5 large males. The groups then become known as breeding, or harem, schools, and there usually is about 1 adult male for every 10 adult females. New evidence suggests that the males may remain only a few hours before moving off in search of another group with estrous females and that there is no long-term control of harems. The exact relationship between the adult males is not well understood. There is apparently some fighting for the right to join a breeding school, and the scars from such battles frequently cover the heads of males. Nonetheless, it may be that several males establish a dominance hierarchy and share the females of one or more mixed schools. Such males may even have composed an organized bachelor school prior to the breeding season. In any event, only 10–25 percent of the fully adult males in a population are able to get into a breeding school. The others move toward the Arctic and the Antarctic. It is not known whether some or all of the adult males in a breeding school also eventually spend some time in polar waters (Best 1979; Ellis 1980; Gaskin 1970; Ohsumi 1971; Whitehead and Arnbom 1987).

The breeding season varies, but in both the Northern and Southern hemispheres mating generally peaks in the spring and calving in the autumn. Females generally give birth at intervals of 5–7 years. Gestation has been variously calculated at 14–19 months. The single newborn is about 400 cm long and weighs nearly 1,000 kg. Some solid food is taken by the age of 1 year, but nursing usually lasts just over 2 years and seems to continue in some instances for up to 13 years. Sexual maturity is attained by females at 7–13 years. Males may have the physiological ability to produce offspring at less than 19 years, but they do not reach social maturity—the ability to enter a breeding school—until they reach 25–27 years. Full physical maturity comes at 35 years in males and 28 years in females. Some individuals are estimated to have lived up to 77 years (Berzin 1972; Best 1968, 1970*b*, 1974, 1979; Best, Canham, and MacLeod 1984; Caldwell, Caldwell, and Rice 1966; Frazer 1973; Haley 1978; Ohsumi 1965, 1966; D. W. Rice 1978*d*, 1989).

The sperm whale has been hunted regularly by people since 1712. The meat is not generally used for human consumption, and of course there is no yield of baleen. The large teeth of *Physeter* were valued, however, as a medium for the artistic form of engraving and carving known as scrimshaw. A product unique to the sperm whale is ambergris, a waxy substance probably formed in the intestines from solid wastes coalescing around a matrix of indigestible material. It is used as a fixative and has the property of retaining the fragrance of perfumes. Apparently the heaviest mass of ambergris from a single animal weighed about 450 grams. Such an amount of the substance would have sold for about $10–$50 in the early 1960s, depending on color, but prices have fallen in recent years because of the development of synthetic fixatives. The most important product of the sperm whale is oil, though unlike that of baleen whales, it cannot be used in the manufacture of margarine. Sperm whale oil was once the major source of fuel for lamps and recently has served as a lubricant and as the base for skin creams and cosmetics. The oil of the spermaceti organ solidifies into a white wax upon exposure to air and is used in making ointments and fine, smokeless candles. It also has served as a high-quality lubricant for precision instruments and machinery and as a component of automatic transmission fluid.

An intensification of sperm whale hunting came around 1750 with the invention of the spermaceti candle and the development of onboard processing facilities, which allowed whaling vessels to remain at sea until their holds were filled with oil (Ellis 1980). For the next 100 years Americans dominated the industry, probably taking up to 5,000 sperm whales annually. This kill may not have been enough to reduce overall populations, but in the second half of the nineteenth century both sperm whale hunting and American whaling declined. Some of the suggested reasons are increasing costs, replacement of whale oil by kerosene, destruction of much of the U.S. fleet in the Civil War, and opening of the western Arctic and North Pacific to hunting of the bowhead whale *(Balaena mysticetus)*. The development of the harpoon gun in the 1860s brought more hunting of the large baleen whales and less emphasis on *Physeter.*

By the 1930s a decline in some of the baleen species had resulted in renewed interest in sperm whale hunting. Floating factory ships could remain for lengthy periods within the prime habitat of *Physeter,* especially the North Pacific. In the 1936–37 season, for the first time in many years, the annual kill rose above 5,000. Subsequent efforts at international regulation were largely unsuccessful. In the 1950–51 season the take was 18,264 sperm whales, in 1963–64 it peaked at 29,255, and it remained above 20,000 in all but one season until 1975–76. About one-fourth of the total catch was made by shore-based operations, and the remainder by pelagic expeditions. Finally, in response to immense scientific and public concern, the International Whaling Commission began to reduce quotas substantially. The kill in the 1978–79 season was 8,536, the quota set for 1980–81 was only 1,849, and no kill was authorized for 1981–82. Moreover, the use of floating factories was banned in the hunting of *Physeter,* and most shore-based sperm whale fisheries closed down. All commercial whaling was halted by 1985 (Committee for Whaling Statistics 1980; Ellis 1980; Gulland 1974; McHugh 1974; Rice 1978*d;* U.S. National Marine Fisheries Service 1978, 1981, 1984, 1987).

There has been considerable controversy regarding the numerical and conservation status of *Physeter.* According to Rice (1989), prior to modern exploitation there were probably close to 3 million sperm whales divided about equally by sex; numbers subsequently fell to fewer than 2 million, with males being reduced by about 45 percent and females by about 17 percent. The U.S. National Marine Fisheries Service (1978, 1981, 1987) indicated that in 1946 the estimated number of males more than 9.2 meters long (the minimum legal limit for catching) and sexually mature females was about 1.1 million and that even into the 1980s the estimate for those categories was just over 700,000. In 1994 that agency listed a worldwide estimate of about 1.8 million for all age and sex categories.

However, at the 1989 meeting of the International Whaling Commission had come the shocking news that direct counts by whaling cruises over the previous eight years had led to new and drastically lower estimates for most of the large whales. Some reports indicated that *Physeter* numbered only 5,000–10,000 individuals in the Southern Hemisphere, having been nearly exterminated by the intensive commercial kills that continued through the 1970s (*Marine Mammal News* 15 [1989]: 5). Such reports apparently referred only to numbers of animals actually seen in limited areas and not to total population size. Recent extrapolation of data indicates that the total number of sperm whales to the south of 30° S is about 128,000 (International Whaling Commission 1994). Information for the Northern Hemisphere is highly uncertain, though there have been estimates of more than 80,000 for the western North Pacific Ocean (Klinowska and Cooke 1991) and about 23,000 for the eastern tropical Pacific (Wade and Gerrodette 1993). *Physeter* now is classified as vulnerable by the IUCN, is listed as endangered by the USDI, and is on appendix 1 of the CITES.

CETACEA; **Family ESCHRICHTIIDAE; Genus ESCHRICHTIUS**
Gray, 1864

Gray Whale

The single known genus and species, *Eschrichtius robustus*, occurs in coastal waters from the Sea of Okhotsk to southern Korea and Japan and from the Chukchi and Beaufort seas to the Gulf of California (Rice 1977). Other populations apparently once existed in the North Atlantic, there being records from the east coast of the United States, western Europe, the Baltic Sea, and Iceland (Fraser 1970; Mead and Mitchell 1984; Mitchell and Mead 1977). Some of the Atlantic records are based on material that is subfossil but from historical times. There also are a few old specimens and capture records indicating that the western Pacific population once extended south along the coast of China as far as Hainan (Omura 1988; Wang 1984). Geologically the family Eschrichtiidae is known only from the Recent and from a single Pleistocene specimen of *E. robustus* about 50,000–120,000 years old found in southern California (Barnes and McLeod 1984). This relatively brief known evolutionary history, together with analysis of mitochondrial DNA, was cited by Arnason and Gullberg (1994) in suggesting that *Eschrichtius* is more closely related to the Balaenopteridae than some species commonly assigned to that family are related to one another. E. R. Hall (1981) used the name *E. gibbosus* in place of *E. robustus*.

At sexual maturity head and body length is about 11.1 meters, at physical maturity the length averages 13 meters in males and 14.1 meters in females, and the maximum reliably recorded length is 14.3 meters in males and 15 meters in females (Rice and Wolman 1971). Pectoral fin length is approximately 200 cm, and the expanse of the tail flukes is about 300 cm. There is no dorsal fin, but there are 8–14 low humps along the midline of the lower back. Adults weigh 20,000–37,000 kg. Coloration is black or slaty gray, with many white spots and skin blotches, some being discolored patches of skin and others being areas of white barnacles.

The snout is high and rigid, and the throat has two or three (rarely four) short, shallow, curved furrows. *Eschrichtius* is a baleen whale and has no teeth. The baleen is yellowish white in color and is arranged in series of 130–80 plates on each side of the mouth. The largest plates are 34–45 cm long. The two rows of plates do not meet in front, as they usually do in the family Balaenopteridae. There are 56 vertebrae in *Eschrichtius*, and the neck vertebrae are separate.

The gray whale is found primarily in shallow water and generally remains closer to shore than any other large cetacean. The eastern Pacific population makes an annual migration of more than 18,000 km. From late May to early October this population concentrates in the shallow waters of the northern and western Bering Sea, the Chukchi Sea (between northern Alaska and Siberia), and the Beaufort Sea (off northeastern Alaska). Some individuals, however, spend the summer farther south, scattered along the coast of Oregon and northern California. From October to January the main part of the population moves down the

Gray whales *(Eschrichtius robustus)*, painting by Richard Ellis.

Gray whale *(Eschrichtius robustus):* Top, photo by David Withrow; Bottom, photo by David Rugh.

east side of the Bering Sea, goes through Unimak Pass in the Aleutians, and proceeds down the west coast of North America. In January and February most whales are along the west coast of Baja California and on the eastern side of the Gulf of California off Sonora and Sinaloa. In these areas are five bays and lagoons, within which the young are born. From late February to June the population migrates northward to its summering sites in the Arctic (Rice and Wolman 1971; Wolman and Rice 1979).

During the southward phase of the migration of the eastern Pacific population the animals are more concentrated and may move somewhat nearer to the shore than during the northward phase. For example, most individuals pass San Diego during a six-week period from late December to early February, and in certain past years 95 percent of the population traveled 3–5 km offshore. At other points the whales sometimes move within 1 km of land. The migration is therefore witnessed by large numbers of people. Unfortunately, perhaps, increasing boat traffic seems to have forced the whales to stay farther from shore (Wolman and Rice 1979). Observations by Leatherwood (1974) indicated that half the population passed more than 64 km off San Diego.

Considerably less is known about the western Pacific population, which has been reduced to very small numbers. Apparently it spent the summer in the Sea of Okhotsk and then migrated through the Sea of Japan and along the Pacific coast of Japan to wintering sites off the southern coast of Korea and in the Inland Sea of Japan. Some animals probably once migrated southwestward across the East

China Sea to breeding waters along the coast of China and around Hainan. The extinct population of the eastern Atlantic probably summered in the Baltic Sea and wintered along the Atlantic and Mediterranean coasts of southern Europe and North Africa. The western Atlantic population may have bred and given birth in the shallow lagoons and bays of south-central and southeastern Florida before migrating northward (Omura 1974, 1984, 1988; Reeves and Mitchell 1988; Rice and Wolman 1971; Wang 1984).

The normal swimming speed of the gray whale is 7–9 km/hr (Rice and Wolman 1971), but when pursued it may reach about 13 km/hr. Migrating individuals usually submerge for four or five minutes and then surface to blow three to five times. The spout rises to a height of about 300 cm. The tail flukes usually appear above the water just before a deep dive but not before a shallow dive. The back is not arched before diving, as in the humpback whale *(Megaptera)*. The gray whale often seems to play in heavy surf and in shallow water along the shore, sometimes throwing itself clear of the water and falling back on its back or side. It occasionally becomes stranded in less than 1 meter of water but refloats on the next tide, without injury. Visual stimuli appear to be involved in orientation during migration. The habit of spyhopping—lifting the head vertically out of the water and appearing to look around—is commonly observed in *Eschrichtius* and may be involved in social interaction (Samaras 1974). A variety of sounds have been reported, including grunts, groans, rumbles, whistles, and clicks. The clicks seem to be of a frequency too low for use in precise echolocation but may possibly assist in sonar navigation or in the detection of dense concentrations of food organisms (Fish, Sumich, and Lingle 1974).

Feeding takes place primarily on the bottom, with individuals apparently sometimes plowing their heads sideways through the mud or sand to stir up prey (Ellis 1980). Water and organisms are sucked into the mouth, and then the water is forced out, leaving the food trapped within the baleen. The diet consists mainly of gammaridean amphipods (small crustaceans), especially *Ampelisca macrocephala*, a creature about 25 mm long and found on sandy bottoms at depths of 5–300 meters (Rice and Wolman 1971). Other crustaceans, certain mollusks and worms, and small fish are also eaten. A captive yearling female consumed and apparently thrived on 900 kg of squid per day (Ray and Schevill 1974). Available evidence suggests that there is little or no feeding except when the animals are in the northern part of their annual migratory route and that fasting lasts up to six months (Rice 1978*c*). Individuals moving south past San Francisco had lost 11–39 percent of their body weight by the time they moved north past the city (Rice and Wolman 1971).

Aggregations of as many as 150 individuals have been seen in the arctic feeding waters (Rice 1967), but *Eschrichtius* is not a particularly social cetacean. It usually migrates alone or in groups of 2–3, though sometimes as many as 16 travel together. Moreover, there is segregation by age and sex during migrations. Generally females precede males and adults precede immature animals. Newly pregnant females are the first to leave on the northward migration (Rice 1978*c;* Rice and Wolman 1971). Segregation also has been reported at the wintering sites, with calving and nursing females remaining well within protected lagoons and males positioning themselves at the entrances (Ellis 1980). Individuals have been observed to aid others that are injured or giving birth by pushing them to the surface so they can breathe. Females are reportedly very protective of their young, even to the point of attacking whaling boats.

Females have a 2-year reproductive cycle. Most enter estrus and mate during a 3-week period in late November and early December, while still migrating south, but some do not mate until they are in the wintering lagoons or even on the northward migration. Births occur mainly from late December to early February. Gestation lasts about 13 months, lactation lasts about 7 months, followed by 3–4 months of anestrus. The single newborn averages about 490 cm in length and 500 kg in weight. It is weaned around August in the summer feeding area, after it has grown to a length of 850 cm. The mean age of puberty is about 8 years in both sexes, and full physical maturity is attained at about 19 years in males and 17 years in females (Rice 1978*c;* Rice and Wolman 1971; Yablokov and Bogoslovskaya 1984). Maximum estimated longevity is 70 years (Haley 1978).

Like various other large cetaceans, the gray whale has been killed by people for its oil, meat, hide, and baleen. It was hunted in ancient times by the native people of northwestern North America and eastern Siberia and probably also by Europeans and Japanese. Because of its migrations close to shore, it was comparatively simple to locate and secure. According to Krupnik (1984), aboriginal whaling began to develop in the northern Pacific nearly 2,000 years ago and eventually became the major economic activity along most coasts of the Chukchi, Bering, and Okhotsk seas. The activity had declined by the mid–nineteenth century in connection with a reduction in the number of native people, cultural changes, and usurpation of whaling by Americans. The European gray whale population is not well known, but there is evidence that it was still being hunted in coastal areas in the Middle Ages, there is a likely record for Iceland in the early seventeenth century, and specimens from England have radiocarbon dates as recent as about 340 years ago. The western Atlantic population apparently survived until about 1700, when it disappeared in the face of intensifying American whaling (Bryant 1995; Fraser 1970; Mead and Mitchell 1984; Mitchell and Mead 1977; Rice and Wolman 1971).

The western Pacific population seems to have always been rather small, with about 50 individuals being taken annually in the eighteenth and nineteenth centuries by Japanese whalers. The branch that bred in the Inland Sea of Japan probably contained fewer than 1,000 whales. It was gone by 1900, the last survivors perhaps being driven away by increasing boat traffic and industrialization (Omura 1974, 1984). The branch that bred off southern Korea may have numbered 1,000–1,500 when modern whaling began in that area around the turn of the century. By 1933 the Korean population appeared to be extinct, but there were 67 known kills from 1948 to 1966 and sightings in Korean waters as late as 1974 (Brownell and Chun 1977; Rice and Wolman 1971; Wolman and Rice 1979). An individual was seen off the Pacific coast of Japan in 1982 (Furuta 1984), and more recently others have been observed in the Sea of Okhotsk and near the Kuril Islands. Some authorities have considered these reports to refer to stray animals from the eastern Pacific population, but Reeves and Mitchell (1988) suggested that they represent survival of the original eastern stock and estimated that this population numbers in the tens or low hundreds.

The wintering lagoons of the eastern Pacific population were discovered by American whalers in 1846. Shore-whaling stations were established in the area, and from 1846 to 1874 the known kill was 10,800 animals. By about the turn of the century regular shore whaling stopped and the population seemed extinct, but there may still have been several thousand individuals left. Shortly thereafter, factory ships came into use for the hunting of the gray whale. From 1921 to 1947, 1,153 individuals are known to have been killed, mainly by Norwegian, Japanese, and Russian vessels. Since 1946 the species has received protection under the International Whaling Convention, except that about 175 individuals have been taken legally each year by Siberian Eskimos, and also a few by Alaskan natives. A substantial number also perish accidentally after becoming entangled in fishing gear. The eastern Pacific population evidently increased steadily in response to protection and currently is estimated to contain 21,000 whales, perhaps not far below the number present before exploitation began. Currently the main problem is disturbance of the animals and their habitat, especially in the calving lagoons, by industrialization, shipping, and even well-meaning tourists, who follow the whales in motorboats (Klinowska and Cooke 1991; Reeves and Mitchell 1988; Rice 1978*c;* Rice and Wolman 1971; Rice, Wolman, and Braham 1984; Storro-Patterson 1977; U.S. National Marine Fisheries Service 1987, 1989; Wolman and Rice 1979).

The gray whale is on appendix 1 of the CITES. The IUCN classifies the western Pacific stock as endangered and the eastern Pacific stock as conservation dependent. Until recently the entire species was listed as endangered by the USDI, but in June 1994 the eastern Pacific population was declassified. The latter measure may have been politically motivated, involving an attempt to demonstrate that a whale population can recover even while undergoing limited exploitation. According to the Marine Mammal Commission (1994), which questioned the USDI declassification, studies by the U.S. National Marine Fisheries Service in 1984 indicated that the eastern Pacific population was threatened by human disruption of its calving waters and migratory corridors. Nonetheless, in response to a 1990 petition by an organization that may have ultimate exploitation interests, the service proposed declassification. Ironically, the proposal identified numerous threats to the population and its habitat but concluded that it was not threatened. That the population still is smaller or not much larger than populations of numerous other still classified cetaceans, that it is concentrated entirely in a limited and environmentally sensitive region, and that it has a demonstrated history of vulnerability to human activity seem to have been disregarded by the USDI.

CETACEA; **Family NEOBALAENIDAE; Genus CAPEREA**
Gray, 1864

Pygmy Right Whale

The single known genus and species, *Caperea marginata*, is known from strandings or sightings in western and southern Australia, Tasmania, New Zealand, the Falkland Islands, the South Atlantic Ocean, South Africa, the Crozet Islands, and possibly Argentina and Chile (Ross, Best, and Donnelly 1975). Many authorities, including Rice (1984), treat the Neobalaenidae as a subfamily of the Balaenidae, but Arnason and Best (1991), Barnes, Domning, and Ray (1985), and Barnes and McLeod (1984) regarded the two groups as separate families. There are no reliably known fossils.

Caperea is the smallest baleen whale. Females apparently are larger than males. According to Baker (1985), the largest female on record was 645 cm long and the largest male measured 609 cm. A female 621 cm long weighed 3,200 kg, and a male 547 cm long weighed 2,850 kg. Another adult female specimen measured as follows: pectoral fin length, 66 cm; dorsal fin height, 25 cm; and expanse of the tail flukes, 181 cm. The coloration is black or dark gray above and paler below. Young animals may be lighter than adults, and one juvenile appears to have been albinistic (Ross, Best, and Donnelly 1975). The tongue and interior part of the mouth are pure white. In profile the ventral outline of the anterior part of the throat is concave (Ross, Best, and Donnelly 1975). There are 213–30 baleen plates on each side of the mouth. They are narrow, have a fringe of fine soft hair, and are colored whitish yellow with a narrow, dark brown marginal band on the external edge (Baker 1985). The plates are up to 70 cm long and are ivory white in color but have a dark outer margin. The sickle-shaped dorsal fin is located on the posterior part of the back.

Caperea resembles *Eubalaena* and *Balaena* in having a relatively large head, about one-fourth of the total length, and a strongly arched lower jaw. Aside from its much smaller size, it differs from these other genera in having a dorsal fin, incipient throat grooves, a different type of baleen, a smaller head size in proportion to the body, a proportionally shorter humerus, and four instead of five digits in the manus. Its skull is very different, having a larger, more anteriorly thrust occipital shield; a shorter, wider, and less arched rostrum; smaller nasal bones; shorter supraorbital processes of the frontals; and orbits and glenoid fossae that are not located as far ventrally (Barnes and McLeod 1984). There are 40–44 vertebrae, fewer than in the other genera (Baker 1985).

Caperea has 34 ribs, more than any other cetacean genus. These ribs also extend farther posteriorly than those of any other cetacean, so only two vertebrae without ribs intervene between those with ribs and the tail. The ribs become increasingly flattened and widened toward the tail and thereby probably provide additional protection to the internal organs.

The pygmy right whale is known only by about 80 specimens and a small number of sightings. Available information on natural history was summarized by Baker (1985), Mitchell (1975*b*), Pavey (1992), and Ross, Best, and Donnelly (1975). The genus appears to be restricted to temperate waters of 5°–20° C in the Southern Hemisphere. Although it is primarily pelagic, stranding records suggest a movement, mainly by juveniles, toward the shore in spring and summer. At such times *Caperea* is frequently found in sheltered, shallow bays. Records indicate that this genus occurs throughout the year around Tasmania but only from August to December off Australia and from December to February off South Africa.

Old reports suggested that *Caperea* dives deeply and may remain for long periods near the bottom. More recent observations indicate that it stays underwater only for three or four minutes at a time, at a depth of two or three meters. It is a relatively slow swimmer, averaging around 7 km/hr, and proceeds with a flexing of the entire body. The diet evidently consists mainly of the minute crustaceans known as copepods.

Schools of as many as eight individuals have been observed. Records of pregnant females and young suggest that mating and calving occur between June and February. At least some young are thought to be born during the austral autumn or winter, apparently well away from land. The gestation period probably is about 10 months. Theoretical length at birth is 190 cm. Nursing is presumed to last about five or six months and to be followed by a movement of the young toward shore in the spring and summer.

Caperea now is on appendix 1 of the CITES. However, there are no records of exploitation or international trade (Klinowska and Cooke 1991).

Pygmy right whale *(Caperea marginata)*, photos by Mr. T. Dicks through Peter B. Best.

CETACEA; **Family BALAENIDAE**

Right Whales

This family of two Recent genera and species, *Eubalaena glacialis* and *Balaena mysticetus*, is found in all oceans and adjoining seas of the world except in tropical and south polar regions (Rice 1984). A third genus, *Caperea* (see account thereof), often has been placed in the Balaenidae.

The two genera are among the largest of baleen whales, and females average larger than males. Both genera have a massive body and a relatively large head, accounting for one-fourth to one-third of the total length. The rostrum is narrow and highly arched, resulting in the cleft of the mouth being curved. The lower jaw is enormous and has a

Dr. Roy Chapman Andrews, 6 ft. 1 in. in height, standing beside the skull of a right whale *(Balaena glacialis).* The plates of baleen are smooth on the outside, but on the inside they are fringed to form an effective strainer. Photo from American Museum of Natural History.

large, fleshy lip. There are no throat and chest grooves, as in the Balaenopteridae, and no dorsal fin. The pectoral fin is short, broad, and rounded. The tongue is heavy and muscular. This family, like the Balaenopteridae and Eschrichtidae, has baleen instead of teeth. The baleen plates of the Balaenidae are the longest in the Cetacea and are usually narrower than in the other two families. There is a row of plates on each side of the upper part of the mouth, and the two rows do not join anteriorly. The plates fold in the floor of the closed mouth and straighten when the mouth opens. The vertebrae number 54–57; the 7 neck vertebrae are fused into a single unit. The manus has five digits. The tail stock is constricted and tapers into the flukes.

The diet consists mainly of the small forms of animal life known collectively as zooplankton (and sometimes called "krill"). These organisms include crustaceans, such as the shrimplike euphausiids and copepods, and the free-swimming mollusks known as pteropods. Unlike most species of the Balaenopteridae, the Balaenidae use the feeding method called "skimming." The whales swim through swarms of prey with their mouths open and their heads above water to just behind the nostrils. When a sufficient mouthful of organisms has been filtered from the water by the inner bristles of the baleen plates, the whales force out the water, dive, and swallow the food.

The common name "right" whale refers to the consideration of *Eubalaena* and *Balaena* in the early days of commercial whaling as the proper or best kind of whales to hunt. They were slower and less active than the sperm whale and most other baleen whales, and they often came closer to land. Unlike the others, they had greater buoyancy and were less likely to sink after being killed. Moreover, they yielded the most valuable products. The amount of oil usually derived from an adult was 80–100 barrels (1 barrel = about 105 liters), more even than from a blue whale *(Balaenoptera musculus).* Some large individual *Eubalaena* and *Balaena* were said to have produced several hundred barrels each. The oil was used mainly as a fuel for lamps and in cooking but also served as a lubricant, in leather tanning, and in the manufacture of soap and paint.

The baleen of *Eubalaena* and *Balaena* was the finest available, and the plates were far longer than those of other species. The usual yield from an adult *Balaena* was 680–760 kg. The hard outer portion of the plates was split into strips that could be used in the manufacture of umbrella ribs, fishing rods, carriage springs, whips, and numerous other items that required a combination of strength and elasticity. Baleen strips were especially useful as a stiffening in fashionable garments, such as the farthingale (a framework for supporting the elaborate dresses of the Elizabethan era) and the hoop skirt of the nineteenth century. The fine inner fibers of the baleen plates were also important in fashion. According to Gilmore (1978), they were woven into fabrics, giving a stiffness and rustle to taffeta and crinoline, and could be washed without softening or loss of elasticity.

The value of whale products fluctuated over the years. Typical nineteenth-century prices in the United States, however, were around $30 per barrel for oil and $10 per kg for baleen. At such rates the average *Balaena* could bring more than $10,000, at a time when the cost of living was about 3 percent of what it is today. By the early twentieth century the last great concentrations of *Eubalaena* and *Balaena* had been eliminated, whale oil was no longer being used for illumination, and spring steel was replacing baleen. Although whaling was about to enter its most intensive era, right whales were not to be of major commercial importance.

The geological range of the family Balaenidae is early Miocene to Pleistocene in South America, middle Miocene to Pleistocene in western North America, late Miocene in Australia, early Pliocene to Pleistocene in Europe, and Recent in all oceans (Rice 1984).

CETACEA; BALAENIDAE; **Genus EUBALAENA**
Gray, 1864

Right Whale

A single species, *E. glacialis,* occurring mainly in temperate parts of the Atlantic, Indian, and Pacific oceans and adjoining seas, was recognized by Banfield (1974), Corbet and Hill (1991, 1992), Gaskin (1982, 1987, 1991), McLeod, Whitmore, and Barnes (1993), Meester et al. (1986), and Rice

Right whale *(Eubalaena glacialis)*, photo by Roger Payne.

(1977, 1984). The populations of the Northern and Southern Hemispheres generally are separated from one another by several thousand kilometers, and some authorities, such as Cummings (1985), Klinowska and Cooke (1991), Mead and Brownell (*in* Wilson and Reeder 1993), Payne and Dorsey (1983), and Reeves and Leatherwood (1994), regard the southern populations as a full species, *E. australis*. Brownell, Best, and Prescott (1986) indicated that the question was unsettled and recommended further systematic work. Recent analysis of mitochondrial DNA suggests that *glacialis* and *australis* diverged about 0.9 million to 1.8 million years ago (Schaeff et al. 1991). *Eubalaena* often is included within the genus *Balaena*, but recently authorities have tended to treat the two as distinct genera (Barnes and McLeod 1984; Brownell, Best, and Prescott 1986; Cummings 1985; Gaskin 1982, 1987; Klinowska and Cooke 1991; McLeod, Whitmore, and Barnes 1993; Mead and Brownell *in* Wilson and Reeder 1993; Payne and Dorsey 1983).

Head and body length at physical maturity is usually 13.6–16.6 meters. Banfield (1974) listed average size as 13.7 meters and 22,000 kg for males and 14 meters and 23,000 kg for females. Rice (1984) stated that maximum length is 18 meters and that one female 17.4 meters long weighed 106,500 kg. Pectoral fin length is 180–210 cm. There is no dorsal fin. The coloration is usually black throughout, but there are sometimes large white patches, especially on the belly. Entirely white calves have been observed (Best 1970*a*). Around the head are series of horny protuberances representing accumulations of cornified layers of skin and commonly infested with barnacles and parasitic crustaceans. The most conspicuous of these callosities is located on the tip of the upper jaw and is known as the "bonnet." The callosities are present from the time of birth, but their exact function is unknown. Payne and Dorsey (1983) found that males have more and larger callosities than do females and suggested that they are used as weapons for intraspecific aggression.

The head of *Eubalaena* is smaller than that of *Balaena*, about one-fourth of the total length, and the upper jaw is not as strongly arched. The narrow upper jaw is practically concealed by the high, massive lower jaw when the mouth is closed. On each side of the upper part of the mouth is a row of 225–50 baleen plates. These plates are usually 180–220 cm in vertical length and range in color from dark gray or brown to black. The two blowholes are set well apart, and thus there are two spouts, which form a V pattern (Gilmore 1978; Leatherwood, Caldwell, and Winn 1976).

The right whale is primarily a species of temperate waters, though some individuals apparently move just north of the Arctic Circle and just south of the Tropic of Cancer. It is usually found closer to land than are most large whales, especially in the breeding season. Although it was largely eliminated by whalers before much scientific information could be gathered, available evidence indicates that there were well-defined migrations to higher latitudes for summer feeding and back to lower latitudes for winter breeding. Populations of the Southern Hemisphere wintered mainly between 30° and 50° S, off the coasts of South America, southern Africa, Australia, and New Zealand, and then moved toward the Antarctic for the summer. The population of the eastern North Atlantic wintered in the Bay of Biscay and summered between Great Britain and Ice-

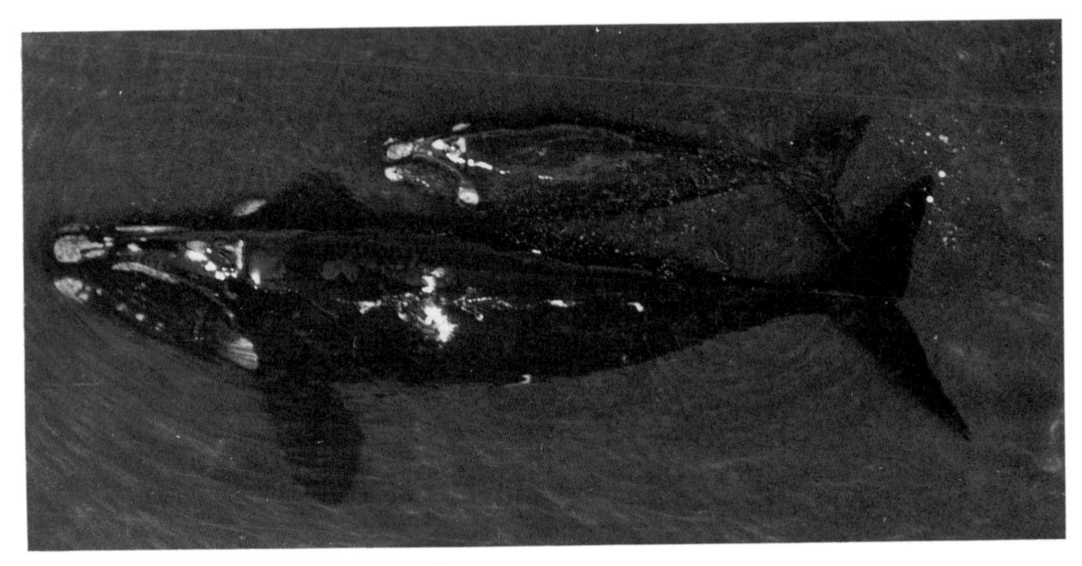

Right whales *(Eubalaena glacialis)*, photo by Roger Payne.

land. In the western North Atlantic from about November to March the whales remained from Cape Cod to Bermuda and the Gulf of Mexico. In late winter they began to move north, and by spring they were concentrated off the northeastern United States. Some remained in the latter region, but the major summer feeding waters were around Newfoundland and in the Labrador Sea. Part of the remnant population in the western North Atlantic calves during winter off Georgia and Florida, feeds off New England in the spring, feeds in the Bay of Fundy and off Nova Scotia in the summer and early autumn, and then migrates southward in late October and November. Pacific populations seem to have passed Japan in April and fed off southern Alaska from May to September; wintering waters are unknown but may possibly have included those around Hawaii and the Ryukyu Islands (Allen 1942; Banfield 1974; Gaskin 1968; Gilmore 1978; Herman et al. 1980; Omura 1986; Reeves, Mead, and Katona 1978; Winn, Price, and Sorensen 1986).

The right whale is a relatively slow swimmer, averaging about 8 km/hr, but frequently leaps clear of the water and engages in other acrobatics. It commonly makes a series of five or six shallow dives and then lifts its tail flukes above the water and submerges for about 20 minutes. It is usually not wary of boats and can be easily approached (Leatherwood, Caldwell, and Winn 1976). It appears to frolic in stormy seas and even to use its tail to sail in the wind (Payne 1976). It employs the skimming method of feeding, generally swimming near the surface with its mouth open to strain out organisms from the water with its baleen (see familial account). Its diet consists mainly of copepods—crustaceans only a few millimeters in diameter—and also includes euphausiids, pteropods, and small fish (Gilmore 1978).

According to Banfield (1974), *Eubalaena* sometimes was found in aggregations of 100 or more. Social activity, however, is generally limited to mating concentrations and the relationship between mother and young. *Eubalaena* emits a number of low-frequency bellows, moans, pulses, and other sounds, mostly during courtship (Cummings 1985; Gilmore 1978). The primary contact call is a low, tonal upsweep that intensifies toward the end (Clark 1982). A male may attempt to mate with several females, and a female may sometimes be courted by several males at once (Ellis 1980; Payne 1976). There is some question whether physical confrontation occurs between males during the breeding season, though sometimes one male supports a female while another male mates with her (Gaskin 1987). Females give birth to a single offspring about every 2–5 years; at that time they separate from the rest of the population and move closer to shore (Kraus et al. 1986). Mothers with young calves in the population that winters off the Valdez Peninsula in Argentina concentrate along the coast in water only about 5 meters deep (Payne 1986). Females in that area usually produce their first calf at 9 years, with most births there occurring from June to December (Payne et al. 1990). The calving season off South Africa lasts from late June to late October, with a peak in August, and observations there indicate a gestation period of 357 or 396 days (Best 1994). In the Southern Hemisphere mating has been reported to peak from August to October (Lockyer 1984). Most births in the western North Atlantic apparently occur off Florida and Georgia between December and March following a gestation period of about 13 months (Gaskin 1991). The young is about 350–550 cm long at birth and nurses for about 7 months. Gaskin (1987) estimated the age of sexual maturity to be about 6 years, and Cummings (1985) estimated it to be about 10 years.

Because of its coastal habitat and economic value (see familial account), *Eubalaena* was among the first of the large whales to be extensively exploited. It may have been hunted as early as the ninth century A.D. off Norway and was regularly taken in the Bay of Biscay in the tenth century. By the late fifteenth century it had become rare in the latter area, and the Basque whalers shifted their emphasis to other waters, eventually reaching Newfoundland. An estimated 25,000–40,000 whales were killed around Newfoundland from 1530 to 1610, after which the Basques began to journey northward in pursuit of the bowhead. The right whale, however, became the basis of a major industry in the American colonies during the seventeenth century, with hunting initially being done from small boats in such places as Delaware Bay and Cape Cod Bay and later involving extended voyages in large ships. By 1700 in Europe and 1800 in the United States the right whale had become too rare to be of commercial interest. Toward the end of the eighteenth century, however, large stocks of the species were discovered in the wintering waters of the Southern Hemisphere and in the summer feeding areas of the North Pacific. Just during the 1840s the right whale in the North Pacific and adjacent seas seems to have gone from a state of great abundance to one of scarcity. By that time the industry was dominated by American companies. On the basis of catch records and oil and baleen imports, it has been estimated that 70,343–74,693 right whales were taken by American vessels from 1805 to 1914 (see Best 1970*a* and 1987 for an explanation of these figures). By the latter year the right whale was very rare throughout its range. Some hunting continued in the early twentieth century, but takes were usually considerably fewer than 100 per year. In 1937, in accordance with the International Agreement for the Regulation of Whaling, *Eubalaena* received complete protection (Aguilar 1986; Allen 1942; Banfield 1974; Reeves, Mead, and Katona 1978; Scarff 1991).

Braham and Rice (1984) estimated the original number of right whales in the world as 100,000–300,000, about two-thirds of them in the Southern Hemisphere. The species now has become one of the rarest of large mammals. According to the U.S. National Marine Fisheries Service (1989, 1994), current estimated numbers are 3,000 in the Southern Hemisphere, 100–200 in the western North Pacific, and 265 in the eastern North Pacific. In the western North Atlantic there are about 200–350 individuals left, with a majority of mothers and calves concentrating in the Bay of Fundy during the summer (Gaskin 1991; Schaeff et al. 1993). Gaskin (1987) stated that the eastern North Atlantic population is virtually extinct. Brown (1986) listed recent records indicating that there still are a few right whales in the eastern North Atlantic but noted that they might be stragglers from the west.

There is evidence that the populations of the Southern Hemisphere have grown since protection was granted. The group wintering off the Valdez Peninsula, Argentina, has been increasing at 7.6 percent annually and in 1986 was estimated to contain 1,200 whales (Payne et al. 1990). There have been suggestions that an increase also has occurred in the North Atlantic, but Reeves, Mead, and Katona (1978) cautioned that recovery has been modest at best and that the species may be jeopardized by such factors as competition for food with the sei whale, accidental trapping in fish nets, pollution, and boat collisions. Braham and Rice (1984) added that the right whale may be the most vulnerable of all great whales, especially because females move into shallow coastal bays to give birth. Klinowska and Cooke (1991) listed numerous threats and saw no sign of any recovery of the western North Atlantic stock. Likewise, Scarff (1986) reported no evidence of recovery in the eastern Pacific. *Eubalaena* is classified as endangered by the USDI and is on appendix 1 of the CITES. The IUCN now recognizes two species (see above) and has designated *E. glacialis* of the Northern Hemisphere as endangered and *E. australis* of the Southern Hemisphere as conservation dependent.

CETACEA; BALAENIDAE; **Genus BALAENA**
Linnaeus, 1758

Bowhead Whale, or Greenland Right Whale

The single species, *B. mysticetus,* is found in the Arctic Ocean and adjoining seas, the Sea of Okhotsk, Hudson Bay, and the Gulf of St. Lawrence (Banfield 1974; Rice 1977). The genus *Eubalaena* (see account thereof) often is included within *Balaena.*

Head and body length at sexual maturity is about 12–13 meters in males and 13.0–13.5 meters in females (Koski et al. 1993). At physical maturity the usual length is 15–18 meters. The largest specimens reach about 19.8 meters in length and weigh more than 100,000 kg (Ellis 1980; Reeves and Leatherwood 1985). Pectoral fin length is about 200 cm, and the expanse of the tail flukes is 550–790 cm. There is no dorsal fin. The adult coloration is black, except that the anterior part of the lower jaw is cream-colored, the belly occasionally has white patches, and the junction of the body and tail is sometimes gray.

The skull is about 40 percent as long as the entire animal (E. R. Hall 1981). Seen from the front, the enormous lower jaw forms a U around the relatively narrow upper jaw. Seen from the side, both jaws are highly arched, and when the mouth is closed the lower jaw may partly conceal the upper. On each side of the mouth is a row of about 300 baleen plates. These plates, the largest of any whale's, are 300–450 cm in vertical length and black in color. As in *Eubalaena,* the two separated blowholes produce a V-shaped spout. According to Ellis (1980), the blubber of *Balaena* is 25–50 cm thick.

Some early commercial and modern subsistence whalers have recognized a relatively short, stocky form of bowhead with shorter, thinner baleen. This form, referred to as the "ingutuk," sometimes even has been considered to be a separate species, but Braham et al. (1980) concluded that it was a morphological variant of *B. mysticetus.*

The bowhead is primarily an arctic species, seldom occurring south of about 45° N, but in 1969 a specimen was discovered off Japan at 33°28′ N (Nishiwaki and Kasuya 1970). Normally the bowhead is found in association with ice floes, appearing to move seasonally in response to the melting and freezing of the ice. During the summer it frequents bays, straits, and estuaries. Prior to exploitation there seem to have been four major populations. One apparently spent the entire year in the Sea of Okhotsk, concentrating in certain coastal areas during winter. Another, the western Arctic stock, wintered in the Bering Sea, migrated north through the Bering Strait in the spring and early summer, remained in the Chukchi and Beaufort seas from late summer to early autumn, and then returned south. A third group apparently wintered along the east coast of Canada as far south as the Gulf of St. Lawrence. In the spring and summer this group moved northward and westward into Davis Strait, Hudson Bay, Baffin Bay, and adjoining waters. The fourth population apparently wintered mostly off southeastern Greenland and migrated to the region from Svalbard (Spitsbergen and associated islands) to Franz Josef Land and beyond during the spring and summer (Banfield 1974; Bockstoce and Botkin 1983; Marquette 1978; Moore and Reeves 1993; Rice 1977; W. G. Ross 1979). There have been some suggestions that the bowheads of Hudson Bay and adjacent Foxe Basin form a fifth population, discrete from that of Baffin Bay and Davis Strait (Braham 1984*a*).

During migration the bowhead swims at about 6 km/hr; its maximum speed is around 15 km/hr. It normally surfaces for up to 2 minutes, blows four to nine times, and then submerges for 5–20 minutes. Harpooned individuals are said to have stayed underwater for more than an hour. Before a dive the bowhead lifts its tail flukes above the surface. It sometimes leaps almost entirely out of the water. To feed, it commonly employs the skimming method (see familial account). It also dives to find prey at various depths and near the bottom. The diet consists mainly of zooplankton—copepods, amphipods, euphausiids, and pteropods.

Right whale *(Eubalaena glacialis):* Top photo shows the baleen in place and the very large tongue. Inset: a smaller piece of baleen of lighter color. Bottom photo shows the tongue and the rough surface of the top of the front portion of the head (the baleen has been removed). Photos by G. C. Pike through I. B. MacAskie.

The estimated daily intake during the summer feeding season is 1,800 kg. At other times of the year the bowhead, like most baleen whales, lives off of stored fat reserves and eats little or nothing (Banfield 1974; Marquette 1978; Würsig and Clark 1993).

Würsig et al. (1985) observed bowheads to sometimes feed in a V-shaped formation, with up to 14 animals, mouths wide open, moving in the same direction at the same speed. Groups consisting of 2 adults and a calf also were seen to persist for at least a few weeks. Previous reports (Banfield 1974; Ellis 1980; Marquette 1978) indicate that *Balaena* usually is found alone or in groups of 2–3 individuals but that larger groups may form during migration, and in the past such schools often included several hundred whales. There is sometimes segregation by age and sex. During the spring migration of the Bering Sea population young animals move north first, followed by large males and females with calves. Clark and Johnson (1984) reported bowhead sounds to vary greatly but to consist mostly of low tones and to include moans, rich purrs, roars, and complex pulsive calls. Würsig and Clark (1993) described various "songs," which may be associated with breeding, and individual "signature" calls, probably for maintaining the cohesion of migrating herds.

Available data (Banfield 1974; Breiwick, Eberhardt, and Braham 1984; Ellis 1980; Koski et al. 1993; Marquette 1978; Nerini et al. 1984; Schell and Saupe 1993; Zeh et al. 1993) indicate that mating occurs mainly in the late winter and spring, and births mostly from April to June, peaking in the western Arctic in May. The female usually gives birth every 3 or 4 years, normally to a single calf. The gestation period is probably about 13–14 months. The young is 400–450 cm long at birth and is weaned 9 to 12 months later. *Balaena* apparently attains sexual maturity at a later age than most whales, probably around 15–20 years. There are questionable reports, based on the finding of old harpoons in newly taken whales, of individuals having lived more than 100 years (Weintraub 1996) and also of individuals having made a transpolar passage from the Atlantic to the Pacific side of the Arctic.

Bowhead whale *(Balaena mysticetus)*, accompanied by belugas, painting by Richard Ellis.

Like *Eubalaena*, the bowhead was one of the "right" whales of commerce. Indeed, because of the length of its baleen and the thickness of its blubber, it was the most economically valuable of all cetaceans (see also familial account). Hunting by the Basques was occurring in the fifteenth century off Iceland and in the sixteenth century off Greenland. They carried out a major commercial whaling operation at the Strait of Belle Isle, Labrador, from about 1540 to 1620. In the early seventeenth century Spitsbergen became the center of the industry. Whaling bases were established there by expeditions from the Netherlands, England, France, Spain, Denmark, and Germany. By the early nineteenth century the bowhead had become so rare in this region that hunting was no longer profitable. The focus of exploitation was then shifting to the west of Greenland. Intensive whaling began in Davis Strait in 1719, Baffin Bay and adjacent sounds in 1818, and Hudson Bay in 1860. In these regions whaling was dominated by the Dutch for most of the eighteenth century and by the British and Americans in the nineteenth century. In some years several hundred ships were involved. By the mid–nineteenth century the bowhead was becoming rare in most areas from Greenland to eastern Canada, and after 1887 not more than 10 ships per year hunted in these waters. The last 2 vessels to try, in 1912 and 1913, did not take a single whale. Meanwhile, however, starting in 1848 American whalers had sailed north through the Bering Strait to exploit the last major summer concentration of bowheads, that of the Beaufort and Chukchi seas. These whales also were exploited in the wintering waters in the Bering Sea and were eliminated there progressively from south to north. For a few years in the 1850s hunting shifted largely to the Sea of Okhotsk, the population there being quickly and drastically reduced. Following a relaxation of the demand for whale oil in the mid–nineteenth century, the intensity of hunting corresponded with the price being paid for baleen for use in fashion. The western Arctic bowhead population probably was approaching extinction by the turn of the century, but the price of baleen per pound fell from $5.00 in 1907 to $0.075 in 1912, and hunting ceased. There is some question whether the present number of bowheads in the western Arctic represents a subsequent moderate increase. The total kill of bowheads had been about 19,000 in the western Arctic since 1848, about 3,500 more in the Sea of Okhotsk, about 37,000 in the waters from northeastern Canada to Greenland since 1719, and about 91,000 in the northeastern Atlantic since 1660 (Allen 1942; Banfield 1974; Bockstoce 1980*a*, 1980*b*, 1986; Bockstoce and Botkin 1983; Bockstoce and Burns 1993; Breiwick, Eberhardt, and Braham 1984; Ellis 1980; W. G. Ross 1974, 1979, 1993).

The bowhead whale received protection under the International Agreement for the Regulation of Whaling in 1937, the International Whaling Convention of 1946, the United States Marine Mammal Protection Act of 1972, and the U.S. Endangered Species Act of 1973. In each case, however, an exception was made for subsistence hunting by aboriginal peoples. Certain groups of Eskimos in northwestern Alaska had been whaling since ancient times. Their cultural and economic status is still associated with the pursuit and kill of the bowhead during its annual migrations.

They once utilized the entire animal, including the blubber for fuel and the baleen for making tools, but now the meat (around 10,000 kg per whale) is the main economic objective. A few bowheads also are taken by natives of northeastern Siberia.

Until about 1970 the usual take by the Alaskan Eskimos was 10–15 whales per year. Subsequently the kill rose, apparently because the petroleum industry had made money available for native Alaskans to purchase the boats, guns, and other equipment necessary for the culturally prestigious venture of whaling. In 1976, 48 bowheads were killed and another 46 were known to have been wounded; in 1977 the respective figures were 29 and 82. Because of growing concern that such exploitation would jeopardize the survival of the remaining bowhead population, the International Whaling Commission decided in 1977 to rescind the exemption for Eskimo whaling. The U.S. government, in a reversal of its general stand against whaling, then urged that some hunting be allowed. The commission thus began to set quotas for the bowhead, just as for commercially harvested species. The quota for 1980 was 18 whales landed or 26 struck, whichever came first ("struck" means either having been landed or hit by a harpoon but then lost). Some Eskimos argued for a higher kill and announced that they would not abide by the commission's quota. Indeed, the quota was substantially exceeded during the 1980 hunts, with 16 whales landed and 18 others struck and lost. Subsequently, however, the take has been kept within or very near the quota. In 1992 and 1993 the annual quotas were 41 for landings and 54 for strikes. In 1992 there were 38 actual landings and 50 strikes; the respective figures for 1993 were 41 and 52 (Bockstoce 1980*a*; Ellis 1980; Evans and Underwood 1978; Marine Mammal Commission 1994; Marquette 1978, 1979; R. Rau 1978; U.S. National Marine Fisheries Service 1978, 1981, 1987). Breiwick, Eberhardt, and Braham (1984) cautioned that any regular kill of more than 22 whales per year might eliminate the hope of a future population increase, though Reeves and Leatherwood (1994) noted that the cooperative management program with the Eskimos seems to have led to improvement in the status of the western arctic stock and to the potential for its slow recovery.

There has been longstanding and intensive disagreement regarding how many bowheads there are and were. Reported estimates of the western Arctic population in the mid-nineteenth century range from 4,000 to 40,000 individuals. The most commonly mentioned range is 10,000–30,000 (Breiwick, Eberhardt, and Braham 1984; Breiwick and Mitchell 1983; Breiwick, Mitchell, and Chapman 1980; Ellis 1980; Evans and Underwood 1978; Marquette 1978). A "best estimate" of 18,000 for the original western Arctic stock recently has been made by the International Whaling Commission, and there also are estimates of 25,000 originally in the eastern Atlantic, about 6,000–12,000 in the western Atlantic, and 3,000–6,500 in the Sea of Okhotsk (Braham 1984*a;* Braham, Krogman, and Carroll 1984; Reeves and Leatherwood 1985; Woodby and Botkin 1993). However, a detailed analysis of logbooks and journals of early whaling cruises led to an estimate for the western Arctic in 1847 of no fewer than 20,000 and no more than 40,000 (Bockstoce and Botkin 1983).

The U.S. National Marine Fisheries Service (1987) gave the estimate of the current population in the western Arctic as 4,417 whales but later (1994) listed an estimate of 7,500, a figure also calculated after extensive analysis by Zeh et al. (1993). Reeves et al. (1983) concluded that even though the western Atlantic population has been severely reduced, it continues to occupy much of its former range and to follow the original migratory schedule. Mitchell and Reeves (1982) warned that low-level but persistent hunting by native peoples and predation by killer whales might be preventing recovery of this population. Finley (1990) estimated that the stock that summers in and to the northeast of Baffin Bay contains at least 250 whales. Reeves and Mitchell (1990) indicated that the possibly separate stock that uses Hudson Bay and adjacent waters contains fewer than 100 animals. McQuaid (1986) suggested that recent sightings indicate the survival of a small relict population in the eastern North Atlantic, and Wiig (1991) observed 7 bowheads off Franz Josef Land in 1990. Zeh et al. (1993) listed estimates of at least 450 for the entire western Atlantic population, "in the tens" for the eastern Atlantic, and 150–200 for the Sea of Okhotsk. In addition to exploitation, *B. mysticetus* may be jeopardized by industrial development in its restricted habitat, disturbance by vessels and machinery, and pollution (Bratton et al. 1993; Richardson and Malme 1993). The species is listed as endangered by the USDI and is on appendix 1 of the CITES. The IUCN classifies the species generally as conservation dependent but specifies the Okhotsk Sea population as endangered and the Baffin Bay–Davis Strait and Hudson Bay–Foxe Basin stocks as vulnerable.

CETACEA; **Family BALAENOPTERIDAE**

Rorquals

This family of two Recent genera and six species occurs in all the oceans and adjoining seas of the world.

Head and body length of sexually mature adults is 6.7–31.0 meters, and weight is estimated to range up to 160,000 kg (Rice 1967). In each species the females are larger than the males. The pectoral fin is tapering, and the dorsal fin is located on the posterior part of the back. The rostrum is broad and flat, and the sides of the lower jaw bow outward. The Balaenopteridae are baleen whales and have no teeth past the embryonic stage. There is one row of baleen plates on each side of the top of the mouth, the two rows usually joining anteriorly. Each row may consist of more than 300 plates, which range in vertical length from 20 to 101 cm. The vertebrae number 42–65; the neck vertebrae are generally separate.

The term *rorqual* is Norwegian and means "tube whale" or "furrow whale" (Ellis 1980). It refers to the longitudinal

Baleen from a whale (*Balaenoptera* sp.), photo by P. F. Wright of specimen in U.S. National Museum of Natural History.

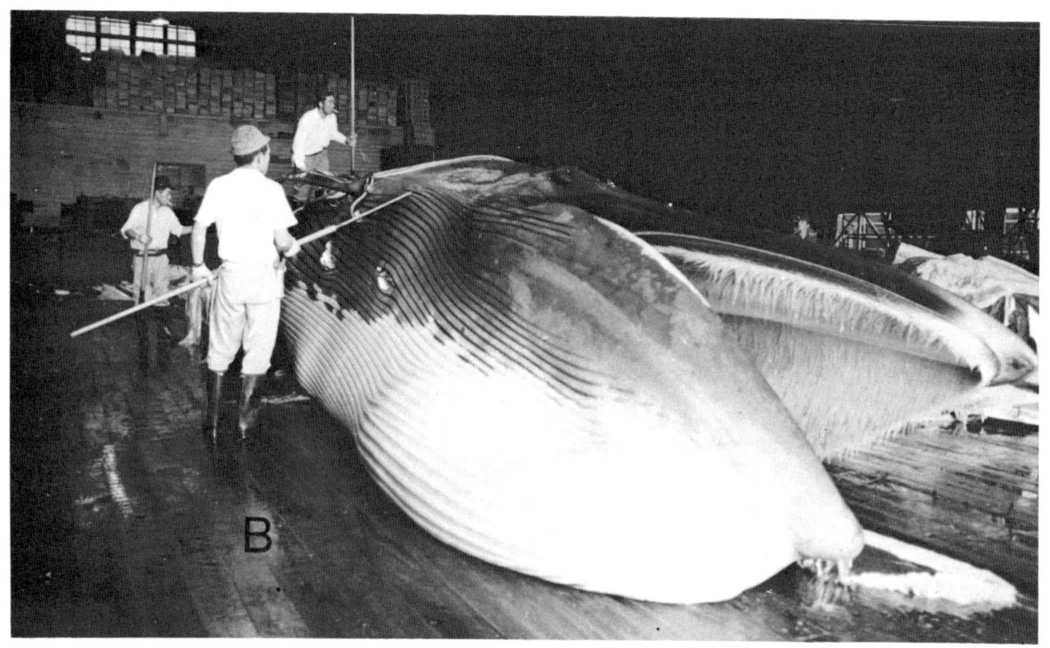

A. Blue whale *(Balaenoptera musculus)*, photo by K. C. Balcomb. B. Sei whale *(B. borealis)*, photo by Warren J. Houck.

folds or pleats, 10–100 in number and 25–50 mm deep, that are present on the throat, the chest, and in some species the belly. These furrows allow the throat to expand enormously, which greatly increases the amount of material the whale can take in when feeding.

The Balaenopteridae are distinguished from the Balaenidae by the presence of the throat and chest furrows, the more elongate and streamlined body form, the relatively smaller head, the more tapering pectoral fin, the softer and less massive tongue, and the usually shorter and less flexible baleen. Unlike the genera *Balaena* and *Eschrichtius,* the Balaenopteridae have a dorsal fin.

The feeding habits and dietary preferences of the Balaenopteridae are associated with the characteristics of the head, mouth, tongue, and baleen plates. The sei whale *(Balaenoptera borealis)* has finely meshed baleen fringes and sometimes uses the feeding method known as "skimming," as do the Balaenidae (Ellis 1980). In skimming the animal swims slowly through swarms of tiny food organisms with its mouth open and its head above water to just behind the nostrils. When a mouthful of organisms has been filtered from the water by the baleen plates, the whale dives, closes its mouth, and swallows the food. The other rorquals use the feeding method known as "swallowing": the animal may turn on its side and often has part of its head above the water; after taking in a great amount of food organisms and water, the whale forces the water out with its tongue, leaving the food trapped in the baleen.

All species are migratory, most giving birth in warm areas and feeding predominantly in colder waters. The diet depends partly on location. The most important foods are shrimplike creatures of the family Euphausiidae (mainly the genera *Euphausia, Thysanoessa,* and *Meganyctiphanes*) and copepods *(Calanus* and *Metrida).* These crustaceans form part of the zooplankton, the mass of minute floating and weak-swimming animals found near the surface of the ocean. In colder areas there are fewer kinds of plankton, but the numbers of each kind are greater. During the summer, shoals of these organisms concentrate in the upper layers of polar waters, and the whales are attracted to them. In the Antarctic the diet of the minke *(B. acutorostrata)*, fin *(B. physalus)*, blue *(B. musculus)*, and humpback *(Megaptera)* whales consists mainly of the shrimp *Euphausia superba* (commonly called "krill"). Outside of the Antarctic the blue whale feeds predominantly on other euphausiids, but the fin, minke, and humpback take substantial amounts of copepods and fish. The fine, fleecy baleen of the sei whale is suited for trapping smaller organisms than those eaten by other rorquals. In the Antarctic the sei feeds on *Euphausia superba* but also on the amphipod *Parathemisto gandichaudi;* in other areas it seems to prefer copepods. In contrast, the Bryde's whale *(B. edeni)* has coarse baleen, does not enter the Antarctic, and feeds mainly on fish (Ellis 1980).

All rorquals have been hunted by people for oil, meat, baleen, and other products. The most intensive exploitation has taken place in antarctic waters, where large numbers of most species gather for part of the year to feed on plankton. The whaling activity in this region followed a systematic pattern (D. G. Chapman 1974; Gulland 1974, 1976; McHugh 1974). The first species to be depleted was the humpback, which has a localized, coastal distribution and could be easily taken from the island whaling stations established in the first years of the twentieth century. The introduction in the 1920s of factory ships with slip sternways, which allowed whalers to operate independently of land facilities, led to increased killing of other species.

The blue, fin, and sei whales, in that order, become smaller in size and concentrate farther from the antarctic ice pack. Exploitation proceeded in this same order. The blue whale, the most valuable species in relation to individual hunting effort, received initial maximum attention and was evidently in decline by the late 1930s. Whalers then centered their activity somewhat farther north, and the fin whale became the mainstay of the industry for the next 30 years. Meanwhile, efforts at protection were largely ineffective. The International Whaling Commission established quotas on the basis of "blue whale units." Each unit equaled 1 blue whale, 2 fin whales, 2.5 humpbacks, or 6 sei whales. Since quotas were assigned by units rather than by species, it remained more profitable to hunt the larger species. By the 1960s, however, the blue whale seemed to be nearly extinct, and the fin whale greatly reduced in numbers, and attention was then directed still farther north, to the smaller sei whale. Another factor in the switch to the sei was a lower demand for whale oil, along with an increase in the value of the meat. The sei has less blubber and relatively more and better meat than the larger species. Concern for the future of the sei, and the setting of lower quotas specifically for it and the larger whales, contributed to an increased kill of the minke whale, smallest of the rorquals, in the 1970s. There also was a rise in the take of Bryde's whale, a species considerably larger than the minke but one that does not migrate to the Antarctic.

All large-scale commercial whaling had ended by the late 1980s, and there was hope that major stocks of most of the rorquals had survived and would form the basis for recovery of the once great Antarctic herds. However, at recent meetings of the International Whaling Commission it was estimated that fewer than 500 blue and 15,000 fin whales survive in the region. These low figures add weight to earlier studies showing that major changes occurred in the Antarctic ecosystem as a result of the decline of the great whales. This reduction meant that more than 100 million metric tons of krill, formerly eaten by the large whales, had become available to other animals (Laws 1985). As a result, there have been remarkable increases in such species as squid, penguins, the crabeater seal *(Lobodon carcinophagus)*, and the Antarctic fur seal *(Arctocephalus gazella)*. As these other species became more abundant, the krill surplus was taken up and the recovery potential for the whales further reduced. A growing human harvest of krill, which reached 500,000 metric tons in 1985–86 (Horwood 1990), is a further threat to the entire ecosystem.

The geological range of the Balaenopteridae is middle Miocene to Pleistocene in North America and Europe, Miocene in eastern Asia, early Pliocene and Pleistocene in South America, and Recent in all oceans (Rice 1984).

CETACEA; BALAENOPTERIDAE; **Genus BALAENOPTERA**
Lacépède, 1804

Minke, Bryde's, Sei, Fin, and Blue Whales

There are five species (Rice 1977):

B. acutorostrata (minke whale), all oceans and adjoining seas;
B. edeni (Bryde's whale), tropical and warm temperate waters of the Atlantic, Indian, and Pacific oceans and adjoining seas;
B. borealis (sei whale), all oceans and adjoining seas except in tropical and polar regions;
B. physalus (fin whale), all oceans and adjoining seas, but rare in tropical waters and in pack ice;
B. musculus (blue whale), all oceans and adjoining seas.

A remarkable recent series of investigations utilizing mainly mitochondrial DNA have identified a female of this genus caught in 1986 as a hybrid between *B. physalus* and *B. musculus* and shown that this female was pregnant with a fetus fathered by a male *B. musculus;* they have also identified two immature males as hybrids of *B. physalus* and *B. musculus* even though the two species are calculated to have diverged about 5 million years ago, far earlier than any other mammals known to have interbred in the wild (Arnason and Gullberg 1993; Arnason et al. 1991; Spilliaert et al. 1991). Other molecular evidence supports recognition of *B. bonaerensis* of the Southern Hemisphere as a species separate from *B. acutorostrata* (Arnason and Gullberg 1994). There also are general morphological differences between the minke whales of the Northern and Southern hemispheres, but Horwood (1990) considered the two populations to be no more than subspecifically distinct.

In addition to the characters set forth in the familial account, *Balaenoptera* is distinguished by a relatively slender body, a compressed tail stock that abruptly joins the flukes, a pointed snout, a relatively short pectoral fin (usually 8–11 percent of head and body length), and a usually sickle-shaped dorsal fin. Additional information is provided separately for each species.

Balaenoptera acutorostrata (minke whale)

In the more common form, head and body length at sexual maturity averages about 7.3 meters in males and 7.9 meters in females. At physical maturity the respective averages are about 8.3 and 8.8 meters (Mitchell 1975*a;* Ohsumi and Masaki 1975; Ohsumi, Masaki, and Kawamura 1970). Maximum length is around 9.8 meters in males and 10.7 meters in females (Klinowska and Cooke 1991). Maximum known weight is 10,000 kg but theoretically could exceed

A. Fin whale *(Balaenoptera physalus)*, photo from Field Museum of Natural History. B & C. Minke whale *(B. acutorostrata)*, photos from Marineland of the Pacific.

13,000 kg. Average size is greater in the population of the Southern Hemisphere, and as in all rorquals, females average slightly larger than males (Horwood 1990). The upper parts are dark gray to black and the underparts are white. There may also be a white patch on the pectoral fin. The rostrum is markedly triangular in shape and shorter than that of any other rorqual. The dorsal fin is high and curved strongly backward. There are 50–70 ventral grooves, and these do not extend as far as the navel. On each side of the upper part of the mouth is a row of about 300 baleen plates mostly yellowish white in color but sometimes asymmetrical in pattern (Ellis 1980; Leatherwood, Caldwell, and Winn 1976). In populations of the Southern Hemisphere (subspecies *bonaerensis*) the posterior baleen plates are usually black on the outer edge (Horwood 1990).

A second form of minke whale has been recognized in waters near Australia, New Zealand, New Caledonia, South Africa (Arnold, Marsh, and Heinsohn 1987; Best 1985), and Brazil and in the southern Indian Ocean (Horwood 1990). Referred to as the dwarf, or diminutive, form, this whale may reach sexual maturity when it is 6–7 meters long, and its full size is 7–8 meters. There is a large white area on the base of the pectoral fin and the adjacent thorax. Coloration of the upper surface of the head and rostrum is paler than in the other form. The baleen plates are predominantly light and do show an asymmetrical pattern. There is uncertainty whether the dwarf minke whale represents a subspecies or even a separate species.

The minke whale is not as coastal an animal as *Eschrichtius* but seldom goes farther than 160 km from land and often enters bays and estuaries. It also moves farther into polar ice fields than any other rorqual. There are a number of discrete populations, at least some of which make extensive migrations. Most populations of the Northern Hemisphere winter in tropical waters, where breeding occurs, but during the spring and summer they spread northward to feeding waters, often beyond the Arctic Circle. While these populations are in the north, those of the Southern Hemisphere are in the tropics for the austral winter. When the northern populations return to the tropics in the boreal autumn, the southern populations set out in the direction of Antarctica (Ellis 1980; Mitchell 1978; U.S. National Marine Fisheries Service 1978).

The minke whale is a fast swimmer and the most acrobatic member of its genus. It has a tendency to approach ships and often leaps completely out of the water (Leatherwood, Caldwell, and Winn 1976; Mitchell 1978). Its spout, however, is invisible. It has a wide variety of vocalizations, including a series of thumps that may be distinctive for each individual (Thompson, Winn, and Perkins 1979). In the Antarctic the diet consists almost exclusively of plankton (see familial account), but in the Northern Hemisphere *B. acutorostrata* also eats squid, herring, cod, sardines, and various other kinds of small fish (Ellis 1980). Like other rorquals, the minke whale feeds mainly from the late spring to early autumn, when it has migrated to high latitudes, and eats very little while in the winter breeding waters (Horwood 1990).

This rorqual is generally seen alone or in groups of two to four individuals, but several hundred sometimes gather in the vicinity of abundant food (Leatherwood, Caldwell, and Winn 1976; Mitchell 1975*b*). During migration there is some segregation by age and sex, adult females tending to move ahead of the others (Horwood 1990). Off Newfoundland adult males evidently remain farther from shore than do females and young (Mitchell and Kozicki 1975). Observations of a population along the coast of Washington revealed the presence of three exclusive adjoining ranges in an area of about 600 sq km, each shared by at least seven whales, though there was no indication of group relationships or territoriality (Dorsey 1983).

Mitchell (1978) gave the calving interval as only 1 year, Mitchell and Kozicki (1975) found more than 85 percent of mature females off Newfoundland to be pregnant, and Horwood (1990) concluded that most individuals breed every year. The breeding period is extensive (Horwood 1990; Klinowska and Cooke 1991; Mitchell 1975*b*). In the Southern Hemisphere mating occurs from June through December, peaking in August and September, and peak numbers of births occur from June to August. In the North Atlantic mating peaks in February and calving in December. In the North Pacific reproduction occurs throughout the year, with mating peaks in January and June and birth peaks in December and June. The gestation period is about 10 months. There is normally a single offspring about 280 cm long. Lactation is thought to last about 6 months. Males have been reported to attain physical maturity at 18–22 years, and females at 20–30 years (Horwood 1990; Ohsumi and Masaki 1975; Ohsumi, Masaki, and Kawamura 1970). There have been varying estimates of age at sexual maturity, but Kato and Sakuramoto (1991) concluded that in the Southern Hemisphere, in association with increasing food availability, this age fell from 12–13 years in the mid-1940s to about 10 years in 1955 and 7–8 years in the early 1970s. Klinowska and Cooke (1991) cited ages of sexual maturity as low as 5 years in males and 6 years in females. Horwood (1990) agreed that a decline in this age probably has occurred but indicated that more research is needed. The natural rate of mortality for a population is about 10 percent. Maximum longevity is estimated at 47 years (Haley 1978).

The minke whale has been hunted in Norway since the Middle Ages and was regularly taken by local fisheries there and in some other countries from the 1920s to 1970s but long was not considered of major commercial importance (Mitchell 1975*a*). With the decline of the larger rorquals in the 1960s, the minke became subject to more intensive hunting, especially in the Antarctic. In the 1976–77 season the worldwide kill peaked at 12,398 individuals, surpassing the take of any other species (Committee for Whaling Statistics 1980). With growing public demands for an end to commercial whaling, and under reduced quotas set by the International Whaling Commission, the kill dropped to 6,055 during the 1983–84 season (Committee for Whaling Statistics 1984). Kills continued at this level for several more years, despite the commission's official moratorium on commercial whaling starting in 1985, because several nations chose to issue a formal objection to the ban. Such measures had been dropped by 1988, however, and the annual kill is now fewer than 1,000 minke whales taken either for aboriginal subsistence or for alleged research purposes. Conrad and Bjorndal (1993) concluded that it would be biologically feasible to resume commercial taking of the minke whale stock that utilizes the Barents Sea and waters off Norway, though they acknowledged that political and humanitarian considerations also would be involved in any decision on the matter.

Ohsumi (1979) gave an estimate of 416,700 minke whales for the Southern Hemisphere and suggested that this figure represented an eightfold increase since the 1930s because the species had multiplied in the absence of competition from the larger rorquals. The U.S. National Marine Fisheries Service (1994) reported the current population estimate of minke whales to be 700,000 in the Southern Hemisphere, 25,000 in the western North Pacific, and 100,000 in the North Atlantic. Notwithstanding these high figures, certain stocks, particularly those of the central and eastern North Atlantic and the western North

Pacific, have been reduced by commercial whaling, and concerns persist about the effects of subsistence whaling and incidental catching by fishing nets (Reeves and Leatherwood 1994). Increasing human pressure on the minke whale's food resources—fish in the Northern Hemisphere and krill in the Southern Hemisphere (see above family account)—also is a concern (Horwood 1990). The IUCN now classifies the subspecies of the Northern Hemisphere, *acutorostrata,* as near threatened and the subspecies of the Southern Hemisphere, *bonaerensis,* as conservation dependent. All populations of the minke whale are on appendix 1 of the CITES, except that off western Greenland, which is on appendix 2.

Balaenoptera edeni (Bryde's whale)

Head and body length at sexual maturity averages 12.2 meters in males and 12.5 meters in females (U.S. National Marine Fisheries Service 1978). Maximum head and body length is 15.5 meters (Klinowska and Cooke 1991), and dorsal fin height is as much as 46 cm (Leatherwood, Caldwell, and Winn 1976). The coloration is dark blue gray above and paler below. *B. edeni* resembles *B. borealis* but can be distinguished by the presence of two dorsal ridges running from the tip of the snout to beside the blowhole, in addition to another ridge down the middle of the snout. The dorsal fin is strongly sickle-shaped. There are approximately 45 ventral grooves, extending to the navel. On each side of the mouth is a row of about 300 baleen plates, and often the rows do not join anteriorly. The plates are relatively short; the front ones are whitish and the rear ones are dark (Ellis 1980).

Certain populations inhabit waters of high productivity near the shore; others are mainly pelagic. Unlike other rorquals, *B. edeni* makes no large-scale movements toward the poles in the spring. Tropical populations may be sedentary, but those in temperate waters seem to make limited migrations in response to shifting food sources. *B. edeni* has a tendency to approach ships and is a relatively deep diver. Its diet consists mainly of small schooling fish, such as anchovies and sardines, but in some areas it also feeds extensively on shrimplike crustaceans of the family Euphausiidae (Ellis 1980; Leatherwood, Caldwell, and Winn 1976; U.S. National Marine Fisheries Service 1978).

Bryde's whale is often seen in groups, usually of 2–10 individuals but sometimes of more than 100. The reproductive season is usually winter, but in some regions, as in the waters off South Africa, it extends throughout the year. Females generally give birth every other year to a single offspring about 335 cm long. The gestation period is 12 months, and lactation probably lasts less than a year. The age of sexual maturity is about 10–13 years and that of physical maturity is 15–18 years. Some individuals are estimated to have lived up to 72 years (Ellis 1980; Haley 1978; Klinowska and Cooke 1991; Lockyer 1984; Mitchell 1978; U.S. National Marine Fisheries Service 1978, 1981).

Bryde's whale long has been hunted along the coasts of Japan, South Africa, and Baja California but did not receive major commercial attention until the 1970s, subsequent to the decline of most other rorquals. The maximum recorded kill was 1,882 in the 1973–74 season (Committee for Whaling Statistics 1980). The 1981–82 quota set by the International Whaling Commission was 1,392. Quotas subsequently dropped and were halted altogether with the formal moratorium on commercial whaling after 1985. Estimates of varying reliability made at various times since 1980 suggest that there may be about 14,000 *B. edeni* in the Indian Ocean, 55,000 in the Pacific, and a small number in the Atlantic (Klinowska and Cooke 1991; Wade and Gerrodette 1993). The species is on appendix 1 of the CITES.

Balaenoptera borealis (sei whale)

Head and body length at sexual maturity averages 13.1 meters in males and 13.7 meters in females (U.S. National Marine Fisheries Service 1978). The biggest specimen on record, a female, was 20 meters long, but most individuals now measure 12.2–15.2 meters. Adults typically reach a weight of about 20,000 kg and rarely as much as 40,000 kg (Horwood 1987). The general coloration is dark steel gray,

Sei whale *(Balaenoptera borealis)*, painting by Richard Ellis.

with irregular white markings ventrally. The dorsal fin, set farther back than on *B. acutorostrata,* is 25–61 cm in height. There are 38–56 ventral grooves, and these end well before the navel. On each side of the upper part of the mouth is a row of 300–380 baleen plates. The color of the plates is ashy black, but the fine inner bristles are whitish (Ellis 1980; Leatherwood, Caldwell, and Winn 1976).

The sei whale is a pelagic species, normally occurring far from shore. Like most rorquals, it feeds in temperate and subpolar regions in summer and migrates to subtropical waters for the winter. It does not, however, go as far into the Antarctic as the blue and fin whales do. It is one of the fastest cetaceans, reportedly able to reach speeds of up to 50 km/hr. It is not usually a deep diver. Periods of submergence generally last 5–10 minutes. When feeding, the sei whale remains near the surface, often twisting on its side as it swims through swarms of prey. Unlike other rorquals, it sometimes obtains food by the skimming method (see familial account). It generally eats smaller organisms than do the other rorquals, mostly copepods and amphipods, but also takes euphausiids and small fish. Estimates of daily consumption per whale are about 200–900 kg (Banfield 1974; Ellis 1980; Horwood 1987; Kawamura 1973, 1974; Mitchell 1978). *B. borealis* is not known to use echolocation to search for prey but has been heard to emit a "sonic burst of 7–10 metallic pulses" (Thompson, Winn, and Perkins 1979).

Groups usually consist of two to five individuals, but sometimes thousands may gather in the vicinity of abundant food (Ellis 1980; Leatherwood, Caldwell, and Winn 1976). According to Banfield (1974), there is a protracted mating season that extends from November to February in the Northern Hemisphere and from May to July in the Southern Hemisphere. Females generally give birth every other year, but Masaki (1978) reported that the percentage of pregnant females being taken has recently increased, perhaps as a natural compensatory response to human exploitation. Horwood (1987) concluded that the pregnancy rate has nearly doubled and the age of sexual maturity has fallen. The gestation period was reported as 10.5 months by Masaki (1976), 11.5 months by Frazer (1973), and 12 months by Ellis (1980). There are rare cases of multiple fetuses, but normally a single offspring about 450 cm long is produced. Lactation lasts 6–7 months. Sexual maturity is attained at around 10 years, and physical maturity at 25 years (Mitchell 1978). Some individuals are estimated to have lived up to 74 years (Haley 1978).

The sei whale has been taken regularly by people since the 1860s but did not achieve major commercial importance for another 100 years. The annual kill in antarctic waters did not exceed 1,000 individuals until 1950. There was a pronounced increase in the kill during the 1950s and 1960s in conjunction with a declining harvest of blue and fin whales (see familial account). The worldwide take of the sei whale peaked at 25,454 in the 1964–65 season (McHugh 1974) and then steadily dropped to only 150 in 1978–79 (Committee for Whaling Statistics 1980). The annual quota set by the International Whaling Commission for most subsequent seasons was 100, and after 1985 all commercial whaling was officially halted. The original number of sexually mature sei whales, exclusive of those in the North Atlantic, is estimated at 200,000; the total population, including immature animals, would have been about 50 percent larger (U.S. National Marine Fisheries Service 1978). Klinowska and Cooke (1991) cited a modern estimate of 40,000 in the Southern Hemisphere but noted that it was based in part on data gathered well before the end of intensive exploitation. More recent estimates are only 8,300 in the Southern Hemisphere (International Whaling Commission 1994), 9,100 in the North Pacific (U.S. National Marine Fisheries Service 1994), and 12,000 in the North Atlantic (Cattanach et al. 1993). *B. borealis* is classified as endangered by the IUCN and USDI and is on appendix 1 of the CITES.

Balaenoptera physalus (fin whale)

Head and body length at sexual maturity averages about 17.7 meters in males and 18.3 meters in females. At physical maturity the respective averages are about 19 and 20 meters. Maximum known length is 25 meters in males and 27 meters in females. Adults have not been weighed, but calculations suggest that a 25-meter animal would weigh about 70,000 kg (Gambell 1985). The strongly curved dorsal fin is as much as 61 cm high, and the expanse of the tail flukes is about 25 percent of the head and body length. The general coloration is brownish gray above and white below; the pattern, however, is asymmetrical, especially in that the lower jaw is white on the right and dark on the left. *B. physalus* is slimmer than *B. musculus;* if two individuals, one of each species, are the same length, the *B. musculus* will be much heavier. Indeed, even though *B. physalus* is the second longest cetacean, it often weighs less than the thickset *Balaena, Eubalaena,* and *Physeter.* The rostrum of *B. physalus* is sharply pointed. There are on average 85 ventral grooves, and these terminate at the level of the navel. On each side of the upper part of the mouth is a row of 350–400 baleen plates. Those plates in the forward third of the right row are whitish, but those in the rear two-thirds of the right row and all of those in the left row are dark grayish blue (Ellis 1980; Leatherwood, Caldwell, and Winn 1976; Mitchell 1978; U.S. National Marine Fisheries Service 1978).

The fin whale is a pelagic species, seldom found in water less than 200 meters deep. Numerous discrete populations have been identified, most of which are known to be highly migratory. Populations generally move into cold temperate and polar waters to feed in the spring and early summer and return to warm temperate and tropical regions in the autumn. Since the seasons in the northern half of the world are opposite those in the southern half, the fin whale populations of one hemisphere do not meet those of the other hemisphere in equatorial waters (Banfield 1974; D. G. Chapman 1974; Ellis 1980).

The fin whale is among the fastest of cetaceans, being able to sustain a speed of around 37 km/hr. It is probably a deeper diver than the blue and sei whales, sometimes reaching depths of at least 230 meters and remaining underwater for up to 15 minutes. It occasionally leaps completely out of the water. A swallowing method of feeding is employed, during which the throat becomes distended to nearly double the normal diameter (see familial account). There is very little, if any, feeding during the autumn and winter, when the whales are in lower latitudes. In the Antarctic the diet consists almost entirely of small, shrimplike crustaceans of the family Euphausiidae. In northern waters *B. physalus* also eats other crustaceans and various kinds of small fish (Ellis 1980; Leatherwood, Caldwell, and Winn 1976). Banfield (1974) indicated that the fin whale uses echolocation to find its food, but this activity has not actually been demonstrated. The species is known to produce a great variety of low-frequency sounds and probably also some high-frequency pulses (Thompson, Winn, and Perkins 1979).

The fin whale may be monogamous and is regularly seen in pairs. Group size is usually 6–7 individuals but often as many as 50, and occasionally as many as 300 travel together on migrations. Both mating and calving occur during the late autumn or winter, when the whales are in warm waters. Each mature female gives birth every 2–3 years. The gestation period is 11.0–11.5 months. There are rare records of as many as six fetuses, but normally a single offspring is produced. The newborn is about 650 cm long and weighs 1,800 kg. It nurses for about 6–7 months, until it is around 11.5–12.2 meters long, and then travels with the female to the polar feeding areas. Sexual maturity is attained at 6–11 years (Banfield 1974; Ellis 1980; Frazer 1973; Lockyer 1984; Mitchell 1978). Some individuals are estimated to have lived as long as 114 years (Haley 1978).

Following development of the harpoon gun in the 1860s, *B. physalus* was regularly hunted by Norwegian whalers in the North Atlantic. Exploitation of the vast antarctic populations began after establishment of the first island whaling station in that region in 1904 and intensified after the introduction of floating factories there in the 1920s. As the blue whale declined, the fin received increased hunting emphasis. In the 1937–38 season the take of *B. physalus* in the Antarctic was 28,009 individuals, nearly twice as great as that of *B. musculus.* After a lull during World War II, large-scale pelagic whaling resumed in the Antarctic and increased in other areas, especially the North Pacific. The worldwide kill of the fin whale exceeded 10,000 animals in every season from 1946–47 to 1964–65 and averaged around 30,000 annually from 1952 to 1962. Such a harvest was far in excess of the sustainable yield, and by 1960 there was evidence of seriously reduced populations. Initial conservation efforts were largely unsuccessful, but by the mid-1970s the International Whaling Commission had drastically lowered annual quotas and had completely banned hunting in some regions. The kill dropped to 5,320 in 1968–69 and 743 in 1978–79 (Committee for Whaling Statistics 1980; Ellis 1980; Gulland 1974; McHugh 1974). Legal quotas fell subsequently and ceased altogether after 1985, except that a small aboriginal subsistence catch (10 in 1987) was allowed off western Greenland (U.S. National Marine Fisheries Service 1987). Some countries, notably Iceland, also continued to hunt *B. physalus* for alleged research purposes.

The number of mature *B. physalus* in the world prior to exploitation is estimated at 470,000, of which 400,000 were in the Southern Hemisphere. Inclusion of immature individuals would increase these figures by roughly 50 percent (U.S. National Marine Fisheries Service 1978). Even after the period of intensive exploitation there were estimates that 105,200–121,900 fin whales still existed, including 85,000 in southern waters (U.S. National Marine Fisheries Service 1989). However, direct counts made by recent whaling cruises, as reported to the International Whaling Commission (1994), now indicate that at most 15,000 fin whales are left in the Southern Hemisphere. This sad figure clearly reflects the commercial slaughter that persisted unchecked through the 1960s and continued even in the 1980s as a supposed research activity. There are another 17,000 in the North Pacific (U.S. National Marine Fisheries Service 1994) and, surprisingly perhaps, at least 46,000 in the North Atlantic (Reeves and Leatherwood 1994). *B. physalus* is classified as endangered by the IUCN and USDI and is on appendix 1 of the CITES.

Balaenoptera musculus (blue whale)

This whale is the largest animal ever known to have existed. There is, however, a pygmy subspecies *(B. m. brevicauda),* described by Ichihara (1966) and reported to inhabit a restricted zone of the Southern Hemisphere to the north of 54° S and between 0° and 80° E. Most authorities now seem to accept this subspecies as valid (Ellis 1980; E. R. Hall 1981; Rice 1977), but others, such as Small (1971), thought it represented only young individuals of another subspecies. That part of the body of the pygmy blue whale posterior to the dorsal fin is described as being relatively shorter than that of other populations, and hence overall head and body length is reduced. Sexual maturity in this subspecies is attained at an average length of 19.2 meters, and the maximum length is about 24.4 meters (Ellis 1980).

Disregarding the pygmy subspecies, head and body length of the blue whale at sexual maturity averages 22.5 meters in males and 24.0 meters in females (U.S. National Marine Fisheries Service 1978). Average length at physical maturity for antarctic specimens is 25 meters in males and 27 meters in females (Banfield 1974). Animals from the Northern Hemisphere are somewhat smaller. There is some question about maximum size, and it is possible that individuals grew larger prior to the period of intensive exploitation that began in the 1920s. According to Yochem and Leatherwood (1985), the longest specimen measured in the scientifically correct manner (in a straight line from the point of the upper jaw to the notch in the tail) was a 33.58-meter female taken in the Antarctic sometime between 1904 and 1920, and the heaviest was a 27.6-meter female that weighed 190,000 kg, taken in the Antarctic in 1947.

The general coloration is slate or grayish blue, mottled with light spots especially on the back and shoulders. Although the underparts have about the same basic color as the upper parts, they sometimes acquire a yellowish coat-

Blue whale *(Balaenoptera musculus)*, painting by Richard Ellis.

ing of microorganisms. Thus, the vernacular name "sulphurbottom" is often applied to *B. musculus*. The dorsal fin is relatively small, usually reaching a height of only 33 cm. The rostrum is less sharply pointed than that of any other member of the genus. There are approximately 90 ventral grooves, and these extend to the navel. On each side of the upper part of the mouth is a row of 300–400 baleen plates. They are black in color and range in length from 50 cm in front to 100 cm in back (Ellis 1980; Leatherwood, Caldwell, and Winn 1976).

Prior to exploitation, at least 90 percent of all blue whales lived in the Southern Hemisphere, but there are populations in the North Atlantic and the North Pacific. Banfield (1974) described the species as a pelagic denizen of polar and temperate seas. Mitchell (1978) noted that whereas the blue whale goes farther into the Antarctic than the other large rorquals, it does not move as close to the northern polar ice fields as the fin whale. Populations generally spend the winter in temperate and subtropical zones, migrate toward the poles in the spring, feed in high latitudes during the summer, and move back toward the equator in the autumn. Because of the difference in seasons between the northern and southern parts of the earth, all migrating populations move in roughly the same direction at the same time, and thus those of the Northern Hemisphere do not meet those of the Southern Hemisphere. There seem to be a number of distinct stocks, and although these may overlap to some extent in the summer feeding areas, they separate and return to their own discrete breeding sites each year (Gulland 1974; Mackintosh 1966). Some individuals are thought to remain at low latitudes yearround in certain parts of the Indian and Pacific oceans (Reilly and Thayer 1990). The stock of the pygmy subspecies around Kerguelen, Crozet, and Heard islands does not appear to migrate to the same extent as other populations (Ellis 1980).

The blue whale normally swims at a speed of around 22 km/hr but may swim as fast as 48 km/hr if alarmed. It usually feeds at depths of less than 100 meters, but harpooned individuals have gone deeper than 500 meters below the surface. Dives normally last 10–20 minutes and are followed by a series of 8–15 blows. The spout reaches a height of 9.1 meters. The blue whale is the only member of its genus that commonly lifts its tail flukes out of the water before a dive. To feed, *B. musculus* takes in large amounts of water and organisms, greatly distending its throat, and then forces the water out, leaving the food trapped in the inner fibers of the baleen (see familial account). The highly restricted diet consists almost entirely of shrimplike crustaceans of the family Euphausiidae. These organisms, some of which are known as krill, are generally less than 5 cm long. When in the summer feeding areas, these whales each have a daily consumption of probably about 40 million individual euphausiids, with a total weight of 3,600 kg. During the rest of the year—a period of up to eight months—the blue whale apparently does not eat at all and lives off of stored fat (Ellis 1980; Leatherwood, Caldwell, and Winn 1976; Mitchell 1978; Rice 1978*b*). *B. musculus* emits deep, low-frequency sounds, as well as series of clicks that may possibly be used in the echolocation of swarms of krill (Thompson, Winn, and Perkins 1979).

Off the coast of California, aggregations of as many as 60 blue whales are common (Rice 1978*b*). Usually, however, the species is seen alone or in groups of 2 or 3 (Banfield 1974). In the Southern Hemisphere mating peaks in the summer (June–July) and births peak in the spring (Lockyer 1984). Females give birth every 2–3 years (U.S. National Marine Fisheries Service 1978). According to Mizroch, Rice, and Breiwick (1984), females appear to be seasonally monestrous, though if they fail to conceive, they may ovulate two or three times during one estrous cycle. The gestation period, usually reported at 10–12 months, is surprisingly short for so large an animal. Frazer (1973) listed gestation as only about 9.6 months, explaining that if the period lasted much longer, the young would be born at a disadvantageous time—just before or during the season spent in cold waters, and before it had accumulated much protective blubber. Twins have been reported on rare occasion, but there is normally a single offspring. At birth it is about 7 meters long and weighs 2,000 kg. It gains 90 kg per day and is weaned after 7–8 months, when it is about 15 meters long. Sexual maturity probably is attained at 10 years (Banfield 1974; Ellis 1980; Klinowska and Cooke 1991; Yochem and Leatherwood 1985). Some individuals are estimated to have lived as long as 110 years (Haley 1978).

Because of its size, speed, strength, and remote habitat, the blue whale was generally considered too difficult a target in the early days of whaling. This situation was changed by a series of developments from the 1860s to 1920s, as described in the account of the order Cetacea. Regular hunting of *B. musculus* began off Norway, steadily spread across the Northern Hemisphere as one population after another was depleted, and finally came to center in the vast feeding waters of the Antarctic. The annual kill increased dramatically following the introduction of factory ships with slip sternways. The total recorded kill in the twentieth century is approximately 350,000 individuals, of which more than 90 percent were taken in the Antarctic. The peak season was 1930–31, when the kill was 29,410 in the Antarctic and 239 in other parts of the world. There subsequently was a general decline in the worldwide seasonal harvest, to 12,559 in 1939–40, 6,313 in 1949–50, and 1,465 in 1959–60. Various international efforts to set size limits, sanctuary areas, and quotas were attempted beginning in the 1930s, but they were inadequate and did not receive full compliance. By the early 1960s it was clear to almost all concerned persons that the blue whale was nearing extinction, and groups of scientists were recommending that the International Whaling Commission establish total protection for the species. Because of a continued lack of cooperation from whaling interests, such protection did not come until the year after the 1965–66 season, when 613 blue whales were killed, only 20 of them in the Antarctic (Allen 1942; Ellis 1980; Gulland 1974; McHugh 1974; Small 1971).

The estimated numbers of blue whales prior to exploitation are: Southern Hemisphere, 150,000–240,000; North Pacific, 4,500–6,000; western North Atlantic, 1,100; and eastern North Atlantic, 6,000–13,000 (Klinowska and Cooke 1991; Yochem and Leatherwood 1985). Although Small (1971), among others, warned that these numbers had been so drastically reduced that recovery might be impossible, there was a general authoritative view through the early 1980s that substantial and viable stocks remained. Numbers estimated by the U.S. National Marine Fisheries Service (1978, 1981, 1987, 1989) were: Southern Hemisphere, 10,000 (about half of which were the pygmy subspecies); North Pacific, 1,600; and North Atlantic, a few hundred. There even had been indications of slowly increasing numbers, and populations off the east and west coasts of North America appeared to be larger than once feared (Berzin 1978; Leatherwood, Caldwell, and Winn 1976). Indeed, in the latter regions international protection against direct killing may be working, with the population of the northeastern Pacific now approaching 4,000 (Reeves and Leatherwood 1994) and that of the North Atlantic estimated at 555 (U.S. National Marine Fisheries Service 1994).

Unfortunately, new data from systematic survey cruises suggest a much grimmer picture in the Southern Hemisphere. Horwood (1986) provided an estimate of only about 1,300–2,000 blue whales left there. More recent analyses of the data indicate that although as many as 6,500 individuals of the pygmy subspecies may still exist (Klinowska and Cooke 1991), only 450 other *B. musculus* survive in all southern waters (International Whaling Commission 1994). The earlier dire predictions now seem to have been justified, and it is feared that remaining populations of the world's largest animal may not be viable. There are other threats to any potential recovery, especially regarding the increasing harvest of antarctic krill for use as human food (Gulland 1974; Laws 1985; McWhinnie and Denys 1980).

The blue whale is classified as endangered by the USDI and is on appendix 1 of the CITES. The IUCN classification is unusual, referring to the entire species as endangered but designating the North Atlantic stock as vulnerable and the North Pacific stock as conservation dependent and indicating that there are insufficient data for an assessment of the pygmy subspecies *(B. m. brevicauda).*

CETACEA; BALAENOPTERIDAE; **Genus MEGAPTERA**
Gray, 1846

Humpback Whale

The single species, *M. novaeangliae,* occurs in all oceans and adjoining seas of the world (Rice 1977).

Head and body length at sexual maturity averages 11.6 meters in males and 11.9 meters in females (U.S. National Marine Fisheries Service 1978). Average length at physical maturity for North Pacific specimens is 12.5 meters in males and 13.0 meters in females (Banfield 1974). Individuals more than 15 meters long are very rare. Average weight is around 30,000 kg. The pectoral fin is about one-third as long as the head and body and is the largest belonging to any cetacean, both absolutely and relatively. The dorsal fin varies from bumplike to strongly curved in shape and from 15 to 60 cm in height. The expanse of the tail flukes also is about one-third the length of the head and body. The general coloration is usually black above and white below, but there is much variation. Clusters of white barnacles are often present. The body is stocky compared with that of *Balaenoptera.* There are 10–36 grooves extending from beneath the tip of the snout to the area of the navel. Irregular knobs and protuberances occur along the snout and lower lip. On each side of the upper part of the mouth is a row of about 340 baleen plates, gray to black in color.

The humpback whale is basically oceanic but enters shallow tropical waters for the winter breeding season. At this time each of the discrete populations utilizes a certain archipelago or stretch of continental coastline. Johnson and Wolman (1984) identified the major wintering areas in the North Pacific as follows: (1) west coast of Baja California, Gulf of California, mainland Mexican coast from southern Sonora to Jalisco, and around Islas Revillagigedo; (2) Hawaiian Islands from Kauai to Hawaii; and (3) around Mariana, Bonin, and Ryukyu islands and Taiwan. During the spring there are movements along well-defined migratory routes toward the high-latitude feeding areas. The Northern Hemisphere migration extends as far as the Chukchi Sea and Spitsbergen. Some of the whales wintering off of Hawaii and Mexico form several summer feeding herds in the waters south of Alaska. There evidently is limited exchange of individuals between these two winter groups, and it may be that all *Megaptera* in the central and eastern North Pacific are part of a single structured stock (Baker et al. 1986; Darling and Jurasz 1983; Payne and Guinee 1983). In the Southern Hemisphere there are summer concentrations off the southern ends of continents and around such subantarctic islands as South Georgia. In the autumn there are return migrations toward the equator. Because of the reversal of seasons, the populations of the Northern Hemisphere are not in equatorial waters at the same time as those of the Southern Hemisphere (Dawbin 1966; Wolman 1978).

Despite its common name, *Megaptera* is a graceful swimmer and among the most acrobatic of large cetaceans. It often makes a somersault by leaping completely out of the water with its belly up, plunging back headfirst, and then circling underwater to the original position. It normally travels at about 7 km/hr, but maximum speed has

Humpback whales *(Megaptera novaeangliae)*, painting by Richard Ellis.

been estimated at 27 km/hr. Just before a dive it emits a series of three to six short, broad spouts and raises its tail flukes high above the water. Feeding is usually accomplished by taking in a large amount of water and organisms and then forcing out the water to strain the food in the baleen (see familial account). There have been observations of one or two whales diving beneath a school of fish and then spiraling upward while emitting bubbles. The bubbles rose in a cylindrical pattern around the fish and seemed to form a barrier, through which they would not pass. The whales then rushed into the school of fish to feed. Small fish evidently form a major part of the diet, but shrimplike crustaceans, especially of the family Euphausiidae, are the most important foods in the Antarctic. As is the general case with rorquals, most if not all feeding is done during the summer in high latitudes (Ellis 1980; Wolman 1978).

Experiments with a captive humpback suggested to Beamish (1978) that the genus does not use echolocation to find food. *Megaptera* is, however, among the most loquacious of cetaceans. It emits a great variety of sounds, certain of which are sometimes combined into an elaborate song. Winn and Winn (1978) reported the song to evolve through "moans and cries," to "yups or ups and snores," to "whos or wos and yups," to "ees and oos," to "cries and groans," and finally to various "snores and cries." These authorities considered the song to be one of the longest, and possibly the most patterned, in the animal kingdom. It lasts 6–35 minutes, but the highly rhythmic pattern is repeated over and over for hours and possibly months. It is heard only in the tropics during the winter breeding season. It is produced only by single individuals, which are thought to be young but sexually mature males. It may serve to space the males, to attract females, or to locate group members. There apparently are dialects, in that the song of the whales in one area differs to some extent from that heard in other areas.

Winn et al. (1981) determined the dialects of the Hawaiian and Mexican breeding groups to be essentially identical, which supports the view that both are part of the same overall population. This dialect differs from one shared by whales of the Cape Verde Islands and West Indies, and the population at Tonga in the South Pacific has still another distinct dialect. Perhaps the most remarkable feature of the song is that in each area it seems to change from year to year and that the new versions are effectively spread throughout the singing population. Payne and Payne (1985) found that songs in successive years overlap in part, they become increasingly different as time elapses, and after three or four years nearly all elements of the song have been modified. Payne, Tyack, and Payne (1983) concluded that the changes are transmitted by listening and learning and that humpback singing seems to fill an evolutionary gap between birds, which increase their repertoire through mimicry, and people, who create new songs. Baker and Herman (1984) showed that some of the same animals known to be singers are also the ones that escort females and young (see below).

There is segregation by age and sex during migrations (Dawbin 1966). In autumn the order of progression is: females with their recently weaned calves, then independent juveniles, then mature males and females that are not reproductively active, and finally females in late pregnancy. During the spring the order is: females in early pregnancy, then independent juveniles, then mature males and females that are not reproductively active, and finally females in the early stage of lactation. This arrangement ensures that pregnant females spend as much time as possible in the feeding waters and that young calves spend the longest possible time in warm regions. In feeding areas *Megaptera* sometimes is seen in aggregations of as many as 150; in the breeding season, however, animals are usually found alone or in groups of 2–15 (most frequently 3–5). The groups seem often to represent a number of males competing for proximity to a female, though they may sometimes involve males seeking to sort out dominance relationships or to form some kind of coalition in the absence of a female (Clapham et al. 1992; Tyack and Whitehead 1983). A female and young calf commonly are accompanied by another adult, evidently a male (Ellis 1980; Herman and Antinoja 1977). This escort male is aggressive toward other males that approach, often inflating its throat with water or air and lunging at its rivals and sometimes striking them with its tail. The pair bonds are not stable, however, and both males and females associate with a number of animals of the opposite sex during the breeding season (Baker and Herman 1984).

The mating and calving season is October–March in the Northern Hemisphere and April–September in the Southern Hemisphere. Females are seasonally polyestrous and usually give birth every 2 years to a single young about 400–500 cm long and weighing about 1,350 kg; some females have been observed to produce calves in successive years, especially in Hawaiian waters. The gestation period is 11.0–11.5 months and lactation lasts about 11 months. Sexual maturity is attained at 4–6 years (Banfield 1974; Clapham 1992; Glockner-Ferrari and Ferrari 1984, 1990; U.S. National Marine Fisheries Service 1978; Winn and Reichley 1985). Some individuals are estimated to have lived as long as 77 years (Haley 1978).

The humpback has been regularly hunted for a much longer period than other large rorquals. It is relatively easy to secure because it tends to stay near the shore in the breeding season, is a slow swimmer, often approaches vessels, and is highly visible. It has been considered a valuable source of oil, meat, and baleen. It was taken by aboriginal peoples in northwestern North America in ancient times and has recently been hunted for subsistence on certain West Indian and Pacific islands and in Greenland. Coastal fisheries began in Japan and Bermuda in the 1600s and in eastern North America by the 1700s. Major commercial exploitation of the genus spread over the North Atlantic and North Pacific in the nineteenth and early twentieth centuries. The most intensive period of hunting, however, began with the establishment of whaling stations on the islands around Antarctica in the early 1900s. The total kill in the Southern Hemisphere from 1904 to 1939 was 102,298 individuals. Populations were quickly and drastically reduced. Humpbacks composed 96.8 percent of the catch of whales at South Georgia in the 1910–11 season but only 9.3 percent in 1916–17. Nonetheless, large-scale exploitation continued for many years. The annual worldwide kill was more than 2,400 in every season from 1948–49 to 1963–64. By this time it was generally realized that the genus was nearing extinction. From 1964 to 1966 the International Whaling Commission extended protection to all populations, except that a small aboriginal quota continued (Allen 1942; Ellis 1980; Gulland 1974; McHugh 1974; Mitchell and Reeves 1983). The quota was 10 whales for the 1981–82 season (*Marine Mammal News* 7 [1981]: 1–3) but zero in 1986–87 (U.S. National Marine Fisheries Service 1987).

Original populations are estimated at 150,000 animals, of which 100,000 were in the Southern Hemisphere, 15,000 in the North Pacific, 10,000 in the northwestern Atlantic, and 5,000 in the eastern North Atlantic. The current worldwide estimate is at least 28,000, of which about 20,000 are in the Southern Hemisphere, 5,500 in the North American Atlantic, a few hundred in the eastern North Atlantic, and

500 in the northern Indian Ocean. Of the approximately 2,000 individuals in the North Pacific about 700 winter in the waters around Hawaii. Interestingly, this population seems to have become established only in the last 200 years. It now may be jeopardized by increasing boat traffic and disturbance from tourists. There are similar problems in the summer waters of Glacier Bay, Alaska. There is evidence that the North Atlantic population has increased since 1915, but individuals sometimes become entangled in fishing nets off eastern Canada (Herman 1980; Johnson and Wolman 1984; Klinowska and Cooke 1991; U.S. National Marine Fisheries Service 1978, 1981, 1987, 1989, 1994; Winn, Edel, and Taruski 1975; Winn and Reichley 1985; Wolman 1978). Because of its size, concentration near the shore, and acrobatic proclivity, the humpback is easily observed and has become a major tourist attraction. Efforts are underway to compile a catalog of photographs that will allow identification of individual whales throughout the world (Marine Mammal Commission 1994; Reeves and Leatherwood 1994). The species may be one of the few cetacean species that is definitively in the process of recovery; however, it still is classified as vulnerable by the IUCN and as endangered by the USDI and is on appendix 1 of the CITES.

Sirenians

Dugong, Sea Cow, and Manatees: Sirenia

This order of aquatic mammals contains two Recent families: the Dugongidae for the genera *Dugong* (dugong, one species) and *Hydrodamalis* (Steller's sea cow, one species, extinct); and the Trichechidae for the single genus *Trichechus* (manatees, three species). The dugong inhabits coastal regions in the tropical parts of the Old World, but some individuals go into the fresh water of estuaries and up rivers. Steller's sea cow occurred in the Bering Sea, being the only Recent member of this order adapted to cold waters. Manatees live along the coast and in coastal rivers in the southeastern United States, Central America, the West Indies, northern South America, and western Africa.

These massive, fusiform (spindle-shaped) animals have paddlelike forelimbs, no hind limbs or dorsal fin, and a tail in the form of a horizontally flattened fin. The adults of the living forms generally are 250–400 cm in length and weigh as much as 908 kg. The skin is thick, tough, often wrinkled, and nearly hairless; stiff, thickened vibrissae (tactile hairs) are present around the lips. The head is rounded, the mouth is small, and the muzzle is abruptly cut off. The nostrils, which are valvular and separate, are located on the upper surface of the muzzle. The eyelids, though small, are capable of contraction, and a well-developed nictitating membrane is present. There is no external ear flap. The neck is short. The females have two pectoral mammae, one on each side in the axilla behind the flipper. The testes in males are abdominal (borne permanently within the abdomen).

The dense, heavy skeleton is characterized by an absence of pneumatic cavities (pachyostosis). The increased specific gravity is probably an adaptation to remaining submerged and maintaining horizontal trim in shallow waters (Domning and de Buffrénil 1991; Kaiser 1974). The skull is large in proportion to the size of the body, though the brain is relatively among the smallest found in mammals (O'Shea and Reep 1990). The nasal opening is located far back on the skull and directed posteriorly. The lower jaws are heavy and united for a considerable distance. The forelimb has a well-developed skeletal support, but there is no trace of the skeletal elements of the hind limb. In *T. senegalensis,* at least, there is no movement in the wrist joint (Kaiser 1974). The pelvis consists of one or two pairs of bones suspended in muscle. The vertebrae are separate and distinct throughout the spinal column; one genus *(Trichechus)* has six neck vertebrae (nearly all other mammals have seven).

The dentition is highly modified and often reduced (functional teeth are lacking in *Hydrodamalis*). Most of the incisor teeth are reduced or absent, and canines are present only in certain fossil species. When the incisors are present, there is a space between them. The cheek teeth, which are arranged in a continuous series, number from 3 to 10 in each half of each jaw. The anterior part of the palate and the corresponding surface of the lower jaw are covered with rough, horny plates, presumably used as an aid in chewing food; the small, fixed tongue is also supplied with rough plates.

The genera differ in a number of ways. In *Dugong* and *Hydrodamalis* the tail fin is deeply notched, but in *Trichechus* the tail fin is more or less evenly rounded. The upper lip is more deeply cleft in manatees than in *Dugong* and *Hydrodamalis.* In *Dugong* there are 3/3 functional cheek teeth, and the males have one pair of tusklike incisors (usually unerupted in females). In *Trichechus* functional incisors are not present, but the cheek teeth are numerous and indefinite in number (up to 10 in each half of each jaw). The cheek teeth of *Trichechus* are replaced consecutively from the rear, as in many proboscideans; an individual cheek tooth is worn down as it moves forward.

Sirenians are solitary, travel in pairs, or associate in groups of three to about six individuals. Generally slow and inoffensive, they spend all their life in the water. They are vegetarians and feed on various water plants. They are the only mammals that have evolved to exploit plant life in the sea margin (Anderson 1979). The ordinal name Sirenia is related to the supposed mermaidlike nursing of dugongs (thought to be the origin of the myths of the sirens) and manatees. The only reliable observations of nursing in manatees, however, have revealed that the young suckle while the mother is underwater in a horizontal position, belly downward. Anderson (1984*b*) reported that suckling in the dugong is somewhat similar but that the calf usually is in an inverted position.

The Sirenia often are classified together with the Proboscidea and the Hyracoidea in the mammalian superorder Paenungulata. The geological range of the order Sirenia is early Eocene to Recent. The earliest fossils are from Hungary and India. By the middle Eocene the order was present in southeastern North America, the West Indies, southern Europe, northern and eastern Africa, and south-central Asia, and three distinct families—the Dugongidae, the Prorastomidae, and the Protosirenidae—had evolved (Dawson and Krishtalka 1984; Domning, Morgan, and Ray 1982). On the basis of morphological studies, Domning (1994) argued that the Trichechidae branched off shortly thereafter, in the late Eocene or early Oligocene, and he criticized molecular analyses suggesting that such divergence did not occur until the early Miocene. Sirenians apparently were more abundant from the Oligocene to the Pliocene than

Left, Manatees *(Trichechus manatus)*, female and calf, photo by James A. Powell, Jr., U.S. Fish and Wildlife Service. Right, dugongs *(Dugong dugon)*, female on right and male on left, photo by T. Kataoka, Toba Aquarium, Japan.

they are now. Their comparative scarcity at the present time probably results from climatic changes in the Pliocene and Pleistocene and, more recently, exploitation by humans for food, hides, and oil. The number of individual sirenians remaining in the world, perhaps 130,000, is smaller than that of any other mammalian order.

SIRENIA; **Family DUGONGIDAE**

Dugong and Sea Cow

This family contains two Recent subfamiles: the Dugonginae, with the single Recent genus *Dugong* (dugong), found in coastal regions of the tropics of the Old World; and the Hydrodamalinae, with the single Recent genus *Hydrodamalis* (Steller's sea cow), formerly found in the Bering Sea (Rathbun 1984).

Dugong is generally 250–350 cm in length, whereas Steller's sea cow measured around 800 cm. The flippers in Steller's sea cow were curiously bent and were said to have been used to pull an individual along the bottom of the ocean as it foraged. Members of the family Dugongidae lack nails on their flippers. The deeply notched tail fin has two pointed, lateral lobes (the tail fin in manatees is more or less evenly rounded). The upper lip, which is more deeply cleft in the adult dugong than in the young, is not as deeply cleft as that of the manatees. The nostrils are located more dorsally than in other sirenians. In the dugong (and apparently all sirenians) the eyelids contain a number of glands that produce an oily secretion to protect the eye against water.

The ribs of *Hydrodamalis* consisted entirely of dense bone; those of *Dugong* contain some porous bone. The skull of a dugong has a downwardly bent rostrum (for holding the incisors), whereas the rostrum of the skull of a Steller's sea cow was only slightly inclined. *Hydrodamalis* lacked functional teeth but did possess functional, rough oral plates. Male *Dugong* have one pair of tusklike incisors (usually unerupted in females), which are directed downward and forward and are partially covered with enamel. *Dugong* has 3/3 functional cheek teeth (as many as 6/6 may be present in an individual's lifetime). The cheek teeth are columnar and covered with cement, lack enamel, and have simple, open roots.

The geological range of this family, aside from the modern distribution, is early Eocene to Pliocene in Europe, middle Eocene and Pliocene in North Africa, Oligocene in Madagascar, late Oligocene in southern Asia and Japan, Miocene in Sri Lanka, middle Eocene to Pleistocene in North America, and early Miocene to early Pliocene in South America (Domning and Ray 1986; De Muzion and Domning 1985; Rathbun 1984).

SIRENIA; DUGONGIDAE; **Genus DUGONG**
Lacépède, 1799

Dugong

The single species, *D. dugon*, originally occurred in the Red Sea and the Gulf of Aqaba, along the eastern coast of Africa as far south as Mozambique and occasionally Natal, along the southern coast of Asia, around Madagascar and many of the islands in the Indian Ocean, throughout the East Indies, off eastern Asia as far north as Taiwan and the Ryukyu Islands, along the coast of Australia except in the south, and in the Pacific Ocean at least as far east as the Caroline Islands and the New Hebrides (Husar 1978*a;* Nishiwaki et al. 1979). According to Nishiwaki and Marsh (1985), individuals occasionally are seen near Sydney, Australia, about 700 km south of the regularly occupied range. Kingdon (1971) indicated that this species also was present within historical time in the eastern Mediterranean Sea.

Adults usually have a total length of 240–70 cm and weigh 230–360 kg, but recorded maximums are 406 cm and 908 kg (Husar 1978*a*). There is little sexual dimorphism, though females may grow to a larger average size than do males (Marsh, Heinsohn, and Marsh 1984). The coloration is variable but is usually dull brownish gray above and somewhat lighter below. The skin is thick, tough, relatively smooth, and covered with widely scattered hairs. The forelimbs are modified into flippers about 35–45 cm in length; they are used for propulsion by the young, but adults propel themselves by means of the flukelike tail, using the flippers only for steering. When the dugong grazes on the ocean floor it uses the flippers for "walking," and not to probe for food, but captives have been seen using the flippers to convey food to the mouth. The dugong has a relatively small and simple stomach and also has unique adaptations for feeding on marine grasses. The upper lip protrudes considerably beyond the lower; it is deeply cleft, forming a large U-shaped, muscular pad that overhangs the small, downwardly opening mouth. On the sides of this fa-

Dugong *(Dugong dugon)*, photos of a live female from the Surabaya Zoological Garden, temporarily stranded while her pool is being cleaned.

cial pad are two ridges that bear short, sturdy, blunt bristles. The lower lip and the distal portion of the palate have rough, horny pads that are used to grasp sea grasses during feeding; the whole upper lip is extended and curved around the base of a plant, which is then grasped with the mouth pads and pulled up by the roots.

The dental formula is: (i 0/0, c 0/0, pm 0/0, m 2–3/2–3) × 2 = 10–14. The upper incisors are rootless and straight, forming short, thick tusks in males more than 12–15 years old. In females the incisors usually do not pierce the gum, which implies that these teeth have a sex-related function rather than a food-gathering function. Adult males invariably carry conspicuous scars that seem to have been made by competing males. The molars are circular in section, thick, rootless, and lacking in enamel. The last molar tooth is a double, rather than a single, cylinder (Lekagul and McNeely 1977). Young animals have as many as six cheek teeth (including three premolars) on each side of each jaw; however, each row of cheek teeth migrates forward, worn teeth dropping out anteriorly, until only two molars remain in each quadrant of old animals (Rathbun 1984).

The dugong occurs in the shallow waters of coastal regions of tropical seas, where there is an abundance of vegetation. It is more strictly marine than manatees and seldom is found in freshwater localities. Long-distance migration is unknown, but off some parts of Africa, southeastern Asia, and Australia seasonal changes in abundance are associated with the monsoons and may reflect movements in response to rough weather and availability of food; in some areas there also seem to be regular daily movements between feeding areas and deeper waters (Husar 1978*a*). The dugong may rest in deep water during the day and move toward the shore to feed at night, but Nishiwaki and Marsh (1985) noted that there is diurnal inshore feeding in some areas. Anderson (1986) determined that a population in Shark Bay, off Western Australia, has cyclic movements, abandoning feeding sites when temperatures fall below 19° C and seeking warmer waters for the winter. Anderson (1984*a*) noted that the distance to these winter waters is more than 160 km. Average swimming speed is 10 km/hr, but animals can nearly double this rate if pressed (Husar 1978*a*). Dives seem to vary in length depending on such factors as water depth and forage species but have been reported normally to last around 1–3 minutes (Anderson 1982*b*; Nishiwaki and Marsh 1985). Feeding generally is at depths of 1–5 meters (Anderson 1984*a*).

Dugong is mostly herbivorous, and historical distribution broadly coincided with the tropical Indo-Pacific distribution of food plants, the phanerogamous sea grasses of the families Potomogetonaceae and Hydrocharitaceae. Such vegetation also is found in the Mediterranean Sea, where the dugong may formerly have occurred (Kingdon 1971). Although sea grasses are the primary food of the dugong, it has been reported to feed on brown algae following a cyclone that damaged the sea grass bed; green algae, marine algae, and some crabs also have been found in dugong stomachs (Husar 1978*a*).

According to Anderson (1982*a*), population densities in Shark Bay range from 0.12 to 12.8 individuals per kilometer of coastline. The dugong is essentially a gregarious animal, though it frequently is solitary. In the past, huge groups, sometimes containing thousands of individuals, were reported in various areas. Herds with hundreds of animals still can be found, but numbers usually are far small-

Dugong *(Dugong dugon)*, photo by T. Preen / National Photographic Index of Australian Wildlife.

er. Anderson suggested that such groups apparently are not simply aggregations that form in response to favorable feeding conditions but may involve deliberate assembly for protection against predators (especially sharks) and a process by which younger animals learn advantageous patterns of movement. Most dugongs in Shark Bay were found in the company of at least one other individual, and about 10–12 percent of the animals there were calves. Additional information, as summarized by Husar (1978*a*) and Lekagul and McNeely (1977), suggests that there may be lasting pair bonds between mated dugongs and that family groups may form within the larger groups. Calves may sometimes separate from the herd to form nursery subgroups of their own. Aggressive behavior evidently is rare. *Dugong* has been reported to make whistling sounds when frightened; calves make a bleating, lamblike cry. Anderson and Barclay (1995) recorded a variety of "chirp-squeaks, barks, trills," and other sounds, some quite complex, that may be associated with identification and territorial advertisement and defense.

Husar (1978*a*) wrote that breeding seems to occur throughout the year, with no well-defined season, though births apparently peak from July to September in Sri Lanka and parts of Australia. Marsh, Heinsohn, and Marsh (1984) reported that most births in northeastern Australia take place from September to December. Nishiwaki and Marsh (1985) stated that females are polyestrous and that the gestation period is 13–14 months. There normally is a single young; twins are rare (Husar 1978*a*). The newborn is 100–120 cm long and weighs 20–35 kg (Nishiwaki and Marsh 1985). It is born underwater and swims immediately to the surface for its first breath of air. The baby clings to its mother's back as she browses through the shoals of sea grass, submerging when the mother submerges and rising when she rises (Lekagul and McNeely 1977). The young have been seen to suckle while underwater in an inverted position behind the mother's axilla, and they sometimes are carried above the water when the mother rolls (Anderson 1984*b*). Lactation may last at least 18 months, though the calves begin to graze within 3 months of birth. Sexual maturity is attained by both males and females at as early as 9–10 years but may be delayed until 15 years. Females have a calving interval of 3–7 years, and the maximum rate of population increase is thought to be about 5 percent (Marsh, Heinsohn, and Marsh 1984). A pair of captives was maintained for 10 years (Husar 1978*a*), but examination of wild specimens indicates that maximum longevity is about 73 years (Marsh 1995).

Sharks are probably the main natural enemy of the dugong, but individuals have been seen to "gang up" on sharks in shallow water and drive them off by butting them with the head (Lekagul and McNeely 1977). Anderson and Prince (1985) related a report of a devastating attack by about 10 killer whales on a tightly bunched group of approximately 40 dugongs. People have had a far more serious long-term effect. Specialized human cultures based on dugong hunting have developed in several areas, such as the Torres Strait, between Australia and New Guinea (Nishiwaki and Marsh 1985). The dugong has been hunted for food throughout its range, its meat being likened to tender veal. The hide has been used to make a good grade of leather. The species also has been taken for its oil (24–56 liters for an average adult) and for its bones and teeth, which have been used to make ivory artifacts and a good grade of charcoal for sugar refining. Commercial dugong fisheries once operated from Sri Lanka, and several Asian cultures have prized dugong products for supposed medicinal and aphrodisiac properties. Large gill nets and harpooning are used to capture the dugong (Husar 1978*a*).

In many areas the dugong has declined greatly in numbers, and fears have been expressed that it may be exterminated as a result of continued hunting pressure. Kingdon (1971) wrote that it was known to the ancient Greeks, Egyptians, and Phoenicians of the Mediterranean; no populations have occurred in that region for centuries, though occasional individuals may have traversed the Suez Canal. According to Nishiwaki and Marsh (1985), populations also have disappeared from the Mascarene, Laccadive, and Maldive islands in the Indian Ocean. They have become very rare around Guam and Yap, in the Ryukyu Islands, and along much of the mainland coast of eastern Asia. Populations along the coasts of India, southwestern Asia, Africa, and Madagascar are thought to be in critical danger. Nishiwaki et al. (1979) estimated that roughly 30,000 dugongs still inhabit the whole Indo-Pacific region. However, recent aerial censuses off Australia suggest that larger numbers survive (Marsh 1989, 1994; Marsh et al. 1994; Marsh and Saalfeld 1989, 1990): at least 10,000 in Shark Bay, Western Australia; nearly 12,000 in the waters of the Great Barrier Reef Marine Park, Queensland; about 17,000 in the Gulf of

Carpentaria; about 14,000 along the rest of the coast of the Northern Territory; and more than 24,000 in Torres Strait, where, however, there is an annual traditional hunting take of about 1,200, which may be more than can be sustained. The total number in Australian waters may exceed 80,000 and is probably more than half of the world's total (Reynolds and Odell 1991). The second largest known concentration, about 7,000 individuals, is in the Persian (or Arabian) Gulf, and it also may be undergoing excessive hunting by villagers (Baldwin and Cockcroft 1995).

The dugong is classified as vulnerable by the IUCN and is on appendix 1 of the CITES, except for the Australian population, which is on appendix 2. The USDI lists the dugong as endangered, but because of an uncorrected technicality this classification does not apply within present or former territory of the United States, such as Guam and Palau. Recent investigations indicate that the population at Palau numbers fewer than 200 individuals and is jeopardized by unsustainable subsistence hunting and habitat loss (Marine Mammal Commission 1994; Marsh et al. 1995; Rathbun et al. 1988). In August 1993 the USDI proposed to extend endangered status to the involved populations, but publication of a final rule to that effect has been continually delayed, in part because of a year-long congressional ban on all listings of endangered and threatened species.

Steller's sea cow *(Hydrodamalis gigas):* A. Drawing from *Extinct and Vanishing Mammals of the Western Hemisphere,* Glover M. Allen; Inset: Steller's sea cow's palate, from *Symbolae Sirenologicae,* drawing by J. F. Brandt; B. Outer surface of skin, photo from Erna Mohr.

SIRENIA; DUGONGIDAE; **Genus HYDRODAMALIS**
Retzius, 1794

Steller's Sea Cow, or Great Northern Sea Cow

The single Recent species, *H. gigas,* occurred in historical time around Bering and Copper islands in the Commander Islands of the western Bering Sea (Rice 1977). Pleistocene remains of the same species have been found on Amchitka Island in the Aleutians and in Monterey Bay, off California. Pliocene remains of a related species, *H. cuestae,* have been found in Japan, California, and Baja California (Domning 1978). Still older fossils indicate that the line leading to *Hydrodamalis* had diverged from the Dugongidae by the late Oligocene (Domning 1994).

This animal was the largest of the sirenians and possibly the largest noncetacean mammal to exist in historical times. Calculations by Scheffer (1972) indicate that the largest specimens were at least 800 cm long and weighed 10,000 kg. Observations by Steller (as reported in Stejneger 1936), the famous German naturalist who accompanied Bering (see below), provide an additional idea of the size and appearance of the live animal. An adult female measured 752 cm from the tip of the nose to the point of the tail flipper. Its greatest circumference measured 620 cm, and its estimated weight was about 4,000 kg. The head was small in proportion to the body. The tail had two pointed lobes forming a caudal flipper. The forelimbs were very small, flipperlike, and somewhat truncated. There were no hind limbs externally, no trace even remaining of the pelvic elements. The skin was naked and was covered with a very thick, barklike, extremely uneven-appearing epidermis, from which the German name *Borkentier* (bark animal) was derived. The skin was dark brown to gray brown in color, occasionally spotted or streaked with white. The forelimbs were covered with short, brushlike hairs. There is a large dried sample of this skin at the Zoological Institute in St. Petersburg, but a reported specimen in the Hamburg Zoological Museum may actually represent a cetacean (Forstén and Youngman 1982). *Hydrodamalis* was parasitized by two or three species of small crustaceans that burrowed into the rough skin. Steller said that he found some of the holes made by the small crablike crustacean *Cyamus* to ooze a thin serum.

According to Forstén and Youngman (1982), the condylobasal length of the skull was 638–722 mm, the zygomatic width was 324–73 mm, the nasal basin on top extended past the eye opening, and the braincase was relatively small. *Hydrodamalis* had no functional teeth; mastication was accomplished by keratinized pads, one each in the rostral areas on the upper and lower jaws. The upper and lower pads had corresponding V-shaped crests and valleys, between which the food was ground.

Hydrodamalis was the only Recent sirenian adapted to cold waters. All that we know about its habits comes from Steller's account. It was quite numerous in the shallow bays and inlets around the coast of Bering Island when first discovered. It fed upon the kelp beds and the extensive growths of various marine algae that grew in the shallower waters. It was slow-moving and utterly fearless of humans and was said to be affectionate toward others of its kind, as individuals apparently tried to aid others that were either wounded or in distress.

The following are excerpts from Steller's own account (from Stejneger 1936):

> Usually entire families keep together, the male with the female, one grown offspring and a little, tender one. To me they appear to be monogamous. They bring forth their young at all seasons, generally however in autumn, judging from the many new-born seen at that time; from the fact that I observed them to mate preferably in the early spring, I conclude that the fetus remains in the uterus more than a year. That they bear not more than one calf I conclude from the shortness of the uterine cornua and the dual number of mammae, nor have I ever seen more than one calf about each cow.
>
> These gluttonous animals eat incessantly, and because of their enormous voracity keep their heads always under water with but slight concern for their life and security, so that one may pass in the very midst of them in a boat even unarmed and safely single out from the herd the one he wishes to hook. All they do while feeding is to lift the nostrils every four or five minutes out of the water, blowing out air and a little water with a noise like that of a horse snorting. While browsing they move slowly forward, one foot after the other, and in this manner half swim, half walk like cattle or sheep grazing. Half the body is always out of water. . . . In winter these animals become so emaciated that not only the ridge of the backbone but every rib shows.
>
> Their capture was effected by a large iron hook, . . . the other end being fastened by means of an iron ring to a very long, stout rope, held by thirty men on shore. . . . The harpooner stood in the bow of the boat with the hook in his hand and struck as soon as he was near enough to do so, whereupon the men on shore grasping the other end of the rope pulled the desperately resisting animal laboriously towards them. Those in the boat, however, made the animal fast by means of another rope and wore it out with continual blows, until, tired and completely motionless, it was attacked with bayonets, knives and other weapons and pulled up on land. Immense slices were cut from the still living animal, but all it did was shake its tail furiously.

In the late Pleistocene, *Hydrodamalis* evidently occurred all around the rim of the North Pacific from Japan to California. Subsequent hunting by primitive peoples probably eventually restricted it to the vicinity of the Commander Islands (Domning 1978). This last population was discovered when a Russian expedition led by Captain Vitus Bering was stranded on Bering Island in 1741. At that time the total number of sea cows probably did not exceed 1,000 to 2,000. Bering's crew, mostly sick with scurvy, and subsequent visitors to the area slaughtered the animals relentlessly. The meat of *Hydrodamalis* was used for food, and the hide for making skin boat covers and shoe leather. The genus probably was extinct by 1768, though there were later reports of its presence in various areas. It is classified as extinct by the IUCN.

SIRENIA; **Family TRICHECHIDAE; Genus TRICHECHUS**
Linnaeus, 1758

Manatees

The single Recent genus, *Trichechus*, contains three species (Caldwell and Caldwell 1985; Domning and Hayek 1986; Husar 1977, 1978*b*, 1978*c*; Rice 1977):

- *T. inunguis* (Amazonian manatee), throughout the Amazon Basin of northern South America;
- *T. manatus* (West Indian manatee), coastal waters and some connecting rivers from Virginia around the Gulf of Mexico and the Caribbean Sea to eastern Brazil, Orinoco Basin, Greater and formerly Lesser Antilles, Bahamas;
- *T. senegalensis* (West African manatee), coastal waters and connecting rivers from Senegal to Angola, occurs as far as 2,000 km up the Niger River and in tributaries of Lake Chad.

The range of *T. inunguis* sometimes is said to include the Orinoco Basin, but such reports apparently are based on misidentification of *T. manatus* in the early nineteenth century. It also is unlikely that *T. inunguis* reaches the waters that connect the Orinoco and Amazon basins. *T. inunguis* occurs at the mouth of the Amazon on the Atlantic coast, where it may be sympatric with *T. manatus*, and the latter species once was present as far south as the state of Espirito Santo at about 20° S (Domning 1981, 1982*a*; Thornback and Jenkins 1982). *T. manatus* was introduced in the Panama Canal around 1960 to aid in weed control, and the population there subsequently increased and extended its range to the Pacific Ocean (Domning and Hayek 1986; Montgomery, Gale, and Murdoch 1982).

These animals have a rounded body, a small head, and a squarish snout. The upper lip is deeply split, and each half is capable of moving independently of the other. The nostrils are borne at the tip of the muzzle, the eyes are small, and there are no external ears. The flattened tail fin is more or less evenly rounded (not notched as in the dugong and sea cow) and is the sole means of propulsion in adults. The flexible flippers are used for aiding motion over the bottom, scratching, touching and even embracing other manatees, and moving food into and cleaning the mouth (Caldwell and Caldwell 1985). Vestigial nails may occur on the flippers. Stout bristles appear on the upper lip, and bristlelike short hairs are scattered singly over the body at intervals of about 1.25 cm. The skin is 5.1 cm thick. Females have two pectoral mammae, one on each side in the axilla behind the flipper.

The skeleton is of extremely dense (pachyostotic) bone, a condition that increases specific gravity and may contribute to neutral buoyancy (Caldwell and Caldwell 1985).

West Indian manatees *(Trichechus manatus)*, mother and calf, photos by James A. Powell, Jr.

The stout ribs consist wholly of dense bone. Manatees have only six neck vertebrae, whereas nearly all other mammals have seven. Nasal bones are present in the skull (these bones are absent or vestigial in the dugong and sea cow). Manatees have 2/2 incisors, which are concealed beneath horny plates and are lost before maturity. The cheek teeth number up to 10 in each half of each jaw (although more than 6/6 are rarely present at any one time). They are low-crowned, enameled, divided into cuspidate cross-crests, lack cement, and have closed roots. These teeth are replaced horizontally; they form at the back of the jaw and wear down as they move forward. This may be an adaptation to eating food mixed with sand or silt; a similar replacement of teeth occurs in many proboscideans and other mammals but not in the Dugongidae.

Rathbun (1984) wrote that the known geological range of the family Trichechidae is early Pliocene to Recent in Atlantic North America, early Miocene to Recent in South America, and Recent in West Africa. Domning (1982*b*, 1994), however, indicated that the Trichechidae arose in the late Eocene or early Oligocene and speculated that the family is descended from protosirenian stock that became isolated during the Eocene in South America. Until the late Miocene, trichechids may have been restricted to the coastal rivers and estuaries of that continent. By the Pliocene a branch of the family leading to modern *T. inunguis* had reached the interior of the Amazon Basin, where it adapted to feeding on the newly available floating meadows of nutrient-rich lakes. Another Pliocene branch, the forerunner of *T. manatus,* spread into the Caribbean, where its wear-resistant dentition proved superior to that of the dugongids, then present in the region, for dealing with a diet modified by the increased silt runoff of the period. A closely related branch dispersed across the Atlantic to West Africa during the Pliocene or Pleistocene and gave rise to *T. senegalensis.* Additional information on each of the three species is given separately below.

Trichechus inunguis (Amazonian manatee)

Husar (1977) wrote that this species is smaller and more slenderly proportioned than either *T. manatus* or *T. senegalensis;* the largest recorded specimen was a 280-cm male. Timm, Albuja V., and Clauson (1986) cited a commercial hunter in Ecuador as saying that the largest manatee he had taken weighed 480 kg. Domning and Hayek (1986) stated that the skin is smooth, slick, and rubberlike, not wrinkled as in the other two species. The general coloration is gray, and most individuals have a distinct white or bright pink patch on the breast. Fine hairs are sparsely distributed over the body, and there are thick bristles on both the upper and lower lips. The flippers, unlike those of the other two species, are elongate and usually lack nails. The skull is relatively long and narrow.

The Amazonian manatee occurs exclusively in fresh water (Thornback and Jenkins 1982). It has been found all around Ilha de Marajó, at the mouth of the Amazon on the Atlantic coast, but even there salt water has little or no influence (Domning 1981). This species favors blackwater lakes, oxbows, and lagoons, and it has been maintained successfully in waters with temperatures of 22°–30° C (Husar 1977). Its key requirements in the wild seem to be large blackwater lagoons with deep connections to large rivers and abundant macrophytic plants such as grasses, bladderworts, and water lilies (Timm, Albuja V., and Clauson 1986). It usually surfaces several times per minute to breathe, but the longest recorded submergence is 14 minutes (Husar 1977). A captive juvenile released and radio-tracked for 20 days was found to be equally active by day and by night, to move about 2.6 km per day, and to spend most of its time where food, especially floating vegetation, was most abundant (Montgomery, Best, and Yamakoshi 1981). The diet of *T. inunguis* consists mostly of vascular aquatic vegetation (Caldwell and Caldwell 1985). Captive adults consume 9–15 kg of leafy vegetables per day (Husar 1977).

Best (1983) found that manatee populations of the central Amazon Basin make an annual movement in July–August, when water levels begin to fall. Some enter rivers, where the animals can continue to find food but must also spend energy to hold their position in the current. Others become restricted to deep parts of larger lakes during the dry season, September–March, where they do not have obvious food sources until water levels rise 1–2 meters. Available evidence suggests that the latter populations fast for nearly seven months, though they may take some vegetation. The manatee's large fat reserves and low metabolic rate, only 36 percent of the usual rate for eutherians, allow the species to survive at this time. Best's study was done at Lago Amana, where 500–1,000 manatees congregate during the dry season.

There have been other reports of large aggregations of manatees in the middle reaches of the Amazon, but such assemblies appear rare today. Loose groups of 4–8 individuals may be present in feeding areas. As with *T. manatus*, estrous females reportedly are pursued by groups of males (Marmontel, Odell, and Reynolds 1992). A few short vocalizations have been noted (Husar 1977). Breeding has been reported to occur throughout the year, at least in some areas, but Timm, Albuja V., and Clauson (1986) received information that births occur mainly in January in one part of Amazonian Ecuador and in June in another. Evidence gathered by Best (1982) indicates that in the central Amazon Basin breeding is seasonal, with nearly all births taking place from December to July, mainly from February to May, the period of rising river levels. Data compiled by Husar (1977) and Marmontel, Odell, and Reynolds (1992) suggest that gestation lasts approximately 1 year, that there normally is a single young weighing 10–15 kg, and that the calving interval may be 2 years. The mother-calf bond seems to be long-lasting; the mother carries her calf on her back or clasped to her side. Two individuals lived 12.5 years in captivity.

Available records (Domning 1982*a*; Thornback and Jenkins 1982) indicate that in Brazil the Amazonian manatee was being exploited commercially for its meat as early as the 1600s and that even by the late eighteenth century there was some concern for its survival. Commercial hunting continued, however, with about 1,000–2,000 individuals being marketed each year from 1780 to 1925. Many other animals were taken for subsistence purposes. The main product during this period was *mixira*, fried meat packed in lard. The commercial kill rose sharply starting in 1935, when there was an increased demand for the manatee's hide to manufacture heavy-duty leather. This industry collapsed about 1954, when synthetic replacements became widely available, but there then was a renewed intensification of meat hunting. About 4,000–10,000 manatees were killed each year for commercial purposes in Brazil from 1935 to the early 1960s. The annual take then fell to about 1,000–2,000, probably reflecting a severe decline of the species. The manatee was protected legally in Brazil in 1973, but subsistence hunting and local commercial killing has continued.

The manatee also has declined in other parts of its range (Thornback and Jenkins 1982). The Siona Indians of Ecuador, who are traditional hunters of the manatee, now have a self-imposed ban on taking the species because of its low numbers. However, Timm, Albuja V., and Clauson (1986) thought that if meat hunting by settlers and the military continued at 1986 levels, the manatee would be eliminated in Ecuador within 10–15 years. Other problems include accidental drowning in commercial fishing nets and degradation of food supplies by soil erosion resulting from deforestation. The large-scale destruction of Amazonian rainforests also would lead to a decline in rainfall, a drop in water levels, and increased mortality of the manatee (Best 1983). The indiscriminate release of mercury in mining activities has put the entire aquatic fauna of the Amazon Basin at risk (Rosas 1994). An overall minimum population estimate of 10,000 individuals was cited by Husar (1977). *T. inunguis* is classified as vulnerable by the IUCN and as endangered by the USDI and is on appendix 1 of the CITES.

Trichechus manatus (West Indian manatee)

According to Husar (1978*a*), total length of adults varies from 250 to more than 450 cm, and weight is 200 to more than 600 kg, but average adults are 300–400 cm long and weigh less than 500 kg. Sexual dimorphism in size has not been documented, though females seem bulkier. The finely wrinkled skin is slate gray to brown in color. Fine, colorless hairs 30–45 mm long, are sparsely distributed over the body, and there are stiff, stout bristles on the muscular, prehensile pads of the upper lip. Nails are present on the dorsal surface of the flippers. The broad skull has a relatively short and downturned snout.

Except as noted, the information for the remainder of this account was taken from Hartman (1979) and refers to *T. manatus* in Florida. This species inhabits shallow coastal waters, bays, estuaries, lagoons, and rivers. It utilizes both salt and fresh water, though it consistently prefers the latter (Lefebvre et al. 1989). It also seems to prefer water above about 20° C but can endure water as cold as 13.5° C. In temperate parts of its range, such as most of Florida, individuals may congregate during the winter at sources of warm water, such as natural hot springs and discharges from power plants. A drop in the air temperature to below 10° C seems to induce congregation in the warm headwaters of the Crystal and Homosassa rivers in central Florida. Some individuals winter in the tropical waters of southern Florida. When air temperatures rise above 10° C, Florida manatees may leave their refugia and wander all along the Gulf Coast and up the Atlantic coast as far as Virginia. One well-known individual reached Chesapeake Bay in 1994,

West Indian manatees *(Trichechus manatus)*, mother and calf, U.S. Fish and Wildlife Service photo by Galen B. Rathbun.

where he was captured and taken back to Florida, but he moved back as far as Rhode Island in 1995, before returning to Florida on his own (Reid 1995).

There is evidence of long-range, offshore migrations between population centers. Individuals have been caught up to 15 km off the coast of Guyana. They also have been reported in rivers 230 km from the sea in Florida and 800 km from the coast in South America (Husar 1978*a*). Most manatees appear to be nomadic and to move hundreds of kilometers, pausing for days, weeks, months, or seasons in estuaries and rivers that supply their needs. They follow established travel routes, preferring channels that are 2 meters or more in depth and shunning those that are less than 1 meter deep. Animals generally swim 1–3 meters below the surface of the water. They use their tails not only to propel themselves through the water but also as rudders by means of which they control roll, pitch, and yaw. The flippers are used in precise maneuvering and in minor corrective movements to stabilize, position, and orient the animal. Speed normally is 3–7 km/hr but can reach 25 km/hr when the animal is pressed. The manatee avoids fast currents, and animals migrating through the Intracoastal Canal were never seen swimming against currents that exceeded 6 km/hr. According to Husar (1978*b*), average submergence time is 259 seconds, but a dive of 980 seconds has been recorded.

T. manatus may be active both by day and by night. During a 24-hour period adults have been observed to feed 6–8 hours in sessions that usually lasted 1–2 hours, to rest 6–10 hours, and to move as much as 12.5 km. The manatee rests by hanging suspended near the surface of the water or by lying prone on the bottom. In both positions the animal lapses into a somnolent state, with eyes closed and body motionless. Sound appears to be the principal sensory mode, but in clear water the preferred method of environmental exploration is visual. In addition, the prevalence of mouthing in social interaction suggests that the species has a chemoreceptive sense by which it can recognize odors in the water. Comfort activities include stretching, using the flippers to scratch and clean the mouth, rubbing against logs and rocks, and rooting in the substrate. Rubbing also may involve deposition of scent, by which an estrous female signals receptivity to wandering males (Rathbun and O'Shea 1984).

The manatee is herbivorous, though some invertebrates are ingested together with vegetation and may provide an important amount of protein. Also, Powell (1978) reported that captives have deliberately eaten dead fish, and wild individuals off Jamaica have been seen to take fish entangled in nets. The diet consists mainly of a wide variety of submerged, vascular plants, but emergent and floating vegetation sometimes is eaten. Water hyacinths *(Hydrilla)* are a staple food in many rivers of Florida. For many years *T. manatus* has been deliberately employed for the control of water hyacinths and other aquatic weeds in waterways in Guyana (Haigh 1991). Sea grasses are taken in salt-water areas. Captives have consumed up to one-fourth of their body weight per day in wet greens. Feeding occurs from the surface to a depth of 4 meters. Individuals are drawn to sources of fresh water, which they appear to require for osmoregulation. However, such a need was questioned by Reynolds and Ferguson (1984), who observed two large manatees 61 km northeast of the Dry Tortugas Islands (about 110 km west of Key West, Florida), which lack fresh water.

In their search for receptive mates males reportedly utilize a much larger home range than do females. Some new information also suggests that individuals come together deliberately in order to learn the location of favorable habitat and migration routes (Thornback and Jenkins 1982). Generally, however, the West Indian manatee is considered to be a weakly social, essentially solitary species. The only lasting association seems to be that between a cow and her calf. Temporary, casual groups form in favorable areas for purposes of migration, feeding, resting, or cavorting. Such groups may be randomly made up of juveniles and adults of both sexes, though Reynolds (1981) suggested a tendency for subadult males to associate. There is no evidence of communal defense or mutual aid and little or no indication of a social hierarchy. Individuals indulge in what appears to be play. They exchange gentle nibbles, kisses, and embraces that are age- and sex-independent. They have exceptional acoustic sensitivity; sound is doubtless the major directional determinant in social interactions. *T. manatus* normally is silent but can emit high-pitched squeals, chirp-squeals, and screams under conditions of fear, aggravation, protest, internal conflict, male sexual arousal, and play. Its vocalizations are probably nonnavigational and lack ultrasonic signals, pulsed emissions, or directional sound fields. Bengtson and Fitzgerald (1985) suggested that certain calls help to relay warnings and to keep groups together during flight. However, the only predictable vocal exchange between manatees involves screams of alarm, by which a cow calls her calf in case of danger, and the responding squeals of the young.

A population wintering in the Crystal River of Florida one year contained 31 adults, 13 juveniles, and 6 calves and was divided about equally between males and females. When a female is in estrus, she may be accompanied for a period of one week to more than a month by as many as 17 males. The courtship of the latter is relentless, but the cow generally seeks to escape and to vigorously repulse their advances. Rathbun and O'Shea (1984) suggested that this process involves mate selection by the female. It also involves constant and sometimes violent pushing and shoving by the bulls as they compete for a position next to the estrous female; such activity seems to be the only display of aggression in the species. Although most calves are born in spring and early summer in Florida (Marmontel 1995), there appears to be no definite breeding season there or elsewhere (Caldwell and Caldwell 1985). Calves have been seen throughout the year in the Dominican Republic and Puerto Rico (Belitsky and Belitsky 1980; Powell, Belitsky, and Rathbun 1981). The interbirth interval probably is normally 2.5–3.0 years but may be shorter if an infant is lost. The gestation period is about 12–13 months. The cow seeks the shelter of a backwater to give birth, normally to a single young but occasionally to twins. The newborn is about 120–30 cm long, weighs 28–36 kg, and is dark in color (Caldwell and Caldwell 1985; Marmontel 1995; Rathbun et al. 1995). Within half a day of birth it is capable of swimming and surfacing on its own, though it occasionally rides on the mother's back. The calf begins to take some vegetation at about 1–3 months, though it continues to suckle until leaving its mother at about 1–2 years. Males appear to reach full reproductive maturity at 9–10 years (Odell, Forrester, and Asper 1978). However, recent laboratory studies indicate that males may be physiologically capable of mating when they are only 2–3 years old (Hernandez et al. 1995). Females are sexually mature at 3–4 years, one has been reproductively active in captivity for 35 years, and one is estimated to have reached an age of 59 years in the wild; a captive specimen of *T. manatus* is 44 years old (Marmontel 1995).

Commercial exploitation of the West Indian manatee began in the sixteenth century, and combined with extensive subsistence hunting, it has resulted in severely reduced populations in most areas. Data compiled by Lefebvre et al. (1989) and Thornback and Jenkins (1982) indicate that *T. manatus* probably had disappeared from the Lesser Antilles by the eighteenth century and is continuing to decline almost everywhere else despite legal protection. Problems include killing for meat, drainage of swamps, accidental drowning in fishing nets and flood control structures, and pollution. The largest remaining population, consisting of thousands of individuals, occurs in Guyana. Numbers may be in the hundreds in Belize, Surinam, French Guiana, and perhaps some of the other South American countries, but there are only around 100 or fewer in each of the other countries of Central America and the Greater Antilles. *T. manatus* is still found in certain wetlands, bays, and rivers along the coast of Mexico from Veracruz to the Yucatan but is being killed off both accidentally and deliberately by fishermen. O'Shea and Salisbury (1991) reported that numbers in Belize have been stable since 1977 and that the number there is greater than the number in any other country bordering the Caribbean, apparently because of a relatively low human population and maintenance of a high-quality habitat, including shallow waters sheltered by the second longest coral reef in the world. The species has been extirpated in most of its former range in Brazil, though there evidently was a small population as far south as Bahia in the 1960s, and there still are several isolated groups, especially at the Rio Mearim in Maranhao and at Barra de Mamanguape in Paraiba (Borobia and Lodi 1992; Domning 1981, 1982*a*; Whitehead 1977). *T. manatus* is classified as vulnerable by the IUCN and as endangered by the USDI and is on appendix 1 of the CITES.

The population in Florida waters originally contained an estimated several thousand animals but was so greatly reduced by 1893 that it was then given legal protection by the state (Caldwell and Caldwell 1985). It now also is covered by the federal Marine Mammal Protection Act and the Endangered Species Act, but many animals are killed and injured by vandals, entrapment in gates of locks and dams, collision with barges, and the propellers of power boats (O'Shea et al. 1985). Indeed it is a sad fact that most manatees in Florida bear scars from propellers or collisions with boats and that such marks are the chief means by which biologists identify the individuals they are studying (Reid, Rathbun, and Wilcox 1991). Aerial surveys in 1992 indicated that at least 1,856 manatees remain in Florida, with a roughly equal number on each coast (Ackerman 1995). There are at least 125–30 deaths annually, at least 30 percent of which are caused by people, and there also are an estimated 120–30 births each year in Florida (Baugh 1987). In the first half of 1996, 304 deaths were recorded and attributed largely to a natural outbreak of the toxic microorganism red tide (*Washington Post*, 8 July 1996, A-2). Despite this problem, aerial counts made in early 1996 indicated that the true number of manatees in Florida may be close to 3,000, though Domning (1996) cautioned that this figure may reflect improved census techniques and ideal survey conditions rather than any significant increase in population size.

During the spring and summer substantial numbers of manatees continue to enter Georgia waters, and some occasionally move farther north. A few individuals also recently have been found in Mississippi, Louisiana, the Galveston area, southern Texas, and Tamaulipas in northeastern Mexico (Fernandez and Jones 1990; Lazcano-Barrero and Packard 1989; Powell and Rathbun 1984). Evidently the animals along most of the Gulf Coast belong to the subspecies *T. manatus latirostris,* which is centered in Florida and represented by wanderers elsewhere in the United States and in the Bahamas. Those in southern Texas and Tamaulipas come from farther south in Mexico and belong to *T. m. manatus,* which is found in the remainder of the range of the species (Domning and Hayek 1986). Occurrences in Texas have declined in association with a drastic reduction of the population in Mexico.

Trichechus senegalensis (West African manatee)

According to Husar (1978*c*), adults are 300–400 cm long and weigh less than 500 kg. The skin is finely wrinkled and grayish brown in color. Fine, colorless hairs are distributed over the body, and there are stout bristles on the upper and lower lip pads. The flippers have nails on the dorsal surface. The skull is broad and has a shortened, slightly deflected snout. Domning and Hayek (1986) found *T. senegalensis* and *T. manatus* to be phenetically similar and noted that the reduced rostral deflection of the former probably is associated with its dependence on emergent or overhanging, rather than submerged, vegetation.

The West African manatee occurs in both shallow coastal waters and freshwater rivers but seems to prefer large, shallow estuaries and weedy swamps; its range apparently is limited to water with a temperature above 18° C (Husar 1978*c*). It seems to avoid salt water (Nishiwaki et al. 1982). An individual may travel 30–40 km per day through lagoons and rivers (Reynolds and Odell 1991). The diet includes aquatic vegetation, of which a free-ranging adult might consume 8,000 kg in a year (Husar 1978*c*). Populations in some rivers depend heavily on overhanging bank growth, and those in estuarine areas feed exclusively on mangroves (Domning and Hayek 1986). In Sierra Leone manatees reportedly remove fish from nets and consume rice in such quantities that they are considered to be pests (Reeves, Tuboku-Metzger, and Kapindi 1988).

Roth and Waitkuwait (1986) reported that *T. senegalensis* lives singly or in family groups of 4, or rarely 6, animals. Husar (1978*c*) related a report of 15 individuals being seen together and indicated that social behavior probably is similar to that of *T. manatus.* The breeding period is uncertain and may last throughout the year. Parturition supposedly occurs in shallow lagoons, usually there is a single calf, and the newborn is about a meter long.

Husar (1978*c*) wrote that there had been no large-scale commercial exploitation of *T. senegalensis* comparable to that of other manatees but that populations had declined markedly in some areas because of local hunting for meat. Reeves, Tuboku-Metzger, and Kapindi (1988) also pointed out such problems as accidental capture in fishing nets and persecution for alleged depredations on netted fish and growing rice. Surveys in 1980–81 (Nishiwaki 1984; Nishiwaki et al. 1982) yielded a rough overall estimate of 9,000–15,000 individual *T. senegalensis,* exclusive of Angola and Congo. The largest numbers, perhaps several thousand in each area, are found in Gabon and the lower reaches of the Niger and Benue rivers. Numbers also are considerable in Cameroon, Ghana, Ivory Coast, and the upper reaches of the Niger but are fewer from Senegal to Liberia. The species is classified as vulnerable by the IUCN and as threatened by the USDI and is on appendix 2 of the CITES.

Other Marine Mammals

CARNIVORA; URSIDAE; Genus URSUS
Linnaeus, 1758

Black, Brown, Polar, Sun, and Sloth Bears

There are six species (Corbet 1978; Corbet and Hill 1992; Ellerman and Morrison-Scott 1966; E. R. Hall 1981; Kurten 1973; Laurie and Seidensticker 1977; Lay 1967; Lekagul and McNeely 1977; Ma 1983; Simpson 1945):

U. thibetanus (Asiatic black bear), Afghanistan, southeastern Iran, Pakistan, Himalayan region, Burma, Thailand, Indochina, China, Manchuria, Korea, extreme southeastern Siberia, Japan (except Hokkaido), Taiwan, Hainan;
U. americanus (American black bear), Alaska, Canada, conterminous United States, northern Mexico;
U. arctos (brown or grizzly bear), western Europe and Palestine to eastern Siberia and Himalayan region, Atlas Mountains of northwestern Africa, Hokkaido, Alaska to Hudson Bay and northern Mexico;
U. maritimus (polar bear), primarily on arctic coasts, islands, and adjacent sea ice of Eurasia and North America;
U. malayanus (Malayan sun bear), Assam southeast of Brahmaputra River, Sichuan and Yunnan in south-central China, Burma, Thailand, Indochina, Malay Peninsula, Sumatra, Borneo;
U. ursinus (sloth bear), India, Nepal, Bangladesh, Sri Lanka.

Each of these species has often been placed in its own genus or subgenus: *Selenarctos* Heude, 1901, for *U. thibetanus; Euarctos* Gray, 1864, for *U. americanus; Ursus* Linnaeus, 1758, for *U. arctos; Thalarctos* Gray, 1825, for *U. maritimus; Helarctos* Horsfield, 1825, for *U. malayanus;* and *Melursus* Meyer, 1793, for *U. ursinus.* E. R. Hall (1981), however, placed all of these names in the synonymy of *Ursus.* The latter arrangement is supported by molecular and karyological data (O'Brien 1993; Wayne et al. 1989) and in part by the captive production of viable offspring through hybridization between several of the above species (Van Gelder 1977). Wozencraft (*in* Wilson and Reeder 1993) treated *Helarctos* and *Melursus* as valid genera.

The systematics of the brown or grizzly bear have caused considerable confusion. Old World populations long have been recognized as composing a single species, with the scientific name *U. arctos* and the general common name "brown bear." In North America the name "grizzly" is applied over most of the range, while the term "big brown bear" is often used on the coast of southern Alaska and nearby islands, where the animals average much larger than those inland. E. R. Hall (1981) listed 77 Latin names that have been used in the specific sense for different populations of the brown or grizzly bear in North America. No one now thinks that there actually are so many species, but some authorities, such as Burt and Grossenheider (1976), have recognized the North American grizzly *(U. horribilis)* and the Alaskan big brown bear *(U. middendorffi)* as species distinct from *U. arctos* of the Old World. Other authorities (Erdbrink 1953; Kurten 1973; Rausch 1953, 1963), based on limited systematic work, have referred the North American brown and grizzly to *U. arctos.* This procedure is being used by most persons now studying or writing about bears and is followed here. Kurten (1973) distinguished three North American subspecies: *U. a. middendorffi,* on Kodiak and Afognak islands; *U. a. dalli,* on the south coast of Alaska and the west coast of British Columbia; and *U. a. horribilis,* in all other parts of the range of the species. Hall (1984), however, recognized nine North American subspecies of *U. arctos.*

From *Tremarctos, Ursus* is distinguished by its masseteric fossa on the lower jaw not being divided by a bony septum into two fossae. From *Ailuropoda* it is distinguished in having an alisphenoid canal (E. R. Hall 1981). Additional information is provided separately for each species.

Ursus maritimus (polar bear)

Head and body length is 2,000–2,500 mm, tail length is 76–127 mm, and shoulder height is up to 1,600 mm. DeMaster and Stirling (1981) gave the weight as 150–300 kg for females and 300–800 kg for males. Banfield (1974), however, wrote that males usually weigh 420–500 kg. The color is often pure white following the molt but may become yellowish in the summer, probably because of oxidation by the sun. The pelage also sometimes appears gray or almost brown depending on the season and light conditions. The neck is longer than that of other bears, and the head is relatively small and flat. The forefeet are well adapted for swimming, being large and oarlike. The soles are haired, probably for insulation from the cold and traction on the ice. Females have four functional mammae (DeMaster and Stirling 1981).

The polar bear is often considered to be a marine mammal. It is distributed mainly in arctic regions around the North Pole. The southern limits of its range are determined by distribution of the pack ice. It has been recorded as far north as 88° N and as far south as the Pribilof Islands in the

Polar bears *(Ursus maritimus):* Top, photo from New York Zoological Society. Insets: A. Forefoot; B. Hind foot; photos from *Proc. Zool. Soc. London;* C. Young, 24 hours old, photo by Ernest P. Walker. Bottom, photo by Sue Ford, Washington Park Zoo, Portland, Oregon.

Bering Sea, the island of Newfoundland, the southern tip of Greenland, and Iceland. There also are permanent populations in James Bay and the southern part of Hudson Bay. Although found generally in coastal areas or on ice hundreds of kilometers from shore, individuals have wandered up to 200 km inland (Stroganov 1969).

According to DeMaster and Stirling (1981), the preferred habitat is pack ice that is subject to periodic fracturing by wind and sea currents. The refreezing of such fractures provides places where hunting by the bear is most successful. Some animals spend both winter and summer along the lower edge of the pack ice, perhaps undergoing extensive north-south migrations as this edge shifts. Others move onto land for the summer and disperse across the ice as it forms along the coast and between islands during winter. The bears of the Labrador coast sometimes move north to Baffin Island, and some individuals have traveled as far as 1,050 km, to the islands of northern Hudson Bay (Stirling and Kiliaan 1980). Individuals in the population of the Beaufort Sea off northern Alaska move several thousand kilometers annually and may use an area exceeding 500,000 sq km over a period of years (Amstrup, Garner, and Durner 1995). The population that summers along the southern shore of Hudson Bay spreads all across the partly ice-covered bay in November and returns to shore in July or August (Stirling et al. 1977). During the latter, ice-free period adult males occupy coastal areas and family groups and pregnant females move farther inland (Derocher and Stirling 1990*a*). Despite such movements, the polar bear is not a true nomad. There are a number of discrete populations, each with its own consistently used areas for feeding and breeding (Stirling, Calvert, and Andriashek 1980).

The polar bear can outrun a reindeer for short distances on land and can attain a swimming speed of about 6.5 km/hr. It swims rather high, with head and shoulders above the water. If killed in the water, it will not immediately sink. According to DeMaster and Stirling (1981), it has been reported to swim for at least 65 km across open water. It is capable of diving under the ice and surfacing in holes utilized by seals. It seems to be most active during the first third of the day and least active in the final third. From July to December in the James Bay region, when a lack of ice prevents seal hunting, *U. maritimus* spends about 87 percent of its time resting, apparently living off of stored fat (Knudsen 1978). The bears sometimes excavate depressions or complete earthen burrows on land during the summer in order to avoid the sun and keep cool (Jonkel et al. 1976).

Any individual bear may make a winter den for temporary shelter during severe weather, but only females, especially those that are pregnant, generally hibernate for lengthy periods. As with other bears, winter sleep involves a depressed respiratory rate and a slightly lowered body temperature but not deep torpor. Most pregnant females evidently do not spend the winter along the pack ice but hibernate on land from October or November to March or April. Maternal dens are usually found within 8 km of the coast, but in the southern Hudson Bay region they are concentrated 30–60 km inland. They are excavated in the snow to depths of 1–3 meters, often on a steep slope. They usually consist of a tunnel several meters long that leads to an oval chamber of about 3 cubic meters. Some dens have several rooms and corridors (DeMaster and Stirling 1981; Harington 1968; Larsen 1975; Stirling, Calvert, and Andriashek 1980; Stirling et al. 1977; Uspenski and Belikov 1976).

The polar bear feeds primarily on the ringed seal *(Phoca hispida)* (DeMaster and Stirling 1981). The bear either remains still until a seal emerges from the water or stealthily stalks its prey on the ice (Stirling 1974). It may also dig out the subnivean dens of seals to obtain the young (Stirling, Calvert, and Andriashek 1980). During summer and autumn in the southern Hudson Bay region *U. maritimus* often swims among sea birds and catches them as they sit on the water (Russell 1975). The diet also includes the carcasses of stranded marine mammals, small land mammals, reindeer, fish, and vegetation. Berries become important for some individuals during summer and autumn (Jonkel 1978).

Reported population densities range from 1/37 sq km to 1/139 sq km (DeMaster and Stirling 1981). Home ranges are not well defined but are thought to vary from 150 to 300 km in diameter and to overlap extensively (Kolenosky 1987). Although *U. maritimus* is generally solitary, large aggregations may form around a major source of food (Jonkel 1978). As many as 40 individuals have been seen at one time in the vicinity of the Churchill garbage dump, on the southern shore of Hudson Bay (Stirling et al. 1977). Of the adult males in the population that inhabits that region, more than half occur in aggregations during the ice-free season. These groups, with an average size of 4 bears, tend to occur at environmentally favorable sites and may help develop familiarity that will avoid future conflicts for resources (Derocher and Stirling 1990*b*). Wintering females evidently tolerate one another well, as dens on Wrangel Island are sometimes found at densities of one per 50 sq meters (Uspenski and Belikov 1976). High concentrations of summer dens also have been reported (Jonkel et al. 1976). Adult females with young are not subordinate to any other age or sex class but tend to avoid interaction with adult males, presumably because the latter are potential predators of the cubs (DeMaster and Stirling 1981).

Estrus lasts 3 days (Hayssen, Van Tienhoven, and Van Tienhoven 1993). The sexes usually come together only briefly during the mating season, March–June. Delayed implantation apparently extends the period of pregnancy to 195–265 days. The young are born from November to January, while the mother is in her winter den. Females give birth every 2–4 years. The number of young per litter averages about two and ranges from one to four. They weigh about 600 grams at birth and are well covered with short white fur, but their eyes are tightly closed. Upon emergence from the den in March or April the cubs weigh 10–15 kg. They usually leave the mother at 24–28 months. The age of sexual maturity averages about 5–6 years. Adult weight is attained at about 5 years by females but not until 10–11 years by males. Wild females apparently have a reduced natality rate after the age of 20. Annual adult mortality in a population is about 8–16 percent. Potential longevity in the wild is estimated at 25–30 years (DeMaster and Stirling 1981; Ramsay and Stirling 1988; Stirling, Calvert, and Andriashek 1980; Ulmer 1966; Uspenski and Belikov 1976). A female at the Detroit Zoo gave birth at 36 years and 11 months (Latinen 1987) and was still living at 45 years (Marvin L. Jones, Zoological Society of San Diego, pers. comm., 1995).

The polar bear often is considered to be dangerous to people, though usually the two species are not found in close proximity. An exception developed during the 1960s in the vicinity of the town of Churchill, on the southern shore of Hudson Bay (Stirling et al. 1977). Bears apparently increased in this area because of a decline in hunting. At the same time, more people moved in and several large garbage dumps were established. A number of persons were attacked and one was killed. Many bears were shot or translocated by government personnel. There also have been a number of confrontations in Svalbard, and one person was killed by a bear there in 1977 (Gjertz and Persen 1987).

The native peoples of the Arctic have long hunted the

polar bear for its fat and fur. Sport and commercial hunting increased in the twentieth century. Of the regularly marketed North American mammal pelts, that of *U. maritimus* is the most valuable. During the 1976–77 season 530 skins from Canada were sold at an average price of $585.22 (Deems and Pursley 1978). Some individual prime pelts have brought more than $3,000 each (Smith and Jonkel 1975). There subsequently appears to have been a decline in demand, and other parts of the polar bear, particularly the gallbladder, seem to lack the market value of those of other bears (Frampton 1995).

The use of aircraft to locate polar bears and to land trophy hunters in their vicinity developed in Alaska in the late 1940s. The annual kill by such means reached about 260 bears by 1972. In that year, however, the killing of *U. maritimus*, except for native subsistence, was prohibited by the United States Marine Mammal Protection Act. Canada and Denmark (for Greenland) also limit hunting to resident natives, and Russia and Norway (for Svalbard) provide complete protection. In 1973 these five nations drafted an agreement calling for the restriction of hunting, the protection of habitat, and cooperative research on polar bears. The agreement was ratified by the United States in 1976. The yearly worldwide kill is now estimated at around 1,000 animals. The total number of polar bears in the wild in 1993 was estimated at 21,470–28,370, and populations are generally thought to be stable or increasing. *U. maritimus*, however, may be threatened by the exploitation of oil and gas reserves in the Arctic, especially by development in the limited areas suitable for denning by pregnant females (DeMaster and Stirling 1981; Stirling and Kiliaan 1980; U.S. Fish and Wildlife Service 1980; Wiig, Born, and Garner 1995).

There now also is concern that the influx of cash from oil and gas development will stimulate increased hunting by native peoples and that such hunting, not being restricted to adult males, could damage the relatively small and vulnerable polar bear populations off northern Alaska and northwestern Canada (Amstrup, Stirling, and Lentfer 1986). Another long-term concern is that global warming, resulting from the greenhouse effect of atmospheric polluting gases, will reduce the southerly extent of sea ice and thereby deny accessibility to seals. Even now the polar bear population of southwestern Hudson Bay is showing signs of nutritional stress (Stirling and Derocher 1993). The species is classified as conservation dependent by the IUCN and is on appendix 2 of the CITES. The provisions of both CITES and the United States Marine Mammal Protection Act do allow for limited importation under certain circumstances. The U.S. Fish and Wildlife Service recently proposed allowing importation of polar bear trophies taken in accordance with carefully regulated sport-hunting programs in the Northwest Territories of Canada (Frampton 1995).

CARNIVORA; **MUSTELIDAE; Genus LONTRA**
Gray, 1843

New World River Otters

There are four species (E. R. Hall 1981; Redford and Eisenberg 1992; Van Zyll de Jong 1972, 1987):

L. canadensis, Alaska, Canada, conterminous United States;
L. longicaudis (neotropical river otter), northwestern Mexico to Uruguay and Buenos Aires Province of Argentina;
L. provocax, central and southern Chile, southern Argentina, Tierra del Fuego;
L. felina (marine otter), Pacific coast from northern Peru to Tierra del Fuego.

Lontra was considered a subgenus or synonym of *Lutra* by E. R. Hall (1981) and Jones et al. (1992) but was treated as a distinct genus by Van Zyll de Jong (1972, 1987) and Wozencraft (*in* Wilson and Reeder 1993). The basis on which various authors rejected generic status for *Lontra* was questioned by Kellnhauser (1983).

Head and body length is 460–820 mm, tail length is 300–570 mm, and weight is 3–15 kg. Males average larger than females (Van Zyll de Jong 1972). The upper parts are various shades of brown, the underparts are light brown or grayish, and the muzzle and throat may be whitish or silvery gray. The fur is short and sleek, with dense underfur overlaid by glossy guard hairs. The general morphology is much like that of *Lutra*, but in the skull the posterior palatine foramina are located more posteriorly, the vomer-ethmoid partition of the nasal cavity extends posteriorly to or behind the first upper molar tooth (in *Lutra* it extends only to between the third and fourth premolars), the first upper molar has a prominent cingulum and expanded talon (in *Lutra* the cingulum is little developed and the talon is small), and the sectorial fourth upper premolar has a talon extending more than two-thirds the length of the tooth (in *Lutra* the talon extends less than two-thirds the length). In both genera the toes are webbed to the terminal digit pads or beyond, the proximal part of the tail is broad and moderately dorsoventrally flattened, and females have four mammae, except in *Lontra provocax*, which has more than four (Van Zyll de Jong 1987).

The natural history of New World river otters, including their aquatic habits and playful behavior, is much like that of *Lutra* (see account thereof). However, there are some particularly different aspects of the ecology of *L. felina* (Estes 1986; Redford and Eisenberg 1992). That species is found largely or exclusively along the exposed seashore, though it may enter freshwater estuaries and large rivers. It stays within about 500 meters of the coast, mainly in areas characterized by rocky outcroppings, heavy seas, and strong winds. It shelters in caves that open at water level, is active mostly in the afternoon, and makes food dives of 15–45 seconds. It feeds mostly on crustaceans, mollusks, and fish. *L. provocax* also sometimes is found along rocky coastlines, whereas *L. longicaudis* depends on permanent streams or lakes with ample riparian vegetation, and shelters in a self-excavated burrow.

According to Melquist and Dronkert (1987), *L. canadensis* is found in both marine and freshwater environments and from coastal areas to high mountains. Densities appear highest in food-rich habitats such as estuaries, lower stream drainages, coastal marshes, and interconnected small lakes and swamps. Favored locations include those with riparian vegetation or rocks that can be used for dens. An individual may use numerous dens and temporary shelters in the course of a year. Beaver lodges are frequently occupied, sometimes simultaneously with the builder. Otters are mainly nocturnal and crepuscular, but daytime activity is not unusual. The diet consists primarily of fish and also includes crustaceans, reptiles, amphibians, and occasionally birds and mammals. Most reported population densities have been about 1/1–10 km of shoreline or waterway length, and home range lengths have been 4–78 km.

In Idaho, Melquist and Hornocker (1983) found *L. canadensis* to have an overall population density of 1/3.9 km of waterway. Seasonal home range length was 8–78 km,

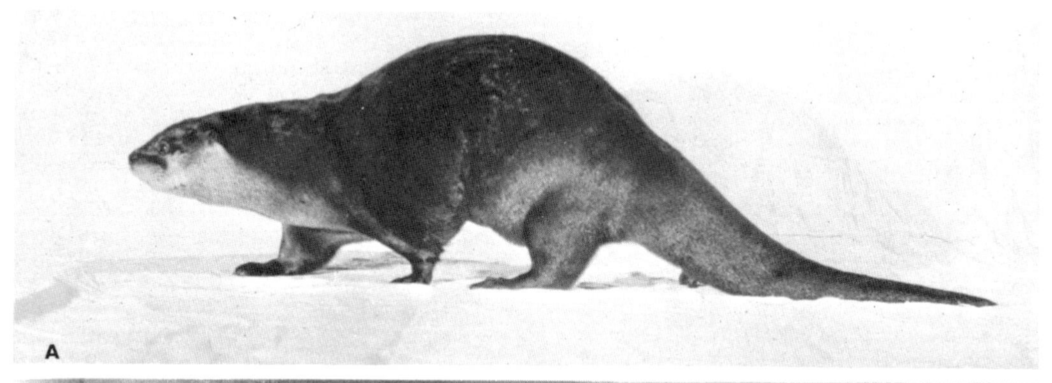

Canadian river otters *(Lontra canadensis)*, photos by Ernest P. Walker.

with males generally having larger ranges than females; however, there was extensive overlap between the ranges of both the same and opposite sexes. There was some mutual avoidance and defense of personal space but no strong territorial behavior. The basic social group consisted of an adult female and her juvenile offspring. Such families broke up before the female again gave birth, though yearlings occasionally associated with the group. Fully adult males were not observed to accompany family groups. Observations in other areas, however, suggest that although the male is excluded from the vicinity of the female when the latter has small young, he joins the family when the cubs are about 6 months old (Banfield 1974; Jackson 1961). *Lontra* has a variety of vocalizations, and like *Lutra*, it communicates through scent marking with urine, feces, and anal gland secretions (Melquist and Dronkert 1987).

In the southern part of the range of *L. felina* there seems to be a birth peak in September and October, the gestation period is 60–65 days, and litter size is two young; they are born in an earthen den or rocky crevice and stay with the female for approximately 10 months (Redford and Eisenberg 1992). Little is known about reproduction in *L. longicaudis*, but recently a captive pair in southern Brazil was observed to produce litters on 1 April 1992, 21 July 1992, and 14 February 1993; the young of the first two litters died shortly after parturition (Blacher 1994). In most populations of *L. canadensis*, of North America, there is delayed implantation of the fertilized eggs in the uterus. Mating occurs in the winter or spring, and births take place the following year, usually from January to May. The total period of pregnancy has been reported to vary from 290 to 380 days, though actual embryonic development is about 60–63 days, the same as in other kinds of river otters. Populations in southern Florida may not experience delayed implantation. The female does not excavate her own den but uses that of another animal or some natural shelter. The male does not assist in rearing the young. Litter size is one to six cubs, usually two or three. They weigh about 120–60 grams at birth, emerge from the den at about 2 months, are weaned at 5–6 months, and usually leave their mother just before she again gives birth. Both females and males attain sexual maturity at about 2 years, but males generally cannot successfully breed until 5–7 years of age. Wild individuals up to 14 years old have been taken, and captives have lived about 25 years.

As a group, river otters have suffered severely through habitat destruction, water pollution, misuse of pesticides, excessive fur trapping, and persecution as supposed predators of game and commercial fish. *L. canadensis* has disappeared or become rare throughout the conterminous United States except in the Northwest, the upper Great Lakes region, New York, New England, and the states along the

Atlantic and Gulf coasts. The southwestern subspecies *L. c. sonora* has nearly disappeared, though there have been several recent reports in Arizona (Polechla *in* Foster-Turley, Macdonald, and Mason 1990). Otters of another subspecies were released in Arizona in 1981, perhaps inadvisably considering the possibility of genetic modification of the native population. Since 1976 there also have been efforts to reintroduce otters in Colorado, Iowa, Kansas, Kentucky, Missouri, Nebraska, Oklahoma, Pennsylvania, and West Virginia (Melquist and Dronkert 1987).

The IUCN classifies *L. felina* as endangered, noting that it has declined by at least 50 percent over the past decade, and *L. provocax* as vulnerable (both are included in the genus *Lutra*). *L. longicaudis, L. provocax,* and *L. felina* are listed as endangered by the USDI and are on appendix 1 of the CITES, and *L. canadensis* is on appendix 2. There may now be fewer than 1,000 individuals of *L. felina* (Estes 1986); it has declined in Chile because of excessive hunting for its fur and in Peru because of persecution for alleged damage to prawn fisheries (Thornback and Jenkins 1982). *L. provocax* remains common at a few isolated sites in extreme southern Chile and Argentina but has disappeared from most of its range because of overhunting and habitat alteration. *L. longicaudis* is still widespread but has disappeared from the highlands of Mexico and is threatened in the rest of that country by habitat destruction and fragmentation (Foster-Turley, Macdonald, and Mason 1990).

The beautiful and durable fur of river otters is used for coat collars and trimming. During the 1976–77 trapping season 32,846 pelts of *L. canadensis* were reported taken in the United States, and the average selling price was $53.00. Respective figures for Canada that season were 19,932 pelts and $69.04 (Deems and Pursley 1978). In 1983–84 the total take was 33,135, and the average selling price was $18.71 (Novak, Obbard, et al. 1987). In the 1991–92 season 10,916 pelts were taken in the United States and sold at an average price of $22.34 (Linscombe 1994). *L. longicaudis* also was taken in large numbers for its valuable skin, with probably about 30,000 killed annually during the early 1970s in Colombia and Peru alone. Continued illegal hunting, along with habitat loss and water pollution, jeopardizes the survival of this species and *L. provocax* (Mason and Macdonald 1986; Thornback and Jenkins 1982).

CARNIVORA; MUSTELIDAE; **Genus ENHYDRA**
Fleming, 1822

Sea Otter

The single species, *E. lutris,* was originally found in coastal waters off Hokkaido, Sakhalin, Kamchatka, the Commander Islands, the Pribilof Islands, the Aleutians, southern Alaska, British Columbia, Washington, Oregon, California, and western Baja California (Estes 1980). There are three subspecies: *E. l. lutris,* Hokkaido to the Commander Islands; *E. l. kenyoni,* Aleutians to Oregon; and *E. l. nereis,* California and Baja California (Wilson et al. 1991). The recent confirmation of the validity of these subspecies is important,

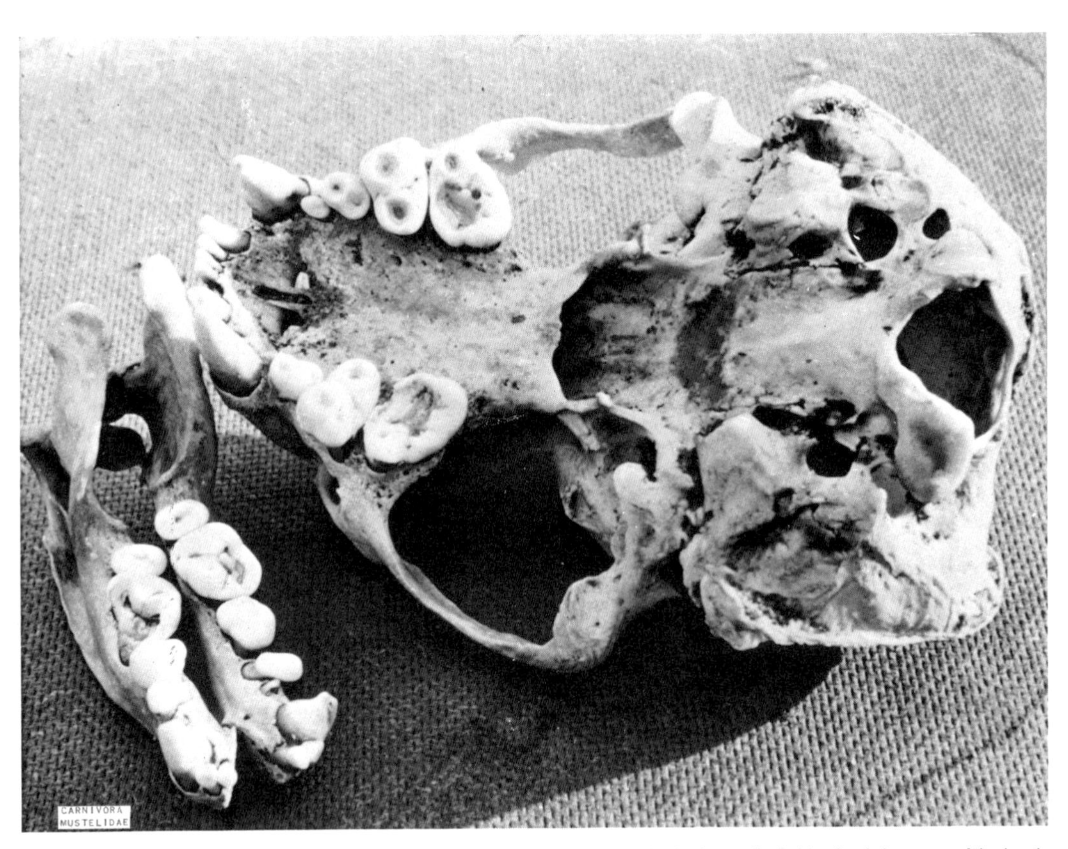

Skull and lower jaw of a sea otter *(Enhydra lutris)*, showing the cavities that develop in the teeth of old animals because of the hard, rough materials that they eat, photo by H. Robert Krear.

both for consideration in reintroduction planning and to dispel claims that certain populations of bioconservation concern are of no systematic significance.

Head and body length is usually 1,000–1,200 mm and tail length is 250–370 mm. Males weigh 22–45 kg and females, 15–32 kg (Estes 1980). The color varies from reddish brown to dark brown, almost black, except for the gray or creamy head, throat, and chest. Albinistic individuals are rare. The head is large and blunt, the neck is short and thick, and the legs and tail are short. The ears are short, thickened, pointed, and valvelike. The hind feet are webbed and flattened into broad flippers; the forefeet are small and have retractile claws. *Enhydra* is the only carnivore with only four incisor teeth in the lower jaw. The molars are broad, flat, and well adapted to crushing the shells of such prey as crustaceans, snails, mussels, and sea urchins. Unlike most mustelids, the sea otter lacks anal scent glands. Females have two abdominal mammae (Estes 1980).

The sea otter differs from most marine mammals in that it lacks an insulating subcutaneous layer of fat. For protection against the cold water it depends entirely on a layer of air trapped among its long, soft fibers of hair. If the hair becomes soiled, as if by oil, the insulating qualities are lost and the otter may perish. The underfur, about 25 mm long, is the densest mammalian fur, averaging about 100,000 hairs per sq cm (Rotterman and Simon-Jackson 1988). It is protected by a scant coat of guard hairs.

Although the sea otter is a marine mammal, it rarely ventures more than 1 km from shore. According to Estes (1980), it forages in both rocky and soft-sediment communities on or near the ocean floor. Off California *Enhydra* seldom enters water of greater depth than 20 meters, but in the Aleutians it commonly forages at depths of 40 meters or more; the maximum confirmed depth of a dive was 97 meters. The usual period of submergence is 52–90 seconds, and the longest on record is 4 minutes and 25 seconds. The sea otter is capable of spending its entire life at sea but sometimes rests on rocks near the water. Such hauling-out behavior is more common in the Alaskan population than in that of California. *Enhydra* walks awkwardly on land. When supine on the surface of the water, it moves by paddling with the hind limbs and sculling with the tail. For rapid swimming and diving it uses dorsoventral undulations of the body. It can attain velocities of up to 1.5 km/hr on the surface and 9 km/hr for short distances underwater. The sea otter is generally diurnal, with crepuscular peaks and a midday period of rest (Riedman and Estes 1990). It often spends the night in a kelp bed, lying under strands of kelp to avoid drifting while sleeping. It sometimes sleeps with a forepaw over the eyes. Daily movements usually extend over a few kilometers, and there may be local seasonal movements but no extensive migrations (Riedman and Estes 1990).

The diet consists mainly of slow-moving fish and marine invertebrates, such as sea urchins, abalone, crabs, and mollusks (Estes 1980). Prey is usually captured with the forepaws, not the jaws. *Enhydra* floats on its back while eating and uses its chest as a "lunch counter." It is one of the few mammals known to use a tool. While floating on its back, it places a rock on its chest and then employs the rock as an anvil for breaking the shells of mussels, clams, and large sea snails in order to obtain the soft internal parts. This activity is most frequent in the population off California, and recent research there has shown considerable variation, such as using the rock as a hammer or using one rock as a hammer and another as an anvil (Riedman and Estes 1990). The sea otter requires a great deal of food: it must eat 20–25 percent of its body weight every day. It obtains about 23 percent of its water needs from drinking sea water and most of the rest from its food.

According to Estes (1980), *Enhydra* is basically solitary but sometimes rests in concentrations of as many as 2,000 individuals. Males and females usually come together only briefly for courtship and mating. At most times there is sexual segregation, with males and females occupying separate sections of coastline. Males usually occur at higher densities. Recent studies (Garshelis and Garshelis 1984; Garshelis, Johnson, and Garshelis 1984; Jameson 1989; Loughlin 1980; Rotterman and Simon-Jackson 1988) indicate that during the breeding season (which may be for most of the year) some males move into the areas occupied by females and establish territories. Such behavior has been documented in both Alaska and California, and males have been observed to return to the same place for up to seven years. The most favorable territories—those that seem to attract the most females—are characterized by availability of food; density of canopy-forming kelp, to which the animals can attach themselves for secure resting; and associated shoreline features that provide a degree of shelter from the open sea (Riedman and Estes 1990). The boundaries of these territories are vigorously patrolled and intruding males are repulsed, but serious fighting is rare. The owner seeks to mate with any female that enters, though sometimes a pair bond is formed for a few days or weeks. Male territories are usually about 20–50 ha. and are smaller than female home ranges in the same area. Annual movements of both sexes frequently cover 50–100 km, considering foraging, breeding, and the passage of males between their territories and the all-male areas. McShane et al. (1995) described 10 vocalizations of *Enhydra*, including screams of distress, heard especially when mothers and young were separated, and coos, heard mostly when individuals were content or in familiar company.

Reproductive data have been summarized by Estes (1980) and Riedman and Estes (1990). Breeding occurs throughout the year, but births peak in late May and June in the Aleutians and from January to March off California; in the latter area there apparently is a secondary peak in late summer and early autumn. Males may mate with more than one female during the season. Pair bonding, and presumably estrus, lasts about 3 days. Females are capable of giving birth every year but usually do so at greater intervals. If a litter does not survive, the female may experience a postpartum estrus. Females are known to adopt and nurse orphaned pups. Reports of the period of pregnancy range from 4 to 12 months, and delayed implantation is probably involved. Estimates of the period of actual implanted gestation vary from about 4.0 to 5.5 months. Births probably occur most often in the water. There is normally a single offspring. About 2 percent of births are multiple, but only one young can be successfully reared. The pup weighs 1.4–2.3 kg at birth. While still small it is carried, nursed, and groomed on the mother's chest as the mother swims on her back. The pup begins to dive in the second month of life. It may take some solid food shortly after birth but may nurse almost until it attains adult size. The period of dependency on the mother is thought to be about 5–8 months. Females become sexually mature at about 4 years. Males are capable of mating at 5–6 years but usually do not become active breeders until several more years have passed. Wendell, Ames, and Hardy (1984) concluded that the reproductive cycle of the California population is shorter than elsewhere, with some females giving birth each year. It now is known that Alaskan females also are capable of annual reproduction (Garshelis, Johnson, and Garshelis 1984). According to Rotterman and Simon-Jackson (1988), a captive male fathered young when at least 19 years old, and maximum estimated longevity for wild females is 23 years. Data from Alaska indicate that sea otter populations have the potential to increase by about 20 percent annually, but the population off California has tended to increase by only 5 percent a year, probably because of high preweaning mortality of young that may be associated with pollutants imparted through lactation (Estes 1990; Riedman et al. 1994).

The fur of the sea otter may be the most valuable of any mammal's. During the 1880s prices on the London market ranged from $105 to $165 per skin. By 1903, when the species had become scarce, large, high-quality skins sold for up to $1,125. Pelts taken in Alaska in the late 1960s, during a brief reopening of commercial activity, sold for an average of $280 each (Kenyon 1969).

Estimates of the original numbers of *Enhydra* are 150,000–300,000 (Riedman and Estes 1990). Intensive exploitation of the genus was begun by the Russians in 1741. Hunting was uncontrolled until 1799, when some conservation measures were established. Unregulated killing resumed in 1867, when Alaska was purchased by the United States. By 1911, when the sea otter was protected by a treaty among the United States, Russia, Japan, and Great Britain, probably only 1,000–2,000 of the animals survived worldwide (Kenyon 1969). Under protection of the treaty, state and national laws, and finally the United States Marine Mammal Protection Act of 1972, the sea otter has steadily increased in numbers and distribution. There are now probably 100,000–150,000 individuals in the major populations off southwestern and south-central Alaska and another 17,000 off Kamchatka and the Kuril and Commander islands in Russia. Alaskan populations are subject to limited killing for native subsistence purposes, may come into conflict with shellfisheries, and are potentially jeopardized by oil spills. Reintroduced populations (from Alaskan stock) apparently have been established off southeastern Alaska (now numbering 4,500 animals), Vancouver Island (about 350), and Washington (280). Reintroduced groups off Oregon and in the Pribilof Islands do not seem to have done well and have all but disappeared (Estes 1980; Jameson 1993; Jameson et al. 1982; Riedman and Estes 1990; Rotterman and Simon-Jackson 1988).

The magnitude of the threat posed by oil spills from damaged tanker ships was tragically demonstrated by the *Exxon Valdez* disaster in 1989. As many as 5,000 sea otters are estimated to have been killed directly (*Oryx* 26 [1992]: 195). Hence, a significant part of the entire world's population of *Enhydra* was destroyed in this single incident. More than 1,000 dead or dying sea otters were actually recovered. However, an intensive effort was made by the U.S. Fish and Wildlife Service and cooperating organizations to save as many of the affected animals as possible. Several hundred were rescued and treated, and 197 were released back into the wild (Bayha and Kormendy 1990).

The southern sea otter (subspecies *E. l. nereis*), which originally ranged from Baja California to at least Oregon was generally considered extinct by 1920. Apparently, however, a group of 50–100 individuals survived off central California in the vicinity of Monterey. In 1938 the presence of this population became generally known. By the 1970s it had grown to include about 1,800 animals, but subsequently numbers stabilized or even declined. The sea otter now regularly occurs along about 350 km of the central California coast, and there have been scattered reports of individuals from southern California and northern Baja California. As this population increased, there was concern that stocks of abalone and other shellfish were being depleted. Some parties with a commercial or recreational interest in these stocks have advocated control of the sea otter population, and there have been cases of illegal killing. Some otters also are being drowned accidentally in fishing nets. Another fear is that an oil spill, associated with either the extensive tanker traffic in the area or offshore drilling, could devastate the population (Armstrong 1979; Carey 1987; Estes and VanBlaricom 1985; Leatherwood, Harrington-Coulombe, and Hubbs 1978; U.S. Fish and Wildlife Service 1980).

There has been concern that the genetic viability of *E. l. nereis* was severely reduced when it approached extinction earlier in the century, but Ralls, Ballou, and Brownell (1983) calculated that the existing population should theoretically retain about 77 percent of the original diversity and that transplanted colonies should also be viable. An effort to establish such a colony was started in 1987 when 63 otters taken from the main California population were released around San Nicolas Island (Brownell and Rathbun 1988). By June of 1990, 137 animals had been brought there, but only 15 were known to have remained in the area (Riedman and Estes 1990) and the experiment was called unsuccessful (*Oryx* 28 [1994]: 95). The southern sea otter is listed as threatened by the USDI and is on appendix 1 of the CITES (other subspecies of *E. lutris* are on appendix 2).

World Distribution of Marine Mammals

For maximum usefulness, it has been necessary to devise the simplest practicable outline of the approximate distribution of the genera in the sequence used in the text. The tabulation should be regarded as an index guide to groups of marine mammals or to geographic regions. At the same time it gives a good overall picture of the general distribution of marine mammals. Generally, seals, sea lions, walruses, fresh-water dolphins, sirenians, otters, and the polar bear frequent coasts and adjacent waters, or rivers, of the lands for which they are recorded. Whales, porpoises, and oceanic dolphins are designated by the water areas they inhabit.

The major geographic distribution of the genera of Recent marine mammals that appears in the tabulation is designed to show their natural distribution at the present time or within comparatively recent times. Most of the animals occupy only a portion of the geographic region that appears at the head of the column. Some are limited to the tropical regions, others to temperate zones, and still others to the colder areas. Also, many restricted ranges cannot be designated either by letters to show the general area or by footnotes because of limited space on the tabulation. *It therefore should not be assumed that a mark indicating that an animal occurs within a geographic region implies that it inhabits that entire area.* For more detailed outlines of the ranges of the respective genera, it is necessary to consult the generic texts.

Explanation of Geographic Column Headings

Europe and Asia constitute a single land mass, but this land mass comprises widely different types of zoogeographic areas created by high mountain ranges, plateaus, latitudes, and prevailing winds. The general distribution of Recent marine mammals can be shown much more accurately by two columns, headed "Europe" and "Asia," than by a single column headed "Eurasia."

Most islands are included with the major land masses nearby unless otherwise specified.

With Europe are included the British Isles and other adjacent islands, including those in the Arctic.

With Asia are included the Japanese Islands, Taiwan, Hainan, Sri Lanka, and other adjacent islands, including those in the Arctic.

With North America are included Mexico and Central America south to Panama, adjacent islands, the Aleutian chain, the islands in the arctic region, and Greenland but not the West Indies.

With South America are included Trinidad, the Netherlands Antilles, and other small adjacent islands but not the Falkland and Galapagos Islands unless named in footnotes.

With Africa are included only Zanzibar Island and small islands close to the continent but not the Cape Verde or Canary Islands.

The island groups treated separately are:

Southeastern Asian islands, in which are included the Andamans, the Nicobars, the Mentawais, Sumatra, Java, the Lesser Sundas, Borneo, Sulawesi, the Moluccas, and the many other adjacent small islands;

New Guinea and small adjacent islands;

the Australian region, in which are included Australia, Tasmania, and adjacent small islands;

the Philippine Islands and small adjacent islands;

the West Indies;

Madagascar and small adjacent islands;

other islands that have only one or a few forms of marine mammals and are named in footnotes.

Footnotes indicate the major easily definable deviations from the distribution indicated in the tables.

Symbols

†	The mammals are extinct.
■	The mammals occur on or adjacent to the land or in the water area.
N	Northern portion
S	Southern portion
E	Eastern portion
W	Western portion
Ne	Northeastern portion
Se	Southeastern portion
Sw	Southwestern portion
Nw	Northwestern portion
C	Central portion

Examples: "N, C" = northern and central; "Nc" = north-central. Numerals refer to footnotes indicating clearly defined limited ranges within the general area.

Genera of Recent Mammals	page	North America	West Indies	South America	Madagascar	Africa	Europe	Asia	Southeast Asia Islands	Philippine Islands	New Guinea	Australian Region	Antarctic Region	Arctic Region	Atlantic Ocean	Indian Ocean	Pacific Ocean
PINNIPEDIA OTARIIDAE																	
Callorhinus	68	■W						■Ne									■N
Arctocephalus	72	■Wc		■1		■S						■S	■		■S	■S	■S
Zalophus	80	■Wc		■2				■3									
Phocarctos	83																■4
Neophoca	85											■S					
Otaria	87			■5													
Eumetopias	88	■W						■Ne									
PINNIPEDIA ODOBENIDAE																	
Odobenus	90													■	■N		■N
PINNIPEDIA PHOCIDAE																	
Monachus	97	■Se	■			■N	■S	■Sw									■6
Lobodon	99			■S		■S						■S	■				■4
Hydrurga	101			■S		■S						■S	■		■S	■S	■S
Leptonychotes	102			■S		■							■		■S	■S	■S
Ommatophoca	103												■				
Mirounga	105	■W		■S									■		■S	■S	■
Erignathus	109	■N					■N	■N						■	■N		■N
Cystophora	110	■N					■N							■	■N		
Halichoerus	112	■Ne					■N								■N		
Phoca	114	■7					■7	■8						■	■N		■N
CETACEA PLATANISTIDAE																	
Platanista	128							■Sc									
CETACEA LIPOTIDAE																	
Lipotes	130							■E									
CETACEA PONTOPORIIDAE																	
Pontoporia	131			■Se													
CETACEA INIIDAE																	
Inia	132			■N,C													
CETACEA MONODONTIDAE																	
Delphinapterus	135													■	■N		■N
Monodon	137													■			
CETACEA PHOCOENIDAE																	
Phocoena	139	■7		■7			■7	■7									
Neophocaena	141							■S,E	■								
Australophocaena	142			■7											■9	■10	■4
Phocoenoides	143																■N
CETACEA DELPHINIDAE																	
Steno	145														■	■	■
Sousa	148					■		■S	■			■					
Sotalia	148			■E													
Lagenorhynchus	151														■	■	■
Grampus	153														■	■	■
Tursiops	153														■	■	■

1. And Galapagos and Falkland Islands. 2. Galapagos Islands only. 3. Japan and Korea only. 4. New Zealand region. 5. And Falkland Islands. 6. Hawaii only. 7. Coastal waters. 8. Northern and eastern coastal waters. 9. Falkland Islands and South Georgia. 10. Kerguelen Islands.

Genera of Recent Mammals	page	North America	West Indies	South America	Madagascar	Africa	Europe	Asia	Southeast Asia Islands	Philippine Islands	New Guinea	Australian Region	Antarctic Region	Arctic Region	Atlantic Ocean	Indian Ocean	Pacific Ocean
CETACEA DELPHINIDAE Continued																	
Stenella	157														■	■	■
Delphinus	160														■	■	■
Lagenodelphis	161														■	■	■
Lissodelphis	163														■S	■S	■
Orcaella	164							■Se	■		■	■N					
Cephalorhynchus	165			■S		■Sw									■1	■2	■3
Peponocephala	167														■	■	■
Feresa	168														■	■	■
Pseudorca	169														■	■	■
Orcinus	170												■	■	■	■	■
Globicephala	172														■	■	■
CETACEA ZIPHIIDAE																	
Berardius	174												■		■S	■S	■
Ziphius	175														■	■	■
Tasmacetus	176												■		■Sw		■S
Hyperoodon	177														■	■	■
Indopacetus	179															■	■
Mesoplodon	179														■	■	■
CETACEA PHYSETERIDAE																	
Kogia	181														■	■	■
Physeter	182												■	■	■	■	■
CETACEA ESCHRICHTIIDAE																	
Eschrichtius	187													■	■4		■N
CETACEA NEOBALAENIDAE																	
Caperea	190														■S	■S	■S
CETACEA BALAENIDAE																	
Eubalaena	192														■	■	■
Balaena	195													■	■N		■N
CETACEA BALAENOPTERIDAE																	
Balaenoptera	200												■	■	■	■	■
Megaptera	207												■	■	■	■	■
SIRENIA DUGONGIDAE																	
Dugong	211					■E		■S	■	■	■	■				■	■5
Hydrodamalis†	214							■6									
SIRENIA TRICHECHIDAE																	
Trichechus	215	■Se	■	■		■W											
CARNIVORA URSIDAE																	
Ursus	221	■				■Nw	■	■	■					■			
CARNIVORA MUSTELIDAE																	
Lontra	224	■		■													
Enhydra	226	■W						■Ne									■N

1. Falkland Islands and South Georgia. 2. Kerguelen Islands. 3. New Zealand region. 4. No longer present. 5. East to Carolines and New Hebrides. 6. Commander Islands only.

Appendix

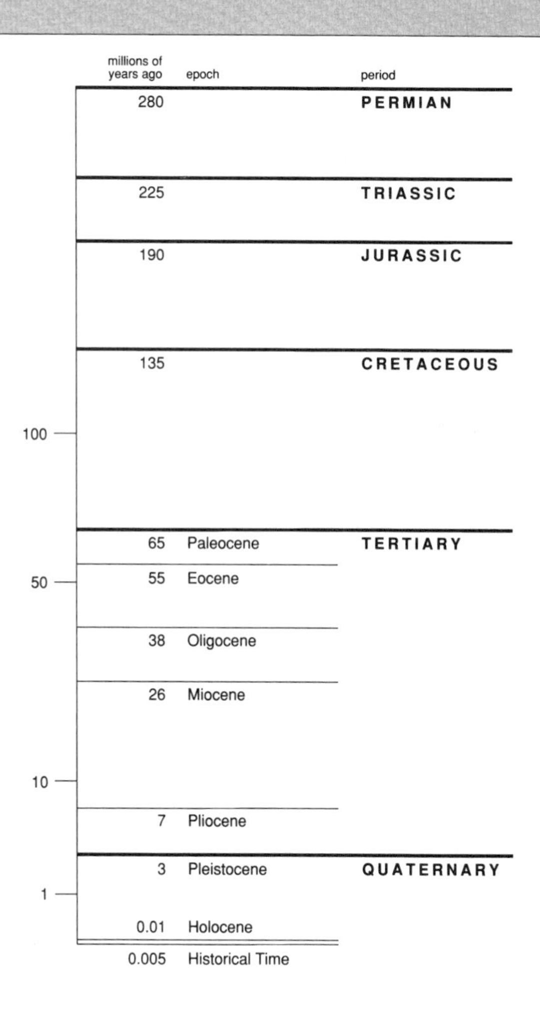
GEOLOGICAL TIME
millions of years ago
epoch
period
280
PERMIAN
225
TRIASSIC
190
JURASSIC
135
CRETACEOUS
100
65
Paleocene
TERTIARY
50
55
Eocene
38
Oligocene
26
Miocene
10
7
Pliocene
3
Pleistocene
QUATERNARY
1
0.01
Holocene
0.005
Historical Time

LENGTH

scales for comparison of metric and U.S. units of measurement

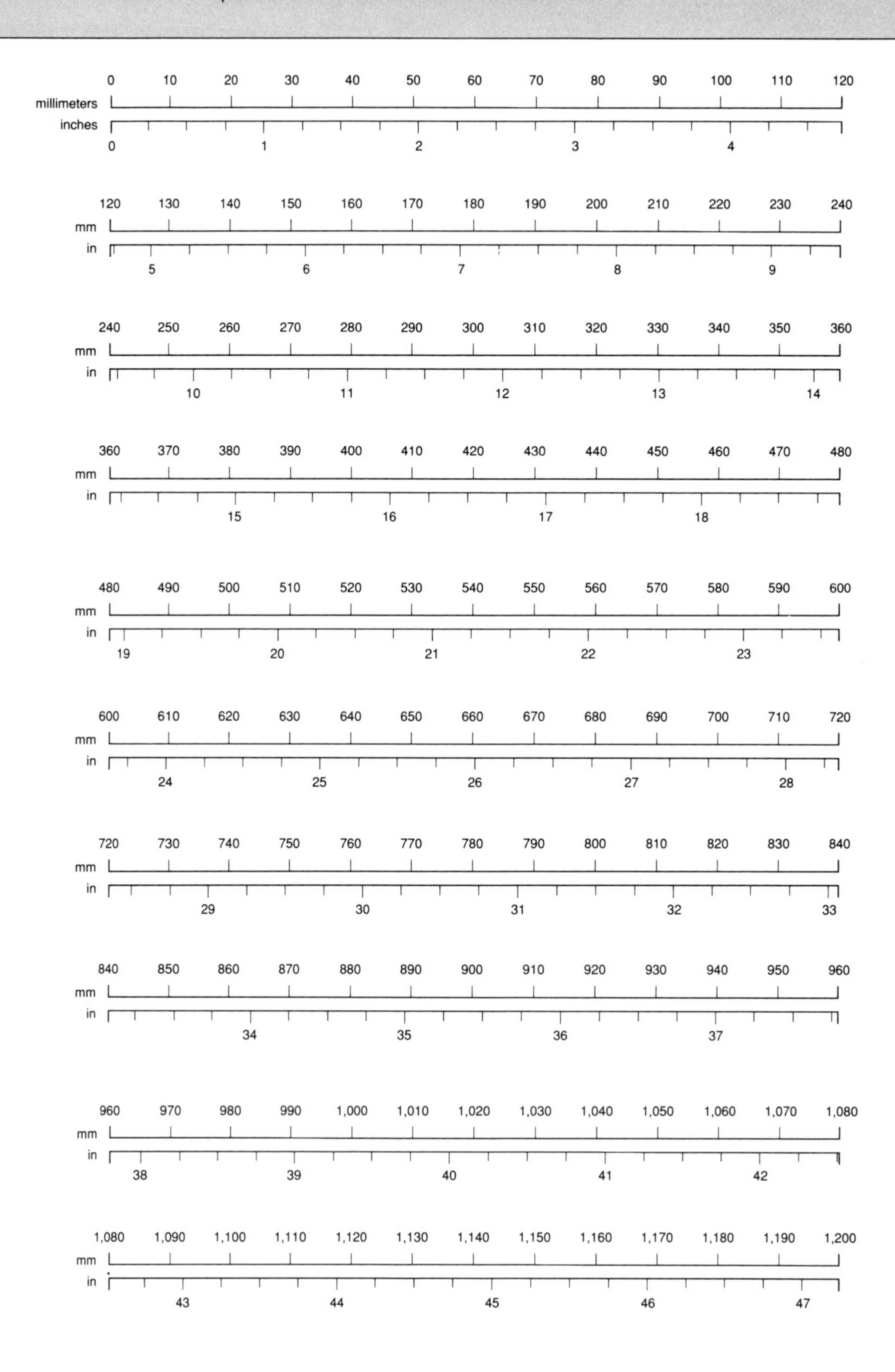

LENGTH

scales for comparison of metric and U.S. units of measurement

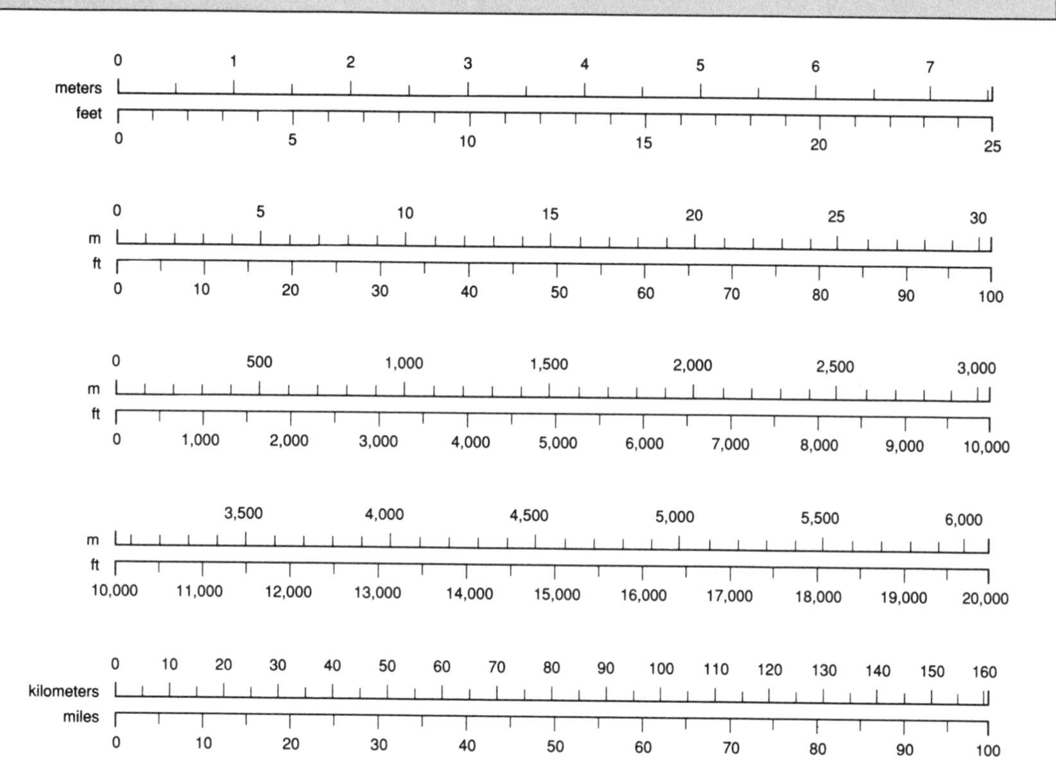

CONVERSION TABLES

	U.S. to Metric *to convert*	*multiply by*	**Metric to U.S.** *to convert*	*multiply by*
LENGTH	in. to mm.	25.4	mm. to in.	0.039
	in. to cm.	2.54	cm. to in.	0.394
	ft. to m.	0.305	m. to ft.	3.281
	yd. to m.	0.914	m. to yd.	1.094
	mi. to km.	1.609	km. to mi.	0.621
AREA	sq. in. to sq. cm.	6.452	sq. cm. to sq. in.	0.155
	sq. ft. to sq. m.	0.093	sq. m. to sq. ft.	10.764
	sq. yd. to sq. m.	0.836	sq. m. to sq. yd.	1.196
	sq. mi. to ha.	258.999	ha. to sq. mi.	0.004
VOLUME	cu. in. to cc.	16.387	cc. to cu. in.	0.061
	cu. ft. to cu. m.	0.028	cu. m. to cu. ft.	35.315
	cu. yd. to cu. m.	0.765	cu. m. to cu. yd.	1.308
CAPACITY (liquid)	fl. oz. to liter	0.03	liter to fl. oz.	33.815
	qt. to liter	0.946	liter to qt.	1.057
	gal. to liter	3.785	liter to gal.	0.264
MASS (weight)	oz. avdp. to g.	28.35	g. to oz. avdp.	0.035
	lb. avdp. to kg.	0.454	kg. to lb. avdp.	2.205
	ton to t.	0.907	t. to ton	1.102
	l. t. to t.	1.016	t. to l. t.	0.984

Abbreviations

avdp.	avoirdupois
cc.	cubic centimeter(s)
cm.	centimeter(s)
cu.	cubic
ft.	foot, feet
g.	gram(s)
gal.	gallon(s)
ha.	hectare(s)
in.	inch(es)
kg.	kilogram(s)
lb.	pound(s)
l. t.	long ton(s)
m.	meter(s)
mi.	mile(s)
mm.	millimeter(s)
oz.	ounce(s)
qt.	quart(s)
sq.	square
t.	metric ton(s)
yd.	yard(s)

WEIGHT

scales for comparison of metric and U.S. units of measurement

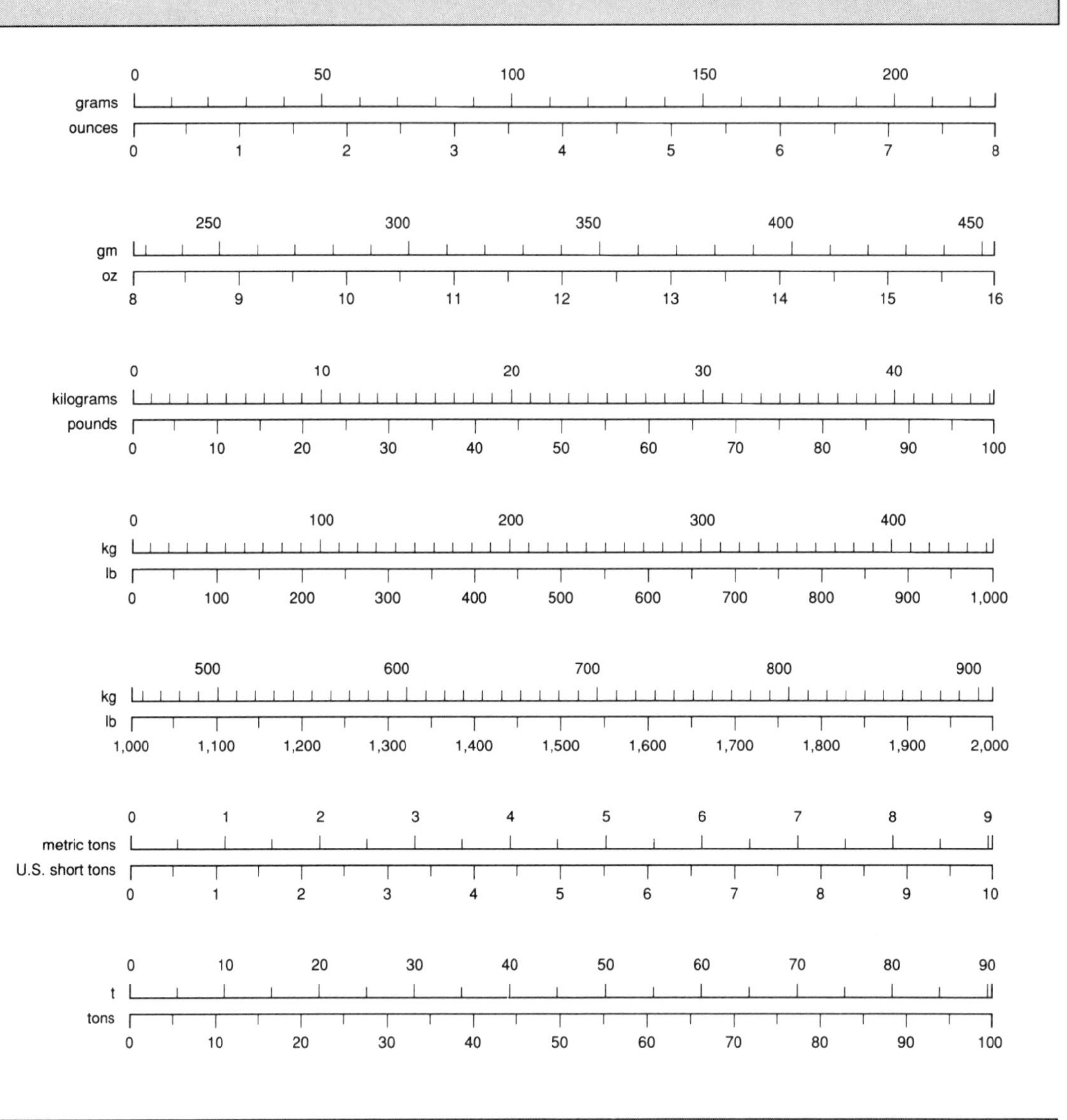

TEMPERATURE

scales for comparison of metric and U.S. units of measurement

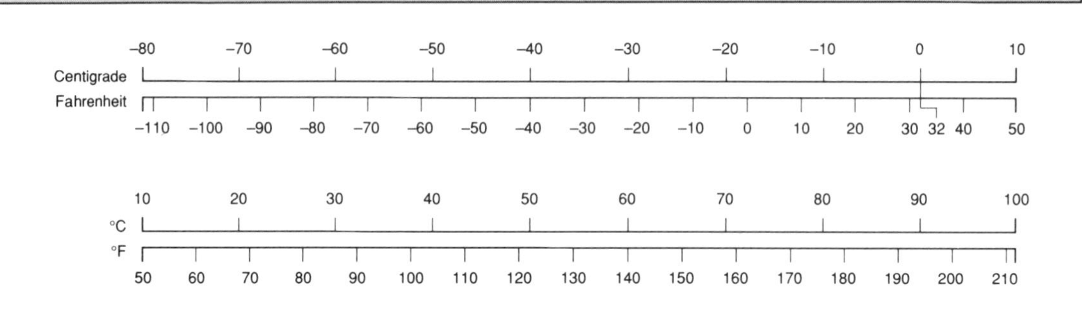

AREA

HECTARE
10,000.0 square meters
107,639.1 square feet

ACRE
4,046.86 square meters
43,560.0 square feet

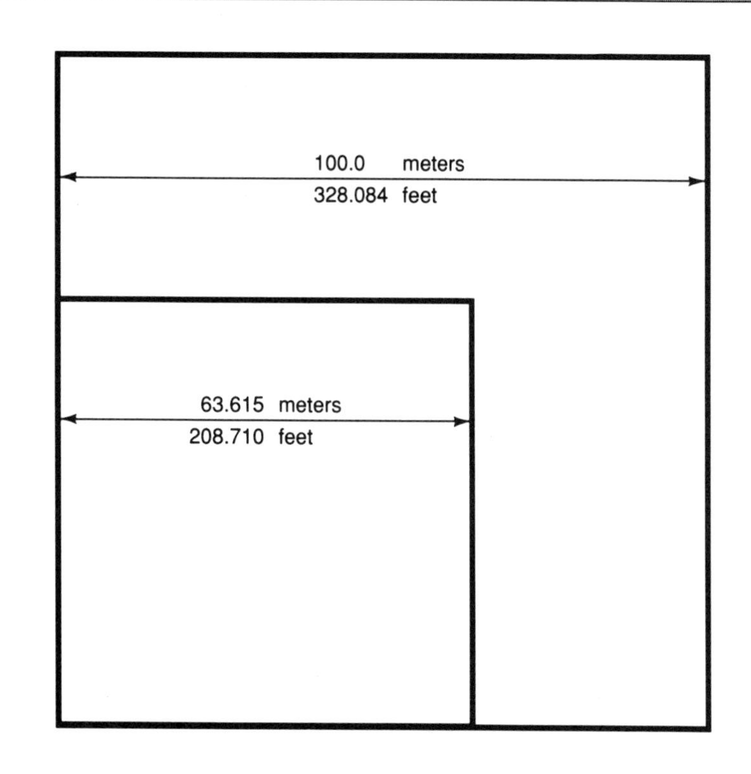

Literature Cited

A

Ackerman, B. B. 1995. Aerial surveys of manatees: a summary and progress report. *In* O'Shea, Ackerman, and Percival (1995), 13–33.

Adler, H. E., and L. L. Adler. 1978. What can dolphins *(Tursiops truncatus)* learn by observation? Cetology, no. 30, 10 pp.

Aguilar, A. 1986. A review of old Basque whaling and its effect on the right whales *(Eubalaena glacialis)* of the North Atlantic. *In* Brownell, Best, and Prescott (1986), 191–99.

Allen, G. M. 1942. Extinct and vanishing mammals of the Western Hemisphere with the marine species of all the oceans. Spec. Publ. Amer. Comm. Internatl. Wildl. Protection, no. 11, xv + 620 pp.

Allen, S. G., S. C. Peaslee, and H. R. Huber. 1989. Colonization by northern elephant seals of the Point Reyes Peninsula, California. Mar. Mamm. Sci. 5:298–302.

Alverson, D. L. 1992. A review of commercial fisheries and the Steller sea lion *(Eumetopias jubatus)*: the conflict arena. Rev. Aquat. Sci. 6:203–56.

Amano, M., and T. Kuramochi. 1992. Segregative migration of Dall's porpoise *(Phocoenoides dalli)* in the Sea of Japan and Sea of Okhotsk. Mar. Mamm. Sci. 8:143–51.

Amano, M., N. Miyazaki, and K. Kureha. 1992. A morphological comparison of skulls of the finless porpoise *Neophocaena phocaenoides* from the Indian Ocean, Yangtze River, and Japanese waters. J. Mamm. Soc. Japan 17:59–69.

Amstrup, S. C., G. W. Garner, and G. M. Durner. 1995. Polar bears in Alaska. *In* LaRoe et al. (1995), 351–53.

Amstrup, S. C., I. Stirling, and J. W. Lentfer. 1986. Past and present status of polar bears in Alaska. Wildl. Soc. Bull. 14:241–54.

Andersen, S. 1976. The taming and training of the harbor porpoise *Phocoena phocoena*. Cetology, no. 24, 9 pp.

Anderson, P. K. 1979. Dugong behavior: on being a marine mammalian grazer. Biologist 61:113–44.

———. 1982*a*. Studies of dugongs at Shark Bay, Western Australia. I. Analysis of population size, composition, dispersion, and habitat use on the basis of aerial survey. Austral. Wildl. Res. 9:69–84.

———. 1982*b*. Studies of dugongs at Shark Bay, Western Australia. II. Surface and subsurface observations. Austral. Wildl. Res. 9:85–99.

———. 1984*a*. Dugong. *In* Macdonald (1984), 298–99.

———. 1984*b*. Suckling in *Dugong dugon*. J. Mamm. 65:510–11.

———. 1986. Dugongs of Shark Bay, Australia—seasonal migration, water temperature, and forage. Natl. Geogr. Res. 2:473–90.

Anderson, P. K., and R. M. R. Barclay. 1995. Acoustic signals of solitary dugongs: physical characteristics and behavioral correlates. J. Mamm. 76:1226–37.

Anderson, P. K., and R. I. T. Prince. 1985. Predation on dugongs: attacks by killer whales. J. Mamm. 66:554–56.

Armstrong, J. J. 1979. The California sea otter: emerging conflicts in resource management. San Diego Law Rev. 16:249–85.

Arnason, U. 1974. Comparative chromosome studies in Pinnipedia. Hereditas 76: 179–225.

Arnason, U., and P. B. Best. 1991. Phylogenetic relationships within the Mysticeti (whalebone whales) based upon studies of highly repetitive DNA in all extant species. Hereditas 114:263–69.

Arnason, U., and A. Gullberg. 1993. Comparison between the complete mtDNA sequences of the blue and the fin whale, two species that can hybridize in nature. J. Molecular Evol. 37:312–22.

———. 1994. Relationship of baleen whales established by cytochrome *b* gene sequence comparison. Nature 367:726–28.

Arnason, U., M. Höglund, and B. Widegren. 1984. Conservation of highly repetitive DNA in cetaceans. Chromosoma 89:238–42.

Arnason, U., and C. Ledje. 1993. The use of highly repetitive DNA for resolving cetacean and pinniped phylogenies. *In* Szalay, Novacek, and McKenna (1993*b*), 74–80.

Arnason, U., R. Spilliaert, A. Pálsdóttir, and A. Arnason. 1991. Molecular identification of hybrids between the two largest whale species, the blue whale *(Balaenoptera musculus)* and the fin whale *(B. physalus)*. Hereditas 115:183–89.

Arnold, P., and G. Heinsohn. 1996. Phylogenetic status of the Irrawaddy dolphin *Orcaella brevirostris* (Owen in Gray): a cladistic analysis. Mem. Queensland Mus. 39: 141–204.

Arnold, P., H. Marsh, and G. Heinsohn. 1987. The occurrence of two forms of minke whale in east Australian waters with a description of external characters and skeleton of the diminutive or dwarf. Sci. Rept. Whales Res. Inst. 38:1–46.

Asper, E. D., W. G. Young, and M. T. Walsh. 1988. Observations on the birth and development of a captive-born killer whale. Internatl. Zoo Yearbook 27:295–304.

Au, D. W. K., and W. L. Perryman. 1985. Dolphin habitats in the eastern tropical Pacific. Fishery Bull. 83:623–43.

B

Baird, R. W., E. L. Walters, and P. J. Stacey. 1993. Status of the bottlenose dolphin, *Tursiops truncatus*, with special reference to Canada. Can. Field-Nat. 107:466–80.

Baker, A. N. 1978. The status of Hector's dolphin, *Cephalorhynchus hectori* (Van Beneden), in New Zealand waters. Rept. Internatl. Whaling Comm. 28:331–34.

———. 1985. Pygmy right whale—*Caperea marginata*. *In* Ridgway and Harrison (1985), 345–54.

Baker, C. S., and L. M. Herman. 1984. Aggressive behavior between humpback whales

(Megaptera novaeangliae) wintering in Hawaiian waters. Can. J. Zool. 62:1922–37.

Baker, C. S., L. M. Herman, A. Perry, W. S. Lawton, J. M. Straley, A. A. Wolman, G. D. Kaufman, H. E. Winn, J. D. Hall, J. M. Reinke, and J. Östman. 1986. Migratory movement and population structure of humpback whales *(Megaptera novaeangliae)* in the central and eastern North Pacific. Mar. Ecol. Prog. Ser. 31:105–19.

Baker, R. C., F. Wilke, and C. H. Baltzo. 1970. The northern fur seal. U.S. Dept. Interior Circ., no. 336, iii + 19 pp.

Baker, R. H. 1977. Mammals of the Chihuahuan Desert region—future prospects. *In* Wauer, R. H., and D. H. Riskind, eds., Transactions of the symposium on the biological resources of the Chihuahuan Desert region, United States and Mexico, U.S. Natl. Park Serv. Trans. Proc. Ser., no. 3, 221–25.

Balcomb, K. C., III. 1989. Baird's beaked whale—*Berardius bairdii* Stejneger, 1883: Arnoux's beaked whale—*Berardius arnuxii* Duvernoy, 1851. *In* Ridgway and Harrison (1989), 261–88.

Balcomb, K. C., III, and M. A. Bigg. 1986. Population biology of the three resident killer whale pods in Puget Sound and off southern Vancouver Island. *In* Kirkevold and Lockard (1986), 85–95.

Baldwin, R., and V. G. Cockcroft. 1995. Is the world's second-largest population of dugongs safe? Sirenews 24:11–13.

Ballance, L. T. 1990. Residence patterns, group organization, and surfacing associations of bottlenose dolphins in Kino Bay, Gulf of California, Mexico. *In* Leatherwood and Reeves (1990), 267–83.

Banfield, A. W. F. 1974. The mammals of Canada. Univ. Toronto Press, xxv + 438 pp.

Banks, R. C., and R. L. Brownell. 1969. Taxonomy of the common dolphins of the eastern Pacific Ocean. J. Mamm. 50:262–71.

Bannister, J. L., J. H. Calaby, L. J. Dawson, J. K. Ling, J. A. Mahoney, G. M. McKay, B. J. Richardson, W. D. L. Ride, and D. W. Walton. 1988. Zoological catalogue of Australia. Volume 5. Mammalia. Australian Government Publ. Serv., Canberra, x + 274 pp.

Bareham, J. R., and A. Furreddu. 1975. Observations on the use of grottoes by Mediterranean monk seals. J. Zool. 175:291–98.

Barnes, L. G. 1985. Evolution, taxonomy, and antitropical distributions of the porpoises (Phocoenidae, Mammalia). Mar. Mamm. Sci. 1:149–65.

———. 1989. A new enaliarctine pinniped from the Astoria Formation, Oregon, and a classification of the Otariidae (Mammalia: Carnivora). Los Angeles Co. Nat. Hist. Mus. Contrib. Sci., no. 403, 26 pp.

———. 1992. A new genus and species of middle Miocene enaliarctine pinniped (Mammalia, Carnivora, Otariidae) from the Astoria Formation in coastal Oregon. Los Angeles Co. Nat. Hist. Mus. Contrib. Sci., no. 431, 27 pp.

Barnes, L. G., D. P. Domning, and C. E. Ray. 1985. Status of studies of fossil marine mammals. Mar. Mamm. Sci. 1:15–53.

Barnes, L. G., and S. A. McLeod. 1984. The fossil record and phyletic relationships of gray whales. *In* Jones, Swartz, and Leatherwood (1984), 3–32.

Barros, N. B., and V. G. Cockcroft. 1991. Prey of humpback dolphins *(Sousa plumbea)* stranded in eastern Cape Province, South Africa. Aquat. Mamm. 17:134–36.

Barzdo, J. 1980. International trade in harp and hooded seals. Oryx 15:275–79.

Baugh, T. 1987. Man and manatee: planning for the future. Endangered Species Tech. Bull. 12(9):7.

Bayha, K., and J. Kormendy, technical coordinators. 1990. Sea Otter Symposium: Proceedings of a symposium to evaluate the response effort on behalf of sea otters after the T/V *Exxon Valdez* oil spill into Prince William Sound, Anchorage, Alaska, 17–19 April 1990. U.S. Fish and Wildl. Serv. Biol. Rept., no. 90(12), x + 485 pp.

Beamish, P. 1978. Evidence that a captive humpback whale *(Megaptera novaeangliae)* does not use sonar. Deep-Sea Res. 25:469–72.

Beaubrun, P.-Ch. 1990. Un cétacé nouveau pour les côtes sud-marocaines: *Sousa teuszii* (Kukenthal, 1892). Mammalia 54:162–64.

Beentjes, M. P. 1990. Comparative terrestrial locomotion of the Hooker's sea lion *(Phocarctos hookeri)* and the New Zealand fur seal *(Arctocephalus forsteri)*: evolutionary and ecological implications. Zool. J. Linnean Soc. 98:307–25.

Belitsky, D. W., and C. L. Belitsky. 1980. Distribution and abundance of manatees *Trichechus manatus* in the Dominican Republic. Biol. Conserv. 17:313–19.

Bengtson, J. L., and S. M. Fitzgerald. 1985. Potential role of vocalizations in West Indian manatees. J. Mamm. 66:816–19.

Bengtson, J. L., and D. B. Siniff. 1981. Reproductive aspects of female crabeater seals *(Lobodon carcinophagus)* along the Antarctic Peninsula. Can. J. Zool. 59:92–102.

Benjaminsen, T., and I. Christensen. 1979. The natural history of the bottlenose whale *Hyperoodon ampullatus* (Forster). *In* Winn and Olla (1979), 143–64.

Berkes, F. 1977. Turkish dolphin fisheries. Oryx 14:163–67.

Berta, A. 1991. New *Enaliarctos* (Pinnipedimorpha) from the Oligocene and Miocene of Oregon and the role of "enaliarctids" in pinniped phylogeny. Smithson. Contrib. Paleontol., no. 69, iii + 33 pp.

———. 1994. What is a whale? Science 263:180–81.

Berta, A., and T. A. Deméré. 1986. *Callorhinus gilmorei* n. sp. (Carnivora: Otariidae) from the San Diego Formation (Blancan) and its implications for otariid phylogeny. Trans. San Diego Soc. Nat. Hist. 21:111–26.

Berta, A., C. E. Ray, and A. R. Wyss. 1989. Skeleton of the oldest known pinniped, *Enaliarctos mealsi*. Science 244:60–62.

Berta, A., and A. R. Wyss. 1990. Oldest pinniped. Science 248:499–500.

———. 1994. Pinniped phylogeny. Proc. San Diego Soc. Nat. Hist. 29:33–56.

Berzin, A. A. 1972. The sperm whale. Israel Progr. Sci. Transl., Jerusalem, v + 394 pp.

———. 1978. Whale distribution in tropical eastern Pacific waters. Rept. Internatl. Whaling Comm. 28:173–77.

Best, P. B. 1968. The sperm whale *(Physeter catodon)* off the west coast of South Africa. 2. Reproduction in the female. S. Afr. Div. Sea Fish. Investig. Rept., no. 66, 32 pp.

———. 1970*a*. Exploitation and recovery of right whales *Eubalaena australis* off the Cape Province. S. Afr. Div. Sea Fish. Investig. Rept., no. 80, 20 pp.

———. 1970*b*. The sperm whale *(Physeter catodon)* off the west coast of South Africa. 5. Age, growth, and mortality. S. Afr. Div. Sea Fish. Investig. Rept., no. 79, 27 pp.

———. 1974. The biology of the sperm whale as it relates to stock management. *In* Schevill (1974), 257–93.

———. 1979. Social organization in sperm whales, *Physeter macrocephalus*. *In* Winn and Olla (1979), 227–89.

———. 1985. External characters of southern minke whales and the existence of a diminutive form. Sci. Rept. Whales Res. Inst. 36:1–33.

———. 1987. Estimates of the landed catch of right (and other whalebone) whales in the American fishery, 1805–1909. Fishery Bull. 85:403–18.

———. 1988. The external appearance of Heaviside's dolphin, *Cephalorhynchus heavisidii* (Gray, 1828). *In* Brownell and Donovan (1988), 279–301.

———. 1994. Seasonality of reproduction and the length of gestation in southern right whales *Eubalaena australis*. J. Zool. 232: 175–89.

Best, P. B., and R. B. Abernethy. 1994. Heaviside's dolphin *Cephalorhynchus heavisidii* (Gray, 1828). *In* Ridgway and Harrison (1994), 289–310.

Best, P. B., P. A. S. Canham, and N. Macleod. 1984. Patterns of reproduction in sperm whales, *Physeter macrocephalus*. *In* Perrin, Brownell, and DeMaster (1984), 51–79.

Best, P. B., and P. D. Shaughnessy. 1981. First record of the melon-headed whale *Peponocephala electra* from South Africa. Ann. S. Afr. Mus. 83:33–47.

Best, R. C. 1981. The tusk of the narwhal (*Monodon monoceros* L.): interpretation of its function (Mammalia: Cetacea). Can. J. Zool. 59:2386–93.

———. 1982. Seasonal breeding in the Amazonian manatee, *Trichechus inunguis* (Mammalia: Sirenia). Biotrópica 14:76–78.

———. 1983. Apparent dry-season fasting in Amazonian manatees (Mammalia: Sirenia). Biotrópica 15:61–64.

Best, R. C., and V. M. F. Da Silva. 1984. Preliminary analysis of reproductive parameters of the boutu, *Inia geoffrensis*, and the tucuxi, *Sotalia fluviatilis*, in the Amazon River system. *In* Perrin, Brownell, and DeMaster (1984), 361–69.

———. 1989*a*. Amazon River dolphin, boto *Inia geoffrensis* (de Blainville, 1817). *In* Ridgway and Harrison (1989), 1–23.

———. 1989*b*. Biology, status and conservation of *Inia geoffrensis* in the Amazon and Orinoco river basins. *In* Perrin et al. (1989), 23–34.

———. 1993. *Inia geoffrensis*. Mammalian Species, no. 426, 8 pp.

Best, R. C., and H. D. Fisher. 1974. Seasonal breeding of the narwhal (*Monodon monoceros* L.). Can. J. Zool. 52:429–31.

Bester, M. N. 1980. Population increase in the Amsterdam Island fur seal *Arctocephalus tropicalis* at Gough Island. S. Afr. J. Zool. 15:229–34.

———. 1981. Seasonal changes in the population composition of the fur seal *Arctocephalus tropicalis* at Gough Island. S. Afr. J. Wildl. Res. 11:49–55.

———. 1982. Distribution, habitat selection, and colony types of the Amsterdam Island fur seal *Arctocephalus tropicalis* at Gough Island. J. Zool. 196:217–31.

———. 1987. Subantarctic fur seal, *Arctocephalus tropicalis*, at Gough Island (Tristan da Cunha group). *In* Croxall and Gentry (1987), 57–60.

———. 1990. Population trends of Subantarctic fur seals and southern elephant seals at Gough Island. S. Afr. J. Antarctic Res. 20:9–12.

Bester, M. N., and G. I. H. Kerley. 1983. Rearing of twin pups to weaning by subantarctic fur seal *Arctocephalus tropicalis*. S. Afr. J. Wildl. Res. 13:86–87.

Bester, M. N., and J.-P. Roux. 1986. Summer presence of leopard seals *Hydrurga leptonyx* at the Courbet Peninsula, Iles Kerguelen. S. Afr. J. Antarctic Res. 16:29–32.

Bigg, M. A. 1981. Harbour seal—*Phoca vitulina* and *P. largha*. *In* Ridgway and Harrison (1981*b*), 1–27.

———. 1988. Status of the Steller sea lion, *Eumetopias jubatus*, in Canada. Can. Field-Nat. 102:315–36.

Bigg, M. A., P. F. Olesiuk, G. M. Ellis, J. K. B. Ford, and K. C. Balcomb, III. 1990. Social organization and genealogy of resident killer whales (*Orcinus orca*) in the coastal waters of British Columbia and Washington state. *In* Hammond, Mizroch, and Donovan (1990), 383–405.

Bigg, M. A., and A. A. Wolman. 1975. Live-capture killer whale (*Orcinus orca*) fishery, British Columbia and Washington, 1962–73. J. Fish. Res. Bd. Can. 32:1213–21.

Blacher, C. 1994. Strategic reproduction of *Lutra longicaudis*. IUCN (World Conservation Union) Otter Specialist Group Bull. 9:6.

Bloch, D., G. Desportes, R. Mouritsen, S. Skaaning, and E. Stefanson. 1993. An introduction to studies of the ecology and status of the long-finned pilot whale (*Globicephala melas*) off the Faroe Islands, 1986–1988. *In* Donovan, Lockyer, and Martin (1993), 1–32.

Bockstoce, J. 1980*a*. Battle of the bowheads. Nat. Hist. 89(5):52–61.

———. 1980*b*. A preliminary estimate of the reduction of the western arctic bowhead whale population by the pelagic whaling industry: 1848–1915. Mar. Fish. Rev. 42 (9–10):20–27.

———. 1986. Whales, ice, and men: the history of whaling in the western Arctic. Univ. Washington Press, Seattle, 394 pp.

Bockstoce, J., and D. B. Botkin. 1983. The historical status and reduction of the western arctic bowhead whale (*Balaena mysticetus*) population by the pelagic whaling industry, 1848–1914. *In* Tillman and Donovan (1983), 107–41.

Bockstoce, J., and J. J. Burns. 1993. Commercial whaling in the North Pacific sector. *In* Burns, Montague, and Cowles (1993), 563–77.

Boness, D. J., W. D. Bowen, and O. T. Oftedal. 1988. Evidence of polygyny from spatial patterns of hooded seals (*Cystophora cristata*). Can. J. Zool. 66:703–6.

Bonner, W. N. 1968. The fur seal of South Georgia. British Antarctic Surv. Sci. Rept., no. 56, 81 pp.

———. 1972. The grey seal and common seal in European waters. Oceanogr. Mar. Biol. Ann. Rev. 10:461–507.

———. 1981*a*. Grey seal—*Halichoerus grypus*. *In* Ridgway and Harrison (1981*b*), 111–14.

———. 1981*b*. Southern fur seals—*Arctocephalus*. *In* Ridgway and Harrison (1981*a*), 161–208.

———. 1982*a*. Seals and man: a study of interactions. Washington Sea Grant, Seattle, xii + 170 pp.

———. 1982*b*. The status of the Antarctic fur seal *Arctocephalus gazella*. Mammals in the Seas, FAO Fish. Ser. No. 5, 4:423–30.

———. 1984*a*. Antarctic renaissance. *In* Macdonald (1984), 260–61.

———. 1984*b*. Eared seals. *In* Macdonald (1984), 253–59.

———. 1985. Impact of fur seals on the terrestrial environment at South Georgia. *In* Siegfried, W. R., P. R. Condy, and R. M. Laws, eds., Antarctic nutrient cycles and food webs, Springer-Verlag, Berlin, 641–46.

Boreman, S. M. 1992. Dolphin-safe tuna: what's in a label? The killing of dolphins in the eastern tropical Pacific and the case for an international legal solution. Nat. Resource J. 32:425–47.

Born, E. W., and I. Gjertz. 1993. A link between walruses (*Odobenus rosmarus*) in northeast Greenland and Svalbard. Polar Record 29:329.

Born, E. W., and L. O. Knutsen. 1992. Satellite-linked radio tracking of Atlantic walruses (*Odobenus rosmarus rosmarus*) in northeastern Greenland, 1989–1991. Z. Saugetierk. 57:275–87.

Borobia, M., and L. Lodi. 1992. Recent observations and records of the West Indian manatee *Trichechus manatus* in northeastern Brazil. Biol. Conserv. 59:37–43.

Borobia, M., S. Siciliano, L. Lodi, and W. Hoek. 1991. Distribution of the South American dolphin *Sotalia fluviatilis*. Can. J. Zool. 69:1025–39.

Borsa, P. 1990. Seasonal occurrence of the leopard seal, *Hydrurga leptonyx*, in the Kerguelen Islands. Can. J. Zool. 68:405–8.

Bowen, W. D., D. J. Boness, and O. T. Oftedal. 1987. Mass transfer from mother to pup and subsequent mass loss by the weaned pup in the hooded seal, *Cystophora cristata*. Can. J. Zool. 65:1–8.

Bowen, W. D., R. A. Myers, and K. Hay. 1987. Abundance estimation of a dispersed, dynamic population: hooded seals (*Cystophora cristata*) in the northwest Atlantic. Can. J. Fish. Aquat. Sci. 44:282–95.

Bowen, W. D., O. T. Oftedal, and D. J. Boness. 1985. Birth to weaning in 4 days: remarkable growth in the hooded seal, *Cystophora cristata*. Can. J. Zool. 63:2841–46.

Boyd, I. L. 1993. Pup production and distribution of breeding Antarctic fur seals (*Arctocephalus gazella*) at South Georgia. Antarctic Sci. 5:17–24.

Boyd, I. L., and J. P. Croxall. 1992. Diving behaviour of lactating Antarctic fur seals. Can. J. Zool. 70:919–28.

Braham, H. W. 1983. Northern records of Risso's dolphin, *Grampus griseus,* in the northeast Pacific. Can. Field-Nat. 97:89–90.

———. 1984*a*. The bowhead whale, *Balaena mysticetus.* Mar. Fish. Rev. 46(4):45–53.

———. 1984*b*. Review of reproduction in the white whale, *Delphinapterus leucas,* narwhal, *Monodon monoceros,* and Irrawaddy dolphin, *Orcaella brevirostris,* with comments on stock assessment. *In* Perrin, Brownell, and DeMaster (1984), 81–89.

Braham, H. W., F. E. Durham, G. H. Jarrell, and S. Leatherwood. 1980. Ingutuk: a morphological variant of the bowhead whale, *Balaena mysticetus.* Mar. Fish. Rev. 42(9–10): 70–73.

Braham, H. W., R. D. Everitt, and D. J. Rugh. 1980. Northern sea lion population decline in the eastern Aleutian Islands. J. Wildl. Mgmt. 44:25–33.

Braham, H. W., B. D. Krogman, and G. M. Carroll. 1984. Bowhead and white whale migration, distribution, and abundance in the Bering, Chukchi, and Beaufort Seas, 1975–1978. U.S. Natl. Mar. Fish. Serv., NOAA Tech. Rept. NMFS SSRF-778, iv + 39 pp.

Braham, H. W., and D. W. Rice. 1984. The right whale, *Balaena glacialis.* Mar. Fish. Rev. 46(4):38–44.

Bratton, G. R., C. B. Spainhour, W. Flory, M. Reed, and K. Jayko. 1993. Presence and potential effects of contaminants. *In* Burns, Montague, and Cowles (1993), 701–44.

Breiwick, J. M., L. L. Eberhardt, and H. W. Braham. 1984. Population dynamics of western arctic bowhead whales *(Balaena mysticetus).* Can. J. Fish. Aquat. Sci. 41:484–96.

Breiwick, J. M., and E. D. Mitchell. 1983. Estimated initial size of the Bering Sea stock of bowhead whales *(Balaena mysticetus)* from logbook and other catch data. *In* Tillman and Donovan (1983), 147–51.

Breiwick, J. M., E. D. Mitchell, and D. G. Chapman. 1980. Estimated initial population size of the Bering Sea stock of bowhead whale, *Balaena mysticetus:* an iterative method. Fishery Bull. 78:843–53.

Broad, S., R. Luxmoore, and M. Jenkins. 1988. Significant trade in wildlife: a review of selected species in CITES appendix II. Volume 1: Mammals. IUCN (World Conservation Union), Gland, Switzerland, xix + 183 pp.

Brodie, P. F. 1971. A reconsideration of aspects of growth, reproduction, and behavior of the white whale *(Delphinapterus leucas),* with reference to the Cumberland Sound, Baffin Island, population. J. Fish. Res. Bd. Can. 28:1309–18.

———. 1989. The white whale *Delphinapterus leucas* (Pallas, 1776). *In* Ridgway and Harrison (1989), 119–44.

Brown, D. H., D. K. Caldwell, and M. C. Caldwell. 1966. Observations on the behavior of wild and captive false killer whales, with notes on associated behavior of other genera of captive delphinids. Los Angeles Co. Nat. Hist. Mus. Contrib. Sci., no. 95, 32 pp.

Brown, S. G. 1986. Twentieth-century records of right whales *(Eubalaena glacialis)* in the northeast Atlantic Ocean. *In* Brownell, Best, and Prescott (1986), 121–27.

Brownell, R. L., Jr. 1975*a*. *Phocoena dioptrica.* Mammalian Species, no. 66, 3 pp.

———. 1975*b*. Progress report on the biology of the Franciscana dolphin, *Pontoporia blainvillei,* in Uruguayan waters. J. Fish. Res. Bd. Can. 32: 1073–78.

———. 1983. *Phocoena sinus.* Mammalian Species, no. 198, 3 pp.

———. 1984. Review of reproduction in platanistid dolphins. *In* Perrin, Brownell, and DeMaster (1984), 149–58.

———. 1986. Distribution of the vaquita, *Phocoena sinus,* in Mexican waters. Mar. Mamm. Sci. 2:299–305.

———. 1989. Franciscana *Pontoporia blainvillei* (Gervais and d'Orbigny, 1844). *In* Ridgway and Harrison (1989), 45–67.

Brownell, R. L., Jr., A. Aguayo L., and D. Torres N. 1976. A Shepherd's beaked whale, *Tasmacetus shepherdi,* from the eastern South Pacific. Sci. Rept. Whales Res. Inst. 28:127–28.

Brownell, R. L., Jr., P. B. Best, and J. H. Prescott, eds. 1986. Right whales: past and present status. Rept. Internatl. Whaling Comm. Spec. Issue, no. 10, vi + 289 pp.

Brownell, R. L., Jr., and C. Chun. 1977. Probable existence of the Korean stock of the gray whale *(Eschrichtius robustus).* J. Mamm. 58:237–39.

Brownell, R. L., Jr., and G. P. Donovan, eds. 1988. Biology of the genus *Cephalorhynchus.* Rept. Internatl. Whaling Comm. Spec. Issue, no. 9, 344 pp.

Brownell, R. L., Jr., L. T. Findley, O. Vidal, A. Robles, and S. Manzanilla N. 1987. External morphology and pigmentation of the vaquita, *Phocoena sinus* (Cetacea: Mammalia). Mar. Mamm. Sci. 3:22–30.

Browneil, R. L., Jr., and E. S. Herald. 1972. *Lipotes vexillifer.* Mammalian Species, no. 10, 4 pp.

Brownell, R. L., Jr., J. E. Heyning, and W. P. Perrin. 1989. A porpoise, *Australophocaena dioptrica,* previously identified as *Phocoena spinipinnis,* from Heard Island. Mar. Mamm. Sci. 5:193–95.

Brownell, R. L., Jr., and R. Praderi. 1976. Records of the delphinid genus *Stenella* in western South Atlantic waters. Sci. Rept. Whales Res. Inst. 28:129–35.

———. 1984. *Phocoena spinipinnis.* Mammalian Species, no. 217, 4 pp.

———. 1985. Distribution of Commerson's dolphin, *Cephalorhynchus commersonii,* and the rediscovery of the type of *Lagenorhynchus floweri.* Sci. Rept. Whales Res. Inst. 36:153–64.

Brownell, R. L., Jr., K. Ralls, and W. F. Perrin. 1989. The plight of the "forgotten" whales. Oceanus 32(1):5–11.

Brownell, R. L., Jr., and G. B. Rathbun. 1988. California sea otter translocation: a status report. Endangered Species Tech. Bull. 13(4): 1, 6.

Brownlee, S. M., and K. S. Norris. 1994. The acoustic domain. *In* Norris et al. (1994), 161–85.

Bryant, P. J. 1995. Dating remains of gray whales from the eastern North Atlantic. J. Mamm. 76:857–61.

Bryden, M. M., R. J. Harrison, and R. J. Lear. 1977. Some aspects of the biology of *Peponocephala electra* (Cetacea: Delphinidae). I. General and reproductive biology. Austral. J. Mar. Freshwater Res. 28:703–15.

Buckland, S. T., D. Bloch, K. L. Cattanach, Th. Gunnlaugsson, K. Hoydal, S. Lens, and J. Sigurjónsson. 1993. Distribution and abundance of long-finned pilot whales in the North Atlantic, estimated from NASS-87 and NASS-89 data. *In* Donovan, Lockyer, and Martin (1993), 33–49.

Buckland, S. T., K. L. Cattanach, and R. C. Hobbs. 1993. Abundance estimates of Pacific white-sided dolphin, northern right whale dolphin, Dall's porpoise, and northern fur seal in the North Pacific, 1987–1990. Internatl. N. Pacific Fish. Comm. Bull. 53:387–407.

Buckland, S. T., T. D. Smith, and K. L. Cattanach. 1992. Status of small cetacean populations in the Black Sea: review of current information and suggestions for future research. Rept. Internatl. Whaling Comm. 42:513–16.

Burns, J. J. 1981*a*. Bearded seal—*Erignathus barbatus. In* Ridgway and Harrison (1981*b*), 145–70.

———. 1981*b*. Ribbon seal—*Phoca fasciata. In* Ridgway and Harrison (1981*b*), 89–109.

Burns, J. J., and F. H. Fay. 1970. Comparative morphology of the skull of the ribbon seal, *Histriophoca fasciata,* with remarks on systematics of the Phocidae. J. Zool. 161:363–94.

Burns, J. J., J. J. Montague, and C. J. Cowles, eds. 1993. The bowhead whale. Spec. Publ. Soc. Marine Mamm., no. 2, xxxvi + 787 pp.

Burt, W. H., and R. P. Grossenheider. 1976. A field guide to the mammals. Houghton Mifflin, Boston, xxv + 289 pp.

Busch, B. C. 1985. The war against the seals. McGill–Queen's Univ. Press, Kingston, Ontario, xviii + 374 pp.

C

Caldwell, D. K., and M. C. Caldwell. 1970. Echolocation-type signals by two dolphins, genus *Sotalia*. Quart. J. Florida Acad. Sci. 33:124–31.

———. 1971*a*. The pygmy killer whale, *Feresa attenuata*, in the western Atlantic, with a summary of world records. J. Mamm. 52:206–9.

———. 1971*b*. Underwater pulsed sounds produced by captive spotted dolphins, *Stenella plagiodon*. Cetology, no. 1, 7 pp.

———. 1977. Cetaceans. *In* Sebeok, T. A., ed., How animals communicate, Indiana Univ. Press, 794–807.

———. 1985. Manatees—*Trichechus manatus, Trichechus senegalensis*, and *Trichechus inunguis*. *In* Ridgway and Harrison (1985), 33–66.

———. 1989. Pygmy sperm whale *Kogia breviceps* (de Blainville, 1838): dwarf sperm whale *Kogia simus* Owen, 1866. *In* Ridgway and Harrison (1989), 235–60.

Caldwell, D. K., M. C. Caldwell, and D. W. Rice. 1966. Behavior of the sperm whale, *Physeter catodon* L. *In* Norris (1966), 678–717.

Caldwell, D. K., M. C. Caldwell, and R. V. Walker. 1976. First records for Fraser's dolphin *(Lagenodelphis hosei)* in the Atlantic and the melon-headed whale *(Peponocephala electra)* in the western Atlantic. Cetology, no. 25, 4 pp.

Caldwell, M. C., and D. K. Caldwell. 1969. The ugly dolphin. Sea Frontiers 15(5):1–7.

———. 1979. The whistle of the Atlantic bottlenosed dolphin *(Tursiops truncatus)*—ontogeny. *In* Winn and Olla (1979), 369–401.

Caldwell, M. C., D. K. Caldwell, and R. L. Brill. 1989. *Inia geoffrensis* in captivity in the United States. *In* Perrin et al. (1989), 35–41.

Caldwell, M. C., D. K. Caldwell, and W. E. Evans. 1966. Sounds and behavior of captive Amazon freshwater dolphins, *Inia geoffrensis*. Los Angeles Co. Nat. Hist. Mus. Contrib. Sci., no. 108, 24 pp.

Caldwell, M. C., D. K. Caldwell, and J. F. Miller. 1973. Statistical evidence for individual whistles in the spotted dolphin, *Stenella plagiodon*. Cetology, no. 16, 21 pp.

Caldwell, M. C., D. K. Caldwell, and P. L. Tyack. 1990. Review of the signature-whistle hypothesis for the Atlantic bottlenose dolphin. *In* Leatherwood and Reeves (1990), 199–234.

Calvert, W., and I. Stirling. 1985. Winter distribution of ringed seals *(Phoca hispida)* in the Barrow Strait area, Northwest Territories, determined by underwater vocalizations. Can. J. Fish. Aquat. Sci. 42:1238–43.

Campagna, C. 1985. The breeding cycle of the southern sea lion, *Otaria byronia*. Mar. Mamm. Sci. 1:210–18.

Campagna, C., and B. J. Le Boeuf. 1988. Reproductive behaviour of southern sea lions. Behaviour 104:233–61.

Campagna, C., B. J. Le Boeuf, and H. L. Cappozzo. 1988. Group raids: a mating strategy of male southern sea lions. Behaviour 105:224–49.

Campbell, R. R. 1987. Status of the hooded seal, *Cystophora cristata*, in Canada. Can. Field-Nat. 101:253–65.

Carey, J. 1987. The sea otter's uncertain future. Natl. Wildl. 25(1):16–20.

Carr, T., N. Carr, and J. H. M. David. 1985. A record of the sub-Antarctic fur seal *Arctocephalus tropicalis* in Angola. S. Afr. J. Zool. 20:77.

Cattanach, K. L., J. Sigurjónsson, S. T. Buckland, and Th. Gunnlaugsson. 1993. Sei whale abundance in the North Atlantic, estimated from NASS-87 and NASS-89 data. Rept. Internatl. Whaling Comm. 43:315–21.

Cawthorn, M. W. 1988. Recent observations of Hector's dolphin, *Cephalorhynchus hectori*, in New Zealand. *In* Brownell and Donovan (1988), 303–14.

Chapman, D. G. 1974. Status of Antarctic rorqual stocks. *In* Schevill (1974), 218–38.

Chen Peixun. 1989. Baiji *Lipotes vexillifer* Miller, 1918. *In* Ridgway and Harrison (1989), 25–43.

Chen Peixun and Hua Yuanyu. 1989. Distribution, population size, and protection of *Lipotes vexillifer*. *In* Perrin et al. (1989), 81–85.

Chen Peixun, Liu Peilin, Liu Renjun, Lin Kejie, and G. Pilleri. 1979. Distribution, ecology, behaviour, and conservation of the dolphins of the middle reaches of Changjiang (Yangtze) River (Wuhan-Yueyang). Investig. Cetacea 10:87–103.

Chen Peixun, Liu Renjun, and R. J. Harrison. 1982. Reproduction and reproductive organs in *Neophocaena asiaeorientalis* from the Yangtse River. Aquat. Mamm. 9:9–16.

Clapham, P. J. 1992. Age at attainment of sexual maturity in humpback whales, *Megaptera novaeangliae*. Can. J. Zool. 70:1470–72.

Clapham, P. J., P. J. Palsboll, D. K. Mattila, and O. Vasquez. 1992. Composition and dynamics of humpback whale competitive groups in the West Indies. Behaviour 122:182–94.

Clark, C. W. 1982. The acoustic repertoire of the southern right whale, a quantitative analysis. Anim. Behav. 30:1060–71.

———. 1985. Economic aspects of marine mammal–fishery interactions. *In* Beddington, Beverton, and Lavigne (1985), 34–38.

Clark, C. W., and J. H. Johnson. 1984. The sounds of the bowhead whale, *Balaena mysticetus*, during the spring migrations of 1979 and 1980. Can. J. Zool. 62:1436–41.

Clarke, M. R. 1979. The head of the sperm whale. Sci. Amer. 240(1):128–41.

Committee for Whaling Statistics. 1980. International whaling statistics. Vols. 85 and 86. Sandefjord, Norway, 67 pp.

———. 1984. International whaling statistics. Vols. 93 and 94. Sandefjord, Norway, 61 pp.

Condit, R., and B. J. Le Boeuf. 1984. Feeding habits and feeding grounds of the northern elephant seal. J. Mamm. 65:281–90.

Conrad, J., and T. Bjorndal. 1993. On the resumption of commercial whaling: the case of the minke whale in the northeast Pacific. Arctic 46:164–71.

Conway, W. 1986. Sultan of South American shores. Anim. Kingdom 89(6):18–26.

Cooke, J., and J. Thomsen. 1993. International Whaling Commission: a report of the 45th annual meeting. Traffic Bull. 14(1):21–24.

Cooper, C. F., and B. S. Stewart. 1983. Demography of northern elephant seals, 1911–1982. Science 219:969–71.

Corbet, G. B. 1978. The mammals of the Palaearctic Region: a taxonomic review. British Mus. (Nat. Hist.), London, 314 pp.

Corbet, G. B., and J. E. Hill. 1991. A world list of mammalian species. Natural History Museum Publ., London, and Oxford Univ. Press, viii + 243 pp.

———. 1992. The mammals of the Indomalayan region: a systematic review. Oxford Univ. Press, viii + 488 pp.

Cousteau, J. Y., and P. Diole. 1972. Killer whales have fearsome teeth and a strange gentleness to man. Smithsonian 3(3):66–73.

Crawley, M. C., and G. J. Wilson. 1976. The natural history and behaviour of the New Zealand fur seal *(Arctocephalus forsteri)*. Tuatara 22:1–29.

Croxall, J. P., and R. L. Gentry, eds. 1987. Status, biology, and ecology of fur seals: proceedings of an international symposium and workshop, Cambridge, England, 23–27 April 1984. Natl. Oceanic Atmos. Admin. Tech. Rept., NMFS 51.

Croxall, J. P., and L. Hiby. 1983. Fecundity, survival, and site fidelity in Weddell seals, *Leptonychotes weddelli*. J. Appl. Ecol. 20:19–32.

Cummings, W. C. 1985. Right whales—*Eubalaena glacialis* and *Eubalaena australis*. *In* Ridgway and Harrison (1985), 275–304.

D

Dahlheim, M. E., and F. Awbrey. 1982. A classification and comparison of vocalizations of captive killer whales *(Orcinus orca)*. J. Accoust. Soc. Amer. 72:661–70.

Darling, J. D., and C. M. Jurasz. 1983. Migratory destinations of North Pacific humpback whales *(Megaptera novaeangliae)*. *In* Payne (1983), 359–68.

Da Silva, V. M. F., and R. C. Best. 1994. Tucuxi *Sotalia fluviatilis* (Gervais, 1853). *In* Ridgway and Harrison (1994), 43–69.

———. 1996. *Sotalia fluviatilis.* Mammalian Species, no. 527, 7 pp.

David, J. H. M., J. Mercer, and K. Hunter. 1993. A vagrant Subantarctic fur seal *Arctocephalus tropicalis* found in the Comores. S. Afr. J. Zool. 28:61–62.

David, J. H. M., and R. W. Rand. 1986. Attendance behavior of South African fur seals. *In* Gentry and Kooyman (1986), 126–41.

Davis, R. A., W. J. Richardson, S. R. Johnson, and W. E. Renaud. 1978. Status of Lancaster Sound narwhal population in 1976. Rept. Internatl. Whaling Comm. 28:209–15.

Dawbin, W. H. 1966. The seasonal migratory cycle of humpback whales. *In* Norris (1966), 145–70.

Dawson, M. R., and L. Krishtalka. 1984. Fossil history of the families of Recent mammals. *In* Anderson and Jones (1984), 11–58.

Dawson, S. M., and E. Slooten. 1988. Hector's dolphin. *Cephalorhynchus hectori:* distribution and abundance. *In* Brownell and Donovan (1988), 315–24.

Deems, E. F., Jr., and D. Pursley, eds. 1978. North American furbearers. Internatl. Assoc. Fish and Wildl. Agencies, Univ. Maryland Press, College Park, x + 165 pp.

Defler, T. R. 1983. Associations of the giant river otter *(Pteronura brasiliensis)* with fresh-water dolphins *(Inia geoffrensis)*. J. Mamm. 64:692.

DeLong, R. L., and B. S. Stewart. 1991. Diving patterns of northern elephant seal bulls. Mar. Mamm. Sci. 7:369–84.

DeLong, R. L., B. S. Stewart, and R. D. Hill. 1992. Documenting migrations of northern elephant seals using day length. Mar. Mamm. Sci. 8:155–59.

DeMaster, D. P., and I. Stirling. 1981. *Ursus maritimus.* Mammalian Species, no. 145, 7 pp.

Deméré, T. A. 1994. The family Odobenidae: a phylogenetic analysis of fossil and living taxa. Proc. San Diego Soc. Nat. Hist. 29:99–123.

De Muzion, C., and D. P. Domning. 1985. The first records of fossil sirenians in the southeastern Pacific Ocean. Bull. Mus. Natl. Hist. Nat. Paris, ser. 4 (Paleontol.-Geol.), 7:189–213.

Derocher, A. E., and I. Stirling. 1990*a*. Distribution of polar bears *(Ursus maritimus)* during the ice-free period in western Hudson Bay. Can. J. Zool. 68:1395–1403.

———. 1990*b*. Observations of aggregating behaviour in adult male polar bears *(Ursus maritimus)*. Can. J. Zool. 68:1390–94.

Desportes, G., D. Bloch, L. W. Andersen, and R. Mouritsen. 1992. The international research program on the ecology and status of the long-finned pilot whale off the Faroe Islands. Frodskaparrit 40:9–29.

Dierauf, L. A. 1984. A northern fur seal, *Callorhinus ursinus,* found in the Sacramento–San Joaquin Delta. California Fish and Game 70:189.

Doidge, D. W. 1987. Rearing of twin offspring to weaning in Antarctic fur seals, *Arctocephalus gazella. In* Croxall and Gentry (1987), 107–11.

Doidge, D. W., T. S. McCann, and J. P. Croxall. 1986. Attendance behavior of Antarctic fur seals. *In* Gentry and Kooyman (1986), 102–14.

Domning, D. P. 1978. Sirenian evolution in the North Pacific Ocean. Univ. California Publ. Geol. Sci. 118:1–176.

———. 1981. Distribution and status of manatees *Trichechus* spp. near the mouth of the Amazon River, Brazil. Biol. Conserv. 19:85–97.

———. 1982*a*. Commercial exploitation of manatees *Trichechus* in Brazil, c. 1785–1973. Biol. Conserv. 22:101–26.

———. 1982*b*. Evolution of manatees: a speculative history. J. Paleontol. 56:599–619.

———. 1994. A phylogenetic analysis of the Sirenia. Proc. San Diego Soc. Nat. Hist. 29: 177–89.

———. 1996. Bad news or good? Sirenews 25:1–2.

Domning, D. P., and V. de Buffrénil. 1991. Hydrostasis in the Sirenia: quantitative data and functional interpretations. Mar. Mamm. Sci. 7:331–68.

Domning, D. P., and L.-A. C. Hayek. 1986. Interspecific and intraspecific morphological variation in manatees (Sirenia: *Trichechus*). Mar. Mamm. Sci. 2:87–144.

Domning, D. P., G. S. Morgan, and C. E. Ray. 1982. North American Eocene sea cows (Mammalia: Sirenia). Smithson. Contrib. Paleobiol., no. 52, 69 pp.

Domning, D. P., and C. E. Ray. 1986. The earliest Sirenian (Mammalia: Dugongidae) from the eastern Pacific Ocean. Mar. Mamm. Sci. 2:263–76.

Dorsey, E. M. 1983. Exclusive adjoining ranges in individually identified minke whales *(Balaenoptera acutorostrata)* in Washington state. Can. J. Zool. 61:174–81.

Douglas, M. E., G. D. Schnell, and D. J. Hough. 1984. Differentiation between inshore and offshore spotted dolphins in the eastern tropical Pacific Ocean. J. Mamm. 65:375–87.

Douglas, M. E., G. D. Schnell, D. J. Hough, and W. F. Perrin. 1992. Geographic variation in cranial morphology of spinner dolphins *Stenella longirostris* in the eastern tropical Pacific Ocean. Fishery Bull. 90:54–76.

Dudley, M. 1992. First Pacific record of a hooded seal, *Cystophora cristata* Erxleben, 1777. Mar. Mamm. Sci. 8:164–68.

Duguy, R. 1986. Observation d'un morse *(Odobenus rosmarus)* sur la côte de Gironde, France. Mammalia 50:563–64.

E

Eibl-Eibesfeldt, I. 1984. The Galapagos seals. Part 1. Natural history of the Galapagos sea lion (*Zalophus californianus wollebaeki,* Sivertsen). *In* Perry (1984), 207–14.

Ellerman, J. R., and T. C. S. Morrison-Scott. 1966. Checklist of Palaearctic and Indian mammals. British Mus. (Nat. Hist.), London, 810 pp.

Ellis, R. 1980. The book of whales. Knopf, New York, xvii + 202 pp.

Erdbrink, D. P. 1953. A review of fossil and Recent bears of the Old World with remarks on their phylogeny based upon their dentition. Deventer, Netherlands, 597 pp.

Estes, J. A. 1980. *Enhydra lutris.* Mammalian Species, no. 133, 8 pp.

———. 1986. Marine otters and their environment. Ambio 15:181–83.

———. 1990. Growth and equilibrium in sea otter populations. J. Anim. Ecol. 59:385–401.

Estes, J. A., and G. R. VanBlaricom. 1985. Sea otters and shellfisheries. *In* Beddington, Beverton, and Levigne (1985), 187–235.

Evans, C. D., and L. S. Underwood. 1978. How many bowheads? Oceanus 21(2):17–23.

Evans, W. E. 1994. Common dolphin, white-bellied porpoise *Delphinus delphis* Linnaeus, 1758. *In* Ridgway and Harrison (1994), 191–224.

F

Fay, F. H. 1978. Belukha whale. *In* Haley (1978), 132–37.

———. 1981. Walrus—*Odobenus rosmarus. In* Ridgway and Harrison (1981*a*), 1–23.

———. 1982. Ecology and biology of the Pacific walrus, *Odobenus rosmarus divergens* Illiger. N. Amer. Fauna, no. 74, vi + 279 pp.

———. 1985. *Odobenus rosmarus*. Mammalian Species, no. 238, 7 pp.

Fay, F. H., B. P. Kelly, and J. L. Sease. 1989. Managing the exploitation of Pacific walruses: a tragedy of delayed response and poor communication. Mar. Mamm. Sci. 5:1–16.

Feldkamp, S. D., R. L. DeLong, and G. A. Antonelis. 1989. Diving patterns of California sea lions, *Zalophus californianus*. Can. J. Zool. 67:872–83.

Fernandez, S., and S. C. Jones. 1990. Manatee stranding on the coast of Texas. Texas J. Sci. 42:103–4.

Finley, K. J. 1990. Isabella Bay, Baffin Island: an important historical and present-day concentration area for the endangered bowhead whale *(Balaena mysticetus)* of the eastern Canadian Arctic. Arctic 43:137–52.

Finley, K. J., J. P. Hickie, and R. A. Davis. 1987. Status of the beluga, *Delphinapterus leucas*, in the Beaufort Sea. Can. Field-Nat. 101:271–78.

Fish, J. F., J. L. Sumich, and G. L. Lingle. 1974. Sounds produced by the gray whale, *Eschrichtius robustus*. Mar. Fish. Rev. 36(4): 38–45.

Fitch, J. E., and R. L. Brownell, Jr. 1968. Fish otoliths in cetacean stomachs and their importance in interpreting feeding habits. J. Fish. Res. Bd. Can. 25:2561–74.

———. 1971. Food habits of the franciscana *Pontoporia blainvillei* (Cetacea: Platanistidae) from South America. Bull. Mar. Sci. 21:626–36.

Fletemeyer, J. 1985. A hooded seal. Florida Nat. 58(2):9.

Ford, J. K. B., and H. D. Fisher. 1978. Underwater acoustic signals of the narwhal *(Monodon monoceros)*. Can. J. Zool. 56:552–60.

———. 1983. Group-specific dialects of killer whales *(Orcinus orca)* in British Columbia. *In* Payne (1983), 129–61.

Fordyce, R. E., and L. G. Barnes. 1994. The evolutionary history of whales and dolphins. Annual Rev. Earth Planet Sci. 22:419–55.

Fordyce, R. E., R. H. Mattlin, and J. M. Dixon. 1984. Second record of spectacled porpoise from subantarctic southwest Pacific. Sci. Rept. Whales Res. Inst. 35:159–64.

Forstén, A., and P. M. Youngman. 1982. *Hydrodamalis gigas*. Mammalian Species, no. 165, 3 pp.

Foster-Turley, P., S. Macdonald, and C. Mason, eds. 1990. Otters: an action plan for their conservation. IUCN (World Conservation Union), Gland, Switzerland, iv + 126 pp.

Fowler, C. W. 1985. An evaluation of the role of entanglement in the population dynamics of northern fur seals on the Pribilof Islands. *In* Shomura, R. S., and H. O. Yoshida, eds., Proceedings of the workshop on the fate and impact of marine debris, Natl. Oceanic Atmos. Admin. Mem., NOAA-TM-NMFS-SWFC-54, 291–307.

———. 1987. Marine debris and northern fur seals: a case study. Marine Pollution Bull. 18:326–35.

Frampton, G. T., Jr. 1995. Importation of polar bear trophies from Canada: proposed rule on legal and scientific findings to implement section 104(c)(5)(A) of the 1994 amendments to the Marine Mammal Protection Act. Federal Register 60:36382–400.

Fraser, F. C. 1970. An early 17th century record of the Californian grey whale in Icelandic waters. Investig. Cetacea 2:13–20.

Frazer, J. F. D. 1973. Specific foetal growth rates of cetaceans. J. Zool. 169:111–26.

Frost, K. J., and L. F. Lowry. 1981. Ringed, Baikal, and Caspian seals—*Phoca hispida, Phoca sibirica*, and *Phoca caspica*. *In* Ridgway and Harrison (1981*b*), 29–53.

Fulton, W. N. 1990. First record of an Australian sea lion on the eastern Australian coast. Victorian Nat. 107:124–25.

Furuta, M. 1984. Note on a gray whale found in the Ise Bay on the Pacific coast of Japan. Sci. Rept. Whales Res. Inst. 35:195–97.

G

Gales, N. J., and H. R. Burton. 1989. The past and present status of the southern elephant seal *(Mirounga leonina* Linn.) in Greater Antarctica. Mammalia 53:35–47.

Gales, N. J., A. J. Cheal, G. J. Pobar, and P. Williamson. 1992. Breeding biology and movements of Australian sea-lions, *Neophoca cinerea*, off the west coast of Western Australia. Wildl. Res. 19:405–16.

Gales, N. J., D. K. Coughran, and L. F. Queale. 1992. Records of Subantarctic fur seals *Arctocephalus tropicalis* in Australia. Austral. Mamm. 15:135–38.

Gales, N. J., P. D. Shaughnessy, and T. E. Dennis. 1994. Distribution, abundance, and breeding cycle of the Australian sea lion *Neophoca cinerea* (Mammalia: Pinnipedia). J. Zool. 234:353–70.

Gallo-R., J. P., and A. Ortega-O. 1986. The first report of *Zalophus californianus* in Acapulco, Mexico. Mar. Mamm. Sci. 2:158.

Gambell, R. 1985. Fin whale—*Balaenoptera physalus*. *In* Ridgway and Harrison (1985), 171–92.

Gao Anli and Zhou Kaiya. 1992. Sexual dimorphism in the baiji, *Lipotes vexillifer*. Can. J. Zool. 70:1484–93.

———. 1993. Growth and reproduction of three populations of finless porpoise, *Neophocaena phocaenoides*, in Chinese waters. Aquat. Mamm. 19:3–12.

Garshelis, D. L., and J. A. Garshelis. 1984. Movements and management of sea otters in Alaska. J. Wildl. Mgmt. 48:665–78.

Garshelis, D. L., A. M. Johnson, and J. A. Garshelis. 1984. Social organization of sea otters in Prince William Sound, Alaska. Can. J. Zool. 62:2648–58.

Gaskin, D. E. 1968. The New Zealand Cetacea. New Zealand Mar. Dept. Fish. Res. Bull., n.s., no. 1, 92 pp.

———. 1970. Composition of schools of sperm whales *Physeter catodon* Linn. east of New Zealand. New Zealand J. Mar. Freshwater Res. 4:456–71.

———. 1976. The evolution, zoogeography, and ecology of Cetacea. Oceanogr. Mar. Biol. Ann. Rev. 14:247–346.

———. 1977. Harbour porpoise *Phocoena phocoena* (L.) in the western approaches to the Bay of Fundy 1969–75. Rept. Internatl. Whaling Comm. 27:487–92.

———. 1982. The ecology of whales and dolphins. Heinemann, London, xii + 459 pp.

———. 1984. The harbour porpoise *Phocoena phocoena* (L.): regional populations, status, and information on direct and indirect catches. Rept. Internatl. Whaling Comm. 34:569–86.

———. 1987. Updated status of the right whale, *Eubalaena glacialis*, in Canada. Can. Field-Nat. 101:295–309.

———. 1991. An update on the status of the right whale, *Eubalaena glacialis*, in Canada. Can. Field-Nat. 105:198–205.

Gaskin, D. E., P. W. Arnold, and B. A. Blair. 1974. *Phocoena phocoena*. Mammalian Species, no. 42, 8 pp.

Gaskin, D. E., and B. A. Blair. 1977. Age determination of harbour porpoise, *Phocoena phocoena* (L.), in the western North Atlantic. Can. J. Zool. 55:18–30.

Gaskin, D. E., G. J. D. Smith, A. P. Watson, W. Y. Yasui, and D. B. Yurick. 1984. Reproduction in the porpoises (Phocoenidae): implications for management. *In* Perrin, Brownell, and DeMaster (1984), 135–48.

Gaskin, D. E., and A. P. Watson. 1984. The harbor porpoise, *Phocoena phocoena*, in Fish Harbour, New Brunswick, Canada: occupancy, distribution, and movements. Fishery Bull. 83:427–42.

Gentry, R. L. 1981. Northern fur seal—*Callorhinus ursinus*. *In* Ridgway and Harrison (1981*a*), 143–60.

Gentry, R. L., D. P. Costa, J. P. Croxall, J. H. M. David, R. W. Davis, G. L. Kooyman, P. Majluf, T. S. McCann, and F. Trillmich. 1986. Synthesis and conclusions. *In* Gentry and Kooyman (1986), 220–64.

Gentry, R. L., and J. R. Holt. 1986. Attendance behavior of northern fur seals. *In* Gentry and Kooyman (1986), 41–60.

Gentry, R. L., and G. L. Kooyman, eds. 1986. Fur seals: maternal strategies on land and at sea. Princeton Univ. Press, xviii + 291 pp.

Gentry, R. L., G. L. Kooyman, and M. E. Goebel. 1986. Feeding and diving behavior of northern fur seals. *In* Gentry and Kooyman (1986), 61–78.

Gerson, H. B., and J. P. Hickie. 1985. Head scarring on male narwhals *(Monodon monoceros)*: evidence for aggressive tusk use. Can. J. Zool. 63:2083–87.

Gewalt, W. 1979*a*. The Commerson's dolphin *(Cephalorhynchus commersonii)*—capture and first experiences. Aquat. Mamm. 7:37–40.

———. 1979*b*. Eine "neue" Delphinart in Delphinarien—der Karib-Delphin *Sotalia guianensis* (Van Beneden, 1864) (?). Saugetierk. Mitt. 27:288–91.

Gilmore, R. M. 1978. Right whale. *In* Haley (1978), 62–69.

Gingerich, P. D., S. M. Raza, M. Arif, M. Anwar, and Zhou Xiaoyuan. 1994. New whale from the Eocene of Pakistan and the origin of cetacean swimming. Nature 368:844–47.

Gjertz, I., and A. Borset. 1992. Pupping in the most northerly harbor seal *(Phoca vitulina)*. Mar. Mamm. Sci. 8:103–9.

Gjertz, I., G. Henriksen, T. Oritsland, and O. Wiig. 1993. Observations of walruses along the Norwegian coast, 1967–1992. Polar Res., n.s., 12:27–31.

Gjertz, I., and E. Persen. 1987. Confrontations between humans and polar bears in Svalbard. Polar Res., n.s., 5:253–56.

Gjertz, I., and O. Wiig. 1994. Past and present distribution of walruses in Svalbard. Arctic 47:34–42.

Glockner-Ferrari, D. A., and M. J. Ferrari. 1984. Reproduction in the humpback whales, *Megaptera novaeangliae*, in Hawaiian waters. *In* Perrin, Brownell, and DeMaster (1984), 237–42.

———. 1990. Reproduction in the humpback whale *(Megaptera novaeangliae)* in Hawaiian waters, 1975–1988: the life history, reproductive rates, and behavior of known individuals identified through surface and underwater photography. *In* Hammond, Mizroch, and Donovan (1990), 163–69.

Godsell, J. 1988. Herd formation and haulout behaviour in harbour seals *(Phoca vitulina)*. J. Zool. 215:83–98.

Goedicke, T. R. 1981. Life expectancy of monk seal colonies in Greece. Biol Conserv. 20:173–81.

Goodall, R. N. P. 1978. Report on the small cetaceans stranded on the coasts of Tierra del Fuego. Sci. Rept. Whales Res. Inst. 30:197–230.

———. 1994. Commerson's dolphin *Cephalorhynchus commersonii* (Lacépède 1804). *In* Ridgway and Harrison (1994), 241–67.

Goodall, R. N. P., A. R. Galeazzi, S. Leatherwood, K. W. Miller, I. S. Cameron, R. K. Kastelein, and A. P. Sobral. 1988. Studies of Commerson's dolphins, *Cephalorhynchus commersonii*, off Tierra del Fuego, 1976–1984, with a review of information on the species in the South Atlantic. *In* Brownell and Donovan (1988), 3–70.

Goodall, R. N. P., K. S. Norris, A. R. Galeazzi, J. A. Oporto, and I. S. Cameron. 1988. On the Chilean dolphin, *Cephalorhynchus eutropia* (Gray, 1846). *In* Brownell and Donovan (1988), 197–257.

Grabert, H. 1984. Migration and speciation of the South American Iniidae (Cetacea, Mammalia). Z. Saugetierk. 49:334–41.

Grzimek, B., ed. 1975. Grzimek's animal life encyclopedia: mammals, I–IV. Van Nostrand Reinhold, New York, vols. 10–13.

———, ed. 1990. Grzimek's encyclopedia of mammals. McGraw-Hill, New York, 5 vols.

Guiler, E. R., H. R. Burton, and N. J. Gales. 1987. On three odontocete skulls from Heard Island. Sci. Rept. Whales Res. Inst. 38:117–24.

Gulland, J. A. 1974. Distribution and abundance of whales in relation to basic productivity. *In* Schevill (1974), 27–52.

———. 1976. Antarctic baleen whales: history and prospects. Polar Record 18:5–13.

H

Haigh, M. D. 1991. The use of manatees for the control of aquatic weeds in Guyana. Irrigation and Drainage Systems 5:339–49.

Hain, J. H. W., and S. Leatherwood. 1982. Two sightings of white pilot whales, *Globicephala melaena*, and summarized records of anomalously white cetaceans. J. Mamm. 63:338–43.

Haley, D., ed. 1978. Marine mammals of eastern North Pacific and Arctic waters. Pacific Search Press, Seattle, 256 pp.

Hall, E. R. 1981. The mammals of North America. John Wiley & Sons, New York, 2 vols.

———. 1984. Geographic variation among brown and grizzly bears *(Ursus arctos)* in North America. Univ. Kansas Mus. Nat. Hist. Spec. Publ., no. 13, 16 pp.

Hammill, M. O., G. B. Stenson, and R. A. Myers. 1992. Hooded seal *(Cystophora cristata)* pup production in the Gulf of St. Lawrence. Can. J. Fish. Aquat. Sci. 49:2546–50.

Harington, C. R. 1968. Denning habits of the polar bear (*Ursus maritimus* Phipps). Can. Wildl. Serv. Rept. Ser., no. 5, 30 pp.

Harrison, C. S. 1979. Sighting of Cuvier's beaked whale *(Ziphius cavirostris)* in the Gulf of Alaska. Murrelet 60:35–36.

Harrison, C. S., and J. D. Hall. 1978. Alaskan distribution of the beluga whale, *Delphinapterus leucas*. Can. Field-Nat. 92:235–41.

Harrison, R. J., and R. L. Brownell, Jr. 1971. The gonads of the South American dolphins, *Inia geoffrensis, Pontoporia blainvillei,* and *Sotalia fluviatilis*. J. Mamm. 52:413–19.

Harrison, R. J., M. M. Bryden, D. A. McBrearty, and R. L. Brownell, Jr. 1981. The ovaries and reproduction in *Pontoporia blainvillei* (Cetacea: Platanistidae). J. Zool. 193:563–80.

Hartman, D. S. 1979. Ecology and behavior of the manatee *(Trichechus manatus)* in Florida. Amer. Soc. Mamm. Spec. Publ., no. 5, viii + 153 pp.

Harwood, J. 1990. The 1988 seal epizootic. J. Zool. 222:349–51.

Hay, K. A., and A. W. Mansfield. 1989. Narwhal *Monodon monoceros* Linnaeus, 1758. *In* Ridgway and Harrison (1989), 145–76.

Hayssen, V., A. Van Tienhoven, and A. Van Tienhoven. 1993. Asdell's patterns of mammalian reproduction: a compendium of species-specific data. Comstock/Cornell Univ. Press, Ithaca, viii + 1023 pp.

Heide-Jorgensen, M.-P., and T. J. Härkönen. 1988. Rebuilding seal stocks in the Kattegat-Skagerrak. Mar. Mamm. Sci. 4:231–46.

Heide-Jorgensen, M. P., T. J. Härkönen, R. Dietz, and P. M. Thompson. 1992. Retrospective of the 1988 seal epizootic. Dis. Aquat. Org. 13:37–62.

Heide-Jorgensen, M. P., and R. R. Reeves. 1993. Description of an anomalous monodontid skull from west Greenland and a possible hybrid? Mar. Mamm. Sci. 9:258–68.

Heimark, R. J., and G. M. Heimark. 1986. Southern elephant seal pupping at Palmer Station, Antarctica. J. Mamm. 67:189–90.

Heimlich-Boran, J. R. 1986. Cohesive relationships among Puget Sound killer whales. *In* Kirkevold and Lockard (1986), 261–84.

Helle, E. 1980. Reproduction, size, and structure of the Baltic ringed seal population of the Bothnian Bay. Acta Univ. Ouluensis, ser. A, no. 106, 47 pp.

Helle, E., H. Hyvärinen, and T. Sipilä. 1984. Breeding habitat and lair structure of the Saimaa ringed seal *Phoca hispida saimensis* Nordq. in Finland. Acta Zool. Fennica 172: 125–27.

Herald, E. S., R. L. Brownell, Jr., F. L. Frye, E. J. Morris, W. E. Evans, and A. B. Scott. 1969. Blind river dolphin: first side-swimming cetacean. Science 166:1408–10.

Herman, L. M. 1980. Humpback whales in Hawaiian waters: a study in historical ecology. Pacific Sci. 33:1–15.

Herman, L. M., and R. C. Antinoja. 1977. Humpback whales in the Hawaiian breeding waters: population and pod characteristics. Sci. Rept. Whales Res. Inst. 29:59–85.

Herman, L. M., C. S. Baker, P. H. Forestell, and R. C. Antinoja. 1980. Right whale *Balaena glacialis* sightings near Hawaii: a clue to the wintering grounds? Mar. Ecol. Prog. Ser. 2:271–75.

Hernandez, P., J. E. Reynolds, III, H. Marsh, and M. Marmontel. 1995. Age and seasonality in spermatogenesis of Florida manatees. *In* O'Shea, Ackerman, and Percival (1995), 84–95.

Hersh, S. L., and D. A. Duffield. 1990. Distinction between northwest Atlantic offshore and coastal bottlenose dolphins based on hemoglobin profile and morphometry. *In* Leatherwood and Reeves (1990), 129–39.

Hersh, S. L., and D. K. Odell. 1986. Mass stranding of Fraser's dolphin, *Lagenodelphis hosei,* in the western North Atlantic. Mar. Mamm. Sci. 2:73–76.

Heyning, J. E. 1984. Functional morphology involved in intraspecific fighting of the beaked whale, *Mesoplodon carlhubbsi.* Can. J. Zool. 62:1645–54.

———. 1989*a*. Comparative facial anatomy of beaked whales (Ziphiidae) and a systematic revision among the families of extant Odontoceti. Los Angeles Co. Nat. Hist. Mus. Contrib. Sci., no. 405, 64 pp.

———. 1989*b*. Cuvier's beaked whale *Ziphius cavirostris* G. Cuvier, 1823. *In* Ridgway and Harrison (1989), 289–308.

Heyning, J. E., and M. E. Dahlheim. 1988. *Orcinus orca.* Mammalian Species, no. 304, 9 pp.

Heyning, J. A., and W. F. Perrin. 1994. Evidence for two species of common dolphins (genus *Delphinus*) from the eastern North Pacific. Los Angeles Co. Nat. Hist. Mus. Contrib. Sci., no. 442, 35 pp.

Hickman, D. L., and E. M. Grigsby. 1978. Comparison of signature whistles in *Tursiops truncatus.* Cetology, no. 31, 10 pp.

Higgins, L. V. 1993. The nonannual nonseasonal breeding cycle of the Australian sea lion, *Neophoca cinerea.* J. Mamm. 74:270–74.

Higgins, L. V., and L. Gass. 1993. Birth to weaning: parturition, duration of lactation, and attendance cycles of Australian sea lions *(Neophoca cinerea).* Can. J. Zool. 71:2047–55.

Hill, D. O. 1975. Vanishing giants. Audubon 77(1):56–107.

Hindell, M. A. 1991. Some life-history parameters of a declining population of southern elephant seals, *Mirounga leonina.* J. Anim. Ecol. 60:119–34.

Hindell, M. A., and H. R. Burton. 1988. The history of the elephant seal industry at Macquarie Island and an estimate of the presealing numbers. Pap. Proc. Roy. Soc. Tasmania 122:159–76.

Hindell, M. A., and G. J. Little. 1988. Longevity, fertility, and philopatry of two female southern elephant seals *(Mirounga leonina)* at Macquarie Island. Mar. Mamm. Sci. 4:168–71.

Hindell, M. A., D. J. Slip, and H. R. Burton. 1991. The diving behaviour of adult male and female southern elephant seals, *Mirounga leonina* (Pinnipedia: Phocidae). Austral. J. Zool. 39:595–619.

Hindell, M. A., D. J. Slip, H. R. Burton, and M. M. Bryden. 1992. Physiological implications of continuous, prolonged, and deep dives of the southern elephant seal *(Mirounga leonina).* Can. J. Zool. 70:370–79.

Hoese, H. D. 1971. Dolphin feeding out of water in a salt marsh. J. Mamm. 52:222–23.

Hollien, H., P. Hollien, D. K. Caldwell, and M. C. Caldwell. 1976. Sound production by the Atlantic bottlenosed dolphin *Tursiops truncatus.* Cetology, no. 26, 8 pp.

Holthuis, L. B. 1987. The scientific name of the sperm whale. Mar. Mamm. Sci. 3:87–90.

Horwood, J. W. 1986. The distribution of the southern blue whale in relation to recent estimates of abundance. Sci. Rept. Whales Res. Inst. 37:155–65.

———. 1987. The sei whale: population biology, ecology, and management. Croom Helm, London, 375 pp.

———. 1990. Biology and exploitation of the minke whale. CRC Press, Boca Raton, Florida, 238 pp.

Hoyt, E. 1977. *Orcinus orca:* separating facts from fantasies. Oceans 10(4):22–26.

Hua Yuanyu, Zhao Qingzhong, and Zhang Guocheng. 1989. The habitat and behavior of *Lipotes vexillifer. In* Perrin et al. (1989), 92–98.

Hubbs, C. L., and K. S. Norris. 1971. Original teeming abundance, supposed extinction, and survival of the Juan Fernandez fur seal. *In* Burt, W. H., ed., Antarctic Pinnipedia, Amer. Geophys. Union, Washington, D.C., 35–51.

Hubbs, C. L., W. F. Perrin, and K. C. Balcomb. 1973. *Stenella coeruleoalba* in the eastern and central tropical Pacific. J. Mamm. 54: 549–52.

Hult, R. W. 1982. Another function of echolocation for bottlenosed dolphins *(Tursiops truncatus).* Cetology, no. 47, 7 pp.

Husar, S. L. 1977. *Trichechus inunguis.* Mammalian Species, no. 72, 4 pp.

———. 1978*a*. *Dugong dugon.* Mammalian Species, no. 88, 7 pp.

———. 1978*b*. *Trichechus manatus.* Mammalian Species, no. 93, 5 pp.

———. 1978*c*. *Trichechus senegalensis.* Mammalian Species, no. 89, 3 pp.

Husson, A. M. 1978. The mammals of Suriname. E. J. Brill, Leiden, xxxiv + 569 pp.

Husson, A. M., and L. B. Holthuis. 1974. *Physeter macrocephalus* Linnaeus, 1758, the valid name for the sperm whale. Zool. Meded. 48:205–17.

Hyvärinen, H. 1989. Diving in darkness: whiskers as sense organs of the ringed seal *(Phoca hispida saimensis).* J. Zool. 218:663–78.

I

Ibáñez, C., M. Delibes, J. Castroviejo, R. Martín, J. F. Beltrán, and S. Moreno. 1988. An unusual record of hooded seal *(Cystophora cristata)* in SW Spain. Z. Saugetierk. 53: 189–90.

Ichihara, T. 1966. The pygmy blue whale *Balaenoptera musculus brevicauda,* a new subspecies from the Antarctic. *In* Norris (1966), 79–113.

International Whaling Commission. 1992. Report of the Sub-Committee on Small Cetaceans. Rept. Internatl. Whaling Comm. 42:178–234.

———. 1994. Report of the Sub-Committee on Southern Hemisphere Baleen Whales, and Report of the Sub-Committee on Aboriginal Subsistence Whaling. Rept. Internatl. Whaling Comm. 45:120–64.

Irvine, A. B., M. D. Scott, R. S. Wells, and J. H. Kaufmann. 1981. Movements and activities of the Atlantic bottlenose dolphin, *Tursiops truncatus,* near Sarasota, Florida. Fishery Bull. 79:671–88.

J

Jackson, H. H. T. 1961. Mammals of Wisconsin. Univ. Wisconsin Press, Madison, xii + 504 pp.

Jameson, R. J. 1989. Movements, home range, and territories of male sea otters off central California. Mar. Mamm. Sci. 5:159–72.

———. 1993. Survey of a translocated sea otter population. Internatl. Union Conserv. Nat. Otter Specialist Group Bull. 8:2–4.

Jameson, R. J., K. W. Kenyon, A. M. Johnson, and H. M. Wight. 1982. History and status of translocated sea otter populations in North America. Wildl. Soc. Bull. 10:100–107.

Jefferson, T. A. 1988. *Phocoenoides dalli.* Mammalian Species, no. 319, 7 pp.

———. 1989. Calving seasonality of Dall's porpoise in the eastern North Pacific. Mar. Mamm. Sci. 5:196–200.

Jefferson, T. A., and B. E. Curry. 1994. A global review of porpoise (Cetacea: Phocoenidae) mortality in gillnets. Biol. Conserv. 67:167–83.

Jefferson, T. A., and S. Leatherwood. 1994.

Lagenodelphis hosei. Mammalian Species, no. 470, 5 pp.

Jefferson, T. A., M. W. Newcomer, S. Leatherwood, and K. Van Waerebeek. 1994. Right whale dolphins *Lissodelphis borealis* (Peale, 1848) and *Lissodelphis peronii* (Lacépède, 1804). *In* Ridgway and Harrison (1994), 335–62.

Jerison, H. J. 1986. The perceptual worlds of dolphins. *In* Schusterman, Thomas, and Wood (1986), 141–66.

Jing Xianying, Xiao Youfu, and Jing Rongcai. 1981. Acoustic signals and acoustic behaviour of Chinese river dolphin *(Lipotes vexillifer)*. Sci. Sinica 24:407–15.

Johnson, D. W. 1990. A southern elephant seal *(Mirounga leonina)* in the Northern Hemisphere (Sultanate of Oman). Mar. Mamm. Sci. 6:242–43.

Johnson, J. H., and A. A. Wolman. 1984. The humpback whale, *Megaptera novaeangliae*. Mar. Fish. Rev. 46(4):30–37.

Johnson, R. H. 1982. Food-sharing behavior in captive Amazon river dolphins *(Inia geoffrensis)*. Cetology, no. 43, 2 pp.

Jones, G. S. 1984. *Phoca* sp. (Mammalia: Carnivora; Pinnipedia) on Taiwan: extra-limital records. J. Taiwan Mus. 37:75–76.

Jones, J. K., Jr., R. S. Hoffmann, D. W. Rice, C. Jones, R. J. Baker, and M. D. Engstrom. 1992. Revised checklist of North American mammals north of Mexico, 1991. Occas. Pap. Mus. Texas Tech Univ., no. 146, 23 pp.

Jones, M. L. 1982. Longevity of captive mammals. Zool. Garten 52:113–28.

Jones, M. L., S. L. Swartz, and S. Leatherwood, eds. 1984. The gray whale *Eschrichtius robustus*. Academic Press, Orlando, xxiv + 600 pp.

Jonkel, C. J. 1978. Black, brown (grizzly), and polar bears. *In* Schmidt and Gilbert (1978), 227–48.

Jonkel, C. J., P. Smith, I. Stirling, and G. B. Kolenosky. 1976. The present status of the polar bear in the James Bay and Belcher Islands area. Can. Wildl. Serv. Occas. Pap., no. 26, 42 pp.

Joseph, J. 1994. The tuna-dolphin controversy in the eastern Pacific Ocean: biological, economic, and political impacts. Ocean Devel. Internatl. Law 25:1–30.

K

Kaiser, H. E. 1974. Morphology of the Sirenia. S. Karger, Basel, 76 pp.

Kamiya, T., and F. Yamasaki. 1974. Organ weights of *Pontoporia blainvillei* and *Platanista gangetica*. Sci. Rept. Whales Res. Inst. 26:265–70.

Kastelein, R. A., and P. Mosterd. 1989. The excavation technique for molluscs of Pacific walruses *(Odobenus rosmarus divergens)* under controlled conditions. Aquat. Mamm. 15:3–5.

Kasuya, T. 1972*a*. Growth and reproduction of *Stenella coeruleoalba* based on the age determination by means of dentinal growth layers. Sci. Rept. Whales Res. Inst. 24:57–79.

———. 1972*b*. Some information on the growth of the Ganges dolphin with a comment on the Indus dolphin. Sci. Rept. Whales Res. Inst. 24:87–108.

———. 1973. Systematic consideration of the Recent toothed whales based on the morphology of tympano-periotic bone. Sci. Rept. Whales Res. Inst. 25:1–103.

———. 1975. Past occurrence of *Globicephala melaena* in the western North Pacific. Sci. Rept. Whales Res. Inst. 27:95–110.

———. 1976. Reconsideration of life history parameters of the spotted and striped dolphins based on cemental layers. Sci. Rept. Whales Res. Inst. 28:73–106.

———. 1977. Age determination and growth of the Baird's beaked whale with a comment on the fetal growth rate. Sci. Rept. Whales Res. Inst. 29:1–20.

———. 1978. The life history of Dall's porpoise with special reference to the stock off the Pacific coast of Japan. Sci. Rept. Whales Res. Inst. 30:1–63.

———. 1985. Effect of exploitation on reproductive parameters of the spotted and striped dolphins off the Pacific coast of Japan. Sci. Rept. Whales Res. Inst. 36:107–38.

———. 1986. Distribution and behavior of Baird's beaked whales off the Pacific coast of Japan. Sci. Rept. Whales Res. Inst. 37:61–83.

Kasuya, T., and A. K. M. Aminul Haque. 1972. Some information on distribution and seasonal movement of the Ganges dolphin. Sci. Rept. Whales Res. Inst. 24:109–15.

Kasuya, T., and R. L. Brownell, Jr. 1979. Age determination, reproduction, and growth of Franciscana dolphin *Pontoporia blainvillei*. Sci. Rept. Whales Res. Inst. 31:45–67.

Kasuya, T., and K. Kureha. 1979. The population of finless porpoise in the Inland Sea of Japan. Sci. Rept. Whales Res. Inst. 31:1–44.

Kasuya, T., and H. Marsh. 1984. Life history and reproductive biology of the short-finned pilot whale, *Globicephala macrorhynchus*, off the Pacific coast of Japan. *In* Perrin, Brownell, and DeMaster (1984), 259–310.

Kasuya, T., H. Marsh, and A. Amino. 1993. Non-reproductive mating in short-finned pilot whales. *In* Donovan, Lockyer, and Martin (1993), 425–37.

Kasuya, T., N. Miyazaki, and W. H. Dawbin. 1974. Growth and reproduction of *Stenella attenuata* in the Pacific Coast of Japan. Sci. Rept. Whales Res. Inst. 26:157–226.

Kasuya, T., and M. Nishiwaki. 1975. Recent status of the population of Indus dolphin. Sci. Rept. Whales Res. Inst. 27:81–94.

Kasuya, T., and S. Tai. 1993. Life history of short-finned pilot whale stocks off Japan and a description of the fishery. *In* Donovan, Lockyer, and Martin (1993), 439–73.

Kato, H., and K. Sakuramoto. 1991. Age at sexual maturity of southern minke whales: a review and some additional analyses. Rept. Internatl. Whaling Comm. 41:331–37.

Kawamura, A. 1973. Food and feeding of sei whale caught in the waters south of 40°N in the North Pacific. Sci. Rept. Whales Res. Inst. 25:219–36.

———. 1974. Food and feeding ecology in the southern sei whale. Sci. Rept. Whales Res. Inst. 26:25–144.

Kellnhauser, J. T. 1983. The acceptance of *Lontra* Gray for the New World river otters. Can. J. Zool. 61:278–79.

Kellogg, R. 1940. Whales, giants of the sea. Natl. Geogr. 77:35–90.

Kenyon, K. W. 1962. History of the Steller sea lion at the Pribilof Islands, Alaska. J. Mamm. 43:68–75.

———. 1969. The sea otter in the eastern Pacific Ocean. N. Amer. Fauna, no. 68, ix + 352 pp.

———. 1981. Monk seals—*Monachus*. *In* Ridgway and Harrison (1981*b*), 195–220.

Kerley, G. I. H. 1983*a*. Comparison haul-out patterns of fur seals *Arctocephalus tropicalis* and *A. gazella* on subantarctic Marion Island. S. Afr. J. Wildl. Res. 13:71–77.

———. 1983*b*. Record for the Cape fur seal *Arctocephalus pusillus pusillus* from subantarctic Marion Island. S. Afr. J. Zool. 18:139–40.

———. 1983*c*. Relative population sizes and trends, and hybridization of fur seals *Arctocephalus tropicalis* and *A. gazella* at the Prince Edward Islands, Southern Ocean. S. Afr. J. Zool. 18:388–92.

———. 1987. *Arctocephalus tropicalis* on the Prince Edward Islands. *In* Croxall and Gentry (1987), 61–64.

Kerley, G. I. H., and T. J. Robinson. 1987. Skull morphometrics of male antarctic and subantarctic fur seals, *Arctocephalus gazella* and *A. tropicalis*, and their interspecific hybrids. *In* Croxall and Gentry (1987), 121–31.

Khan, K. M., and M. S. Niazi. 1989. Distribution and population status of the Indus dolphin, *Platanista minor*. *In* Perrin et al. (1989), 77–80.

King, J. E. 1983. Seals of the world. British Mus. (Nat. Hist.), London, 240 pp.

Kingdon, J. 1971. East African mammals: an atlas of evolution in Africa. I. Academic Press, London, ix + 446 pp.

Kingsley, M. C. S., and M. A. Ramsay. 1988. The spiral in the tusk of the narwhal. Arctic 41:236–38.

Kinzelbach, R., and J. Boessneck. 1992. Vorkommen der Mönchsrobbe *Monachus monachus* auf Sal (Kapverdische Inseln). Z. Saugetierk. 57:58–59.

Kleinenberg, S. E., A. V. Yablokov, B. M. Bel'Kovich, and M. N. Tarasevich. 1969. Beluga *(Delphinapterus leucas):* investigation of the species. Israel Progr. Sci. Transl., Jerusalem, vi + 376 pp.

Klinowska, M., and J. Cooke. 1991. Dolphins, porpoises, and whales of the world: the IUCN red data book. IUCN (World Conservation Union), Gland, Switzerland, viii + 429 pp.

Knudsen, B. 1978. Time budgets of polar bears *(Ursus maritimus)* on North Twin Island, James Bay, during summer. Can. J. Zool. 56:1627–28.

Kolenosky, G. B. 1987. Polar bear. *In* Novak, Baker, et al. (1987), 474–87.

Kooyman, G. L. 1981*a*. Crabeater seal—*Lobodon carcinophagus*. *In* Ridgway and Harrison (1981*b*), 221–35.

———. 1981*b*. Leopard seal—*Hydrurga leptonyx*. *In* Ridgway and Harrison (1981*b*), 261–74.

———. 1981*c*. Weddell seal: consummate diver. Cambridge Univ. Press, viii + 135 pp.

———. 1981*d*. Weddell seal—*Leptonychotes weddelli*. *In* Ridgway and Harrison (1981*b*), 275–96.

Kooyman, G. L., R. W. Davis, and J. P. Croxall. 1986. Diving behaviour of Antarctic fur seals. *In* Gentry and Kooyman (1986), 115–25.

Kooyman, G. L., and R. L. Gentry. 1986. Diving behavior of South African fur seals. *In* Gentry and Kooyman (1986), 145–52.

Kooyman, G. L., and F. Trillmich. 1986. Diving behavior of Galapagos sea lions. *In* Gentry and Kooyman (1986), 209–19.

Koski, L. F., R. A. Davis, G. W. Miller, and D. E. Withrow. 1993. Reproduction. *In* Burns, Montague, and Cowles (1993), 239–74.

Kovacs, K. M., and D. M. Lavigne. 1986. *Cystophora cristata*. Mammalian Species, no. 258, 9 pp.

Kraus, S. D., J. H. Prescott, A. R. Knowlton, amd G. S. Stone. 1986. Migration and calving of right whales *(Eubalaena glacialis)* in the western North Atlantic. *In* Brownell, Best, and Prescott (1986), 139–44.

Krupnik, I. I. 1984. Gray whales and the aborigines of the Pacific Northwest: the history of aboriginal whaling. *In* Jones, Swartz, and Leatherwood (1984), 103–20.

Kurten, B. 1973. Transberingean relationships of *Ursus arctos* Linne (brown and grizzly bears). Commentat. Biol. 65:1–10.

L

Lander, R. H., and H. Kajimura. 1982. Status of northern fur seals. Mammals in the Seas, FAO Fish. Ser. No. 5, 4:319–45.

Larsen, T. 1975. Polar bear den surveys in Svalbard in 1973. Norsk Polarinst. Arbok, 1973, 101–12.

Latinen, K. 1987. Longevity and fertility of the polar bear, *Ursus maritimus* Phipps, in captivity. Zool. Garten 57:197–99.

Laurie, A., and J. Seidensticker. 1977. Behavioural ecology of the sloth bear *(Melursus ursinus)*. J. Zool. 182:187–204.

Laws, R. M. 1984. The krill-eating crabeater. *In* Macdonald (1984), 280–81.

———. 1985. Animal conservation in the Antarctic. *In* Hearn, J. P., and J. K. Hodges, eds., Advances in animal conservation, Symp. Zool. Soc. London 54:3–23.

———. 1994. History and present status of southern elephant seal populations. *In* Le Boeuf and Laws (1994*b*), 49–65.

Lay, D. M. 1967. A study of the mammals of Iran. Fieldiana Zool. 54:1–282.

Lazcano-Barrero, M. A., and J. M. Packard. 1989. The occurrence of manatees *(Trichechus manatus)* in Tamaulipas, Mexico. Mar. Mamm. Sci. 5:202–5.

Leatherwood, J. S. 1974. Aerial observations of migrating gray whales, *Eschrichtius robustus*, off southern California, 1969–72. Mar. Fish. Rev. 36(4):45–49.

Leatherwood, S., D. K. Caldwell, and H. E. Winn. 1976. Whales, dolphins, and porpoises of the western North Atlantic: a guide to their identification. U.S. Natl. Mar. Fish. Serv., NOAA Tech. Rept. NMFS Circ. 396, iv + 176 pp.

Leatherwood, S., J. S. Grove, and A. E. Zuckerman. 1991. Dolphins of the genus *Lagenorhynchus* in the tropical South Pacific. Mar. Mamm. Sci. 7:194–97.

Leatherwood, S., L. J. Harrington-Coulombe, and C. L. Hubbs. 1978. Relict survival of the sea otter in central California and evidence of its recent dispersal south of Point Conception. Bull. S. California Acad. Sci. 77:109–15.

Leatherwood, S., R. A. Kastelein, and K. W. Miller. 1988. Observations of Commerson's dolphin and other cetaceans in southern Chile, January–February 1984. *In* Brownell and Donovan (1988), 71–83.

Leatherwood, S., and R. R. Reeves. 1978. Porpoises and dolphins. *In* Haley (1978), 97–111.

———. 1983. The Sierra Club handbook of whales and dolphins. Sierra Club Books, San Francisco, xviii + 302 pp.

———, eds. 1990. The bottlenose dolphin. Academic Press, San Diego, xviii + 653 pp.

———. 1994. River dolphins: a review of activities and plans of the Cetacean Specialist Group. Aquat. Mamm. 20:137–54.

Leatherwood, S., R. R. Reeves, W. F. Perrin, and W. E. Evans. 1982. Whales, dolphins, and porpoises of the eastern North Pacific and adjacent arctic waters. U.S. Natl. Mar. Fish. Serv., NOAA Tech. Rept. NMFS Circ. 444, v + 245 pp.

Leatherwood, S., and W. A. Walker. 1979. The northern right whale dolphin *Lissodelphis borealis* Peale in the eastern North Pacific. *In* Winn and Olla (1979), 85–141.

Le Boeuf, B. J. 1994. Variation in the diving pattern of northern elephant seals with age, mass, sex, and reproductive condition. *In* Le Boeuf and Laws (1994*b*), 237–52.

Le Boeuf, B. J., D. Aurioles, R. Condit, C. Fox, R. Gisiner, R. Romero, and F. Sinsel. 1983. Size and distribution of the California sea lion population in Mexico. Proc. California Acad. Sci. 43:77–85.

Le Boeuf, B. J., D. P. Costa, A. C. Huntley, and S. D. Feldkamp. 1988. Continuous deep diving in female northern elephant seals, *Mirounga angustirostris*. Can. J. Zool. 66: 446–58.

Le Boeuf, B. J., K. W. Kenyon, and B. Villa-Ramirez. 1986. The Caribbean monk seal is extinct. Mar. Mamm. Sci. 2:70–72.

Le Boeuf, B. J., and R. M. Laws. 1994*a*. Elephant seals: an introduction to the genus. *In* Le Boeuf and Laws (1994*b*), 1–26.

———, eds. 1994*b*. Elephant seals: population ecology, behavior, and physiology. Univ. California Press, Berkeley, xviii + 414 pp.

Le Boeuf, B. J., Y. Naito, T. Asaga, D. Crocker, and D. P. Costa. 1992. Swim speed in a female northern elephant seal: metabolic and foraging implications. Can. J. Zool. 70:786–95.

Le Boeuf, B. J., Y. Naito, A. C. Huntley, and T. Asaga. 1989. Prolonged, continuous, deep diving by northern elephant seals. Can. J. Zool. 67:2514–19.

Lefebvre, L. W., T. J. O'Shea, G. B. Rathbun, and R. C. Best. 1989. Distribution, status, and biogeography of the West Indian manatee. *In* Woods (1989*a*), 567–610.

Lekagul, B., and J. A. McNeely. 1977. Mammals of Thailand. Sahakarnbhat, Bangkok, li + 758 pp.

Lichter, A. A. 1986. Records of beaked whales (Ziphiidae) from the western South Atlantic. Sci. Rept. Whales Res. Inst. 37:109–27.

Lichter, A. A., F. Fraga, and H. P. Castello. 1990. First record of the pygmy killer whale, *Feresa attenuata*, in the southwest Atlantic. Mar. Mamm. Sci. 6:85–86.

Lilly, J. C. 1977. The cetacean brain. Oceans 10(4):4–7.

Lima, M., and E. Páez. 1995. Growth and reproductive patterns in the South American fur seal. J. Mamm. 76:1249–55.

Ling, J. K. 1987. New Zealand fur seal, *Arctocephalus forsteri,* in South Australia. *In* Croxall and Gentry (1987), 53–55.

Ling, J. K., and M. M. Bryden. 1981. Southern elephant seal—*Mirounga leonina. In* Ridgway and Harrison (1981*b*), 297–327.

———, eds. 1985. Studies of sea mammals in south latitudes. S. Austral. Mus., Sydney, vii + 132 pp.

Linscombe, G. 1994. U.S. fur harvest (1970–1992) and fur value (1974–1992) statistics by state and region. Louisiana Department of Wildlife and Fisheries, Baton Rouge, 29 pp.

Lint, D. W., J. W. Clayton, W. R. Lillie, and L. Postma. 1990. Evolution and systematics of the beluga whale, *Delphinapterus leucas,* and other odontocetes: a molecular approach. Can. Bull. Fish. Aquat. Sci. 224:7–22.

Lloyd, D. S., C. P. McRoy, and R. H. Day. 1981. Discovery of northern fur seals *(Callorhinus ursinus)* breeding on Bogoslof Island, southeastern Bering Sea. Arctic 34:318–20.

Lo, N. C. H., and T. D. Smith. 1986. Incidental mortality of dolphins in the eastern tropical Pacific, 1959–72. Fishery Bull. 84:27–33.

Lockyer, C. 1984. Review of baleen whale (Mysticeti) reproduction and implications for management. *In* Perrin, Brownell, and DeMaster (1984), 27–50.

———. 1990. Review of incidents involving wild, sociable dolphins, worldwide. *In* Leatherwood and Reeves (1990), 337–53.

Lockyer, C., R. N. P. Goodall, and A. R. Galeazzi. 1988. Age and body-length characteristics of *Cephalorhynchus commersonii* from incidentally-caught specimens off Tierra del Fuego. *In* Brownell and Donovan (1988), 103–18.

Lopez, J. C., and D. Lopez. 1985. Killer whales *(Orcinus orca)* of Patagonia, and their behavior of intentional stranding while hunting near shore. J. Mamm. 66:181–83.

Loughlin, T. R. 1980. Home range and territoriality of sea otters near Monterey, California. J. Wildl. Mgmt. 44:576–82.

Loughlin, T. R., C. H. Fiscus, A. M. Johnson, and D. J. Rugh. 1982. Observations of *Mesoplodon stejnegeri* (Ziphiidae). J. Mamm. 63: 697–700.

Loughlin, T. R., and M. A. Perez. 1985. *Mesoplodon stejnegeri.* Mammalian Species, no. 250, 6 pp.

Loughlin, T. R., M. A. Perez, and R. L. Merrick. 1987. *Eumetopias jubatus.* Mammalian Species, no. 283, 7 pp.

Loughlin, T. R., A. S. Perlov, and V. A. Vladimirov. 1992. Range-wide survey and estimation of total number of Steller sea lions in 1989. Mar. Mamm. Sci. 8:220–39.

Loughlin, T. R., D. J. Rugh, and C. H. Fiscus. 1984. Northern sea lion distribution and abundance: 1956–80. J. Wildl. Mgmt. 48: 729–40.

M

Ma Yi-ching. 1983. The status of bears in China. Acta Zool. Fennica 174:165–66.

Maas, M. C., and J. G. M. Thewissen. 1995. Enamel microstructure of *Pakicetus* (Mammalia: Archaeoceti). J. Paleontol. 69:1154–63.

Mackintosh, N. A. 1966. The distribution of southern blue and fin whales. *In* Norris (1966), 125–44.

Majluf, O. 1987. South American fur seal, *Arctocephalus australis,* in Peru. *In* Croxall and Gentry (1987), 33–35.

Mansfield, A. W., T. G. Smith, and B. Beck. 1975. The narwhal, *Monodon monoceros,* in eastern Canadian waters. J. Fish. Res. Bd. Can. 32:1041–46.

Marchessaux, D. 1989. Distribution et statut des populations du phoque moine *Monachus monachus* (Hermann, 1779). Mammalia 53: 621–42.

Marchessaux, D., and C. Pergent-Martini. 1991. Biologie de la reproduction et developpement des nouveaux nes chez le phoque moine *Monachus monachus. In* Boudouresque, C. F., M. Avon, and V. Gravez, eds., Les espèces marines à protéger en Méditerranée, GIS Posidonie, France, 349–58.

Marine Mammal Commission. 1994. Annual Report to Congress 1993. Marine Mammal Commission, Washington, D.C., xv + 240 pp.

Marmontel, M. 1995. Age and reproduction in female Florida manatees. *In* O'Shea, Ackerman, and Percival (1995), 98–119.

Marmontel, M., D. K. Odell, and J. E. Reynolds, III. 1992. Reproductive biology of South American manatees. *In* Hamlett, W. C., ed., Reproductive biology of South American vertebrates, Springer-Verlag, New York, 295–312.

Marquette, W. M. 1978. Bowhead whale. *In* Haley (1978), 70–81.

———. 1979. The 1977 catch of bowhead whales *(Balaena mysticetus)* by Alaskan Eskimos. Rept. Internatl. Whaling Comm. 29:281–89.

Marsh, H. 1989. Dugong numbers in Western Australia. Sirenews 12:3–4.

———. 1994. Is the catch of dugongs in Torres Strait sustainable? Sirenews 22:1–2.

———. 1995. The life history, pattern of breeding, and population dynamics of the dugong. *In* O'Shea, Ackerman, and Percival (1995), 75–83.

Marsh, H., G. E. Heinsohn, and L. M. Marsh. 1984. Breeding cycle, life history, and population dynamics of the dugong, *Dugong dugon* (Sirenia: Dugongidae). Austral. J. Zool. 32:767–88.

Marsh, H., R. Lloze, G. E. Heinsohn, and T. Kasuya. 1989. Irrawaddy dolphin *Orcaella brevirostris* (Gray, 1866). *In* Ridgway and Harrison (1989), 101–18.

Marsh, H., R. I. T. Prince, W. K. Saalfeld, and R. Sheperd. 1994. The distribution and abundance of the dugong in Shark Bay, Western Australia. Wildl. Res. 21:149–61.

Marsh, H., G. B. Rathbun, T. J. O'Shea, and A. R. Preen. 1995. Can dugongs survive in Palau? Biol. Conserv. 72:85–89.

Marsh, H., and W. K. Saalfeld. 1989. Distribution and abundance of dugongs in the northern Great Barrier Reef Marine Park. Austral. Wildl. Res. 16:429–40.

———. 1990. The distribution and abundance of dugongs in the Great Barrier Reef Marine Park south of Cape Bedford. Austral. Wildl. Res. 17:511–24.

Martin, A. R., M. C. S. Kingsley, and M. A. Ramsay. 1994. Diving behaviour of narwhales *(Monodon monoceros)* on their summer grounds. Can. J. Zool. 72:118–25.

Martin, A. R., P. Reynolds, and M. G. Richardson. 1987. Aspects of the biology of pilot whales *(Globicephala melaena)* in recent mass strandings on the British coast. J. Zool. 211:11–23.

Martin, A. R., and P. Rothery. 1993. Reproductive parameters of female long-finned pilot whales *(Globicephala melas)* around the Faroe Islands. *In* Donovan, Lockyer, and Martin (1993), 263–304.

Martin, A. R., and T. G. Smith. 1992. Deep diving in wild, free-ranging beluga whales, *Delphinapterus leucas.* Can. J. Fish. Aquat. Sci. 49:462–66.

Martin, P. S., and H. E. Wright, eds. 1967. Pleistocene extinctions. Yale Univ. Press, New Haven, x + 453 pp.

Masaki, Y. 1976. Biological studies on the North Pacific sei whale. Bull. Far Seas Fish. Res. Lab. 14:1–104.

———. 1978. Yearly change in the biological parameters of the Antarctic sei whale. Rept. Internatl. Whaling Comm. 28:421–29.

Mason, C. F., and S. M. Macdonald. 1986. Otters: ecology and conservation. Cambridge Univ. Press, vii + 236 pp.

Mattlin, R. H. 1987. New Zealand fur seal, *Arctocephalus forsteri,* within the New Zealand region. *In* Croxall and Gentry (1987), 49–51.

McCann, C. 1975. A study of the genus *Berardius* Duvernoy. Sci. Rept. Whales Res. Inst. 27:111–37.

McCann, T. S. 1980. Territoriality and breeding behaviour of adult male Antarctic fur seal, *Arctocephalus gazella*. J. Zool. 192: 295–310.

———. 1985. Size, status, and demography of southern elephant seal *(Mirounga leonina)* populations. *In* Ling and Bryden (1985), 1–17.

McCann, T. S., and D. W. Doidge. 1987. Antarctic fur seal, *Arctocephalus gazella*. *In* Croxall and Gentry (1987), 5–8.

McGinnis, S. M., and R. J. Schusterman. 1981. Northern elephant seal—*Mirounga angustirostris*. *In* Ridgway and Harrison (1981*b*), 329–49.

McHugh, J. L. 1974. The role and history of the International Whaling Commission. *In* Schevill (1974), 305–35.

McLaren, I. A. 1960. Are the Pinnipedia biphyletic? Syst. Zool. 9:18–28.

———. 1984. True seals. *In* Macdonald (1984), 270–79.

McLeod, S. A., F. C. Whitmore, Jr., and L. G. Barnes. 1993. Evolutionary relationships and classification. *In* Burns, Montague, and Cowles (1993), 45–70.

McNally, R. 1977. Echolocation: cetaceans' sixth sense. Oceans 10(4):27–33.

McQuaid, C. D. 1986. Post-1980 sightings of bowhead whales *(Balaena mysticetus)* from the Spitsbergen stock. Mar. Mamm. Sci. 2:316–18.

McShane, L. J., J. A. Estes, M. L. Riedman, and M. M. Staedler. 1995. Repertoire, structure, and individual variation of vocalizations in the sea otter. J. Mamm. 76:414–27.

McWhinnie, M., and C. J. Denys. 1980. The high importance of the lowly krill. Nat. Hist. 89(3):66–73.

Mead, J. G. 1975. Preliminary report on the former net fisheries for *Tursiops truncatus* in the western North Atlantic. J. Fish. Res. Bd. Can. 32:1155–62.

———. 1984. Survey of reproductive data for the beaked whales (Ziphiidae). *In* Perrin, Brownell, and DeMaster (1984), 91–96.

———. 1989*a*. Beaked whales of the genus *Mesoplodon*. *In* Ridgway and Harrison (1989), 349–430.

———. 1989*b*. Bottlenose whales *Hyperoodon ampullatus* (Forster, 1770) and *Hyperoodon planifrons* Flower, 1882. *In* Ridgway and Harrison (1989), 321–48.

Mead, J. G., and A. N. Baker. 1987. Notes on the rare beaked whale, *Mesoplodon hectori* Gray). J. Roy. Soc. New Zealand 17:303– 12.

Mead, J. G., and E. D. Mitchell. 1984. Atlantic gray whales. *In* Jones, Swartz, and Leatherwood (1984), 33–53.

Mead, J. G., D. K. Odell, R. S. Wells, and M. D. Scott. 1980. Observations on a mass stranding of spinner dolphin, *Stenella longirostris*, from the west coast of Florida. Fishery Bull. 78:353–60.

Mead, J. G., and R. S. Payne. 1975. A specimen of the Tasman beaked whale, *Tasmacetus shepherdi*, from Argentina. J. Mamm. 56: 213–18.

Mead, J. G., and C. W. Potter. 1990. Natural history of bottlenose dolphins along the central Atlantic coast of the United States. *In* Leatherwood and Reeves (1990), 165–95.

———. 1995. Recognizing two populations of the bottlenose dolphin *(Tursiops truncatus)* off the Atlantic coast of North America: morphologic and ecologic considerations. IBI Reports 5:31–44.

Mead, J. G., W. A. Walker, and W. J. Houck. 1982. Biological observations on *Mesoplodon carlhubbsi* (Cetacea: Ziphiidae). Smithson. Contrib. Zool., no. 344, iii + 25 pp.

Meade, R. H., and L. Koehnken. 1991. Distribution of the river dolphin, tonina *Inia geoffrensis*, in the Orinoco River Basin of Venezuela and Colombia. Interciencia 16: 300–312.

Medway, Lord. 1977. Mammals of Borneo. Monogr. Malaysian Branch Roy. Asiatic Soc., no. 7, xii + 172 pp.

Meester, J., I. L. Rautenbach, N. J. Dippenaar, and C. M. Baker. 1986. Classification of southern African mammals. Transvaal Mus. Monogr., no. 5, x + 359 pp.

Melquist, W. E., and A. E. Dronkert. 1987. River otter. *In* Novak, Baker, et al. (1987), 626–41.

Melquist, W. E., and M. G. Hornocker. 1983. Ecology of river otters in west central Idaho. Wildl. Monogr., no. 83, 60 pp.

Méndez, E., and B. Rodriguez. 1984. A southern sea lion *Otaria flavescens* (Shaw) found in Panama. Caribbean J. Sci. 20:105–8.

Merrick, R. L., T. R. Loughlin, and D. G. Calkins. 1987. Decline in abundance of the northern sea lion, *Eumetopias jubatus*, in Alaska, 1956–86. Fishery Bull. 85:351–65.

Mignucci Giannoni, A. A. 1986. Caribbean monk seals in Puerto Rico. J. Colorado-Wyoming Acad. Sci. 18:54.

Milinkovitch, M. C. 1995. Molecular phylogeny of cetaceans prompts revision of morphological transformations. Tree 10:328–34.

Milinkovitch, M. C., A. Meyer, and J. R. Powell. 1994. Phylogeny of all major groups of cetaceans based on DNA sequences from three mitochondrial genes. Molecular Biol. Evol. 11:939–48.

Milinkovitch, M. C., G. Ortí, and A. Meyer. 1993. Revised phylogeny of whales suggested by mitochondrial ribosomal DNA sequences. Nature 361:346–48.

———. 1995. Novel phylogeny of whales revisited but not revised. Molecular Biol. Evol. 12:518–20.

Miller, E. H. 1975*a*. Body and organ measurements of fur seals, *Arctocephalus forsteri* (Lesson), from New Zealand. J. Mamm. 56: 511–13.

———. 1975*b*. Walrus ethology. I. The social role of tusks and applications of multidimensional scaling. Can. J. Zool. 53:590–613.

Miller, E. H., and D. J. Boness. 1979. Remarks on display functions of the snout of the grey seal, *Halichoerus grypus* (Fab.), with comparative notes. Can. J. Zool. 57:140–48.

Mitchell, E. D. 1975*a*. Porpoise, dolphin, and small whale fisheries of the world. Internatl. Union Conserv. Nat. Monogr., no. 3, 129 pp.

———. 1975*b*. Report of the meeting on smaller Cetaceans, Montreal, April 1–11, 1974. J. Fish. Res. Bd. Can. 32:889–983.

———. 1978. Finner whales. *In* Haley (1978), 36–45.

Mitchell, E. D., and V. M. Kozicki. 1975. Supplementary information on minke whale *(Balaenoptera acutorostrata)* from Newfoundland fishery. J. Fish. Res. Bd. Can. 32: 985–94.

Mitchell, E. D., and J. G. Mead. 1977. History of the gray whale in the Atlantic Ocean (abstr.). Proc. 2nd Conf. Biol. Mar. Mamm., San Diego.

Mitchell, E. D., and R. R. Reeves. 1982. Factors affecting abundance of bowhead whales *Balaena mysticetus* in the eastern Arctic of North America, 1915–1980. Biol. Conserv. 22:59–78.

———. 1983. Catch history, abundance, and present status of northwest Atlantic humpback whales. *In* Tillman and Donovan (1983), 153–212.

Miyazaki, N. 1984. Further analyses of reproduction in the striped dolphin, *Stenella coeruleoalba*, off the Pacific coast of Japan. *In* Perrin, Brownell, and DeMaster (1984), 343–54.

Miyazaki, N., L. L. Jones, and R. Beach. 1984. Some observations on the schools of *dalli*- and *truei*-type Dall's porpoises in the northwestern Pacific. Sci. Rept. Whales Res. Inst. 35:93–105.

Miyazaki, N., and W. F. Perrin. 1994. Rough-toothed dolphin *Steno bredanensis* (Lesson, 1828). *In* Ridgway and Harrison (1994), 1–21.

Mizroch, S. A., D. W. Rice, and J. M. Breiwick. 1984. The blue whale, *Balaenoptera musculus*. Mar. Fish. Rev. 46(4):15–19.

Mizue, K., M. Nishiwaki, and A. Takemura. 1971. The underwater sounds of Ganges River dolphins *(Platanista gangetica)*. Sci. Rept. Whales Res. Inst. 23:123–28.

Montgomery, G. G., R. C. Best, and M. Yamakoshi. 1981. A radio-tracking study of the Amazonian manatee *Trichechus inunguis*. Biotrópica 13:81–85.

Montgomery, G. G., N. B. Gale, and W. P. Murdoch, Jr. 1982. Have manatee entered the eastern Pacific Ocean? Mammalia 46:257–58.

Moore, J. C. 1968. Relationships among the living genera of beaked whales with classifications, diagnoses, and keys. Fieldiana Zool. 53:209–98.

———. 1972. More skull characters of the beaked whale *Indopacetus pacificus* and comparative measurements of austral relatives. Fieldiana Zool. 62:1–19.

Moore, S. E., and R. R. Reeves. 1993. Distribution and movement. *In* Burns, Montague, and Cowles (1993), 313–86.

Morejohn, G. V. 1979. The natural history of Dall's porpoise in the North Pacific Ocean. *In* Winn and Olla (1979), 45–83.

Mouchaty, S., J. A. Cook, and G. F. Shields. 1995. Phylogenetic analysis of northern hair seals based on nucleotide sequences of the mitochondrial cytochrome *b* gene. J. Mamm. 76:1178–85.

Mullen, D. A. 1977. The striped dolphin, *Stenella coeruleoalba*, in the Gulf of California. Bull. S. California Acad. Sci. 76:131–32.

Murray, M. D. 1981. The breeding of the southern elephant seal, *Mirounga leonina*, on the Antarctic continent. Polar Record 20:370–71.

Myrick, A. C., Jr., A. A. Hohn, J. Barlow, and P. A. Sloan. 1986. Reproductive biology of female spotted dolphins, *Stenella attenuata*, from the eastern tropical Pacific. Fishery Bull. 84:247–59.

N

Nagorsen, D. W. 1985. *Kogia simus*. Mammalian Species, no. 239, 6 pp.

Nerini, M. K., H. W. Braham, W. M. Marquette, and D. J. Rugh. 1984. Life history of the bowhead whale, *Balaena mysticetus* (Mammalia: Cetacea). J. Zool. 204:443–68.

Newman, M. A. 1978. Narwhal. *In* Haley (1978), 138–44.

Nishiwaki, M. 1966. A discussion of rarities among the smaller cetaceans caught in Japanese waters. *In* Norris (1966), 192–204.

———. 1975. Ecological aspects of smaller cetaceans, with emphasis on the striped dolphin *(Stenella coeruleoalba)*. J. Fish. Res. Bd. Can. 32:1069–72.

———. 1984. Current status of the African manatee. Acta Zool. Fennica 172:135–36.

Nishiwaki, M., and T. Kasuya. 1970. A Greenland right whale caught at Osaka Bay. Sci. Rept. Whales Res. Inst. 22:45–62.

Nishiwaki, M., T. Kasuya, N. Miyazoki, T. Tobayama, and T. Kataoka. 1979. Present distribution of the dugong in the world. Sci. Rept. Whales Res. Inst. 31:133–41.

Nishiwaki, M., and H. Marsh. 1985. Dugong— *Dugong dugon*. *In* Ridgway and Harrison (1985), 1–31.

Nishiwaki, M., and K. S. Norris. 1966. A new genus, *Peponocephala*, for the odontocete cetacean species *Electra electra*. Sci. Rept. Whales Res. Inst. 20:95–100.

Nishiwaki, M., and N. Oguro. 1971. Baird's beaked whales caught on the coast of Japan in recent ten years. Sci. Rept. Whales Res. Inst. 23:111–22.

Nishiwaki, M., M. Yamaguchi, S. Shokita, S. Uchida, and T. Kataoka. 1982. Recent survey on the distribution of the African manatee. Sci. Rept. Whales Res. Inst. 34:137–47.

Nores, C., and C. Pérez. 1988. The occurrence of walrus *(Odobenus rosmarus)* in southern Europe. J. Zool. 216:593–96.

Norris, K. S., ed. 1966. Whales, dolphins, and porpoises. Univ. California Press, Berkeley, xv + 789 pp.

Norris, K. S., and G. W. Harvey. 1972. A theory for the function of the spermaceti organ of the sperm whale (*Physeter catodon* L.). *In* Galler, S. R., K. Schmidt-Koenig, G. J. Jacobs, and R. E. Belleville, eds., Animal orientation and navigation, U.S. Natl. Aeronautics and Space Admin., Washington, D.C., 397–417.

Norris, K. S., B. Würsig, and R. S. Wells. 1994. The spinner dolphin. *In* Norris et al. (1994), 14–30.

Norris, K. S., B. Würsig, R. S. Wells, M. Würsig, S. M. Brownlee, C. M. Johnson, and J. Solow. 1994. The Hawaiian spinner dolphin. Univ. California Press, Berkeley, xxiii + 408 pp.

Novak, M., M. E. Obbard, J. G. Jones, R. Newman, A. Booth, A. J. Satterthwaite, and G. Linscombe. 1987. Furbearer harvests in North America, 1600–1984. Ontario Ministry Nat. Res., Toronto, xvi + 270 pp.

O

O'Brien, S. J. 1993. The molecular evolution of bears. *In* Stirling (1993), 26–35.

Odell, D. K. 1975. Status and aspects of the life history of the bottlenose dolphin, *Tursiops truncatus*, in Florida. J. Fish. Res. Bd. Can. 32:1055–58.

———. 1981. California sea lion—*Zalophus californianus*. *In* Ridgway and Harrison (1981*a*), 67–97.

———. 1984. The fight to mate. *In* Macdonald (1984), 262–63.

Odell, D. K., D. Forrester, and E. Asper. 1978. Growth and sexual maturation in the West Indian manatee. Amer. Soc. Mamm., Abstr. Tech. Pap., 58th Ann. Mtg., 7–8.

Ohsumi, S. 1965. Reproduction of the sperm whale in the north-west Pacific. Sci. Rept. Whales Res. Inst. 19:1–35.

———. 1966. Sexual segregation of the sperm whale in the North Pacific. Sci. Rept. Whales Res. Inst. 20:1–16.

———. 1971. Some investigations on the school structure of sperm whale. Sci. Rept. Whales Res. Inst. 23:1–25.

———. 1979. Population assessment of the Antarctic minke whale. Rept. Internatl. Whaling Comm. 29:407–20.

Ohsumi, S., and Y. Masaki. 1975. Biological parameters of the Antarctic minke whale at the virginal population level. J. Fish. Res. Bd. Can. 32:995–1004.

Ohsumi, S., Y. Masaki, and A. Kawamura. 1970. Stock of the Antarctic minke whale. Sci. Rept. Whales Res. Inst. 22:75–110.

Ohtaishi, N., and M. Yoneda. 1981. A thirty four years old male Kuril seal from Shiretoko Pen., Hokkaido. Sci. Rept. Whales Res. Inst. 33:131–35.

Olesiuk, P. F., M. A. Bigg, and G. M. Ellis. 1990. Life history and population dynamics of resident killer whales *(Orcinus orca)* in the coastal waters of British Columbia and Washington state. *In* Hammond, Mizroch, and Donovan (1990), 209–43.

Oliva, D. 1988. *Otaria byronia* (de Blainville, 1820), the valid scientific name for the southern sea lion (Carnivora: Otariidae). J. Nat. Hist. 22:767–72.

Omura, H. 1974. Possible migration route of the gray whale on the coast of Japan. Sci. Rept. Whales Res. Inst. 26:1–14.

———. 1984. History of gray whales in Japan. *In* Jones, Swartz, and Leatherwood (1984), 57–77.

———. 1986. History of right whale catches in the waters around Japan. *In* Brownell, Best, and Prescott (1986), 35–41.

———. 1988. Distribution and migration of the western Pacific stock of the gray whale. Sci. Rept. Whales Res. Inst. 39:1–9.

O'Shea, T. J., B. B. Ackerman, and H. F. Percival, eds. 1995. Population biology of the Florida manatee. Natl. Biol. Serv. Inform. Tech. Rept., no. 1, vi + 289 pp.

O'Shea, T. J., C. A. Beck, R. K. Bonde, H. I. Kochman, and D. K. Odell. 1985. An analysis of manatee mortality patterns in Florida, 1976–81. J. Wildl. Mgmt. 49:1–11.

O'Shea, T. J., and R. L. Reep. 1990. Encephalization quotients and life-history traits in the Sirenia. J. Mamm. 71:534–43.

O'Shea, T. J., and C. A. L. Salisbury. 1991.

Belize—a last stronghold for manatees in the Caribbean. Oryx 25:156–64.

P

Pagnoni, G., and S. Saba. 1989. New record of the spectacle porpoise. Mar. Mamm. Sci. 5:201–2.

Panou, A., J. Jacobs, and D. Panos. 1993. The endangered Mediterranean monk seal *Monachus monachus* in the Ionian Sea, Greece. Biol. Conserv. 64:129–40.

Pascual, M. A., and M. D. Adkison. 1994. The decline of the Steller sea lion in the northeast Pacific: demography, harvest, or environment. Ecol. Applic. 4:393–403.

Paul, J. R. 1968. Risso's dolphin, *Grampus griseus*, in the Gulf of Mexico. J. Mamm. 49:746–48.

Pavey, C. R. 1992. The occurrence of the pygmy right whale, *Caperea marginata* (Cetacea: Neobalaenidae), along the Australian coast. Austral. Mamm. 15:1–6.

Payne, K., and R. Payne. 1985. Large scale changes over nineteen years in songs of humpback whales in Bermuda. Z. Tierpsychol. 68:89–114.

Payne, K., P. Tyack, and R. Payne. 1983. Progressive changes in the songs of humpback whales *(Megaptera novaeangliae)*: a detailed analysis of two seasons in Hawaii. *In* Payne (1983), 9–57.

Payne, P. M., and D. W. Heinemann. 1993. The distribution of pilot whales (*Globicephala* spp.) in shelf/shelf-edge and slope waters of the northeastern United States, 1978–1988. *In* Donovan, Lockyer, and Martin (1993), 51–68.

Payne, R. 1976. At home with right whales. Natl. Geogr. 149:322–39.

———, ed. 1983. Communication and behavior of whales. Westview Press, Boulder, 643 pp.

———. 1986. Long term behavioral studies of the southern right whale *(Eubalaena australis)*. *In* Brownell, Best, and Prescott (1986), 161–67.

Payne, R., and E. M. Dorsey. 1983. Sexual dimorphism and aggressive use of callosities in right whales *(Eubalaena australis)*. *In* Payne (1983), 295–329.

Payne, R., and L. N. Guinee. 1983. Humpback whale *(Megaptera novaeangliae)* songs as an indicator of "stocks." *In* Payne (1983), 333–58.

Payne, R., V. Rowntree, J. S. Perkins, J. G. Cooke, and K. Lankester. 1990. Population size, trends, and reproductive parameters of right whales *(Eubalaena australis)* off Peninsula Valdes, Argentina. *In* Hammond, Mizroch, and Donovan (1990), 271–78.

Pemberton, D., and I. J. Skira. 1989. Elephant seals in Tasmania. Victorian Nat. 106:202–4.

Perrin, W. F. 1975*a*. Distribution and differentiation of populations of dolphins of the genus *Stenella* in the eastern tropical Pacific. J. Fish. Res. Bd. Can. 32:1059–67.

———. 1975*b*. Variation of spotted and spinner porpoise (genus *Stenella*) in the eastern tropical Pacific and Hawaii. Bull. Scripps Inst. Oceanogr. Univ. California 21:1–206.

———. 1976. First record of the melon-headed whale, *Peponocephala electra*, in the eastern Pacific, with a summary of world distribution. Fishery Bull. 74:457–58.

———. 1990. Subspecies of *Stenella longirostris* (Mammalia: Cetacea: Delphinidae). Proc. Biol. Soc. Washington 103:453–63.

Perrin, W. F., P. A. Akin, and J. V. Kashiwada. 1991. Geographic variation in external morphology of the spinner dolphin *Stenella longirostris* in the eastern Pacific and implications for conservation. Fishery Bull. 89: 411–28.

Perrin, W. F., P. B. Best, W. H. Dawbin, K. C. Balcomb, R. Gambell, and G. J. B. Ross. 1973. Rediscovery of Fraser's dolphin *Lagenodelphis hosei*. Nature 241:345–50.

Perrin, W. F., and R. L. Brownell, Jr. 1989. Report of the workshop. *In* Perrin et al. (1989), 1–22.

Perrin, W. F., R. L. Brownell, Jr., and D. P. DeMaster, eds. 1984. Reproduction in whales, dolphins, and porpoises. Rept. Internatl. Whaling Comm. Spec. Issue, no. 6, xii + 495 pp.

Perrin, W. F., R. L. Brownell, Jr., Zhou Kaiya, and Liu Jiankang, eds. 1989. Biology and conservation of river dolphins. Occas. Pap. IUCN (World Conservation Union) Species Survival Comm., no. 3, v + 173 pp.

Perrin, W. F., D. K. Caldwell, and M. C. Caldwell. 1994. Atlantic spotted dolphin *Stenella frontalis* (G. Cuvier, 1829). *In* Ridgway and Harrison (1994), 173–90.

Perrin, W. F., J. M. Coe, and J. R. Zweifel. 1976. Growth and reproduction of the spotted porpoise, *Stenella attenuata*, in the offshore eastern tropical Pacific. Fishery Bull. 74:229–69.

Perrin, W. F., and J. W. Gilpatrick, Jr. 1994. Spinner dolphin *Stenella longirostris* (Gray, 1828). *In* Ridgway and Harrison (1994), 99–128.

Perrin, W. F., and J. R. Henderson. 1984. Growth and reproductive rates in two populations of spinner dolphins, *Stenella longirostris*, with different histories of exploitation. *In* Perrin, Brownell, and DeMaster (1984), 417–30.

Perrin, W. F., and A. A. Hohn. 1994. Pantropical spotted dolphin *Stenella attenuata*. *In* Ridgway and Harrison (1994), 71–98.

Perrin, W. F., D. B. Holts, and R. B. Miller. 1977. Growth and reproduction of the eastern spinner dolphin, a geographical form of *Stenella longirostris* in the eastern tropical Pacific. Fishery Bull. 75:725–50.

Perrin, W. F., S. Leatherwood, and A. Collet. 1994. Fraser's dolphin *Lagenodelphis hosei* Fraser, 1956. *In* Ridgway and Harrison (1994), 225–40.

Perrin, W. F., R. B. Miller, and P. A. Sloan. 1977. Reproductive parameters of the offshore spotted dolphin, a form of *Stenella attenuata*, in the eastern tropical Pacific, 1973–75. Fishery Bull. 75:629–33.

Perrin, W. F., E. D. Mitchell, J. G. Mead, D. K. Caldwell, M. C. Caldwell, P. J. H. Van Bree, and W. H. Dawbin. 1987. Revision of the spotted dolphins, *Stenella* spp. Mar. Mamm. Sci. 3:99–170.

Perrin, W. F., E. D. Mitchell, J. G. Mead, D. K. Caldwell, and P. J. H. Van Bree. 1981. *Stenella clymene*, a rediscovered tropical dolphin of the Atlantic. J. Mamm. 62:583–98.

Perrin, W. F., N. Miyazuki, and T. Kasuya. 1989. A dwarf form of the spinner dolphin *(Stenella longirostris)* from Thailand. Mar. Mamm. Sci. 5:213–27.

Perrin, W. F., and S. B. Reilly. 1984. Reproductive parameters of dolphins and small whales of the family Delphinidae. *In* Perrin, Brownell, and DeMaster (1984), 97–133.

Perrin, W. F., M. D. Scott, G. J. Walker, and V. L. Cass. 1985. Review of geographical stocks of tropical dolphins *(Stenella* spp. and *Delphinus delphis)* in the eastern Pacific. U.S. Natl. Mar. Fish. Serv., NOAA Tech. Rept. NMFS 28, iv + 28 pp.

Perrin, W. F., and W. A. Walker. 1975. The rough-toothed porpoise, *Steno bredanensis*, in the eastern tropical Pacific. J. Mamm. 56:905–9.

Perrin, W. F., C. E. Wilson, and F. I. Archer, II. 1994. Striped dolphin *Stenella coeruleoalba* (Meyen, 1833). *In* Ridgway and Harrison (1994), 129–59.

Perry, E. A., S. M. Carr, S. E. Bartlett, and W. S. Davidson. 1995. A phylogenetic perspective on the evolution of reproductive behavior in pagophilic seals of the northwest Atlantic as indicated by mitochondrial DNA sequences. J. Mamm. 76:22–31.

Perryman, W. L., D. W. K. Au, S. Leatherwood, and T. A. Jefferson. 1994. Melon-headed whale *Peponocephala electra* Gray, 1846. *In* Ridgway and Harrison (1994), 363–86.

Peterson, R. S., and G. A. Bartholomew. 1967. The natural history and behavior of the California sea lion. Amer. Soc. Mamm. Spec. Publ., no. 1, xi + 79 pp.

Peterson, R. S., C. L. Hubbs, R. L. Gentry, and R. L. DeLong. 1968. The Guadalupe fur seal: habitat, behavior, population size, and field identification. J. Mamm. 49: 665–75.

Pilleri, G. 1971. On the La Plata dolphin

Pontoporia blainvillei off the Uruguayan coast. Investig. Cetacea 3:59–67.

———. 1979. Observations on the ecology of *Inia geoffrensis* from the Rio Apure, Venezuela. Investig. Cetacea 10:137–43.

Pilleri, G., and Chen Peixun. 1979. How the finless porpoise *(Neophocaena asiaeorientalis)* carries its calves on its back, and the function of the denticulated area of skin, as observed in the Changjiang River, China. Investig. Cetacea 10:105–8.

Pilleri, G., and M. Gihr. 1971. Zur Systematik der Gattung *Platanista* (Cetacea). Rev. Suisse Zool. 78:746–59.

———. 1975. On the taxonomy and ecology of the finless black porpoise, *Neophocaena* (Cetacea, Delphinidae). Mammalia 39:657–73.

———. 1977*a*. Neotype for *Platanista indi* Blyth, 1859. Investig. Cetacea 8:77–81.

———. 1977*b*. Observations on the Bolivian (*Inia boliviensis* d'Orbigny, 1834) and the Amazonian bufeo (*Inia geoffrensis* de Blainville, 1817) with description of a new subspecies *(Inia geoffrensis humboldtiana)*. Investig. Cetacea 8:11–76.

———. 1977*c*. Radical extermination of the South American sea lion *Otaria byronia* (Pinnipedia, Otariidae) from Isla Verde, Uruguay. Brain Anat. Inst., Univ. Berne, Switzerland, 15 pp.

———. 1981*a*. Additional considerations on the taxonomy of the genus *Inia*. Investig. Cetacea 11:15–24.

———. 1981*b*. Checklist of the cetacean genera *Platanista, Inia, Lipotes, Pontoporia, Sousa,* and *Neophocaena*. Investig. Cetacea 11:33–36.

Pilleri, G., M. Gihr, P. E. Purves, K. Zbinden, and C. Kraus. 1976. On the behaviour, bioacoustics, and functional morphology of the Indus River dolphin (*Platanista indi* Blyth, 1859). Investig. Cetacea 6:1–151.

Pilleri, G., and O. Pilleri. 1979*a*. Observations on the dolphins in the Indus Delta *(Sousa plumbea* and *Neophocaena phocaenoides)* in winter 1978–1979. Investig. Cetacea 10:129–35.

———. 1979*b*. Precarious situation of the dolphin population (*Platanista indi* Blyth, 1859) in the Punjab, upstream from the Taunsa Barrage, Indus River. Investig. Cetacea 10:121–27.

Pilleri, G., K. Zbinden, and M. Gihr. 1976. The "black finless porpoise" (*Neophocaena phocaenoides* Cuvier, 1829) is not black. Investig. Cetacea 7:161–64.

Pinedo, M. C. 1987. First record of a dwarf sperm whale from southwest Atlantic, with reference to osteology, food habits, and reproduction. Sci. Rept. Whales Res. Inst. 38:171–86.

Pitman, R. L., A. Aguayo L., and J. Urbán R. 1987. Observations of an unidentified beaked whale (*Mesoplodon* sp.) in the eastern tropical Pacific. Mar. Mamm. Sci. 3:345–52.

Powell, J. A., Jr. 1978. Evidence of carnivory in manatees *(Trichechus manatus)*. J. Mamm. 59:442.

Powell, J. A., Jr., D. W. Belitsky, and G. B. Rathbun. 1981. Status of the West Indian manatee *(Trichechus manatus)* in Puerto Rico. J. Mamm. 62:642–46.

Powell, J. A., Jr., and G. B. Rathbun. 1984. Distribution and abundance of manatees along the northern coast of the Gulf of Mexico. Northeast Gulf Sci. 7:1–28.

Praderi, R., M. C. Pinedo, and E. A. Crespo. 1989. Conservation and management of *Pontoporia blainvillei* in Uruguay, Brazil, and Argentina. *In* Perrin et al. (1989), 52–56.

Prescott, J. 1991. The Saint Lawrence beluga: a concerted effort to save an endangered isolated population. Environ. Conserv. 18:351–55.

Prestrud, P. 1990. The most northerly harbor seal, *Phoca vitulina*, at Prins Karls Forland, Svalbard. Mar. Mamm. Sci. 6:215–20.

Pryor, K., J. Lindbergh, S. Lindbergh, and R. Milano. 1990. A dolphin-human fishing cooperative in Brazil. Mar. Mamm. Sci. 6:77–82.

Pryor, K., and I. K. Shallenberger. 1991. Social structure in spotted dolphins *(Stenella attenuata)* in the tuna purse seine fishery in the eastern tropical Pacific. *In* Pryor, K., and K. S. Norris, eds., Dolphin societies: discoveries and puzzles, Univ. California Press, Berkeley, 161–96.

Purves, P. E., and G. Pilleri. 1974–75. Observations on the ear, nose, throat, and eye of *Platanista indi*. Investig. Cetacea 5:13–57.

———. 1978. The functional anatomy and general biology of *Pseudorca crassidens* (Owen) with a review of the hydrodynamics and acoustics in Cetacea. Investig. Cetacea 9:68–227.

———. 1983. Echolocation in whales and dolphins. Academic Press, London, xiv + 261 pp.

R

Ralls, K., J. Ballou, and R. L. Brownell, Jr. 1983. Diversity in California sea otters: theoretical considerations and management implications. Biol. Conserv. 25:209–32.

Ralls, K., and R. L. Brownell, Jr. 1991. A whale of a new species. Nature 350:560.

Ralls, K., P. Fiorelli, and S. Gish. 1985. Vocalizations and vocal mimicry in captive harbor seals, *Phoca vitulina*. Can. J. Zool. 63:1050–56.

Ramsay, M. A., and I. Stirling. 1988. Reproductive biology and ecology of female polar bears *(Ursus maritimus)*. J. Zool. 214:601–34.

Rathbun, G. B. 1984. Sirenians. *In* Anderson and Jones (1984), 537–48.

Rathbun, G. B., R. L. Brownell, Jr., K. Ralls, and J. Engbring. 1988. Status of dugongs in waters around Palau. Mar. Mamm. Sci. 4:265–70.

Rathbun, G. B., and T. J. O'Shea. 1984. The manatee's simple social life. *In* Macdonald (1984), 300–301.

Rathbun, G. B., J. P. Reid, R. K. Bonde, and J. A. Powell. 1995. Reproduction in free-ranging Florida manatees. *In* O'Shea, Ackerman, and Percival (1995), 135–56.

Rau, R. 1978. The end of the hunt? Natl. Wildl. 16(3):4–11.

Rausch, R. L. 1953. On the status of some arctic mammals. Arctic 6:91–148.

———. 1963. Geographic variation in size in North American brown bears, *Ursus arctos* L., as indicated by condylobasal length. Can. J. Zool. 41:33–45.

Ray, C. E., and A. E. Spiess. 1981. The bearded seal, *Erignathus barbatus*, in the Pleistocene of Maine. J. Mamm. 62:423–27.

Ray, G. C. 1981. Ross seal—*Ommatophoca rossi*. *In* Ridgway and Harrison (1981*b*), 237–60.

Ray, G. C., and W. E. Schevill. 1974. Feeding of a captive gray whale, *Eschrichtius robustus*. Mar. Fish. Rev. 36(4):31–38.

Ray, G. C., W. A. Watkins, and J. J. Burns. 1969. The underwater song of *Erignathus* (bearded seal). Zoologica 54:79–83.

Read, A. J. 1990. Reproductive seasonality in harbour porpoises, *Phocoena phocoena*, from the Bay of Fundy. Can. J. Zool. 68:284–88.

Read, A. J., and D. E. Gaskin. 1988. Incidental catch of harbor porpoises by gill nets. J. Wildl. Mgmt. 52:517–23.

Read, A. J., K. Van Waerebeek, J. C. Reyes, J. S. McKinnon, and L. C. Lehman. 1988. The exploitation of small cetaceans in coastal Peru. Biol. Conserv. 46:53–70.

Redford, K. H., and J. F. Eisenberg. 1992. Mammals of the neotropics: the southern cone. Univ. Chicago Press, ix + 430 pp.

Reeves, R. R. 1977. Hunt for the narwhal. Oceans 10(4):50–57.

———. 1978. Atlantic walrus *(Odobenus rosmarus rosmarus)*: a literature survey and status report. U.S. Fish and Wildl. Serv. Wildl. Res. Rept., no. 10, 41 pp.

———. 1993. Domestic and international trade in narwhal products. Traffic Bull. 14(1): 13–20.

Reeves, R. R., and R. L. Brownell, Jr. 1989. Susu *Platanista gangetica* (Roxburgh, 1801) and *Platanista minor* Owen, 1853. *In* Ridgway and Harrison (1989), 69–99.

Reeves, R. R., and S. Leatherwood. 1985. Bowhead whale—*Balaena mysticetus. In* Ridgway and Harrison (1985), 305–44.

———. 1994. Dolphins, porpoises, and whales: 1994–1998 action plan for the conservation of cetaceans. IUCN (World Conservation Union), Gland, Switzerland, viii + 92 pp.

Reeves, R. R., and J. K. Ling. 1981. Hooded seal—*Cystophora cristata. In* Ridgway and Harrison (1981*b*), 171–94.

Reeves, R. R., J. G. Mead, and S. Katona. 1978. The right whale, *Eubalaena glacialis*, in the western North Atlantic. Rept. Internatl. Whaling Comm. 28:303–12.

Reeves, R. R., and E. Mitchell. 1981. The whale behind the tusk. Nat. Hist. 90(8): 50–57.

———. 1984. Catch history and initial population of white whales *(Delphinapterus leucas)* in the River and Gulf of St. Lawrence, eastern Canada. Naturaliste Canadien 111: 63–121.

———. 1987*a*. Catch history, former abundance, and distribution of white whales in Hudson Strait and Ungava Bay. Naturaliste Canadien 114:1–65.

———. 1987*b*. Distribution and migration, exploitation, and former abundance of white whales *(Delphinapterus leucas)* in Baffin Bay and adjacent waters. Can. Spec. Publ. Fish. Aquat. Sci., no. 99, vi + 34 pp.

———. 1987*c*. History of white whale *(Delphinapterus leucas)* exploitation in eastern Hudson Bay and James Bay. Can. Spec. Publ. Fish. Aquat. Sci., no. 95, vi + 45 pp.

———. 1988. Current status of the gray whale, *Eschrichtius robustus*. Can. Field-Nat. 102:369–90.

———. 1989. Status of white whales, *Delphinapterus leucas*, in Ungava Bay and eastern Hudson Bay. Can. Field-Nat. 103:220–39.

———. 1990. Bowhead whales in Hudson Bay, Hudson Strait, and Foxe Basin: a review. Nat. Can. (Rev. Ecol. Syst.) 117: 25–43.

Reeves, R. R., E. Mitchell, A. Mansfield, and M. McLaughlin. 1983. Distribution and migration of the bowhead whale, *Balaena mysticetus*, in the eastern North American Arctic. Arctic 36:5–64.

Reeves, R. R., E. Mitchell, and H. Whitehead. 1993. Status of the northern bottlenose whale, Hyperoodon ampullatus. Can. Field-Nat. 107:490–508.

Reeves, R. R., and S. Tracey. 1980. *Monodon monoceros*. Mammalian Species, no. 127, 7 pp.

Reeves, R. R., D. Tuboku-Metzger, and R. A. Kapindi. 1988. Distribution and exploitation of manatees in Sierra Leone. Oryx 22:75–84.

Reid, J. 1995. Chessie's most excellent adventure: the 1995 east coast tour. Sirenews 24:9–11.

Reid, J. P., G. B. Rathbun, and J. R. Wilcox. 1991. Distribution patterns of individually identifiable West Indian manatees *(Trichechus manatus)* in Florida. Mar. Mamm. Sci. 7: 180–90.

Reijnders, P. J. H. 1982. Observation, catch, and release of a walrus, *Odobenus rosmarus*, on the Isle of Ameland, Dutch Wadden Sea. Zool. Anz., 209:88–90.

Reijnders, P. J. H., S. Brasseur, J. Van Der Toorn, P. Van Der Wolf, I. Boyd, J. Harwood, D. Lavigne, and L. Lowry. 1993. Seals, fur seals, sea lions, and walrus: status survey and conservation action plan. IUCN (World Conservation Union), Gland, Switzerland, vii + 88 pp.

Reilly, S. B. 1978. Pilot Whale. *In* Haley (1978), 112–19.

Reilly, S. B., and V. G. Thayer. 1990. Blue whale *(Balaenoptera musculus)* distribution in the eastern tropical Pacific. Mar. Mamm. Sci. 6:265–77.

Reiner, F. 1986. First record of Sowerby's beaked whale from Azores. Sci. Rept. Whales Res. Inst. 37:103–7.

Reiter, J., K. J. Panken, and B. J. Le Boeuf. 1981. Female competition and reproductive success in northern elephant seals. Anim. Behav. 29:670–87.

Repenning, C. A. 1990. Oldest pinniped. Science 248: 499.

Repenning, C. A., and R. H. Tedford. 1977. Otarioid seals of the Neogene. U.S. Geol. Surv. Prof. Pap., no. 992, vi + 93 pp.

Reyes, J. C. 1990. Gray's beaked whale *Mesoplodon grayi* in the south east Pacific. Z. Saugetierk. 55:139–41.

Reyes, J. C., J. G. Mead, and K. Van Waerebeek. 1991. A new species of beaked whale, *Mesoplodon peruvianus* sp. n. (cetacea: Ziphiidae), from Peru. Mar. Mamm. Sci. 7:1–24.

Reynolds, J. E., III. 1981. Aspects of the social behaviour and herd structure of a semi-isolated colony of West Indian manatees, *Trichechus manatus*. Mammalia 45:431–51.

Reynolds, J. E., III, and J. C. Ferguson. 1984. Implications of the presence of manatees *(Trichechus manatus)* near the Dry Tortugas Islands. Florida Sci. 47:187–89.

Reynolds, J. S., III, and D. K. Odell. 1991. Manatees and dugongs. Facts on File, New York, xiv + 192 pp.

Rice, D. W. 1967. Cetaceans. *In* Anderson and Jones (1967), 291–324.

———. 1977. A list of the marine mammals of the world (third edition). U.S. Natl. Mar. Fish. Serv., NOAA Tech. Rept. NMFS SSRF-711, iii + 15 pp.

———. 1978*a*. Beaked whales. *In* Haley (1978), 88–95.

———. 1978*b*. Blue whale. *In* Haley (1978), 30–35.

———. 1978*c*. Gray whale. *In* Haley (1978), 54–61.

———. 1978*d*. Sperm whales. *In* Haley (1978), 82–87.

———. 1984. Cetaceans. *In* Anderson and Jones (1984), 447–90.

———. 1989. Sperm whale *Physeter macrocephalus* Linnaeus, 1758. *In* Ridgway and Harrison (1989), 177–233.

Rice, D. W., and A. A. Wolman. 1971. The life history and ecology of the gray whale *(Eschrichtius robustus)*. Amer. Soc. Mamm. Spec. Publ., no. 3, viii + 142 pp.

Rice, D. W., A. A. Wolman, and H. W. Braham. 1984. The gray whale, *Eschrichtius robustus*. Mar. Fish. Rev. 46(4):7–14.

Richard, P. R. 1991*a*. Abundance and distribution of narwhals *(Monodon monoceros)* in northern Hudson Bay. Can. J. Fish. Aquat. Sci. 48:276–83.

———. 1991*b*. Status of the belugas, *Delphinapterus leucas*, of southeast Baffin Island, Northwest Territories. Can. Field-Nat. 105:206–14.

———. 1993. Status of the beluga, *Delphinapterus leucas*, in western and southern Hudson Bay. Can. Field-Nat. 107:524–32.

Richard, P. R., and R. R. Campbell. 1988. Status of the Atlantic walrus, *Odobenus rosmarus rosmarus*, in Canada. Can. Field-Nat. 102:337–50.

Richard, P. R., and D. G. Pike. 1993. Small whale co-management in the eastern Canadian Arctic: a case history and analysis. Arctic 46:138–43.

Richards, R. 1994. "The uplands seal" of the Antipodes and Macquarie Islands: a historian's perspective. J. Roy. Soc. New Zealand 24:289–95.

Richardson, W. J., and C. I. Malme. 1993. Man-made noise and behavioral responses. *In* Burns, Montague, and Cowles (1993), 631–700.

Ridgway, S. H. 1986*a*. Diving by cetaceans. *In* Brubakk, A. O., J. W. Kanwisher, and G. Sundness, eds., Diving in animals and man, Roy. Norwegian Soc. Sci. Letters, Trondheim, 33–62.

———. 1986*b*. Physiological observations on dolphin brains. *In* Schusterman, Thomas, and Wood (1986), 31–59.

Ridgway, S. H., and R. Harrison, eds. 1981*a*. Handbook of marine mammals. Volume 1. The walrus, sea lions, fur seals, and sea otter. Academic Press, London, xiv + 235 pp.

———, eds. 1981*b*. Handbook of marine mammals. Volume 2. Seals. Academic Press, London, xv + 359 pp.

———, eds. 1985. Handbook of marine mammals. Volume 3. The sirenians and baleen whales. Academic Press, London, xviii + 362 pp.

———, eds. 1989. Handbook of marine mammals. Volume 4. River dolphins and the larger toothed whales. Academic Press, London, xix + 442 pp.

———, eds. 1994. Handbook of marine mammals. Volume 5. The first book of dolphins. Academic Press, London, xx + 416 pp.

Ridgway, S. H., and C. C. Robinson. 1985. Homing by released captive California sea lions, *Zalophus californianus,* following release on distant islands. Can. J. Zool. 63: 2162–64.

Riedman, M. L., and J. A. Estes. 1990. The sea otter *(Enhydra lutris):* behavior, ecology, and natural history. U.S. Fish and Wildl. Serv. Biol. Rept., no. 90(14), iii + 126 pp.

Riedman, M. L., J. A. Estes, M. M. Staedler, A. A. Giles, and D. R. Carlson. 1994. Breeing patterns and reproductive success of California sea otters. J. Wildl. Mgmt. 58:391–99.

Robineau, D., and M. Vely. 1993. Stranding of a specimen of Gervais' beaked whale *(Mesoplodon europaeus)* on the coast of West Africa (Mauritania). Mar. Mamm. Sci. 9:438–40.

Rodriguez, D. H., and R. O. Bastida. 1993. The southern sea lion, *Otaria byronia* or *Otaria flavescens.* Mar. Mamm. Sci. 9:372–81.

Roff, D., and W. D. Bowen. 1986. Further analysis of population trends in the northwest Atlantic harp seal *(Phoca groenlandica)* from 1967 to 1985. Can. J. Fish. Aquat. Sci. 43:553–64.

Ronald, K., and P. J. Healey. 1981. Harp seal—*Phoca groenlandica. In* Ridgway and Harrison (1981*b*), 55–87.

Ronald, K., J. Selley, and P. Healey. 1982. Seals. *In* Chapman, J. A., and G. A. Feldhamer, eds., Wild mammals of North America, Johns Hopkins Univ. Press, Baltimore, 769–827.

Rosas, F. C. W. 1994. Biology, conservation, and status of the Amazonian manatee *Trichechus inunguis.* Mamm. Rev. 24: 49–59.

Rosas, F. C. W., M. Haimovici, and M. C. Pinedo. 1993. Age and growth of the South American sea lion, *Otaria flavescens* (Shaw, 1800), in southern Brazil. J. Mamm. 74:141–47.

Rosel, P. E., A. E. Dizon, and J. E. Heyning. 1994. Genetic analysis of sympatric morphotypes of common dolphins (genus *Delphinus*). Mar. Biol. 119:159–67.

Ross, G. J. B. 1977. The taxonomy of bottlenosed dolphins *Tursiops* species in South African waters, with notes on their biology. Ann. Cape Prov. Mus. (Nat. Hist.) 11:135–94.

———. 1979. Records of pygmy and dwarf sperm whales, genus *Kogia,* from southern Africa, with some biological notes and some comparisons. Ann. Cape Prov. Mus. (Nat. Hist.) 11:259–327.

———. 1984. The smaller cetaceans of the south east coast of southern Africa. Ann. Cape Prov. Mus. (Nat. Hist.) 15:173–410.

Ross, G. J. B., P. B. Best, and B. G. Donnelly. 1975. New records of the pygmy right whale *(Caperea marginata)* from South Africa, with comments on distribution, migration, appearance, and behavior. J. Fish. Res. Bd. Can. 32:1005–17.

Ross, G. J. B., and V. G. Cockcroft. 1990. Comments on Australian bottlenose dolphins and the taxonomic status of *Tursiops aduncus* (Ehrenberg, 1832). *In* Leatherwood and Reeves (1990), 101–28.

Ross, G. J. B., G. E. Heinsohn, and V. G. Cockcroft. 1994. Humpback dolphins *Sousa chinensis* (Osbeck, 1765), *Sousa plumbea* (G. Cuvier, 1829), and *Sousa teuszii* (Kukenthal, 1892). *In* Ridgway and Harrison (1994), 23–42.

Ross, G. J. B., and S. Leatherwood. 1994. Pygmy killer whale *Feresa attenuata* Gray, 1874. *In* Ridgway and Harrison (1994), 387–404.

Ross, W. G. 1974. Distribution, migration, and depletion of bowhead whales in Hudson Bay, 1860 to 1915. Arctic and Alpine Res. 6:85–98.

———. 1979. The annual catch of Greenland (bowhead) whales in waters north of Canada, 1719–1915: a preliminary compilation. Arctic 32:91–121.

———. 1993. Commercial whaling in the North Atlantic sector. *In* Burns, Montague, and Cowles (1993), 511–61.

Roth, H. H., and E. Waitkuwait. 1986. Répartition et statut des grandes espèces de mammifères en Côte-d'Ivoire. III. Lamantins. Mammalia 50:227–42.

Rotterman, L. M., and T. Simon-Jackson. 1988. Sea otter: *Enhydra lutris. In* Lentfer, J. W., ed., Selected marine mammals of Alaska, Mar. Mamm. Comm., Washington, D.C., 239–75.

Roux, J.-P. 1987. Recolonization process in the subantarctic fur seal, *Arctocephalus tropicalis,* on Amsterdam Island. *In* Croxall and Gentry (1987), 189–94.

Roux, J.-P., and A. D. Hes. 1984. The seasonal haul-out cycle of the fur seal *Arctocephalus tropicalis* (Gray, 1872) on Amsterdam Island. Mammalia 48:377–89.

Russell, R. H. 1975. The food habits of polar bears of James Bay and southwest Hudson Bay in summer and autumn. Arctic 28:117–29.

S

Saayman, G. S., and C. K. Tayler. 1973. Social organisation of inshore dolphins *(Tursiops aduncus* and *Sousa)* in the Indian Ocean. J. Mamm. 54:993–96.

———. 1979. The socioecology of humpback dolphins (*Sousa* sp.). *In* Winn and Olla (1979), 165–226.

Samaras, W. F. 1974. Reproductive behavior of the gray whale *Eschrichtius robustus* in Baja California. Bull. S. California Acad. Sci. 73:57–64.

Sarich, V. M. 1969. Pinniped phylogeny. Syst. Zool. 18:416–22.

Scarff, J. E. 1986. Historic and present distribution of the right whale *(Eubalaena glacialis)* in the eastern North Pacific south of 50° N and east of 180° W. *In* Brownell, Best, and Prescott (1986), 43–63.

———. 1991. Historic distribution and abundance of the right whale *(Eubalaena glacialis)* in the North Pacific, Bering Sea, Sea of Okhotsk, and Sea of Japan from the Maury whale charts. Rept. Internatl. Whaling Comm. 41:467–89.

Schaeff, C., S. Kraus, M. Brown, J. Perkins, R. Payne, D. Gaskin, P. Boag, and B. White. 1991. Preliminary analysis of mitochondrial DNA variation within and between the right whale species *Eubalaena glacialis* and *Eubalaena australis. In* Hoelzel, R., ed., Genetic ecology of whales and dolphins: incorporating the Proceedings of the Workshop on the Genetic Analysis of Cetacean Populations, Rept. Internatl. Whaling Comm. Spec. Issue, no. 13, 217–23.

Schaeff, C., S. Kraus, M. Brown, and B. N. White. 1993. Assessment of the population structure of western North Atlantic right whales *(Eubalaena glacialis)* based on sighting and mtDNA data. Can. J. Zool. 71:339–45.

Scheffer, V. B. 1972. The weight of the Steller sea cow. J. Mamm. 53:912–14.

———. 1978*a*. False killer whale. *In* Haley (1978), 128–31.

———. 1978*b*. Killer whale. *In* Haley (1978), 120–27.

Schell, D. M., and S. M. Saupe. 1993. Feeding and growth as indicated by stable isotopes. *In* Burns, Montague, and Cowles (1993), 491–509.

Schevill, W. E. 1974. The whale problem: a status report. Harvard Univ. Press, Cambridge, x + 419 pp.

———. 1986. The International Code of Zoological Nomenclature and a paradigm: the name *Physeter catodon* Linnaeus 1758. Mar. Mamm. Sci. 2:153–57.

Schmidly, D. J., M. H. Beleau, and H. Hildebran. 1972. First record of Cuvier's dolphin from the Gulf of Mexico with comments on the taxonomic status of *Stenella frontalis.* J. Mamm. 53:625–28.

Schnell, G. D., M. E. Douglas, and D. J. Hough. 1985. Sexual dimorphism in spotted dolphins *(Stenella attenuata)* in the eastern tropical Pacific Ocean. Mar. Mamm. Sci. 1:1–14.

———. 1986. Geographic patterns of variation in offshore spotted dolphins *(Stenella attenuata)* of the eastern tropical Pacific Ocean. Mar. Mamm. Sci. 2:186–213.

Schusterman, R. J. 1981. Steller sea lion—*Eumetopias jubatus. In* Ridgway and Harrison (1981*a*), 119–41.

Schusterman, R. J., J. A. Thomas, and F. G. Wood, eds. 1986. Dolphin cognition and behavior: a comparative approach. Lawrence Erlbaum Associates, Hillsdale, New Jersey, xv + 393 pp.

Scott, M. D., and S. J. Chivers. 1990. Distribution and herd structure of bottlenose dolphins in the eastern tropical Pacific Ocean. *In* Leatherwood and Reeves (1990), 387–402.

Scott, M. D., R. S. Wells, and A. B. Irvine. 1990. A long-term study of bottlenose dolphins on the west coast of Florida. *In* Leatherwood and Reeves (1990), 235–44.

Sergeant, D. E. 1973. Biology of white whales *(Delphinapterus leucas)* in western Hudson Bay. J. Fish. Res. Bd. Can. 30:1065–90.

———. 1976. History and present status of populations of harp and hooded seals. Biol. Conserv. 10:96–117.

Sergeant, D. E., and P. F. Brodie. 1975. Identity, abundance, and present status of populations of white whales, *Delphinapterus leucas,* in North America. J. Fish. Res. Bd. Can. 32:1047–54.

Sergeant, D. E., D. K. Caldwell, and M. C. Caldwell. 1973. Age, growth, and maturity of bottlenosed dolphin *(Tursiops truncatus)* from northeast Florida. J. Fish. Res. Bd. Can. 30:1009–11.

Sergeant, D. E., and W. Hoek. 1988. An update of the status of white whales, *Delphinapterus leucas,* in the Saint Lawrence Estuary, Canada. Biol. Conserv. 45:287–302.

Sergeant, D. E., D. J. St. Aubin, and J. R. Geraci. 1980. Life history and northwest Atlantic status of the Atlantic white-sided dolphin, *Lagenorhynchus acutus.* Cetology, no. 37, 12 pp.

Shane, S. H. 1990. Behavior and ecology of the bottlenose dolphin at Sanibel Island, Florida. *In* Leatherwood and Reeves (1990), 245–65.

Shane, S. H., R. S. Wells, and B. Würsig. 1986. Ecology, behavior, and social organization of the bottlenose dolphin: a review. Mar. Mamm. Sci. 2:34–63.

Shaughnessy, P. D., and F. H. Fay. 1977. A review of the taxonomy and nomenclature of North Pacific harbour seals. J. Zool. 182:385–419.

Shaughnessy, P. D., and L. Fletcher. 1987. Fur seals, *Arctocephalus* spp., at Macquarie Island. *In* Croxall and Gentry (1987), 177–88.

Shaughnessy, P. D., and S. D. Goldsworthy. 1990. Population size and breeding season of the Antarctic fur seal *Arctocephalus gazella* at Heard Island—1987/88. Mar. Mamm. Sci. 6:292–304.

Shaughnessy, P. D., and G. J. B. Ross. 1980. Records of the subantarctic fur seal *(Arctocephalus tropicalis)* from South Africa with notes on its biology and some observations of captive animals. Ann. S. Afr. Mus. 82:71–89.

Shipley, C., M. Hines, and J. S. Buchwald. 1986. Vocalizations of northern elephant seal bulls: development of adult call characteristics during puberty. J. Mamm. 67:526–36.

Shirakihara, M., A. Takemura, and K. Shirakihara. 1993. Age, growth, and reproduction of the finless porpoise, *Neophocaena phocaenoides,* in the coastal waters of western Kyushu, Japan. Mar. Mamm. Sci. 9: 392–406.

Silber, G. K. 1988. Recent sightings of the Gulf of California harbor porpoise, *Phocoena sinus.* J. Mamm. 69:430–33.

Silverman, H. B., and M. J. Dunbar. 1980. Aggressive tusk use by the narwhal (*Monodon monoceros* L.). Nature 284:57–58.

Simmonds, M. 1991. What future for European seals now that the epidemic is over? Oryx 25:27–32.

———. 1994. Saving Europe's dolphins. Oryx 28:238–48.

Simons, L. S. 1984. Seasonality of reproduction and dentinal structures in the harbor porpoise *(Phocoena phocoena)* of the North Pacific. J. Mamm. 65:491–95.

Simpson, G. G. 1945. The principles of classification and a classification of the mammals. Bull. Amer. Mus. Nat. Hist. 85:i–xvi + 1–350.

Siniff, D. B. 1991. An overview of the ecology of Antarctic seals. Amer. Zool. 31:143–49.

Siniff, D. B., I. Stirling, J. L. Bengtson, and R. A. Reichle. 1979. Social and reproductive behavior of crabeater seals *(Lobodon carcinophagus)* during the austral spring. Can. J. Zool. 57:2243–55.

Siniff, D. B., and S. Stone. 1985. The role of the leopard seal in the tropho-dynamics of the Antarctic marine ecosystem. *In* Siegfried, W. R., P. R. Condy, and R. M. Laws, eds., Antarctic nutrient cycles and food webs, Springer-Verlag, Berlin, 555–60.

Sipilä, T. 1990. Lair structure and breeding habitat of the Saimaa ringed seal *(Phoca hispida saimensis* Nordq.) in Finland. Finnish Game Res. 47:11–20.

Skinner, J. D., and R. H. N. Smithers. 1990. The mammals of the southern African subregion. Univ. Pretoria, xxxii + 771 pp.

Slip, D. J., M. A. Hindell, and H. R. Burton. 1994. Diving pattern of southern elephant seals from Macquarie Island: an overview. *In* Le Boeuf and Laws (1994*b*), 253–70.

Slooten, E. 1991. Age, growth, and reproduction in Hector's dolphins. Can. J. Zool. 69: 1689–1700.

Slooten, E., and S. M. Dawson. 1988. Studies on Hector's dolphin, *Cephalorhynchus hectori:* a progress report. *In* Brownell and Donovan (1988), 325–38.

———. 1994. Hector's dolphin *Cephalorhynchus hectori* (van Beneden, 1881). *In* Ridgway and Harrison (1994), 311–33.

Small, G. L. 1971. The blue whale. Columbia Univ. Press, New York, xiii + 248 pp.

Smeenk, C. 1987. The harbour porpoise, *Phocoena phocoena* (L., 1758), in the Netherlands: stranding records and decline. Lutra 30:77–90.

Smit, C. J., and A. Van Wijngaarden. 1981. Threatened mammals in Europe. Akademische Verlagsgesellschaft, Wiesbaden, 259 pp.

Smith, P. A., and C. J. Jonkel. 1975. Résumé of the trade in polar bear hides in Canada, 1973–74. Can. Wildl. Serv. Progress Notes, no. 48, 5 pp.

Smith, R. I. L. 1988. Destruction of Antarctic terrestrial ecosystems by a rapidly increasing fur seal population. Biol. Conserv. 45:55–72.

Smith, R. J., and D. M. Lavigne. 1994. Subspecific status of the freshwater harbor seal *(Phoca vitulina mellonae):* a re-assessment. Mar. Mamm. Sci. 10:105–10.

Smith, T. G., and M. O. Hammill. 1981. Ecology of the ringed seal, *Phoca hispida,* in its fast ice breeding habitat. Can J. Zool. 59:966–81.

———. 1986. Population estimates of white whale, *Delphinapterus leucas,* in James Bay, eastern Hudson Bay, and Ungava Bay. Can. J. Fish. Aquat. Sci. 43:1982–87.

Smith, T. G., M. O. Hammill, D. J. Burrage, and G. A. Sleno. 1985. Distribution and abundance of belugas, *Delphinapterus leucas,* and narwhals, *Monodon monoceros,* in the Canadian high Arctic. Can. J. Fish. Aquat. Sci. 42:676–84.

Smith, T. G., M. O. Hammill, and G. Taugbol. 1991. A review of the developmental, behavioural, and physiological adaptations of the ringed seal, *Phoca hispida,* to life in the arctic winter. Arctic 44:124–31.

Smith, T. G., and I. Stirling. 1975. The breeding habitat of the ringed seal *(Phoca hispida)*: the birth lair and associated structures. Can. J. Zool. 53:1297–1305.

Smithers, R. H. N. 1983. The mammals of the southern African subregion. Univ. Pretoria, xxii + 736 pp.

Spilliaert, R., G. Vikingsson, U. Arnason, A. Palsdottir, J. Sigurjonsson, and A. Arnason. 1991. Species hybridization between a female blue whale *(Balaenoptera musculus)* and a male fin whale *(B. physalus)*: molecular and morphological documentation. J. Hered. 82:269–74.

Spotte, S. 1982. The incidence of twins in pinnipeds. Can. J. Zool. 60:2226–33.

Spotte, S., C. W. Radcliffe, and J. L. Dunn. 1979. Notes on Commerson's dolphin *(Cephalorhynchus commersonii)* in captivity. Cetology, no. 35, 9 pp.

Stacey, P. J., S. Leatherwood, and R. W. Baird. 1994. *Pseudorca crassidens*. Mammalian Species, no. 456, 6 pp.

Stains, H. J. 1984. Carnivores. *In* Anderson and Jones (1984), 491–522.

Stein, J. L., M. Herder, and K. Miller. 1986. Birth of a northern fur seal on the mainland California coast. California Fish and Game 72:179–81.

Stejneger, L. 1936. Georg Wilhelm Steller. Harvard Univ. Press, Cambridge, 623 pp.

Steltner, H., S. Steltner, and D. E. Sergeant. 1984. Killer whales, *Orcinus orca*, prey on narwhals, *Monodon monoceros*: an eyewitness account. Can. Field-Nat. 98:458–62.

Stewart, B. S., and R. L. DeLong. 1995. Double migrations of the northern elephant seal, *Mirounga longirostris*. J. Mamm. 76:196–205.

Stewart, B. S., and W. T. Everett. 1983. Incidental catch of a ribbon seal *(Phoca fasciata)* in the central North Pacific. Arctic 36:369.

Stewart, B. S., and H. R. Huber. 1993. *Mirounga angustirostris*. Mammalian Species, no. 449, 10 pp.

Stewart, B. S., P. K. Yochem, H. R. Huber, R. L. DeLong, R. J. Jameson, W. J. Sydeman, S. G. Allen, and B. J. Le Boeuf. 1994. History and present status of the northern elephant seal population. *In* Le Boeuf and Laws (1994*b*), 29–48.

Stirling, I. 1971. *Leptonychotes weddelli*. Mammalian Species, no. 6, 5 pp.

———. 1973. Vocalization in the ringed seal *(Phoca hispida)*. J. Fish. Res. Bd. Can. 30: 1592–94.

———. 1974. Midsummer observations on the behavior of wild polar bears *(Ursus maritimus)*. Can. J. Zool. 52:1191–98.

Stirling, I., W. Calvert, and D. Andriashek. 1980. Population ecology studies of the polar bear in the area of southeastern Baffin Island. Can. Wildl. Serv. Occas. Pap., no. 44, 33 pp.

Stirling, I., and A. E. Derocher. 1993. Possible impacts of climatic warming on polar bears. Arctic 46:240–45.

Stirling, I., C. Jonkel, P. Smith, R. Robertson, and D. Cross. 1977. The ecology of the polar bear *(Ursus maritimus)* along the western coast of Hudson Bay. Can. Wildl. Serv. Occas. Pap., no. 33, 64 pp.

Stirling, I., and H. P. L. Kiliaan. 1980. Population ecology studies of the polar bear in northern Labrador. Can. Wildl. Serv. Occas. Pap., no. 42, 21 pp.

Stirling, I., and G. L. Kooyman. 1971. The crabeater seal *(Lobodon carcinophagus)* in McMurdo Sound, Antarctica, and the origin of mummified seals. J. Mamm. 52:175–80.

Stirling, I., and D. B. Siniff. 1979. Underwater vocalizations of leopard seals *(Hydrurga leptonyx)* and crabeater seals *(Lobodon carcinophagus)* near the South Shetland Islands, Antarctica. Can. J. Zool. 57:1244–48.

Stoett, P. J. 1993. International politics and the protection of great whales. Environ. Politics 2:277–302.

Storro-Patterson, R. 1977. Gray whale protection: how well is it working? Oceans 10(4): 44–49.

Strahan, R., ed. 1983. The Australian Museum complete book of Australian mammals. Angus & Robertson, London, xxi + 530 pp.

Stroganov, S. U. 1969. Carnivorous mammals of Siberia. Israel Progr. Sci. Transl., Jerusalem, x + 522 pp.

Strong, J. T. 1988. Status of the narwhal, *Monodon monoceros*, in Canada. Can. Field-Nat. 102:391–98.

Sullivan, R. M. 1982. Agonistic behavior and dominance relationships in the harbor seal, *Phoca vitulina*. J. Mamm. 63:554–69.

Sylvestre, J.-P. 1983. Review of *Kogia* specimens (Physeteridae, Kogiinae) kept alive in captivity. Investig. Cetacea 15:201–19.

Sylvestre, J.-P., and S. Tasaka. 1985. On the intergeneric hybrids in cetaceans. Aquat. Mamm. 11:101–8.

Szczepaniak, I. D., M. A. Webber, and T. A. Jefferson. 1992. First record of a *truei*-type Dall's porpoise from the eastern North Pacific. Mar. Mamm. Sci. 8:425–28.

T

Taruski, A. G. 1979. The whistle repertoire of the North Atlantic pilot whale *(Globicephala melaena)* and its relationship to behavior and environment. *In* Winn and Olla (1979), 345–68.

Tedford, R. H. 1976. Relationships of pinnipeds to other carnivores (Mammalia). Syst. Zool. 25:363–74.

Teilmann, J., and R. Dietz. 1994. Status of the harbour seal, *Phoca vitulina*, in Greenland. Can. Field-Nat. 108:139–55.

Terry, R. P. 1986. The behaviour and trainability of *Sotalia fluviatilis guianensis* in captivity: a survey. Aquat. Mamm. 12:71–79.

Testa, J. W. 1987. Long-term reproductive patterns and sighting bias in Weddell seals *(Leptonychotes weddelli)*. Can. J. Zool. 65: 1091–99.

Testa, J. W., and D. B. Siniff. 1987. Population dynamics of Weddell seals *(Leptonychotes weddelli)* in McMurdo Sound, Antarctica. Ecol. Monogr. 57:149–65.

Testaverde, S. A., and J. G. Mead. 1980. Southern distribution of the Atlantic whitesided dolphin, *Lagenorhynchus acutus*, in the western North Atlantic. Fishery Bull. 78:167–69.

Thein, U. T. 1977. The Burmese freshwater dolphin. Mammalia 41:233–34.

Thewissen, J. G. M. 1994. Phylogenetic aspects of cetacean origins: a morphological perspective. J. Mamm. Evol. 2:157–84.

Thewissen, J. G. M., S. T. Hussain, and M. Arif. 1994. Fossil evidence for the origin of aquatic locomotion in archaeocete whales. Science 263:210–12.

Thomas, J., V. Pastukhov, R. Elsner, and E. Petrov. 1982. *Phoca sibirica*. Mammalian Species, no. 188, 6 pp.

Thomas, J. A., and I. Stirling. 1983. Geographic variation in the underwater vocalizations of Weddell seals *(Leptonychotes weddelli)* from Palmer Peninsula and McMurdo Sound, Antarctica. Can. J. Zool. 61:2203–12.

Thompson, T. J., H. E. Winn, and P. J. Perkins. 1979. Mysticete sounds. *In* Winn and Olla (1979), 403–31.

Thornback, J., and M. Jenkins. 1982. The IUCN mammal red data book. Part 1: Threatened mammalian taxa of the Americas and the Australasian zoogeographic region (excluding Cetacea). Internatl. Union Conserv. Nat., Gland, Switzerland, xl + 516 pp.

Timm, R. M., L. Albuja V., and B. L. Clauson. 1986. Ecology, distribution, harvest, and conservation of the Amazonian manatee *Trichechus inunguis* in Ecuador. Biotrópica 18:150–56.

Torres N., D. 1987. Juan Fernandez fur seal, *Arctocephalus philippii*. *In* Croxall and Gentry (1987), 37–41.

Trebbau, P. 1975. Measurements and some observations on the freshwater dolphin, *Inia geoffrensis*, in the Apure River, Venezuela. Zool. Garten 45:153–67.

Trebbau, P., and P. J. H. Van Bree. 1974. Notes concerning the freshwater dolphin *Inia geoffrensis* (de Blainville, 1817) in Venezuela. Z. Saugetierk. 39:50–57.

Trillmich, F. 1986. Attendance behavior of Galapagos sea lions. *In* Gentry and Kooyman (1986), 196–208.

———. 1987*a*. Galapagos fur seal, *Arctocephalus galapagoensis. In* Croxall and Gentry (1987), 23–27.

———. 1987*b*. Seals under the sun. Nat. Hist. 96(10):42–49.

Trillmich, F., G. L. Kooyman, P. Majluf, and M. Sanchez-Griñan. 1986. Attendance and diving behavior of South American fur seals during El Niño in 1983. *In* Gentry and Kooyman (1986), 153–67.

Trillmich, F., and P. Majluf. 1981. First observations on colony structure, behavior, and vocal repertoire of the South American fur seal (*Arctocephalus australis* Zimmermann, 1783) in Peru. Z. Saugetierk. 46:310–22.

Trites, A. W. 1992. Northern fur seals: why have they declined? Aquat. Mamm. 18:3–18.

Trites, A. W., and P. A. Larkin. 1989. The decline and fall of the Pribilof fur seal *(Callorhinus ursinus)*: a simulation study. Can. J. Fish. Aquat. Sci. 46:1437–45.

Tyack, P., and H. Whitehead. 1983. Male competition in large groups of wintering humpback whales. Behaviour 83:132–54.

U

Ulmer, F. A., Jr. 1966. Hand-rearing polar bear cubs. America's First Zoo 18(1):3–5.

U.S. (United States) Fish and Wildlife Service. 1980. Administration of the Marine Mammal Protection Act of 1972, April 1, 1979 to March 31, 1980. Washington, D.C., v + 86 pp.

———. 1993. Administration of the Marine Mammal Protection Act of 1972, January 1, 1991 to December 31, 1991. Washington, D.C., 33 pp.

U.S. (United States) National Marine Fisheries Service. 1978. The Marine Mammal Protection Act of 1972: Annual Report, 1977–78. Washington, D.C., v + 183 pp.

———. 1981. Marine Mammal Protection Act of 1972: Annual Report, 1980/81. Washington, D.C., v + 143 pp.

———. 1984. Marine Mammal Protection Act of 1972: Annual Report, 1983/84. Washington, D.C., 146 pp.

———. 1985. Marine Mammal Protection Act of 1972: Annual Report, 1984/85. Washington, D.C., 50 pp.

———. 1986. Marine Mammal Protection Act of 1972: Annual Report, 1985/86. Washington, D.C., 57 pp.

———. 1987. Marine Mammal Protection Act of 1972: Annual Report, 1986/87. Washington, D.C., 47 pp.

———. 1989. Marine Mammal Protection Act of 1972: Annual Report, 1987/88. Washington, D.C., 68 pp.

———. 1992. Listing of eastern spinner dolphin as a threatened species. Federal Register 57:47620–26.

———. 1993*a*. Listing of eastern spinner dolphin as depleted. Federal Register 58: 45066–74.

———. 1993*b*. Listing of the northeastern offshore spotted dolphin as depleted. Federal Register 58:58285–97.

———. 1994. Marine Mammal Protection Act of 1972: Annual Report, 1992/93. Washington, D.C., 136 pp.

Urbán-Ramírez, J., and D. Aurioles-Gamboa. 1992. First record of the pygmy beaked whale, *Mesoplodon peruvianus,* in the North Pacific. Mar. Mamm. Sci. 8:420–25.

Urian, K. W., D. A. Duffield, A. J. Read, R. S. Wells, and E. D. Shell. 1996. Seasonality of reproduction in bottlenose dolphins, *Tursiops truncatus.* J. Mamm. 394–403.

Uspenski, S. M., and S. E. Belikov. 1976. Research on the polar bear in the USSR. *In* Pelton, Lentfer, and Folk (1976), 321–23.

V

Van Bree, P. J. H. 1971. On *Globicephala sieboldii* Gray, 1846, and other species of pilot whales (Notes on Cetacea, Delphinoidea III). Beaufortia 19:79–87.

———. 1976. On the correct Latin name of the Indus susu (Cetacea, Platanistoidea). Bull. Zool. Mus. Univ. Amsterdam 5:139–40.

Van Bree, P. J. H., and M. D. Gallagher. 1978. On the taxonomic status of *Delphinus tropicalis* Van Bree, 1971 (Notes on Cetacea, Delphinoidea IX). Beaufortia 28:1–8.

Van Bree, P. J. H., and P. E. Purves. 1972. Remarks on the validity of *Delphinus bairdii* (Cetacea, Delphinidae). J. Mamm. 53:372–74.

Van Gelder, R. G. 1977. Mammalian hybrids and generic limits. Amer. Mus. Novit., no. 2635, 25 pp.

Van Waerebeek, K., and A. J. Read. 1994. Reproduction of dusky dolphins, *Lagenorhynchus obscurus,* from coastal Peru. J. Mamm. 75:1054–62.

Van Zyll de Jong, C. G. 1972. A systematic review of the nearctic and neotropical river otters (genus *Lutra,* Mustelidae, Carnivora). Roy. Ontario Mus. Life Sci. Contrib., no. 80, 104 pp.

———. 1987. A phylogenetic study of the Lutrinae (Carnivora; Mustelidae) using morphological data. Can J. Zool. 65:2536–44.

Vaz-Ferreira, R. 1981. South American sea lion—*Otaria flavescens. In* Ridgway and Harrison (1981*a*), 39–65.

Villa, B. 1976. Report on the status of *Phocoena sinus,* Norris and McFarland 1958, in the Gulf of California. An. Inst. Biol. Univ. Nac. Autón. México, Ser. Zool., 47:203–8.

W

Wachtel, P. S. 1986. Silja saves the grey Baltic seal. World Wildl. Fund News, no. 43, 7.

Wade, P. R. 1993. Estimation of historical population size of the eastern spinner dolphin *(Stenella longirostris orientalis).* Fishery Bull. 91:775–87.

Wade, P. R., and T. Gerrodette. 1993. Estimates of cetacean abundance and distribution in the eastern tropical Pacific. Rept. Internatl. Whaling Comm. 43:477–93.

Walker, G. E., and J. K. Ling. 1981*a*. Australian sea lion—*Neophoca cinerea. In* Ridgway and Harrison (1981*a*), 99–118.

———. 1981*b*. New Zealand sea lion—*Phocarctos hookeri. In* Ridgway and Harrison (1981*a*), 25–38.

Waller, G. N. H. 1983. Is the blind river dolphin sightless? Aquat. Mamm. 10:106–8.

Wang Ding, Lu Wenxiang, and Wang Zhifan. 1989. A preliminary study of the acoustic behavior of the baiji, *Lipotes vexillifer. In* Perrin et al. (1989), 137–40.

Wang Peilie. 1984. Distribution of the gray whale *(Eschrichtius gibbosus)* off the coast of China. Acta Theriol. Sinica 4:21–26.

Warneke, R. M. 1982. The distribution and abundance of seals in the Australasian region, with summaries of biology and current research. Mammals in the Seas, FAO Fish. Ser. No. 5, 4:431–75.

Warneke, R. M., and P. D. Shaughnessy. 1985. *Arctocephalus pusillus,* the South African and Australian fur seal: taxonomy, evolution, biogeography, and life history. *In* Ling and Bryden (1985), 53–77.

Watkins, W. A. 1976. A probable sighting of a live *Tasmacetus shepherdi* in New Zealand waters. J. Mamm. 57:415.

Watkins, W. A., M. A. Daher, K. Fristrup, and G. Notarbartolo di Sciara. 1994. Fishing and acoustic behavior of Fraser's dolphin *(Lagenodelphis hosei)* near Dominica, southeast Caribbean. Caribbean J. Sci. 30:76–82.

Watkins, W. A., W. E. Schevill, and P. B. Best. 1977. Underwater sounds of *Cephalorhynchus heavisidii* (Mammalia: Cetacea). J. Mamm. 58:316–18.

Watkins, W. A., P. Tyack, K. E. Moore, and G. Notarbartolo di Sciara. 1987. *Steno bredanensis* in the Mediterranean Sea. Mar. Mamm. Sci. 3:78–82.

Wayne, R. K., R. E. Benveniste, D. N. Janczewski, and S. J. O'Brien. 1989. Molecular and biochemical evolution of the Carnivora. *In* Gittleman (1989), 465–94.

Webber, M. A., and J. Roletto. 1987. Two recent occurrences of the Guadalupe fur seal, *Arctocephalus townsendi,* in central California. Bull. S. California Acad. Sci. 86:159–63.

Weintraub, B. 1996. Harpoon blades point to long-lived whales. Natl. Geogr. 189(3):xix.

Wellington, G. M., and Tj. De Vries. 1976. The South American sea lion, *Otaria byronia,* in the Galapagos Islands. J. Mamm. 57: 166–67.

Wells, R. S. 1991. The role of long-term study in understanding the social structure of a bottlenose dolphin community. *In* Pryor, K., and K. S. Norris, eds., Dolphin societies: discoveries and puzzles, Univ. California Press, Berkeley, 199–225 pp.

Wendell, F. E., J. A. Ames, and R. A. Hardy. 1984. Pup dependency period and length of reproductive cycle: estimates from observations of tagged sea otters, *Enhydra lutris,* in California. California Fish and Game 70: 89–100.

Whitehead, H., and T. Arnbom. 1987. Social organization of sperm whales off the Galapagos Islands, February–April 1985. Can. J. Zool. 65:913–19.

Whitehead, P. J. P. 1977. The former southern distribution of New World manatees (*Trichechus* spp.). Biol. J. Linnean Soc. 9:165–89.

Wiig, O. 1983. On the relationship of pinnipeds to other carnivores. Zool. Scripta 12: 225–27.

———. 1991. Seven bowhead whales (*Balaena mysticetus* L.) observed at Franz Josef Land in 1990. Mar. Mamm. Sci. 7:316–19.

Wiig, O., E. W. Born, and G. W. Garner, eds. 1995. Polar bears: proceedings of the eleventh working meeting of the IUCN/SSC Polar Bear Specialist Group, 25–27 January 1993, Copenhagen, Denmark. IUCN (World Conservation Union), Gland, Switzerland, v + 192 pp.

Wiig, O., and R. W. Lie. 1984. An analysis of the morphological relationships between the hooded seals *(Cystophora cristata)* of Newfoundland, the Denmark Strait, and Jan Mayen. J. Zool. 203:227–40.

Wilkinson, I. S., and M. N. Bester. 1990. Continued population increase in fur seals, *Arctocephalus tropicalis* and *A. gazella,* at the Prince Edward Islands. S. Afr. J. Antarctic Res. 20:58–63.

Williamson, G. R. 1988. Seals in Loch Ness. Sci. Rept. Whales Res. Inst. 39:151–57.

Wilson, D(on). E., M. A. Bogan, R. L. Brownell, Jr., A. M. Burdin, and M. K. Maminov. 1991. Geographic variation in sea otters, *Enhydra lutris.* J. Mamm. 72:22–36.

Wilson, D(on). E., and D. M. Reeder, eds. 1993. Mammal species of the world: a taxonomic and geographic reference. Smithsonian Inst. Press, Washington, D.C., xviii + 1206 pp.

Winn, H. E., R. K. Edel, and A. G. Taruski. 1975. Population estimate of the humpback whale *(Megaptera novaeangliae)* in the West Indies by visual and acoustic techniques. J. Fish. Res. Bd. Can. 32:499–506.

Winn, H. E., and B. L. Olla, eds. 1979. Behavior of marine animals. Volume 3: cetaceans. Plenum Press, New York, xix + 438 pp.

Winn, H. E., P. J. Perkins, and L. Winn. 1970. Sounds and behavior of the northern bottle-nosed whale. Proc. Ann. Conf. Biol. Sonar and Diving Mammals 7:53–59.

Winn, H. E., C. A. Price, and P. W. Sorensen. 1986. The distributional biology of the right whale *(Eubalaena glacialis)* in the western North Atlantic. *In* Brownell, Best, and Pres-cott (1986), 129–37.

Winn, H. E., and N. E. Reichley. 1985. Humpback whale—*Megaptera novaeangliae. In* Ridgway and Harrison (1985), 241–73.

Winn, H. E., T. J. Thompson, W. C. Cummings, J. Hain, J. Hudnall, H. Hays, and W. W. Steiner. 1981. Song of the humpback whale—population comparisons. Behav. Ecol. Sociobiol. 8:41–46.

Winn, H. E., and L. K. Winn. 1978. The song of the humpback whale *(Megaptera novaeangliae)* in the West Indies. Mar. Biol. 47:97–114.

Wolman, A. A. 1978. Humpback whale. *In* Haley (1978), 46–53.

Wolman, A. A., and D. W. Rice. 1979. Current status of the gray whale. Rept. Internatl. Whaling Comm. 29:275–79.

Woodby, D. A., and D. B. Botkin. 1993. Stock sizes prior to commercial whaling. *In* Burns, Montague, and Cowles (1993), 387–409.

Worthy, T. H. 1992. Fossil bones of Hooker's sea lions in New Zealand caves. New Zealand Nat. Sci. 19:31–39.

Wozencraft, W. C. 1989. The phylogeny of the Recent Carnivora. *In* Gittleman (1989), 495–535.

Würsig, B. 1982. Radio tracking dusky porpoises in the South Atlantic. Mammals in the Seas, FAO Fish. Ser. No. 5, 4:145–60.

———. 1986. Delphinid foraging strategies. *In* Schusterman, Thomas, and Wood (1986), 347–59.

Würsig, B., and R. Bastida. 1986. Long-range movement and individual associations of two dusky dolphins *(Lagenorhynchus obscurus)* off Argentina. J. Mamm. 67:773–74.

Würsig, B., and C. Clark. 1993. Behavior. *In* Burns, Montague, and Cowles (1993), 157–99.

Würsig, B., E. M. Dorsey, M. A. Fraker, R. S. Payne, and W. J. Richardson. 1985. Behavior of bowhead whales, *Balaena mysticetus,* summering in the Beaufort Sea: a description. Fishery Bull. 83:357–77.

Würsig, B., R. S. Wells, K. S. Norris, and M. Würsig. 1994. A spinner dolphin's day. *In* Norris et al. (1994), 63–102.

Würsig, B., and M. Würsig. 1980. Behavior and ecology of the dusky dolphin, *Lagenorhynchus obscurus,* in the South Atlantic. Fishery Bull. 77:871–90.

Wyss, A. R. 1987. The walrus auditory region and the monophyly of pinnipeds. Amer. Mus. Novit., no. 2871, 31 pp.

———. 1988*a*. Evidence from flipper structure for a single origin of pinnipeds. Nature 334:427–28.

———. 1988*b*. On "retrogression" in the evolution of the Phocinae and phylogenetic affinities of the monk seals. Amer. Mus. Novit., no. 2924, 38 pp.

Wyss, A. R., and J. J. Flynn. 1993. A phylogenetic analysis and definition of the Carnivora. *In* Szalay, Novacek, and McKenna (1993*b*), 32–52.

Y

Yablokov, A. V., and L. S. Bogoslovskaya. 1984. A review of Russian research on the biology and commercial whaling of the gray whale. *In* Jones, Swartz, and Leatherwood (1984), 465–85.

Yochem, P. K., and S. Leatherwood. 1985. Blue whale—*Balaenoptera musculus. In* Ridgway and Harrison (1985), 193–240.

Yurick, D. B., and D. E. Gaskin. 1988. Asymmetry in the skull of the harbour porpoise *Phocoena phocoena* (L.) and its relationship to sound production and echolocation. Can. J. Zool. 66:399–402.

Z

Zachariassen, P. 1993. Pilot whale catches in the Faroe Islands, 1709–1992. *In* Donovan, Lockyer, and Martin (1993), 69–88.

Zeh, J. E., C. W. Clark, J. C. George, D. Withrow, G. M. Carroll, and W. R. Koski. 1993. Current population size and dynamics. *In* Burns, Montague, and Cowles (1993), 409–89.

Zhou Kaiya. 1982. Classification and phylogeny of the superfamily Platanistoidea, with notes on evidence of the monophyly of the Cetacea. Sci. Rept. Whales Res. Inst. 34: 93–108.

———. 1986. The ringed seal and other pinnipeds wandering off the coast of China. Acta Theriol. Sinica 6:107–13.

Zhou Kaiya, Gao Anli, and Sun Jiang. 1993. Notes on the biology of the finless porpoise in Chinese waters. IBI Reports 4:69–74.

Zhou Kaiya and Li Yuemin. 1989. Status and aspects of the ecology and behavior of the baiji, *Lipotes vexillifer,* in the lower Yangtze River. *In* Perrin et al. (1989), 86–91.

Zhou Kaiya, G. Pilleri, and Li Yuemin. 1979.

Observations on the baiji *(Lipotes vexillifer)* and the finless porpoise *(Neophocaena asiaeorientalis)* in the Changjiang (Yangtze) River between Nanjing and Taiyangzhou, with remarks on some physiological adaptations of the baiji to its environment. Investig. Cetacea 10:109–20.

———. 1980. Observations on baiji *(Lipotes vexillifer)* and finless porpoise *(Neophocaena asiaeorientalis)* in the lower reaches of the Chang Jiang. Sci. Sinica 23:785–94.

Zhou Kaiya and Qian Weijuan. 1985. Distribution of the dolphins of the genus *Tursiops* in the China Seas. Aquat. Mamm. 11:16–19.

Zhou Kaiya, Qian Weijuan, and Li Yuemin. 1977. Studies on the distribution of baiji, *Lipotes vexillifer* Miller. Acta Zool. Sinica 23:72–79.

———. 1979. The osteology and the systematic position of the baiji, *Lipotes vexillifer.* Acta Zool. Sinica 25:58–74

entific names of orders, families, and genera that have titled accounts in the text are in boldfaced type, as are the page numbers on which such nts begin. Other scientific names and vernacular names appear in roman.